QUALITATIVE RESEARCH

We dedicate this book to

Our Children:

*Maxwell and Evelyn, who were born during the writing of this book.
Ari and Lev, who continued to grow, teach, and learn as we wrote.
We hope this book will contribute to your generation and beyond
through the good work of those who read this and use research as a tool
to help heal our world.*

And to Our Students:

*Whose questions, passions, concerns, and commitments have shaped
this book beyond words. Thank you for your engaging learning,
your curiosity, and for being our teachers.*

*Curiosity as restless questioning, as movement toward the revelation of something hidden,
as a question verbalized or not, as search for clarity, as a moment of attention, suggestion,
and vigilance, constitutes an integral part of the phenomenon of being alive. There could be no
creativity without the curiosity that moves us and sets us patiently impatient before a world
that we did not make, to add to it something of our own making.*

—Paulo Freire, *Pedagogy of Freedom*

QUALITATIVE RESEARCH

Bridging the Conceptual, Theoretical, and Methodological

SHARON M. RAVITCH

University of Pennsylvania

NICOLE MITTENFELNER CARL

University of Pennsylvania

Los Angeles | London | New Delhi
Singapore | Washington DC

Los Angeles | London | New Delhi
Singapore | Washington DC

FOR INFORMATION:

SAGE Publications, Inc.

2455 Teller Road

Thousand Oaks, California 91320

E-mail: order@sagepub.com

SAGE Publications Ltd.

1 Oliver's Yard

55 City Road

London EC1Y 1SP

United Kingdom

SAGE Publications India Pvt. Ltd.

B 1/I 1 Mohan Cooperative Industrial Area

Mathura Road, New Delhi 110 044

India

SAGE Publications Asia-Pacific Pte. Ltd.

3 Church Street

#10-04 Samsung Hub

Singapore 049483

Printed in the United States of America

ISBN 978-1-4833-5174-2

This book is printed on acid-free paper.

Certified Chain of Custody
Promoting Sustainable Forestry
www.sfiprogram.org
SFI-01268

SFI label applies to text stock

Acquisitions Editor: Vicki Knight

Editorial Assistant: Yvonne McDuffee

eLearning Editor: Katie Bierach

Production Editor: Libby Larson

Copy Editor: Gillian Dickens

Typesetter: C&M Digitals (P) Ltd.

Proofreader: Dennis W. Webb

Indexer: Wendy Allex

Cover Designer: Leonardo March

Marketing Manager: Nicole Elliott

17 18 19 10 9 8 7 6 5 4 3

Brief Contents

Detailed Contents

3 CRITICAL QUALITATIVE RESEARCH DESIGN   65

4 DESIGN AND REFLEXIVITY IN DATA COLLECTION 111

5   METHODS OF DATA COLLECTION   145

11   RESEARCH ETHICS AND THE RELATIONAL QUALITY OF RESEARCH   343

Foreword

Frederick Erickson

Qualitative research is not magic, but it is sometimes done as sleight of hand, as if to pull a rabbit out of a hat. And like the stage performance of a skilled magician, each successive flourish can appear easy to accomplish, from the point of view of the audience. Yet much is going on behind the scenes that the audience does not see. Explaining the backstage aspects of qualitative inquiry—its fundamental aims and conduct—can be done either simplistically—cookbook fashion, which makes it look deceptively simple and straightforward—or with elaborate complexity—invoking unexplained distinctions in epistemology, ontology, and axiology in impenetrable discourse that mystifies the reader. What this book achieves is something quite different: it presents qualitative inquiry as a deliberative enterprise that is careful in its rationales, in its ethics, and in its handling of evidence and representation, discussing these matters with clarity, cogency, and subtlety.

"Bridging," a term used in the subtitle of the book, is an image of connection, and this book certainly accomplishes that, making connections horizontally and vertically. Horizontal connections manifest across successive chapters, from initial consideration of the purposes and intellectual foundations of qualitative inquiry (essays in "methodology" as distinct from method), through discussion of methods—various research tools, and various aspects of relationship between researchers and those who are studied, concluding with discussion of data discovery, data analysis, and writing, with thoughtful consideration of choices involved in narrative reporting; representations of the everyday practices and meaning perspectives of those who are being studied.

Vertical connections also manifest within each chapter of the book, across differing expository voices. There is the voice of the authors, who write from within their considerable experience in conducting qualitative inquiry and teaching others how and why to do it. There are also voices of students, who write from within their experience of beginning to learn qualitative inquiry and finding that this is not as easy as it might have looked at first glance. In addition, there are the voices of other scholars who do qualitative inquiry and write about how it is done well. This three-layered combination of voices, some very junior in experience and others very senior, constructs an especially rich conversation among diverse participants in differing communities of the practices of qualitative inquiry—a conversation that draws the reader into dialogue with the text as a fourth voice; a fourth layer in the conversation.

The extensive presentation of instructional exercises and examples of students' work makes this book pedagogically unusual, and perhaps unique. The breadth and currency of literature review, connecting this book's discussion with wider conversations among classic and contemporary writers who address qualitative inquiry approaches, make this book substantively rich.

Just as I do when I review fieldnotes or watch video clips or listen to audio recordings of interviews again and again, so as to apprehend, reflect upon, and digest their contents more and more thoroughly, so I read a book like this iteratively. I invite the reader to do the same. Read the whole book once. Then read it again. After that, begin to review single chapters repeatedly. Take some time to hunt down the writing of various authors who have been cited. Then go back to the chapter in which that citation appeared. Whether you are an experienced qualitative researcher or a beginner, you will find that this ruminative approach pays off, in fresh ideas and in deeper insights into what you may already know. After a few years, come back to this book again. You don't get it all at once.

Preface

READING THIS BOOK: A NOTE TO STUDENTS

Welcome students.

This book is written with you in mind—your learning goals and needs, your backgrounds, interests, the contexts of your learning, and possibly your anxieties, concerns, and desire for balance between theory and action as you learn about qualitative research. We seek to support you as you endeavor to learn about qualitative research and about becoming a qualitative researcher. Our overarching goal is to provide you with a solid understanding of what qualitative research is and how to do it with focus and rigor. This book seeks to help you cultivate and integrate theoretical, methodological, and conceptual knowledge; to help you see their interaction; and to help you understand the central concepts, topics, and skills you need to engage in rigorous, valid, and respectful qualitative research.

This book comes out of Sharon's decade and a half of teaching qualitative methods to students across programs, disciplines, fields, and continents, as well as Sharon and Nicole's individual and shared work as applied researchers in a variety of contexts, including education, community, and corporate contexts across the globe. It also emerged from our co-teaching of an introductory qualitative methods course at the University of Pennsylvania over the period of 2 years, during which we paid careful attention to students' questions, interests, goals, and confusions about qualitative research and developed creative ways to respond to these. Throughout the book, we bring our experiences of learning with and from our students about what they are interested in, what they care about, and what they need to make sense of in the field of qualitative research. We hope that you will find that the book supports you in your learning about qualitative research as a field as well as in your *doing* of qualitative research in ways that keep it lively, engaging, and meaningful.

This book is an answer to the many questions we have gotten over the years about how to learn the basics of qualitative research in a way that captures and teaches its complexity and nuances. Balancing these two goals—of communicating the foundations and processes of qualitative research with clarity and simplicity while at the same time capturing its complexity and multilayeredness—has been the central challenge of writing this book. We have endeavored to write this book both for new researchers as well as those who have experience with qualitative research. Our hope is that whether you are new to qualitative research or not, reading this book will develop and deepen your understanding of an approach to research that seeks, designs for, and engages what we think of as *criticality in qualitative research* (defined and described in depth in Chapter One), which we believe bridges the theoretical, methodological, and conceptual aspects of qualitative research.

This book is unique in its approach to conceptualizing and explaining qualitative research in a number of ways. First, while we want the content of the book to be accessible, we do not want it to be watered down. Throughout the book, we discuss qualitative research in complex, yet straightforward, ways that seek to keep it complicated but not overwhelming. Second, we have designed the book, to the extent possible, to be used interactively, that is, as a guide for an experiential and dialogic approach to learning about and engaging in qualitative research. To ensure that the text is both accessible and complex and that it supports an active engagement in your learning, we include a variety of features throughout the book, which we describe below.

- We define central concepts throughout the book as we explain ideas and processes so that you do not have to interrupt your reading to look up too many terms.

- We include a variety of perspectives, concepts, and terms from the key research methods scholars about specific topics so as to familiarize you with a range of viewpoints and help you understand the lay of the land.

- We include an array of student work with multiple examples of each kind of researcher-generated text as a way to *show* qualitative research and how it is constructed as well as to shed light on some of the thought processes that students make transparent in their work. In some cases, we annotate aspects of students' work to highlight key features and explain choices they have made.

- We include a variety of specific recommended practices, exercises, and thought-provoking questions to help you engage in and think actively about the concepts and processes we describe.

- We share many real-life examples of questions, issues, and situations that stem from our research and that of our students to humanize the research process and offer thoughts about possible scenarios that arise through the research process.

- At the end of each chapter, we include questions for further reflection to help you integrate the learning from each chapter.

- We include resources for further reading (at the end of each chapter, throughout some chapters, and in notes) that include texts on specific topics that are beyond the scope of this book but that you will likely need as you develop your specific research studies.

The organization of the book, and of each chapter, reflects our desire to balance rigor, engagement, and accessibility. Each chapter begins with a chapter overview and goals that set the stage for the chapter and list key objectives that you should understand and be able to discuss with thought partners after you finish the chapter. The chapters are arranged thematically, and the headings are structured so that you can also easily refer to specific topics and the practices that you can engage in as you learn and relearn them over time. We provide multiple examples, which include a range of memos, sample conceptual framework narratives and graphics, a variety of research proposals, a sample research report, sample data collection instruments, consent forms, samples of qualitative coding and data analysis, and concept mapping exercises, which are both embedded in the chapters as well as in the appendixes.

We hope this book will help make your journey into qualitative research enjoyable and that you will engage and reengage with this book as you move through the phases of your own research, now and in the future. We believe strongly in the power of qualitative research to inform and change the world, and we hope our passion and grounded optimism will carry with you as you move into your own research.

USING THIS BOOK: A NOTE TO INSTRUCTORS

As you know, a major issue in teaching introductory-level qualitative research courses is that there are few texts that both help students understand the processes and procedures of research design and the conduct of research and concurrently speak to methodology as it relates to reflexivity, positionality, ethics, collaboration, and rigor from a more critical angle; thus, the readings in such courses tend to be divided between more functional/transactional guides to doing research and more theoretical/conceptual pieces on research epistemology and ontology, which are typically not as accessible to novice researchers or those less interested in postmodern and poststructural language, which tends to alienate and obscure. We consider the separation of these aspects of qualitative research to be quite problematic at this early developmental stage of researcher identity, knowledge, and skills formation. This book seeks to integrate these approaches so that students gain insight into the overarching frames and values of qualitative research in direct relation to methods choices and procedures.

The book is meant to be engaging and lively while also thought-provoking and instructional. We hope that you and your students will find the book useful, engaging, challenging, and supportive to an energized teaching and learning about qualitative research.

OUR APPROACH: CRITICALITY, REFLEXIVITY, COLLABORATION, AND RIGOR

Many of the most commonly used books on qualitative research focus on qualitative research design in a fairly traditional and narrow sense, meaning that they typically address the pragmatic aspects of research, focusing primarily on research questions, data collection, and writing up findings. This book seeks to complicate and contextualize qualitative research in a more complex (yet still accessible) manner. We emphasize an approach to qualitative research design and to the entire research process that is based on the themes of *criticality, reflexivity, collaboration,* and *rigor.* These themes are the cornerstones of this book, as they highlight the ethical, contextual, and relational nature of qualitative research that is not only ideological but also deeply methodological.

Throughout the book, we discuss how the paradigmatic values of qualitative research must be actively considered throughout every stage of the research process with a focus on the importance of critically conceptualizing context and issues of representation. Importantly, the book does this in a purposefully accessible way intended for novice researchers as well as more experienced researchers interested in understanding (and teaching) research design in more methodologically sophisticated and critical ways.

In this book, we discuss specific research methods—and the concepts that guide them—that help you, as new or somewhat experienced researchers, to approach research participants as "experts of their own experiences" (Jacoby & Gonzales, 1991; van Manen, 1990) and to consider the methodological issues, concerns, and processes related to the subjectivity of your decisions and sense-making as researchers throughout the research process. We do so by examining specific methods that centralize the iterative, reflexive, systematic, and recursive ways that you cultivate your approach to developing your research design (and your broader methodological approach) to achieve a more refined and specific approach to research rigor and validity. Throughout each chapter, we animate the qualitative research process as decidedly ideological, political, and subjective, and we underscore the ethical, relational, and critical stance(s) that qualitative researchers can cultivate to engage in research that embodies the foundational values of the qualitative paradigm.

OVERVIEW OF THE BOOK

In the first three chapters, we establish our approach to qualitative research and introduce what we consider to be horizontal values of qualitative research—criticality, reflexivity, collaboration, and rigor. We also provide specific procedures and strategies for achieving these ideals. In Chapter One, we define and situate qualitative research as a field of inquiry and approach to research. Chapter Two discusses the role of conceptual frameworks in all aspects and phases of qualitative research. Chapter Three offers an ecological approach to qualitative research design that engages and supports criticality in qualitative research.

We address data collection processes and methods in Chapters Four and Five. Chapter Four focuses on the processes of qualitative data collection and emphasizes considering data from a holistic perspective that views research participants as experts of their own experiences. In Chapter Five, we provide a concrete discussion of the primary modes and methods of qualitative data collection as well as how to approach these methods, their strengths, and potential challenges. Chapter Six defines, reframes, and complicates the concept and related processes of validity as more than a set of procedures and approaches validity as both an active methodological process and a research goal. Chapter Seven details approaches to qualitative data analysis and describes an integrative approach to data analysis that critically explores and addresses the interpretative power in/of analysis and the need for an ethical and critical stance on data analysis. In Chapter Eight, we describe a three-pronged approach to data analysis that includes the ongoing processes of *data organization and management, immersive engagement,* and *writing and representation.* In these five chapters, we underscore the iterative, reflexive, recursive, and formative aspects of collecting and analyzing qualitative data as well as conducting valid research. We offer multiple recommended practices to help you apply what we have discussed as well as to push you to critically consider what can sometimes seem like technical acts.

Chapters Nine and Ten focus on writing in qualitative research. Chapter Nine specifically discusses the final research report and addresses the power inherent in qualitative research and representation; we argue for respectful, authentic, and ethical representations of participants and settings. Chapter Ten addresses the many considerations related to writing qualitative research proposals and includes an annotated example of an exemplary research proposal.

In Chapter Eleven, we discuss the complexities and intricacies of taking a relational and reciprocal approach to research in terms of the processes and frameworks that can inform authentic engagement while keeping research boundaries well defined. In the Epilogue, we revisit the horizontal values of qualitative research and provide examples of alternative ways to think of and enact these important ideals.

We have created a robust Appendix to provide additional examples that will illuminate the processes of conducting qualitative research. The appendixes offer examples of memos, conceptual frameworks, instruments, a full pilot study, analysis plans, data displays, coding schemes, conference and grant proposals, consent forms, and assent forms.

We hope that our passion for research and belief in the power of equitable research to contribute to the world will inspire you to retain an appreciation for the real considerations and steps necessary to engage in rigorous and ethical qualitative research. For us, the possibility born out of equitable research that seeks local and contextualized knowledge generation through an attention to rigor, validity, and ethics is at the heart of qualitative research and is a goal toward which we should all strive.

SAGE EDGE

http://edge.sagepub.com/ravitchandcarl

SAGE edge for Students provides a personalized approach to help students accomplish their coursework goals in an easy-to-use learning environment.

- Mobile-friendly eFlashcards strengthen understanding of key terms and concepts.

- Mobile-friendly practice quizzes allow students to independently assess mastery of course material.

- A customized online action plan enhances students' learning experience.

- Learning objectives help reinforce the most important material.

- Meaningful multimedia content facilitates further exploration of topics.

- EXCLUSIVE! Access to full-text SAGE journal articles that have been carefully selected to support and expand on the concepts presented in each chapter.

SAGE edge for Instructors supports teaching by making it easy to integrate quality content and create a rich learning environment for students.

- A sample course syllabus provides suggested models for structuring one's course.

- Editable, chapter-specific PowerPoint® slides offer complete flexibility for creating multimedia presentations for the course.

- EXCLUSIVE! Access to full-text SAGE journal articles that have been carefully selected to support and expand on the concepts presented in each chapter to encourage students to think critically.

- Multimedia content appeals to students with different learning styles.

Acknowledgments

Books are always the result of many minds and much collaboration. In that spirit, we have many people to thank for their support of and input into this book. We would like to thank the following people for their direct hand in helping us with this book and for their support throughout the book-writing process:

At SAGE, Vicki Knight, for your vision, tenacity, and support throughout the book development and writing process and for pushing us to make the book stronger. Yvonne McDuffee, for all of your support throughout this process. Gillian Dickens, for your thorough copyediting and help clarifying the manuscript. Katie Bierach for coordinating the development of the ancillary material, and Libby Larson for managing the copyediting and production of the book. And a special thank you to Jim Strandberg, for your sage (no pun intended) editorial work, guidance, and good humor.

We want to thank Frederick Erickson for writing the Foreword to the book and inspiring so much of it as well. And we want to sincerely thank our reviewers for your excellent reviews and critiques of earlier drafts; you helped us improve our thinking and the book in so many ways:

- Ifeoma Amah, The University of Texas at Arlington
- Janet M. Duncan, SUNY Cortland
- Frederick Erickson, University of California
- Kathryn G. Herr, Montclair State University
- Lydia Kyei-Blankson, Illinois State University
- Miriam Levitt, University of Ottawa
- Carole L. Lund, Alaska Pacific University
- Penny A. Pasque, University of Oklahoma
- Kimberly M. Sheridan, George Mason University
- Ronald J. Shope, University of Nebraska—Lincoln
- Jianzhong Xu, Mississippi State University

We thank the student contributors, who are yourselves researchers and professionals doing amazing work in the world. You have shared your work and reflections on your research processes with us in ways that helped us conceptualize this book. Your research exemplifies the goals and values of criticality in qualitative research. We thank you for your generosity in sharing your work and your reflections with us and the readers: Mustafa Abdul-Jabbar, Susan Bickerstaff, Ceci Cardesa-Lusardi Laura Colket, Susan Feibelman, Adrianne Flack, Shaun Harper, Charlotte Jacobs, Brandi Jones, Sarah Klevan, Ceci Cardesa-Lusardi, Keon McGuire, Demetri Morgan, Jaime Nolan, Cecilia Orphan, Arjun Shankar, Casey Stokes-Rodriguez, Matthew Tarditi, Tanner Terrell, and Hillary Zimmerman.

At PennGSE, Patricia Friess, for all of your help on permissions, sources, and your steadfast support and kindness all around! Justin Jimenez, for your background research. David Schor, for your transcriptions of our course lectures. Alison Diefenderfer, for help developing ancillary materials.

Our research associates from the Center for the Study of Boys' and Girls' Lives (CSBGL): Jeremy Cutler, Nora Gross, Charlotte Jacobs, Michael Kokozos, and Joseph Nelson for helping us continuously learn about how to engage in research that can push stakeholder-driven change in schools. And we want to thank Peter Kuriloff and Michael Reichert for helping us to think about applied research and research rigor in ways that have helped us refine our thoughts on teaching and doing youth participatory action research (YPAR) and research more broadly and for being thought partners, mentors, and true collaborators.

Sharon would like to thank:

Nicole, for being my true partner in this book. You are such a powerful thinker and researcher and a truly skilled writer. Our partnership is so nourishing and vibrant, and you teach me a great deal. I stand in awe of how you have managed to be such a powerful thinker and focused author amid having newborn twins, being a doctoral student, leading the impact evaluation component of CSBGL, and working on your own research. In addition to being a brilliant student, you are an incredible woman, friend, colleague, and coauthor. I adore, value, and respect you so much.

Matthew Riggan, whose ideas continue to teach me and shape my thinking about all facets of research. Your way of thinking about and doing research (and life) continues to inspire me and deeply informs the work in this book.

Matthew Tarditi, my comrade, thought partner, teacher, and student. Our work together in Nicaragua has been the most exciting and challenging of my career. Your relational integrity, passion, commitment to *el bien común* and your desire to push against normative modes of representation in research are so inspiring. I am deeply grateful for these years of collaboration, mutual growth, and powerful engagement.

My many wonderful colleagues at Penn: Mike Nakkula, whose influence on my thinking about the world and about theory, research, and practice integration is beyond estimation. Everything I think has its roots in your mentorship and modeling. Susan Lytle, for your grounded, world-changing, inspiring irreverence, for pushing into hierarchy and speaking truth to power in ways that are heroic and provide a sense of true possibility in and for research. And for your friendship, which I treasure. Howard Stevenson, for being a north for me and for generations of students; you are a powerful truth teller and truly skilled facilitator of generative introspection and paradigm shifts. Dana Kaminstein, with your quiet, unwavering brilliance, generous thought partnership, and high standards for feedback to students. Your knowledge is as vast as your heart is deep. Annie McKee, for being a role model of powerful emotional intelligence and humility and for your rigorous commitment to quality teaching and learning. And to my colleagues at Penn with whom I teach and do research: Susan Yoon, Elliot Weinbaum, Mike Johanek, Matt Hartley, Janine Remillard, Stanton Wortham, and Torch Lytle. Each of you has a real impact on how I teach and conduct research and I appreciate your colleagueship and friendship very much.

My colleagues around the world: In Nicaragua: Duilio, Ernesto, Adriana, and Tono Baltodano, Rosa Rivas, Nayibe Montenegro, and Eveling Estrada from the Seeds for Progress Foundation. Kevin Maranacci from The Fabretto Foundation. And the ever-dynamic and engaging Kenneth Urbina. In India: Gowri Ishwaran and Shiv Khemka from The Global Education and Leadership Foundation. At the Africa-America Institute, Amini Kajunju and Melissa Howell. Thank you each and all for helping me understand the role of context and relationships in applied research. In the U.S. government: Tim Sheeran, unparalleled strategy advisor, for teaching me what co-creating the conditions for strategic and relational integrity in international applied development work means and entails.

My current research and teaching assistants: Kelsey Jones, Demetri Morgan, and Adrianne Flack, for helping me to understand students' questions, for your amazing research spirits, and for being thought partners about teaching and doing qualitative research. And my students turned colleagues with whom I am in close touch: Laura Colket, Arjun Shankar, Bill Dunworth, Jaime Nolan, Dave Almeda, Sarah Klevan, Dave DeFilippo, Mustafa Abdul-Jabbar, Susan Fiebelman, Monica Clark, Yvonne McCarthy, Chris Steel, Sherry Coleman, Marti Richmond, Raj Ramachandran, Yve-Car Momperousse, Kelsey Jones, Tony Sinanis, and Irene Greaves Jaimes. Each of you continues to inform my thinking, research, and teaching.

My mentors whom I feel walk with me through my career: Carol Gilligan, Fred Erickson, Joseph Maxwell, Sarah Lawrence-Lightfoot, and Meg Turner; each and all of you have taught me to view the world in ways that elevate me, my work, and, I hope, the people around me. Ruthy Kaiser, for sharing the concept of inclining the mind toward the wholeness of life and Frankl's conceptualization of the space between stimulus and response. And Kathy Schultz, for being an intellectual and feminist role model for me and for being so passionate about teaching, learning, and teacher inquiry. I cherish you and our friendship dearly.

My family: Andy, Ari, and Lev, who supported me through writing this book, talked with me about it a lot as I worked on it, and kept me laughing all the way through. Ari and Lev, I have really appreciated talking about the book with you on our morning walks; you help me keep it real. My parents, Arline and Carl Ravitch, whom I love, appreciate, respect, and find comfort in beyond words; you are a pillar of strength for me. To my Uncle Gary, who embodies the kindness of my grandparents, Edith and Albert Karp (z"1), reminding me where my roots are and for how you and Aunt Mindy always support and cheer me on; you are both such a gift. My support network of dear friends, Laura Hoffman, Deborah Melincoff, Jen Finkelstein, Stefanie Gabel, Amy Leventhal, Peter Siskind, Alyssa Levy, and Wendy McGrath: Each and all of you help lift, inspire, and sustain me.

All of my students, you really teach and nourish me more than words could ever express. Some garden, I teach. Thank you for your questions, your humor, your engagement, your resonance, your strong questioning spirits.

Nicole would like to thank:

Sharon, for being an incredible adviser, mentor, colleague, and friend. Working with you in multiple capacities throughout these past 4 years has made me a better scholar, researcher, student, teacher, and person. I consider myself very lucky to have such a wonderful thought partner and adviser who pushes me, challenges me, and engages so thoughtfully in my work. I have learned and continue to learn so much from you about what it means to do justice to people's lived experiences. Writing this book with you has been a true pleasure. I admire and respect you immensely and look up to you as a scholar, researcher, teacher, activist, mother, and woman.

I have been blessed with many wonderful mentors, including Roger Platizky, Rand Quinn, Peter Kuriloff, Philippe Bourgois, and Torch Lytle. All of you have shaped and continue to shape the way I think about and see the world. Thank you for your thoughtful critiques, support, guidance, and encouragement.

Thank you to my mother, Iva Linda Baird, for raising me to embrace difference and for helping us care for our children and to Robert and Jennifer Carl for all of your help and continued support. I also thank my father, Nicholas Mittenfelner; grandmother, Lois Bishop; brothers, Matthew and Thomas Mittenfelner; and sisters-in-law, Yating Mittenfelner and Jenna, Jillian, and Jordan Carl for your encouragement. My dear friends, Abby Miller and Gillian Kamata, you are always a great source of comfort and support.

I especially want to thank my husband, Jason Carl, for your support and understanding throughout the book-writing process amid much other change in our lives. You continue to keep me grounded. Maxwell and Evelyn, thank you for the joy you bring me each and every day and for all that you teach me. I love both of you beyond words.

About the Authors

Sharon M. Ravitch, PhD, is a Senior Lecturer at the University of Pennsylvania's Graduate School of Education, where she is Research Co-Director at the Center for the Study of Boys' and Girls' Lives and a Founding Co-Director of Penn's Inter-American Educational Leadership Network. She serves as the Principal Investigator of *Semillas Digitales* (Digital Seeds), a multiyear applied development research initiative in Nicaragua (http://www2.gse.upenn.edu/nicaragua/). Ravitch's research integrates across the fields of qualitative research, education, international development, cultural anthropology, and human development and has four main strands: (1) practitioner research as a means to engendering sustainable professional and institutional development and innovation, (2) international applied development research that works from participatory and action research approaches (projects currently in the United States, Nicaragua, and India), (3) ethnographic and participatory evaluation research, and (4) leader education and professional development. Ravitch has published three books: *Reason and Rigor: How Conceptual Frameworks Guide Research* (with Matthew Riggan, Sage Publications, 2012), *School Counseling Principles: Diversity and Multiculturalism* (American School Counselor Association Press, 2006), and *Matters of Interpretation: Reciprocal Transformation in Therapeutic and Developmental Relationships With Youth* (with Michael Nakkula, Jossey-Bass, 1998). Ravitch earned two master's degrees from Harvard University in Human Development and Psychology and in Education and a doctorate from the University of Pennsylvania in an interdisciplinary program that combined education, anthropology, and sociology.

Nicole Mittenfelner Carl is a doctoral candidate in the teaching, learning, and leadership division at the University of Pennsylvania Graduate School of Education (PennGSE). She is Director of Impact Assessment for the Center for the Study of Boys' and Girls' Lives (CSBGL) and is conducting a multisited, multiyear participatory evaluation of the center's impact. Carl facilitates and teaches youth participatory action research and researches ways that schools can be more equitable and humane places as a part of her work with CSBGL. Carl is also a graduate research associate at PennGSE, where she is part of a research team that examines ways that teachers and parents can organize and act collectively as a means of educational problem solving. Carl has conducted qualitative research for over a decade. She has published articles about new teachers' experiences with standardized curricula as well as about teacher activism and organizing. Her current research focus includes critical approaches to qualitative research, participatory action research, educational equity, urban education, and teacher education. Carl is a former middle school language arts and lead teacher in the School District of Philadelphia as well as a former mentor to and supervisor of first-year teachers. Carl's research interests stem from her experiences as an educator, and she focuses her work on ways that students, parents, and teachers can work together to address educational inequities.

Qualitative Research

An Opening Orientation

CHAPTER OVERVIEW AND GOALS

This text is written to provide you with a better understanding of what qualitative research is and with the tools you can use to effectively practice qualitative research. As the title of the book implies, it is our goal to bridge the methodological (how to design and conduct qualitative research), theoretical (the theoretical and philosophical underpinnings of phenomena), and conceptual (the ways the researcher shapes and conceives of the study and its multiple contexts) while teaching you the technical aspects of qualitative research, including data collection and analysis. This book seeks to help you to cultivate and integrate theoretical, methodological, and conceptual knowledge; to help you see their important and generative interaction; and to help you appreciate and understand the central concepts, topics, and skills you need to engage in rigorous, valid, and respectful research.

We begin this chapter with an overview of the qualitative research process so that you will have an understanding of the broad components and processes that comprise a qualitative study. After highlighting the specific processes of qualitative research, we delve into the history of qualitative research, its key values, assumptions, and components. We then introduce and define four key pillars of qualitative research (what we refer to as "horizontals")—criticality, reflexivity, collaboration, and rigor, which we emphasize throughout the book. Next, we briefly overview the main approaches to qualitative research. The chapter ends with a discussion about the power and possibilities for qualitative research.

BY THE END OF THIS CHAPTER YOU WILL BETTER UNDERSTAND

- The overall processes of a qualitative research study
- The history and foundations of qualitative research

(Continued)

(Continued)

- The key components of qualitative research
- The core values, beliefs, and assumptions upon which qualitative research is based
- The role of the researcher in qualitative research
- A range of key research concepts and terms, including iterative, recursive, positionality, social location, emergent, epistemology, ontology, methodology, criticality in qualitative research, binaries, and dialogic engagement
- The way the horizontal values of qualitative research—criticality, reflexivity, collaboration, and rigor—influence, shape, and guide all aspects of qualitative research
- An overview of some of the more commonly used approaches to qualitative research

AN OVERVIEW OF THE PROCESSES OF QUALITATIVE RESEARCH

It is our goal that, after reading this book, you will have a better understanding not only of how to conduct qualitative research but also of the theoretical, methodological, and conceptual complexities that comprise qualitative research. Broadly defined, qualitative research attempts to understand individuals, groups, and phenomena in their natural settings in ways that are contextualized and reflect the meaning that people make out of their own experiences. Before providing more specific definitions and discussions of qualitative research, its history, and its values, we begin this chapter with a visual and narrative overview of the processes of conducting qualitative research.

Qualitative research is not a linear process. Figure 1.1 overviews (and admittedly oversimplifies) the dynamic and interactive processes of qualitative research. While we try to indicate the dynamic aspects with the multiple arrows in Figure 1.1, it is difficult to demonstrate the fluidity of qualitative research in graphic form. The processes of qualitative research are continuously interacting and building off one another in cyclical fashion. For example, the process of developing research questions stems from an interest, problem, identification of a gap in literature, or some combination of these. However, once you develop your research question(s), you will continue to consult theory throughout your study. During data analysis, you will again revisit the literature to help you understand the relationship of your data to other research that relates to yours. The arrows in the center of the graphic illustrate how all of the different processes relate to one another in dynamic ways.

A qualitative study usually begins with an interest, problem, or question, as indicated in the top center of Figure 1.1. To develop this interest, you may seek out a variety of sources to get a "lay of the land" on the topic, including reading a range of texts and talking with others who know something about the topic or setting. This is represented in the graphic as building a theoretical framework and reviewing literature. As you become more familiar with the literature relevant to your topic, you develop a primary research question (and possibly a set of research questions) that will guide your study. As the arrows indicate, this is often a back-and-forth process, and you will continue to review and consult literature throughout your study.

Figure 1.1 The Dynamic Elements of Qualitative Research

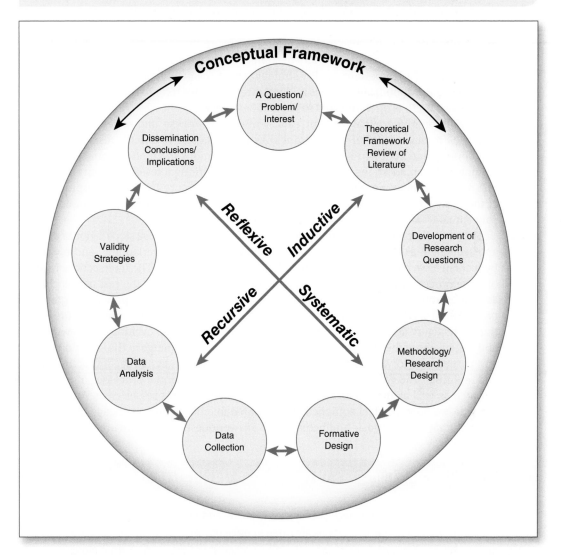

After developing research questions, you will begin to design your study. This includes determining which methods will best help you to answer the research questions. The selection of methods will often necessitate consulting literature as well. During the design process, you will select a research setting, determine the participant selection criteria, develop a research design, and plan for data collection and analysis. This research design is often developed through the creation of a research proposal to be vetted by others.

Methods and research instruments (the tools you use to collect data) are often piloted (or tested) as well as rehearsed and vetted to ensure that they are going to generate the data necessary to answer your research question(s); this is illustrated in what we call *formative design* in the graphic. Formative design may lead you to revise your research questions, methods, and/or study instruments as well as revisit the literature. After making necessary adjustments, you collect data. As you analyze your data by the means detailed in your research design, you will also include efforts to ensure that your data are valid and trustworthy. These efforts include a variety of strategies that we discuss in depth in Chapter Six. One example of an important validity strategy is to check in with participants to determine what they think about your analysis and interpretations; we refer to this strategy as participant validation (it is often referred to as member checks).

As you continue to analyze your data, you will also revisit and review literature that helps you to make sense of your findings. As your data analysis yields research findings, you develop and concertize your findings or key learnings in an effort to respond to the guiding research questions and goals of the study by writing a research report or another means of disseminating your findings. As represented in Figure 1.1, the development of all of these aspects and phases of research is integrated within and through the building of a conceptual framework, which is the focus of the next chapter.

The multidirectional arrows in Figure 1.1 signify that each of these steps is not as discrete and sequential as they may seem but are rather intersecting, overlapping, and recursive. By recursive, we mean that each of the steps informs other steps. The intersectional nature of these processes will become more apparent throughout reading the rest of this chapter as well as subsequent chapters in this book. Also, while Figure 1.1 provides a graphical presentation of what the processes of qualitative research look like, it cannot capture everything. The figure does not depict how qualitative research is also exciting, nerve-wracking, and messy—from the confusions of design to the issues that emerge in the selection of sites and participants to various aspects of conducting fieldwork and struggling with analysis and reporting. The goal of this book is to make each and all of the processes of qualitative research as well as the values and priorities underlying the processes clear. After reading this text, we hope you feel equipped and prepared to embark on and engage in qualitative research that fits your goals and helps you to answer questions that are important to you.

Terms and Concepts Often Used in Qualitative Research

Recursive: Qualitative research is recursive in that it builds and depends on all of its component parts. For example, your research questions are often informed by your personal and/or professional experiences, literature you have read, and the way you understand the world. Furthermore, as you begin to implement your research, the preliminary data you collect will also inform (and possibly lead you to refine) your research questions.

Positivism: Hughes (2001) describes the key aspects of the positivist paradigm through its view of the world as comprising unchanging, universally applicable laws and the belief that life events and social phenomena are/can be explained by knowledge of these universal laws and immutable truths. Within this paradigm, the belief is that understanding these universal laws requires observation and recording of social events and phenomena in systematic ways that allow the "knower" to define the underlying principle or truth that is the "cause" for the event(s) to occur. Positivist research also assumes that researchers are able to be objective and neutral.

SITUATING QUALITATIVE RESEARCH

Qualitative research, as a formalized field, emerged in part as a critique of, and alternative paradigm to, the **positivist** tradition (defined in the Terms and Concepts box above) that dominated research across most disciplines. With its focus on context, interpretation, subjectivity, representation, and the nonneutrality of the researcher as key aspects of the inquiry process, qualitative research emerged as a formal field in the late 1960s (Schwandt, 2015).

It is important to note that qualitative inquiry certainly existed and generated scholarship prior to the 1960s. For example, already widely known and practiced before the 1960s were the ethnographic tradition from cultural anthropology (Atkinson, Coffey, & Delamont, 2001; Hammersley & Atkinson, 2007) and the action-based research and participatory action research (described in Table 1.2) traditions that developed in the 1940s and 1950s (Fals-Borda & Rahman, 1991; Lewin, 1946).

A range of philosophical, **epistemological** (how knowledge is developed or constructed), and methodological beliefs informs qualitative research (Erickson, 2011; Hammersley, 2008; Mason, 2002). Denzin and Lincoln (2011a), premiere scholars in the field of qualitative research, describe the field as encompassing eight historical moments that overlap and exist presently.[1] Despite its association with specific disciplines and theories, qualitative research should not be constrained to specific traditions, disciplines, or methods; it is an "umbrella term that encompasses many approaches" (Atkinson et al., 2001, p. 7). To achieve a working definition of qualitative research, Mason (2002) describes three unifying principles of qualitative research; she states that it (a) is based on interpretivist assumptions, which she defines as "how the social world is interpreted, understood, experienced, produced or constituted" (p. 3); (b) includes data collection methods that are context specific and flexible (so that they can be adapted in real time based on emerging understandings and needs); and (c) uses methods of analysis that "involve understandings of complexity, detail and context" so that findings are properly contextualized (p. 3).

Expanding upon Mason's (2002) and others' (e.g., Denzin & Lincoln, 2011a; Erickson, 2011; Maxwell, 2013; Patton, 2015) conceptions of qualitative research, we position qualitative research as a mode of inquiry that centralizes the complexity and subjectivity of lived experience and values these aspects of human *being* and meaning making through methodological means. Qualitative researchers do not believe or claim that there are universal, static "Truths" but rather assert that there are multiple, situated truths and perspectives. We also view context and contextualization as central to understanding any person, group, experience, or phenomena. We question the interpretive role and authority of the researcher and acknowledge the subjectivity of all researchers. More broadly, we are ever mindful of what is referred to as **positionality,** which refers to a researcher's role and identity in relationship to the context and setting of the research. In these ways, qualitative research has changed how many researchers think about issues such as objectivity, interpretation, and the relationship between methods and study findings (Emerson, Fretz, & Shaw, 2011; Tetreault, 2012; Tisdell, 2008). Qualitative researchers also pay close attention to the relational aspects of research, including how interpersonal dynamics and broader issues of power and identity shape and mediate all aspects of the research process and ultimately the data and findings (Josselson, 2013; Steinberg & Cannella, 2012).

Terms and Concepts Often Used in Qualitative Research

Positionality: Positionality refers to the researcher's role and social location/identity in relationship to the context and setting of the research. For example, you could be a practitioner in the setting, located as an expert, an insider or outsider to the setting, a supervisor of employees, a member of community involved in the research, share a cultural or ethnic relationship with the participants, and so on.

Epistemology: Epistemology concerns the nature of knowledge, including how it is constructed and how it can be acquired. The epistemological assumption underlying qualitative research is that knowledge is developed "through the subjective experiences of people. It becomes important, then, to conduct studies in the 'field,' where the participants live and work" (Creswell, 2013, p. 20).

Ontology: Ontology concerns the nature of being or reality. In qualitative research, an ontological assumption is that there is not a single Truth or reality. Researchers, participants, and readers have differing realities, and a goal of qualitative research is to engage with, understand, and report these multiple realities (Creswell, 2013).

Methodology: Qualitative methodology refers to where ideology and epistemology meet research approach, design, methods, and implementation and shape the overall approach to the methods in a study, including the related processes, understandings, theories, values, and beliefs that inform them. This also refers to the ways in which your overall stance and approach to research broadly and your study specifically shape your specific research methods to collect and analyze data (e.g., interviews, focus groups, specific analytic processes).

DEFINING QUALITATIVE RESEARCH

The field of qualitative research has evolved into a vibrant, multifaceted, complex range of approaches and methodologies that are not easily grouped or defined. As noted by Denzin and Lincoln (2011a),

> Qualitative research, as a set of interpretive activities, privileges no single methodological practice over another. As a site of discussion or discourse, qualitative research is difficult to define clearly. It has no theory or paradigm that is distinctly its own . . . . Qualitative research is used in many separate disciplines . . . It does not

belong to a single discipline. . . . [Q]ualitative research is a set of complex interpretive practices. As a constantly shifting historical formation, it embraces tensions and contradictions, including disputes over its methods and the form its findings and interpretations take. (p. 6)

We, learning much from Denzin and Lincoln and other key scholars who have helped to develop, critique, and refine the field of qualitative research,[2] view qualitative research as a site of immense possibility precisely because it is not limited to any one discipline, theoretical perspective, or approach. Furthermore, its generative tensions and conflicts serve to continually push qualitative researchers to examine their assumptions, blind spots, and the ways that they reproduce dominance and privilege in and through research. Clearly, there is not one singular way to define or engage in qualitative research.

There are, however, shared perspectives and sensibilities across qualitative researchers and studies that help frame the broad array of approaches to it. To state it another way, qualitative research is not a monolith; despite sharing certain foundational ideas, there is great range and variation in approaches to qualitative research (Denzin & Lincoln, 2011a; Hammersley, 2008). See Table 1.2 for a summary of some of the more commonly used approaches in qualitative research. We now turn our attention to key characteristics of this dynamic field to create a shared understanding of qualitative research while still retaining the sense of its complexity, diversity, and multiplicity.

Qualitative research, broadly, is based on the methodological pursuit of understanding the ways that people see, view, approach, and experience the world and make meaning of their experiences as well as specific phenomena within it. Erickson (2011) describes it this way:

Qualitative inquiry seeks to discover and to describe in narrative reporting what particular people do in their everyday lives and what their actions mean to them. It identifies meaning-relevant *kinds* of things in the world—kinds of people, kinds of actions, kinds of beliefs and interests—focusing on differences in forms of things that make a difference for meaning. (p. 43)

Stated another way, and building on the work of Cochran-Smith and Lytle (1993a, 1993b, 1999, 2001, 2009), qualitative research is ideally built upon a **methodological** stance of inquiry on/in your practice as a researcher. A primary goal of this book is to complicate (in an accessible way) qualitative research and its attendant methods in ways that help new qualitative researchers understand the values and goals that must guide those who engage in critical and reflexive research that honors and respects the entire research endeavor. But in complicating qualitative research, we also seek to make clear that it is doable and that while subjective, deeply contextual, and therefore not generalizable (in a quantitative sense of that term), qualitative research is incredibly valuable to knowledge construction in a variety of ways that we discuss throughout the book.

Qualitative research as a field and as a set of methodologies has developed significantly over time. A scan of the editions of *The SAGE Handbook of Qualitative Research* (spanning from the first edition in 1984 to its fourth edition in 2011) shows the growth of the field, including the development of multiple theoretical orientations and conceptual frameworks that guide qualitative research, improved and more sophisticated methodological frameworks and methods of data collection and analysis, and engagement with the representational aspects of qualitative inquiry and its relationship to issues of equity, discrimination, marginalization, and social transformation (Kincheloe & McLaren, 2000). Denzin and Lincoln (2011a) describe the multiplicity of qualitative research well:

> Qualitative research is a field of inquiry in its own right. It crosscuts disciplines, fields, and subject matter. A complex, interconnected family of terms, concepts, and assumptions surrounds the term . . . . There are separate and detailed literatures on the many methods and approaches that fall under the category of qualitative research, such as case study, politics and ethics, participatory inquiry, interviewing, participant observation, visual methods, and interpretive analysis . . . . *Qualitative research* is a situated activity that locates the observer in the world. Qualitative research consists of a set of interpretive, material practices that make the world visible. These practices transform the world. They turn the world into a series of representations, including fieldnotes, interviews, conversations, photographs, recordings, and memos to the self. At this level, qualitative research involves an interpretive, naturalistic approach to the world. This means that qualitative researchers study things in their natural settings, attempting to make sense of or interpret phenomena in terms of the meanings people bring to them. (p. 3)

We take much from this characterization of qualitative research, and we too make connections across the theoretical (formal theories that frame an inquiry), epistemological (how we gain knowledge and know what we know), ontological (the nature of being), and methodological (the meeting place of ideology, inquiry, and methods) aspects of the field.

Looking across a wide range of texts devoted to describing qualitative research, there is deep and wide diversity in interpretive frames and approaches to qualitative inquiry.[3] Beyond this diversity, qualitative research is often defined and described in relationship to quantitative research. As you may know, quantitative research is associated with positivism and involves data that are analyzed numerically through statistical or other mathematical means. We do not think that comparing quantitative and qualitative research is necessary and often creates a false dichotomy. Many scholars describe qualitative research in relation to quantitative research to point out its underlying values and epistemologies. While for decades (roughly 1970s–1990s) people spoke of "the paradigm wars"—meaning in the broadest sense that there was tension between qualitative and quantitative researchers and then between those engaged in different forms of qualitative research[4]—to many, the historic sense of acrimony between qualitative and quantitative researchers is no longer active since we know that certain kinds of research questions require one or the other approach—or their strategic combination (what is known as mixed-methods research)—to gather the data they need to respond to research questions (Creswell & Plano Clark, 2011; Thomas, 2003). However, it is important to note that this tension between the paradigms can also be understood as an artifact of the ongoing efforts of many researchers to keep such conversations and tensions alive to generate knowledge that is methodologically appropriate and sophisticated and to challenge norms that seek to constrain and delegitimize qualitative ways of knowing.

While some qualitative researchers apply qualitative criteria and standards to quantitative research and some quantitative researchers apply their standards and validity criteria to qualitative research in ways that generate defensiveness and misunderstanding, the truth is that many researchers work from the understanding that each paradigm has different goals and each research approach serves different purposes. Simply criticizing one approach or the other does not generate knowledge or support methodological appropriateness or sophistication.

As we discuss in subsequent chapters, you might choose to use quantitative methods alongside qualitative methods for a variety of reasons given the goals of a specific study and the domains of inquiry delimited by the guiding research questions. Researchers should find/develop the methods that map onto—in the sense that they are the right ones to generate appropriate data—their research questions to get the most relevant and useful data. We are neither wedded to the sole use of quantitative or qualitative methods nor wedded to a particular qualitative approach since, as we argue throughout this book, the chosen approach and related methods depend on the research questions and goals of a study as well as on other contextual variables. The methodological approach should be organic and emerge from research questions, study goals, the context in which the research is being carried out, and other mediating variables such as researcher identity and positionality. We will discuss this in more depth as we explore the roles of conceptual frameworks in research in Chapter Two and research design in Chapter Three.

KEY COMPONENTS OF QUALITATIVE RESEARCH

For the reasons described above, we are hesitant to provide broad generalizations of qualitative research or to simply compare it with quantitative research. However, to provide you an orientation, in Table 1.1, we describe what we view as the key components of qualitative research. Some of the shared values and epistemological stances of qualitative researchers include conducting fieldwork using naturalistic engagement, focusing on both describing and analyzing, seeking complexity and contextualization, situating the researcher as the primary instrument in the study, paying careful attention to process and relationships, maintaining fidelity to participants, focusing on meaning making, and placing primacy on inductive understandings and processes.

Of necessity, the components described in Table 1.1 are not exhaustive (to say the least) and are also a bit overgeneralized. However, this table highlights some of the central aspects of qualitative research in terms of its foundational beliefs, assumptions, and methodological dimensions and approaches. Later in this chapter, we discuss specific approaches to qualitative research that highlight some of the differences even in the face of these ideally shared underlying values and foci.

To thread the above points together, qualitative research is at its core about viewing, understanding, and engaging with people as having expertise broadly and specifically in relation to their own experiences (Jacoby & Gonzales, 1991; van Manen, 1990). As we emphasized above, there is no goal of finding an objective or immutable "Truth" in qualitative research. Within qualitative research, people's experiences and perspectives are deeply embedded in the contexts that shape their lives, and how people experience aspects of their lives and the world is subjective and can change over time. Qualitative researchers are precisely interested in people's subjective interpretations of their experiences, events, and other inquiry domains.

Table 1.1 Components of Qualitative Research

Fieldwork and naturalistic engagement	Qualitative research involves *fieldwork and naturalistic engagement,* which means that the researcher is physically present with the people, in a community and/or institution to engage, observe, and record experience and behavior within a natural setting (Glesne, 2016; Merriam, 2009; Robson, 2011).
Descriptive and analytic	Qualitative research is both *descriptive and analytic* in that researchers are interested in understanding, describing, and ultimately analyzing, in detailed and deeply contextualized ways, the complex processes, meanings, and understandings that people have and make within their experiences, contexts, and milieu (Maxwell, 2013; Merriam, 2009; Robson, 2011).
Seeks complexity and contextualization	Qualitative research *seeks complexity and contextualization* in terms of how reality exists and unfolds in ways that are temporal, contextual, and highly individualized even as participants may share certain experiences and perspectives (Alvesson & Sköldberg, 2009; Maxwell, 2013).
Researcher as instrument	In qualitative research, the researcher is considered the *primary instrument* (Lofland, Snow, Anderson, & Lofland, 2006; Porter, 2010) of the research throughout the research process, meaning that the subjectivity, social location/identity, positionality, and meaning making of the researcher shape the research in terms of its processes and methods and therefore shape the data and findings. Thus, the identity of the researchers is viewed as a central and vital part of the inquiry itself.
Process and relationships	Qualitative researchers pay careful attention to *process and relationships,* meaning that there is an intentional focus on how the research process—including procedures, methods, and interpersonal dynamics—itself generates meaning and important frames for understanding data (Hammersley & Atkinson, 2007; Maxwell, 2013).
Fidelity to participants	Qualitative research shows a *fidelity to participants* and their experiences rather than a strict adherence to methods and research design and in that sense can take an emergent approach to research design and implementation (Emerson, Fretz, & Shaw, 2011; Glesne, 2016; Hammersley, 2008; Hammersley & Atkinson, 2007; Maxwell, 2013).
Meaning and meaning making	Qualitative researchers are interested in *meaning and meaning making*, which entails a deep investment in understanding how people make sense of their lives and experiences, as well as how the meanings people make of/in their lives are socially and individually constructed within and directly in relation to social and institutional structures (Cannella & Lincoln, 2012; Kincheloe & McLaren, 2000).
Inductive	The process of qualitative research is largely *inductive* in that the researcher builds concepts, hypotheses, and theories from data that are contextualized and that emerge from engagement with research participants (Maxwell, 2013; Merriam, 2009).

The Role of the Researcher in Qualitative Research

Because the researcher is the primary instrument in qualitative research, the *role of the researcher* is a central consideration in qualitative research. Drawing on the concept that a researcher's identity is central in qualitative research design (Maxwell, 2013), we assert that positionality and social location are two central components of researcher identity and, furthermore, that both positionality and social location are central to understanding the researcher's role in every stage of the research process. Positionality, which we introduced earlier in this chapter, is the researcher's role and identity as they intersect and are in relationship to the context and setting of the research. Positionality consists of the multitude of roles and relationships that exist between the researcher and the participants within and in relation to the research setting, topic, and broader contexts that shape it. To fully understand researcher positionality, researchers must consider **social location** (identity markers of researchers; see the full definition in the box on page 12).

Sometimes researcher social location/identities and positionalities are discussed in ways that create **binaries** (such as polarized notions of insider and outsider positionalities, of practitioners and scholars, or binaried racial, cultural, or gender categories). Realistically, there is great range and variation in the roles and positions that researchers take up and embody in relation to research participants and settings (positionality) and the ways a researcher's social identity/social location is interpreted as well as how researchers interpret themselves.. This might mean that researchers can be considered as both insider and outsider; as scholar and practitioner; as supervisor and employee; as teacher and student; as a member of multiple cultural, social, or thought communities; as multiracial and multicultural; as not having a binaried gender identity; and so on (Henslin, 2013; Neyrey, 1991; Tetreault, 2012; Tisdell, 2008).

Each researcher has a set of roles and identities, and these roles and identities can also shift and change over time. Part of taking a methodological approach to engaging **criticality in qualitative research**, which we define in the box below and discuss throughout the book, is understanding these complexities and not seeing them as either/or identities but rather as both/and, meaning that roles and identities are always in complex interaction. As we discuss in Chapter Two, positionality and social location should be thought of in complex relationship to the researcher, the setting, and participants. Macro-sociopolitical contexts shape social locations and positionalities, and these relationships are temporal, dynamic, and contextual. From our experience, researchers often approach the consideration of social location and positionality as a kind of checklist of things to do at the outset of a study or as a "mea culpa" that seemingly absolves the researcher of actually engaging with issues that the confluence of identities and roles can create in research in deep and ongoing ways throughout all aspects and phases of the research process. Thinking about and addressing issues such as positionality and the identity of the researcher more broadly should not be a checklist; these issues should constitute a vibrant source of inquiry and generative tension as a researcher is reflexively engaged in every research aspect. When examining social location/identity and positionality with respect to the research context, it is reflected in all aspects of the research process (e.g., developing research questions, engaging with [or excluding] theories, selecting and recruiting research participants, structuring interview protocols and other data collection instruments, interacting with research participants, analyzing data, and sharing [or not] aspects of data and analyses with research participants). Throughout this book, we argue that reflexively considering positionality and **social location** should be a complex, multifaceted, and systematic process in qualitative research. Building on this premise, we discuss methodological ways to engage this approach throughout each chapter of this book.

Terms and Concepts Often Used in Qualitative Research

Emergent: This term is often used in relationship to qualitative research design to signify that qualitative research does not typically follow a fixed design. Based on multiple factors, qualitative research can evolve and change. Researchers can refine and revise their research questions, data collection methods, and other aspects of a qualitative study. This aspect of qualitative research is often described as being emergent. Qualitative researchers also use the term *emergent* to mean aspects or understandings that arise from data. For example, an emergent theory is a working assumption that you are building from analysis of your data.

Social location: In qualitative research, social location is often used synonymously with social identity. Social location/identity includes the researcher's gender, social class, race, sexual identity/orientation, culture, and ethnicity as well as the intersections of these and other identity markers such as national origin, language communities, and so on (Henslin, 2013; Neyrey, 1991).

Criticality in qualitative research: Criticality in qualitative research, as we define it, is the meeting place of the theoretical (formal theory), the conceptual and contextual (conceptualizations and enactments of everyday life and social arrangements), and the methodological (where ideology and epistemology meet research methods). It is the researcher who acts as the translator and mediator of the spaces between the realms of the theoretical, conceptual, and methodological. One aspect of criticality in qualitative research is a recognition and address of the power inherent in research and in society more broadly. Criticality in qualitative research also involves transparency, intentionality, and striving to present as complex and contextualized a picture as possible. Criticality is achieved through the methods you use as well as how you engage with all aspects of research. We discuss other aspects of criticality in qualitative research throughout this chapter and the entire book.

Binaries: In qualitative research, binaries can refer to dichotomies such as polarized notions of insider and outsider positionalities, of practitioners and scholars, or of racial, cultural, or gender categories, and so on. Binaries serve to reduce complexity and impose an either/or frame to typically more complex and multifaceted lived experiences.

Hegemony: The concept of hegemony, developed by scholar and activist Antonio Gramsci, refers to the social, cultural, ideological, and economic influence imposed by dominant groups in society. The dissemination of dominant ideologies is enacted and maintained through ideological, social, cultural, and institutional means in such a way that dominant ideas, values, and beliefs appear to be "normal" and neutral (Agnew & Corbridge, 2002) because "the ideas, values, and experiences of dominant groups are validated in public discourse" and represented in and through public process and structures, including education, politics, law, and social institutions (Lears, 1985, p. 574).

Iterative: Qualitative research is often described as iterative, signifying that it (a) involves a back and forth of processes and (b) changes and evolves over time as you engage in these processes. Ideally, these back-and-forth processes lead to a progressive, evolutionary refinement of your research at conceptual, theoretical, and methodological levels.

HORIZONTALS IN QUALITATIVE RESEARCH: CRITICALITY, REFLEXIVITY, COLLABORATION, AND RIGOR

Working from a hermeneutic perspective,[5] which we view as a framework for conceptualizing and understanding the iterative processes of interpretation of everyday experience, including interpretation of yourself, those around us (near and far), and human interaction in ways that are contextualized and that shift and change in context over time, we argue that bias exists in all research (quantitative, qualitative, and mixed methods) and that the multitude of choices researchers make—within and across paradigms—is a result of underlying beliefs and assumptions. In qualitative research, understanding and confronting the values and beliefs underlying decisions and approaches is vital and at the heart of the inquiry itself. This examination of biases, we argue, becomes an ethical responsibility for researchers since it has both indirect and direct implications for other people's lives (Nakkula & Ravitch, 1998). Thus, qualitative researchers should make deliberate methodological choices to acknowledge, account for, and approach researcher bias. Related to these methodological choices and the processes that stem from them, we assert that *criticality, reflexivity, collaboration,* and *rigor* are vital to conducting ethical and valid qualitative research.

We refer to and discuss these as horizontals not because there is anything linear about them but because we believe that these are crucial aspects of qualitative research that should run throughout all phases and processes of qualitative research. As we discuss throughout the book, the research questions, goals, and purposes of a study guide and inform the choices researchers make. Thus, while we believe that criticality, reflexivity, collaboration, and rigor are important to all aspects of qualitative research, we acknowledge that the kind of research topics and goals that researchers have shape the degree to which they engage these ideals within the actual research design, implementation, analysis, and written reports. Furthermore, it takes a particular kind of focus and intentionality to conduct qualitative research that is critical, reflexive, collaborative, and rigorous. We describe this focus as we revisit these horizontals throughout the book. Finally, we argue that to conduct ethical qualitative research, researchers should address criticality, reflexivity, collaboration, and rigor in specific rather than general ways, that is, to engage them methodologically, not just conceptually or ideologically. We operationally define and describe these horizontal aspects in the sections that follow.

Criticality

We argue for what we term criticality in qualitative research. The word *critical* has a range of meanings, both broadly in academia and specifically in relation to qualitative research.[6] Our conception of criticality in qualitative research entails the wedding of the ideological and theoretical with the methodological and looks holistically at how theory, methods, process, context (macro and micro), and researcher identity and positionality intersect with and shape qualitative work. We seek to create a more inclusive and broad definition of criticality in qualitative research rather than one that situates it solely within critical social theory.[7]

From Cannella and Lincoln's (2012) *Critical Qualitative Research Reader*, we adopt (and adapt) an umbrella definition of critical qualitative research that includes key characteristics and beliefs in that it (a) recognizes power and power relationships; (b) de-normalizes reified social norms and assumptions; (c) asks questions about asymmetries and power relations; (d) interrogates who benefits from and/or is marginalized by society (and research); (e) challenges the hegemony of dominant, grand narratives; (f) illuminates hidden power structures and deeply considers language usage and the circulation of discourses that organize everyday life; (g) is concerned with issues of race, gender, and social class; and (h) resists colonialism, neocolonialism, and hegemony more broadly (pp. 105–106). Like Cannella and Lincoln's description of critical qualitative research, our framing of criticality includes the conceptual, ideological, theoretical, and epistemological. In contrast, we do not view critical social theory as synonymous with criticality in qualitative research. In this sense, criticality refers to a broader conceptualization of what it means to critically approach qualitative research other than solely examining its relationship to critical social theory. Considering issues such as power and inequity, which includes impositions of social hierarchy and issues of structural inequity in everyday life, is central to taking a more critical approach to qualitative research (Kincheloe & McLaren, 2000).

It is certainly true that an aspect of what we consider criticality in qualitative research stems from our engagement and alignment with critical social theory. Specific examples of critical theories that we have engaged in our research include critical multicultural theory, postcolonial theory, critical hermeneutics, and critical race theory. However, we argue that researchers need not strictly adhere to a specific critical social theory framework to critically approach qualitative research. Whether you study and frame your research within specific critical social theories directly, issues of power and equity are inherent in all research and research studies, and thinking (and acting) reflexively (in the methodological sense of that term) about these issues in systematic and ongoing ways that situate the researcher as fallible, as a learner, and as deeply embedded in the process and findings of research is what engenders what we consider to be criticality in qualitative research.[8]

Criticality in qualitative research centralizes the adoption of a critical methodological approach to your research, which we believe is about seeking out and creating the conditions that allow us, as researchers, to see, engage, contextualize, and make meaning of the complexity of people's lives, society, and the social, political, institutional, and economic forces that shape them. This includes maintaining a fidelity to the complexity and layeredness of people's complicated experiences, identifying and resisting hegemonic hierarchy and power asymmetries, working against binaries and deficit thinking,[9] and engaging in a specific methodological process that engages these issues with intentionality, through an inquiry stance that centralizes thinking about critical self-reflection and social reflection as a part of an iterative methodology that attends to complexity and context in a systematic, reflexive, and transparent way (Alvesson & Sköldberg, 2009; Chilisa, 2012; Nakkula & Ravitch, 1998). This methodological process, which we term *critical research design,* is described in Chapter Three. Throughout the book, we discuss how researchers can cultivate and engage in certain methodological processes to achieve the complexity we believe reflects a more critical approach to research design and implementation.

Taking an inquiry stance (Cochran-Smith & Lytle, 2001, 2009) in your research is central to criticality in qualitative research. Taking an inquiry stance on research requires that researchers cultivate and refine understandings of the active role of reflection in research. Furthermore, we view critical inquiry as a research ethic as well as a vital aspect of understanding ourselves; it means that researchers must be committed to rigorous processes of self-reflection and continual investigation into and systematic, data-based critique of research practices and the macro and micro contexts that shape them (Tierney & Sallee, 2008). An inquiry stance on research translates into more person-centered, systematic, and proactive approaches to understanding people in context (Cochran-Smith & Lytle, 2001, 2009; Ravitch, 2006a, 2006b, 2014; Ravitch & Tillman, 2010). This practice of research seeks to resist the current confines, norms, and challenges of research—and the contexts in which it is carried out. Criticality in qualitative research seeks to develop and serve as a counternarrative to dominant cultural knowledge and normative narratives that circulate in everyday life.

Reflexivity

A central aspect in qualitative research is *researcher reflexivity*. Broadly, researcher reflexivity is the systematic assessment of your identity, positionality, and subjectivities. It is an active and ongoing awareness and address of the researcher's role and influence in the construction of and relational contribution to meaning and interpretation throughout the research process (L. Anderson, 2008; Denzin, 1994; Guba & Lincoln, 2005; Steedman, 1991). Methodologically, this entails a self-reflection of biases, theoretical preferences, research settings, the selection of participants, personal experiences, relationships with participants, the data generated, and analytical interpretations (Schwandt, 2015). Reflexivity, then, requires you to be vigilant and to frequently reassess your positionality and subjectivities.

Considering that the researcher is the primary instrument of qualitative research (Lofland et al., 2006; Porter, 2010) and therefore the importance of systematically considering and methodologically addressing social location and positionality (Henslin, 2013; Neyrey, 1991; Tetreault, 2012; Tisdell, 2008), it is clear that the values and epistemologies of the researcher are vitally important to the research design, implementation, and findings of any study (Alvesson & Sköldberg, 2010). Related to this, the researcher's beliefs, socialization experiences, and understandings of concepts and experiences (e.g., emotion, culture, experiences in schooling, social engagement) might seem neutral but are actually quite subjective, political, and value laden.

Given that qualitative research is focused on and steeped in an appreciation of subjectivity and interpretation, understanding personal subjectivities is of vital importance. Acknowledging our subjectivities as researchers, which some refer to as a "disciplined subjectivity" (Erickson, 1973; LeCompte & Goets, 1982; McMillan & Schumacher, 1997), is central to rigorous and valid research. Reflexivity requires systematic attention to our subjectivity; this entails that subjectivities and biases are engaged with and scrutinized in systematic ways (L. Anderson, 2008; Denzin, 1994; Guba & Lincoln, 2005; Steedman, 1991). It is one of the responsibilities of the researcher to seek to understand the nature of those subjectivities as they directly relate to the construction, design, and enactment of the research.

Throughout the book, we discuss the inextricable connection between methods and findings in qualitative inquiry, but it is important to underscore here that as with all research, you, as a qualitative researcher, shape research in ways that reflect your values and assumptions about the world. This, in turn, shapes how studies are designed, how the data are collected, how such data are interpreted and analyzed, and therefore what you argue and represent in your texts. For this reason, careful attention to who you are; what you think of and assume about yourself, other people, and the world; and how you view the role of research in understanding human *being* is vital to rigorous, valid research. The way that you approach engaging in critical reflection about all of these aspects of who you are and how that figures into and shapes myriad aspects of your research are important aspects of researcher reflexivity.

Terms and Concepts Often Used in Qualitative Research

Dialogic engagement: We refer to the collaborative, dialogue-based processes that qualitative researchers engage in throughout a research study as dialogic engagement. These processes, which we discuss and provide examples of throughout the book, focus on pushing yourself to think about various aspects of the research process (and products) through talking about them with strategically selected thought partners. Thought partners are people who can challenge you to see your self and your research from a variety of angles at various stages throughout the research process. These people can include colleagues, advisers, peers, research team members, inquiry group members, and/or research participants.

Collaboration

In addition to adopting a systematic practice of reflexivity, engaging with participants, colleagues, advisers, and mentors in thoughtful and deliberate ways is crucial to conducting good qualitative research. This is the third "horizontal" value of qualitative research that we revisit often in this text: *collaboration*. There is a wide range of possibilities for what collaboration can look like in qualitative research. Regardless if you are a lone researcher, member of a research team, or involved in a participatory study that is co-constructed among participants, we argue that collaboration should be an active part of your qualitative research study throughout all stages of the research process.

Qualitative research should not be understood or approached as an isolated endeavor but as an endeavor that needs collaboration. Researchers should actively seek out interlocutors with whom we explore and challenge ourselves in ways impossible in the solitude and safety of our own minds. To this end, and based on the ideas of many before us, we argue that dialogic engagement is a requirement of rigorous, reflexive research and constitutes an approach to qualitative research that engenders and supports criticality (Bakhtin, 1981, 1984; Lillis, 2003; Rule, 2011; Tanggaard, 2009). Dialogic engagement processes allow you to co-create the conditions of collaboration by deliberately engaging thought partners, critical friends, and/or research participants to challenge your biases and interpretations.[10] About this relational aspect of reflexivity and collaboration, Ravitch and Riggan (2012) offer this:

We strongly encourage you to engage in [reflexive] exercises individually as well as in dialogue and collaboration with others who will engage thoughtfully and critically with you as you design and carry out your research, pushing you to examine parts of yourself and your research that you might otherwise take for granted or leave unexamined. Dialogue and exchange are essential to the trustworthiness of your empirical work and we strongly encourage an approach to research that is dialogical and relational as well as internally engaged. The two go hand in hand as a means of conducting the most rigorous, credible research possible. Conceptualizing and carefully documenting these processes is an important part of your methodological approach. (p. 147)

In the subsequent chapters, we discuss dialogic engagement as a necessary component of qualitative research, and we suggest specific ways to collaboratively engage with participants, colleagues, advisers, and mentors in and throughout the research process. We also highlight ways to think about, document, and engage in collaboration that relate to some of its possibilities and challenges.

Rigor

To study individuals' lived experiences and understand them in truly complex and contextualized ways requires a faithful attention to methodological rigor. In *The SAGE Encyclopedia of Qualitative Research Methods,* Saumure and Given (2008) offer this insight about rigor in qualitative research:

As a concept, rigor is perhaps best thought of in terms of the quality of the research process. In essence, a more rigorous research process will result in more trustworthy findings. A number of features are thought to define rigorous qualitative research: transparency, maximal validity or credibility, maximal reliability or dependability, comparativeness, and reflexivity. . . . Using these criteria for building a rigorous research study will enable qualitative researchers to report results that are considered as both useful and credible by their peers. (pp. 795–796)

Using the above definition as a broad umbrella, we conceive of *rigor in qualitative research* to encompass a variety of concepts, considerations, and actions, including the following:

- Developing and engaging a research design that seeks complexity and contextualization through the structure and strategic sequencing of methods and mapping of research methods onto the guiding research questions

- Maintaining a fidelity to participants' experiences through engaging in inductive (or what we think of as emergent design) research that is responsive to emerging meanings while at the same time ensures a systematic approach to data collection and analysis (Cavallo, 2000)

- Seeking to understand and represent as complex and contextualized a picture of people, contexts, events, and experiences as possible

- Transparently addressing the processes, challenges, and limitations of your study

As we discuss throughout the book, engaging in rigorous qualitative research entails designing a study that is responsive not only to the research questions and goals but also to the participants and emerging learnings throughout the research. It is as much about strategic and appropriate research design as it is about the implementation and conduct of research.

An additional way that rigor is achieved in qualitative studies is through the reflexive engagement processes described above. Achieving rigor in this manner includes a systematic uncovering of and attention to yourself and your views, assumptions, and biases and how they play out in and shape all aspects of your research. This engagement leads qualitative researchers to an understanding of the subjectivity of individual experience and of not only intergroup variability (differences across cultural and social groups) but also intragroup variability, which Erickson (2004) defines as "variability of culture within social groups and the continual presence of cultural change as well as cultural continuity across time" (p. 46). The concept of intragroup variability suggests that there is incredible diversity not only across various groups of people (be they cultural or social) but within any group of people. Interrupting normative approaches to research that are typically steeped in oversimplified thinking about culture is an important aspect of conducting rigorous qualitative research that resists essentializing (and therefore denying individuality to) individuals. Furthermore, rigor involves paying careful attention throughout the research process to context and complexity; without this attention, qualitative research can simply serve to reinscribe reductionist, essentializing, disrespectful, and ultimately, we argue, unethical interpretations and representations of people's experiences and lives.

We discuss rigor throughout the book and specifically in Chapter Three, which focuses on research design, and Chapter Six, which focuses on validity in qualitative research.

APPROACHES TO QUALITATIVE RESEARCH: AN OVERVIEW

One goal of this introductory text is to orient you to the qualitative research paradigm. Within that goal is the desire to explicate the specific approaches within the qualitative paradigm. The choice of methodological approach is primarily guided by the study's research questions and aims; it also stems from various contextual influences, the researcher's epistemological leanings and beliefs, and existing theory and research. Thus, the methodological approach is a part of—shaping of and derived from—the conceptual framework of a study (described in depth in Chapter Two) and is variable in terms of each study since some researchers will work *from* an approach and others will arrive *at* an approach. There are a multitude of approaches to qualitative research; in this section, we briefly define some of the commonly used approaches and refer you to different texts for additional exposition of approaches.[11]

While there are more approaches to qualitative research than chapters in this book, we briefly define 10 main approaches: action research, case study research, ethnography and critical ethnography, evaluation research, grounded theory, narrative research, participatory action research, phenomenology, and practitioner research since these are the more common of the specific approaches to qualitative research, and therefore our students typically need to be familiar with these approaches to consider their options and develop their methodological approaches to whatever it is they are studying. We do not review these in depth since that is beyond the scope of this book but rather provide overview definitions to orient the reader in Table 1.2. We also provide additional reading resources at the end of the chapter.

Before including overviews to these 10 approaches, it is important to note that the majority of qualitative research studies, in terms of approach, remain unnamed/unspecified and are referred to as "general qualitative research." Since many qualitative studies do not situate themselves within a specific approach and since even when using different approaches there is much shared across qualitative methods, this book explores and describes qualitative research in general rather than within one or another approach. It is also important to note that in addition to these approaches, there are also multiple interpretive frameworks including (but not limited to) feminist theory, hermeneutics, critical race theory, anti/postcolonial theory, queer theory, disability theories, Black feminist epistemology, poststructuralism, and critical realist theories. While we do not go into detail on these here because this is beyond the scope of this book, we refer you to several helpful sources on this, which include *Conceptualizing Qualitative Inquiry: Mindwork for Fieldwork in Education and the Social Sciences* (Schram, 2003), *Qualitative Inquiry and Research Design: Choosing Among Five Approaches* (Creswell, 2013), *The SAGE Handbook of Qualitative Research* (Denzin & Lincoln, 2011c), and *Critical Qualitative Research Reader* (Steinberg & Cannella, 2012).

To be clear, there are many more qualitative approaches than defined here, but in an effort not to overwhelm and to provide working definitions of the most commonly used approaches, we have simply summarized these more commonly used approaches here. There are numerous additional and valuable approaches, including (but still not limited to) appreciative inquiry, autoethnography, indigenous research methodologies, social science portraiture, teacher research, and many of the approaches listed here that have more "critical" forms of the approach as well. We encourage you to refer to the references we provide at the end of the chapter and in the chapter notes.

These approaches are important to carefully consider because they have not only ideological and conceptual implications but methodological ones as well. It is important to note that at times, people can combine these approaches, for example, engaging in a case study that employs participatory methods or using ethnographic methods to inform the design of an evaluation study. It is also important to note that you can use elements of some of these approaches, for example, using some level of participatory methods at various points of a study (e.g., to collaboratively construct guiding research questions with a community) but not engaging in an entirely participatory process throughout all stages of research or using ethnographic methods of participant observation and fieldnote writing without engaging in an entire ethnography. Within and across these approaches are specific values and meanings. The researcher must carefully consider how these relate to the study goals, driving questions, setting, guiding theories, and commitments.

Table 1.2 Approaches to Qualitative Research

Action research	Action research can be understood as a practice of situated, interpretive, reflexive, collaborative, ethical, democratic, and practical research. Because problems and issues that action research deals with derive from the lived experiences of everyday life, it does not view theory and practice as separate entities but as an integral part of social interactions. In his seminal book, *Action Research*, Stringer (2014) defines action research as
	a systematic approach to investigation that enables people to find effective solutions to problems they confront in their everyday lives. Unlike experimental or quantitative research that looks for generalizable explanations related to a small number of variables, action research seeks to engage the complex dynamics involved in any social context. It uses continuing cycles of investigation designed to reveal effective solutions to issues and problems experienced in specific situations and localized settings, providing the means by which people in schools, businesses, community agencies and organizations, and health and human services may increase the effectiveness and efficiency of their work. In doing so it also seeks to build a body of knowledge that enhances professional and community practices and works to increase the well-being of the people involved. (p. 1)
	Action research is the meeting place of research and action, and the two are intertwined in generative ways that inform each other. The process of action research is central to its core value and stance as a collaborative and democratic process within and through which researchers are co-inquirers who take shared responsibility for the overall research endeavor and who share a goal of applying insights gained through systematic inquiry to the context at the heart of the investigation.
Case study research	Case study research methods involve studying a case (or multiple cases) of contemporary, real-life events (Yin, 2009). A case is typically understood as bounded by time and place (Creswell, 2013). Stake (1995) further explains what makes something a case:
	A child may be a case. A teacher may be a case. But her teaching lacks the specificity, the boundedness, to be called a case. An innovative program may be a case. All the schools in Sweden can be a case. But a relationship among schools, the reasons for innovative teaching, or the policies of school reform are less commonly considered a case. These topics are generalities rather than specifics. The case is a specific, a complex, a functioning thing. (p. 2)
	Miles and Huberman (1994) consider the case to be one's unit of analysis. Case study research methods tend to employ a variety of data sources, including direct observations, interviews, documents, artifacts, and other sources (Eisenhardt, 1989; Yin, 2009). Furthermore, case study research is not exclusive to qualitative research methods.[12] Building theory from case study research involves an iterative process, and there are specific steps that researchers can follow (Dooley, 2002; Eisenhardt, 1989). Dooley (2002) explains that often in case study research new theory is developed through a process in which findings are "extended from one case to the next and more and more data are collected and analyzed. This form of reiteration and continuous refinement, more commonly referred to as the multiple case study, occurs over an extended period of time" (p. 336).[13]

Ethnography and critical ethnography	Ethnography, like many other forms of qualitative research, places an emphasis on in-person field study and includes immersion, through participant observation, in a setting to decipher cultural meaning and generate rich, descriptive data that emerge in relation to the "development of rapport and empathy with respondents, the use of multiple data sources, the making of fieldnotes.
	. . . Ethnography unites process and product, fieldwork, and written text. Fieldwork, undertaken as participant observation, is the process by which the ethnographer comes to know a culture; the ethnographic text is how culture is portrayed" (Schwandt, 2015, pp. 98–99).
	Ethnography stems from anthropology and has a complex history; it is diverse and variable and even contested in terms of its definition and what constitutes immersion, culture, and participant observation (Hammersley & Atkinson, 2007).
	There are multiple forms of ethnography, including holistic, semiotic, and critical. We describe ethnography broadly above and also highlight critical ethnography here because they are the most common approaches. *Critical ethnography*, which is viewed as related to but departing from general ethnography,
	aim[s] to criticize the taken-for-granted social, economic, cultural, and political assumptions and concepts . . . of Western, liberal, middle-class, industrialist, capitalist societies. Critical ethnographies are focused, theorized studies of specific social institutions or practices that aim to change awareness and/or life itself. . . . While difficult to characterize in terms of a single set of features, critical ethnographies in the main are marked by several shared dispositions: a disavowal of the model of ethnographer as detached, neutral participant observer; a focus on specific practices and institutions more so than holistic portraits of an entire culture; an emancipatory versus a solely descriptive intent; and a self-referential form of reflexivity that aims to criticize the ethnographer's own production of an account. (Schwandt, 2015, pp. 47–48)
	This critical form of ethnography is based on critique of normative research as well as of the representation of hegemony in institutions and society more broadly. While it shares methods of data collection and analysis with traditional forms of ethnography, its guiding ideology and attendant methodology, as well as its goals and processes, differ from more traditional forms of ethnography in important ways.

(Continued)

Table 1.2 (Continued)

Evaluation research	Evaluation research, broadly, "can include any effort to judge or enhance human effectiveness through systematic data-based inquiry" (Patton, 2015, p. 18). Such research is used "to provide accountability; for analysis and learning; to facilitate funding allocation; and for advocacy" (Guthrie, Wamae, Diepeveen, Wooding, & Grant, 2013, p. 1). Evaluation research can be either/both quantitative and qualitative; the goal of qualitative data in evaluation research is to create greater understanding and to contextualize and humanize statistics and numbers. Qualitative research methods can contribute to multiple kinds of evaluations, including program evaluation, which focuses on the processes and outcomes of a program, and quality assurance, which focuses on how processes and outcomes affect individuals (Patton, 2015). The criteria used in evaluation research depend on the specific type of evaluation being conducted. For example, in a program evaluation, evaluators may consider the expressed goals of the program, historical data, and a variety of other factors.[14] In addition, in goal-free evaluation, researchers deliberately avoid studying the expressed goals of the program and instead focus on the effects and outcomes of participants' needs (Patton, 2015). Not only can evaluation research methods include quantitative and qualitative methods, but researchers may employ a variety of qualitative approaches such as phenomenology, grounded theory, ethnography, and action research to conduct an evaluation (Patton, 2015).[15] Despite the different kinds of evaluation research,[16] researchers need not follow strict methodological guidelines as the context greatly influences the type of research that will be conducted (Clark & Dawson, 1999).
Grounded theory research	Grounded theory is an approach to qualitative research that attempts to develop theory that comes from data or the field. Grounded theory involves specific procedures,[17] which Schwandt (2015) summarizes well: [Grounded theory methodology is] often used in a non-specific way to refer to any approach to developing theoretical ideas (concepts, models, formal theories) that begins with data. But grounded theory methodology is a specific, rigorous set of procedures for analyzing qualitative data to produce formal, substantive theory of social phenomena . . . [it] requires a concept-indicator model of analysis that, in turn, employs the method of constant comparison. Empirical indicators from the data (actions and events observed, recorded, or described in documents in the words of interviewees and respondents) are compared looking for similarities and differences. From this process, the analyst identifies underlying uniformities in the indicators and processes a coded category or concept. Concepts are compared with more empirical indicators and with each other to sharpen the definition of the concept and to define its properties. Theories are formed from proposing plausible relationships among concepts and sets of concepts. Tentative theories or theoretical propositions are further explored through additional instances of data. The testing of the emergent theory is guided by theoretical sampling. (pp. 62–63) Data for grounded theory studies can come from a variety of sources such as interviews, observations, documents, and other sources. Important to grounded theory is the premise that data analysis begins as soon as the first piece of data is collected (Corbin & Strauss, 1990, 2015). As we assert the importance of memoing throughout the book, it is also an especially important analytical tool throughout all aspects of grounded theory research studies (Corbin & Strauss, 1990, 2015).

Narrative research/inquiry	Narrative research can refer to a phenomenon under study or the specific methods used to investigate the phenomenon; narrative researchers describe individuals' storied lives and write about the experience (Connelly & Clandinin, 1990). Schwandt (2015) states,

This is a broad term encompassing the interdisciplinary study of the activities involved in generating and analyzing stories of life experiences (e.g., life histories, narrative interviews, journals, diaries, memoirs, autobiographies, biographies) and reporting that kind of research. Narrative inquiry or research also includes examining the methodology and aim of research in the form of personal narrative and autoethnography. (p. 211)

Narrative research methodologically gives primacy to the lived experiences of individuals as expressed in their stories. Connelly and Clandinin (1990) posit, "The main claim for the use of narrative in educational research is that humans are storytelling organisms who, individually and socially, lead storied lives" (p. 2). Narrative research holds that individuals construct reality through the narration of their stories (Lichtman, 2014). Narrative research typically includes a focus on "one or two individuals, gathering data through the collection of their stories, reporting individual experiences, and chronologically ordering the meaning of those experiences (or using life course stages)" (Creswell, 2013, p. 70). It is important to note that narrative research techniques can be used in conjunction with other qualitative approaches such as during in-depth interviews (Lichtman, 2014). |
| **Participatory action research (PAR)** | *Participatory action research* (PAR) is an umbrella term for a variety of participatory approaches to action-oriented research that focus on challenging hierarchical and asymmetrical relationships between research and action and between researchers and members of minoritized or marginalized communities (Kindon, Pain, & Kesby, 2007). According to Schwandt (2015), PAR involves working with groups and communities that experience the effects of hegemony through social control, oppression, or colonization; it places a priority on "the politics and power of knowledge production and use" (p. 229) and concerns itself with action toward the promotion of social transformation. Schwandt (2015) states that

three characteristics appear to distinguish the forms of this practice from other forms of social inquiry: (1) its participatory character—cooperation and collaboration between the researcher(s) and other participants in problem definition, choice of methods, data analysis, and use of findings . . . (2) its democratic impulse—PAR embodies democratic ideals or principles but it is not necessarily a recipe for bringing about democratic change; (3) its objective of producing both useful knowledge and action as well as consciousness raising—empowering people through the process of constructing and using their own knowledge. PAR is also marked by tension surrounding the simultaneous realization of the aims of participant involvement, social improvement, and knowledge production. (p. 229)

PAR, at its core, is about local knowledge generation and dissemination toward the accomplishment of stakeholder-driven goals for change and transformation. |

(Continued)

Table 1.2 (Continued)

Phenomenological research	Phenomenology is considered both a research method as well as a philosophy (Goulding, 2005), and it is largely attributed to the philosophy of Edmund Husserl (Creswell, 2013; Moustakas, 1994; Schwandt, 2015). Researchers employing phenomenological research methods tend to be interested in individuals' lived experiences of a phenomenon (such as homeless parenting or crisis leadership). A phenomenon does not need to be bounded by space and time; being a parent is an example of such a phenomenon. Alternatively, a phenomenon can be a specific event. The purpose of phenomenological research is "to identify phenomena through how they are perceived by the actors in a situation" (Lester, 1999, p. 1). Phenomenological research methods often include exploring a phenomenon with a group of individuals, and data collection tends to include interviews (Creswell, 2013). Interviews are not always the only source of data collection; data may also include participant observation or other sources such as documents and poems (Creswell, 2013; Lester, 1999). To understand individuals' lived experiences, phenomenological researchers often employ the process of bracketing. This bracketing process, often also referred to as epoche or phenomenological reduction (Gearing, 2004), involves that researchers bracket, or set aside, their everyday assumptions. Gearing (2004) provides a helpful definition of this process: Phenomenological reduction is the scientific process in which a researcher suspends or holds in abeyance his or her presuppositions, biases, assumptions, theories, or previous experiences to see and describe the phenomenon. Bracketing, as in a mathematical equation, suspends certain components by placing them outside the brackets, which then facilitates a focusing in on the phenomenon within the brackets. (pp. 1430–1431)[18] The goal of phenomenological research methods is to "obtain comprehensive descriptions that provide the basis for a reflective structural analysis that portrays the essences of the experience" (Moustakas, 1994, p. 13).
Practitioner research	Practitioner research constitutes a range of systematic, inquiry-based research efforts that are directed toward creating and extending professional knowledge, skills, ideas, and practices. In practitioner research, questions emerge from practice, and then practitioners design research studies to collect and analyze practice-based data that respond to these questions within their organizational or communal contexts. Practitioner research is undertaken by practitioners who seek to improve their own practice through the purposeful and critical examination of and reflection on aspects of their work, on the experiences of their colleagues and constituencies, and on institutional cultures, policies, and practices that shape these realities. Such systematic examination is designed to increase awareness of the contexts that shape professional actions, decisions, and judgments, enabling practitioners to see their practices anew; to recognize and articulate the complexities of their work; and to discover the values and choices at the core of professional practice. Practitioner research enables practitioners to engage in structured inquiries that are directed toward knowledge generation; it helps practitioners to gain formative insight into what concerns or confuses them; about what aspects of practice are most challenging and rewarding; about their roles as supporters, advocates, collaborators, and change agents; and about the parameters, possibilities, and constraints of their work settings (G. Anderson, Herr, & Nihlen, 2007; Cochran-Smith & Lytle, 2009; Ravitch, 2014).

A NOTE ON THE POSSIBILITIES OF QUALITATIVE RESEARCH

We argue that criticality in qualitative research (and all research in general) is vital at this historical moment with so many powerful logics, ideologies, and policies that work against individuals and groups having a stake and say in aspects that affect their daily lives, thereby serving to marginalize and minoritize their voices and options (Jackson & Mazzei, 2009; Steinberg & Cannella, 2012). Within the contexts of globalization, neoliberalism, and market-driven philosophies, as well as top-down education and social policy and mandates, this book conceptualizes and positions qualitative research as a powerful stance, as a possible set of methodological counteractions that can generate counternarratives upon which local, data-based resistance(s) can be cultivated and made public (Kincheloe & McLaren, 2000). The book seeks to make a case for nonhegemonic ways of thinking about and approaching theory-research-action connections and the transformative possibilities of criticality in qualitative research that work from decidedly relational, contextualized, person-centered, equity-oriented, inter- and transdisciplinary perspectives and methodologies.

As researchers, practitioners, scholars, teachers, and learners, we believe in the transformative possibilities of qualitative research. We acknowledge, however, that even making such an assertion about what constitutes transformation (i.e., According to whom? Evaluated by whom and with what criteria?) involves power asymmetries that must be thoughtfully considered, such as issues of reciprocity and for whom is research transformative (Steinberg & Cannella, 2012). That is part of why we advocate for criticality in qualitative research and present specific processes throughout this book and specifically in our discussion of critical research design in Chapter Three. We argue for an approach to qualitative research that resists reinscribing—to the extent possible—inequity, power, asymmetries, hegemony, and the co-opting of others people's experiences. We believe that qualitative research has the potential to provide interruptive and ultimately transformative experiences (as defined by the people whom it affects) as a result of its ability to generate local knowledge, its potential for informed action, and how it attends to the complexity of lived experience (Kincheloe & McLaren, 2000). While much qualitative research is not explicitly or intentionally driving at social change or transformation, which is of course understandable and fine, we believe at the very least it should resist reinscribing inequity in its methods and articulations.

QUESTIONS FOR REFLECTION

- What are the general processes of a qualitative research study?
- How does the history of qualitative research affect its practice?
- What are the key components of qualitative research?
- What are core values, beliefs, and assumptions on which qualitative research is based?

- What role does the researcher play in qualitative research?

- How are the terms *iterative, recursive, positionality, social location, emergent, epistemology, ontology, methodology, criticality in qualitative research, binaries,* and *dialogic engagement* used in qualitative research?

- How are criticality, reflexivity, collaboration, and rigor integral to qualitative research?

- What stands out about the different approaches to qualitative research?

- What do you consider to be the possibilities of qualitative research?

In the next chapter, we build on the book's premise that the conceptual aspects of qualitative research cannot be separated from the theoretical and methodological. Specifically, we show how theory, methods, goals, research questions, micro and macro contexts, reflexivity, dialogic engagement, and you (as the researcher) come together to form an evolving conceptual framework that simultaneously informs and is informed by the research study.

RESOURCES FOR FURTHER READING

Criticality in Qualitative Research

Bhabha, H. K. (1990). The other question: Difference, discrimination and the discourse of colonialism. In R. Ferguson, M. Gever, T. T. Minh-ha, & C. West (Eds.), *Out there: Marginalization and contemporary cultures* (pp. 71–88). New York, NY: The New Museum of Contemporary Art.

Denzin, N. K., Lincoln, Y. S., & Smith, L. T. (2008). *Handbook of critical and indigenous methodologies*. Thousand Oaks, CA: Sage.

Jackson, A. Y., & Mazzei, L. A. (2009). *Voice in qualitative inquiry: Challenging conventional, interpretive, and critical conceptions in qualitative research*. London, UK: Routledge.

Kincheloe, J. (1995). Meet me behind the curtain: The struggle for a critical postmodern action research. In P. McLaren & J. Giarelli (Eds.), *Critical theory and educational research*. Albany: State University of New York Press.

Kincheloe, K., & Tobin, J.L. (2009). The much exaggerated death of positivism. *Cultural Studies of Science Education, 4*(3), 513–528.

Lather, P. (1992). Critical frames in educational research: Feminist and post-structural perspectives. *Theory Into Practice, 2,* 87–98.

Smith, L. T. (1999). *Decolonizing methodologies: Research and indigenous peoples*. London, UK: Zed Books, Ltd.

Steinberg, S. R., & Cannella, G. S. (Eds.). (2012). *Critical qualitative research reader*. New York, NY: Peter Lang.

Zuberi, T., & Bonilla-Silva, E. (Eds.). (2008). *White logic, white methods: Racism and methodology*. London, UK: Rowman and Littlefield.

Action Research

Greenwood, D. J., & Levin, M. (2006). *Introduction to action research*. Thousand Oaks, CA: Sage.

McNiff, J., Lomax, P., & Whitehead, J. (2003). *You and your action research project*. London, UK: Routledge.

Noffke, S. (1997). Professional, personal, and political dimensions of action research. In M. Apple (Ed.), *Review of research in education* (Vol. 22, pp. 305–343). Washington, DC: American Educational Research Association.

Reason, P., & Bradbury, H. (Eds.). (2008). *The SAGE handbook of action research: Participative inquiry and practice*. Thousand Oaks, CA: Sage.

Stringer, T. (2014). *Action research* (4th ed.). Thousand Oaks, CA: Sage.

Case Study Research

Dooley, L. M. (2002). Case study research and theory building. *Advances in Developing Human Resources, 1*(4), 3.

Eisenhardt, K. M. (1989). Building theories from case study research. *The Academy of Management Review, 14*(4), 532–550.

Stake, R. E. (1995). *The art of case study research*. Thousand Oaks, CA: Sage.

Yin, R. K. (Ed.). (2004). *The case study anthology*. Thousand Oaks, CA: Sage.

Yin, R. K. (2013). *Case study research: Design and methods* (5th ed.). Thousand Oaks, CA: Sage.

Ethnography and Critical Ethnography

Atkinson, P., Delamont, S., Coffey, A. J., Lofland, J., & Lofland, L. H. (Eds.). (2005). *Handbook of ethnography*. Thousand Oaks, CA: Sage.

Emerson, R. M., Fretz, R. I., & Shaw, L. L. (2011). *Writing ethnographic fieldnotes* (2nd ed.). Chicago, IL: University of Chicago Press.

Hammersley, M. (1992). *What's wrong with ethnography?* London: Routledge.

Hammersley, M., & Atkinson, P. (2007). *Ethnography: Principles in practice* (3rd ed.). Hoboken, NJ: Taylor & Francis.

Lareau, A., & Shultz, J. (Eds.). (1996). *Journeys through ethnography: Realistic accounts of fieldwork*. Boulder, CO: Westview.

Lave, J. (2011). *Apprenticeship in critical ethnographic practice*. Chicago, IL: University of Chicago Press.

Madison, D. S. (Ed.). (2011). *Critical ethnography: Method, ethics, and performance* (2nd ed.). Thousand Oaks, CA: Sage.

Evaluation Research

Bamberger, M., Rugh, J., & Mabry, L. (2012). *RealWorld evaluation: Working under budget, time, data, and political constraints* (2nd ed.). Thousand Oaks, CA: Sage.

Clarke, A., & Dawson, R. (1999). *Evaluation research: An introduction to principles, methods, and practice*. London, UK: Sage.

Guba, E. G., & Lincoln, Y. S. (1981). *Effective evaluation: Improving the usefulness of evaluation results through responsive and naturalistic approaches.* San Francisco, CA: Jossey-Bass.

Patton, M. Q. (2015). *Qualitative research and evaluation methods: Integrating theory and practice* (4th ed.). Thousand Oaks, CA: Sage.

Rossi, P. H., Lipsey, M. W., & Freeman, H. E. (2004). *Evaluation: A systematic approach.* Thousand Oaks, CA: Sage.

Shaw, I. F. (1999). *Qualitative evaluation.* Thousand Oaks, CA: Sage.

Grounded Theory

Corbin, J., & Strauss, A. (2014). *Basics of qualitative research: Techniques and procedures for developing grounded theory* (4th ed.). Thousand Oaks, CA: Sage.

Charmaz, K. (2006). *Constructing grounded theory: A practical guide through qualitative analysis.* Thousand Oaks, CA: Pine Forge Press.

Gasson, S. (2004). Rigor in grounded theory research: An interpretive perspective on generating theory from qualitative field studies. In M. E. Whitman & A. B. Woszczynski (Eds.), *The handbook of information systems research* (pp. 79–102). Hershey, PA: Idea Group.

Glaser, B. G., & Strauss, A. L. (1967) *The discovery of grounded theory.* New York, NY: Aldine.

Narrative Research/Inquiry

Clandinin, J. D., & Rosiek, J. (2007). Mapping a landscape of narrative inquiry: Borderland spaces and tensions. In D. J. Clandinin (Ed.), *Handbook of narrative inquiry: Mapping a methodology* (pp. 35–75). Thousand Oaks, CA: Sage.

Connelly, F. M., & Clandinin, D. J., (2006). Narrative inquiry. In J. Green, G. Camilli, & P. Elmore (Eds.), *Handbook of complementary methods in education research* (pp. 375–385). Mahwah, NJ: Lawrence Erlbaum.

Lyons, N., & LaBoskey, V. K. (Eds.). (2002). *Narrative inquiry in practice: Advancing the knowledge of teaching.* New York, NY: Teachers College Press.

Pavlenko, A. (2002). Narrative study: Whose story is it, anyway? *TESOL Quarterly, 36*(2), 213–218.

Participatory Action Research (PAR)

Chevalier, J. M., & Buckles, D. J. (2013). *Participatory action research: Theory and methods for engaged inquiry.* London, UK: Routledge.

Fals-Borda, O., & Rahman, M. A. (1991). *Action and knowledge.* Lanham, MD: Rowman & Littlefield.

Freire, P. (2000). *Pedagogy of the oppressed.* New York, NY: Continuum. (Original work published 1970)

Kindon, S. L., Pain, R., & Kesby, M. (2007). *Participatory action research approaches and methods: Connecting people, participation and place.* London, UK: Routledge.

Mukherjee, N. (2002). *Participatory learning and action: With 100 field methods.* New Delhi, India: Concept Publishing Company.

Ozerdem, A., & Bowd, R. (Eds.). (2010). *Participatory research methods in development and post-disaster reconstruction*. Surrey, UK: Ashgate.

Tolman, D. L., & Brydon-Miller, M. (Eds.). (2001). *From subjects to subjectivities: A handbook of interpretive and participatory methods*. New York: New York University Press.

Phenomenology

Lester, S. (1999). *An introduction to phenomenological research*. Retrieved from www.devmts.demon.co.uk/resmethy.htm

Moustakas, C. E. (1994). *Phenomenological research methods*. Thousand Oaks, CA: Sage.

Smith, J., Flowers, P., & Larkin, M. (2009). *Interpretive phenomenological analysis: Theory, method and research*. London, UK: Sage.

Starks, H., & Brown Trinidad, S. (2008). Choose your method: A comparison of phenomenology, discourse analysis, and grounded theory. *Qualitative Health Research, 17*(10), 1372–1380.

Practitioner Research

Anderson, G., Herr, K., & Nihlen, A. (2007). *Studying your own school: An educator's guide to practitioner action research* (2nd ed.). Thousand Oaks, CA: Corwin.

Cochran-Smith, M., & Lytle, S. (2009). *Inquiry as stance: Practitioner research for the next generation*. New York, NY: Teachers College Press.

Noffke, S. (1999). What's a nice theory like yours doing in a practice like this? And other impertinent questions about practitioner research. *Change: Transformations in Education, 2*(1), 25–35.

Zeichner, K., & Noffke, S. (2001). Practitioner research. In V. Richardson (Ed.), *Teaching* (4th ed.). New York, NY: Macmillan.

Zeni, Z. (Ed.). (2001). *Ethical issues in practitioner research*. New York, NY: Teachers College Press.

Overview of the Range of Qualitative Research Approaches

Bansal, P., & Corley, K. (2011). The coming of age for qualitative research: Embracing the diversity of qualitative methods. *Academy of Management Journal, 54*(2), 233–237.

Creswell, J. W. (2013). *Qualitative inquiry and research design: Choosing among five approaches* (3rd ed.). Thousand Oaks, CA: Sage.

Denzin, N. K., & Lincoln, Y. S. (2003). *The landscape of qualitative research: Theories and issues*. Thousand Oaks, CA: Sage.

Denzin, N. K., & Lincoln, Y. S. (Eds.). (2011). *The SAGE handbook of qualitative research*. Thousand Oaks, CA: Sage.

Guba, E. G., & Lincoln, Y. S. (2003). Paradigmatic, controversies, contradictions, and emerging confluences. In N. Denzin & Y. Lincoln (Eds.), *The SAGE handbook of qualitative research* (pp. 191–216). Thousand Oaks, CA: Sage.

ONLINE RESOURCES

Sharpen your skills with SAGE edge

Visit edge.sagepub.com/ravitchandcarl for mobile-friendly chapter quizzes, eFlashcards, multimedia resources, SAGE journal articles, and more.

ENDNOTES

1. Denzin and Lincoln (2011a) describe these moments "as the traditional (1900–1950), the modernist or golden age (1950–1970), blurred genres (1970–1986), the crisis of representation (1986–1990), the postmodern, a period of experimental and new ethnographies (1990–1995), post-experimental inquiry (1995–2000), the methodologically contested present (2000–2010), and the future (2010–), which is now" (p. 3). For more information, see Denzin and Lincoln (2011a) in *The SAGE Handbook of Qualitative Research*.

2. Examples of these scholars include Bogdan and Biklen (2006), Corbin and Strauss (2008), Creswell (2013), Erickson (1998, 2011), Fine (1994), Hammersley and Atkinson (2007), Lincoln and Guba (2003), Maxwell (2013), Marshall and Rossman (2016), Merriam (2009), Olsen (2004), Patton (2015), and Richardson (1990, 1997).

3. Interpretive frames include (but are not limited to) feminist theory, hermeneutics, critical race theory, anti/postcolonial theory, queer theory, disability theories, Black feminist epistemology, and critical realist theories.

4. There are also within-paradigm arguments that existed at this time and persist to this day. For a great discussion of this, see Denzin and Lincoln (2011b).

5. Hermeneutics can be broadly defined as "a framework for conceptualizing the circular process of interpretation in a manner that contextualizes ourselves, others, and our interactions more fully" (Nakkula & Ravitch, 1998, p. xi). For a more detailed and historically situated definition and a deeper conceptualization of hermeneutics, see Nakkula and Ravitch (1998).

6. For one of the best descriptions of the origins of critical research, see Kincheloe and McLaren (2000). See also Denzin and Lincoln (2011a, 2011b), Cannella and Lincoln (2012), and Steinberg and Cannella (2012) for rich discussions of critical social theory in qualitative research.

7. See Denzin and Lincoln (2011a, 2011b), Cannella and Lincoln (2012), and Steinberg and Cannella (2012) for rich discussions of these.

8. We, like many others (e.g., Anfara & Mertz, 2015; Maxwell, 2013; C. Robson, 2011), assume that every researcher brings his or her own theories, both formal and informal, to the research process. Despite the unique frameworks that all researchers bring, both theoretical and conceptual, we argue that taking a critical stance on qualitative research highlights multiple layers of complexity that are needed to conduct ethical research.

9. For a valuable conceptualization and exploration of deficit thinking, see Valencia (2010).

10. The poet, Rainer Maria Rilke, writes that there is great value in choosing to "unfold" aspects of ourselves that remain hidden, even to us, and therefore conceal parts of us, constraining our ability to live honestly and in more authentic relationship with others. In his poem, "I Am Much Too Alone in This World, Yet Not Alone," Rilke (2001) states, "I want to unfold. Nowhere I wish to stay crooked, bent; for there I would be dishonest, untrue" (p. 17). We consider dialogic engagement part of this "unfolding." Source: Rilke, R. M. (2001). (A. S. Kidder, Trans.). *The book of hours: Prayers to a lowly god*. Evanston, IL: Northwestern University Press.

11. For discussions of the different approaches to qualitative research, see Creswell (2013), Marshall and Rossman (2016), and Patton (2015).

12. For further discussion of case studies, see Yin (2009).

13. See Eisenhardt (1989) and Dooley (2002) for descriptions and processes of building theory from case study research.

14. For a detailed discussion of the criteria used in evaluation research, see Rossi, Lipsey, and Freeman (2004).

15. See Patton (2015) for a description of goal-free evaluation research as well as other types of evaluation research.

16. See Clarke and Dawson (1999) for a description of the many kinds of qualitative research.

17. For more information about grounded theory and more detailed procedures, see Strauss and Corbin (1998) and Corbin and Strauss (1990, 2015).

18. For further discussion of phenomenological bracketing, see Gearing (2004).

Using Conceptual Frameworks in Research

CHAPTER OVERVIEW AND GOALS

In this chapter, we emphasize the importance of conceptual frameworks in research and discuss the important role they play in all aspects and phases of qualitative research. This chapter begins with a discussion of what constitutes a conceptual framework. After establishing a working sense of what a conceptual framework is and a beginning roadmap of its uses, we focus on what it does—or, more precisely, what you do with it. Following this, we discuss how to construct a conceptual framework and provide recommended practices and examples of how researchers describe their own conceptual frameworks.

BY THE END OF THIS CHAPTER YOU WILL BETTER UNDERSTAND

- What a conceptual framework is

- The roles and uses of a conceptual framework

- How a conceptual framework guides and grounds empirical research

- How you, as a researcher, can use a conceptual framework to build and refine your study

- How to construct your own conceptual framework

- The relationship between a conceptual and theoretical framework

- The difference between a literature review and a theoretical framework

- Each of the component parts of a conceptual framework and how they fit together

- The dynamic nature of conceptual frameworks

- The multiple possibilities for developing and representing conceptual frameworks in narrative and graphical form

DEFINING AND UNDERSTANDING CONCEPTUAL FRAMEWORKS AND THEIR ROLE IN RESEARCH

It is vital to understand what a conceptual framework is, what its components are and how they interact, and how it is used to guide and nourish sound qualitative research.

What Is a Conceptual Framework?

Conceptual frameworks have historically been a somewhat confusing aspect of qualitative research design because there is relatively little written on them, and various terms, including *conceptual framework, theoretical framework, theory, idea context, logic model,* and *concept maps,* are used somewhat interchangeably and often in unclear ways in the methods literature. *Reason and Rigor* by Ravitch and Riggan (2012) emerged as a response to this conceptual and definitional murkiness; it focuses on articulating what comprises a conceptual framework and how conceptual frameworks guide research from its inception to its completion. Because that book goes into great detail about how a conceptual framework is used in actual studies (with multiple empirical studies used to exemplify its role at various stages of the research process), it is the best current source for seeing how the conceptual framework guides every facet of research. In this chapter, we build on that text and the work it builds on (discussed below) and seek to conceptualize the term and highlight the roles and uses of the conceptual framework, as well as the process of developing one, since we believe a conceptual framework is a generative source of thinking, planning, conscious action, and reflection throughout the research process.

A conceptual framework makes the case for why a study is significant and relevant and for how the study design (including data collection and analysis methods) appropriately and rigorously answers the research questions. In addition, a conceptual framework situates a study within multiple contexts, including how you, as the researcher, are located in relationship to the research (Ravitch & Riggan, 2012). The conceptual framework consists of multiple parts and serves a variety of intersecting, ongoing, and iterative functions for researchers embarking on and engaging in research and the scholarship it produces. As Figure 2.1 illustrates, a conceptual framework includes multiple components that intersect, inform, and influence each other. For example, the methods that you use to answer your research questions are informed by (in conscious and unconscious ways) by your tacit theories, theoretical framework, the personal and professional goals of your study, ways that you systematically reflect on multiple aspects of your research (structured reflexivity), the intentional interactions you have with others (dialogic engagement),[1] the macro and micro contexts of your research setting, and *you* (as the researcher), including your social location and positionality (which reflects the macro contexts of the research as well). We describe these components and how they come together throughout this chapter and overview them here to help you begin to visualize and understand what comprises a conceptual framework.

Terms and Concepts Often Used in Qualitative Research

Tacit theories: Tacit theories refer to the informal and even unconscious ways that you understand or make sense of the world that are not explicit, spoken, or possibly even known to you without intentional reflection. These might include working hypotheses, assumptions, or conceptualizations you have about why things occur and how they operate. Unlike formal theories, tacit theories are not directly situated in academic literatures. Rather, they are an outgrowth of the attitudes, perspectives, ideologies, and values into which you have been socialized, often without knowing it. Everyone operates from tacit theories about people and the world, and without attention and reflection, they can constrain your research in a variety of ways. One example of this could be if you have a tacit theory that respecting authority is always good and therefore that all challenges to authority reflect badly on people. This theory might lead you to deem anyone who confronts authority as problematic versus seeing that context mediates whether an authority figure should be respected and what that means and looks like. We discuss how tacit theories relate to your conceptual framework throughout this chapter.

The Components of a Conceptual Framework[2]

As noted above, the roles of a conceptual framework include arguing for the significance of a topic, grounding the topic in its multiple contexts (theoretical and actual), guiding the development and iteration of research questions, the selection of theories and methods within a methodological framework that makes sense for the topic and allows for rigor, and providing you, as the researcher, with a framework and set of contexts in which you can examine your social location/identity and positionality as it relates to and shapes methodology overall and methods choices specifically. The conceptual framework creates a bridge between the context, theory, both formal and tacit, and the way that the study is structured and conducted in relation to all of these contextualizing and mediating influences. When conceptualized holistically, a conceptual framework serves as the "connective tissue" of a research study in that it helps you to integrate and mobilize your understanding of the various influences on and aspects of a specific research study in ways that create a more intentional and systematic process of explicitly connecting the various parts of the study. It is important to point out that the conceptual framework should be defined not just by its constituent parts, or even by its collective, but by what it *does*, that is, how these parts, or processes, are intentionally placed in relationship to each other by the researcher and work together in an integrated way to guide research.

What Does a Conceptual Framework Help You Do?

A conceptual framework can serve as a means of explaining why your topic is important practically and theoretically as well as detailing how your methods will answer your research questions. It serves as a thought integration mechanism for learning from existing

Figure 2.1 The Components of a Conceptual Framework[2]

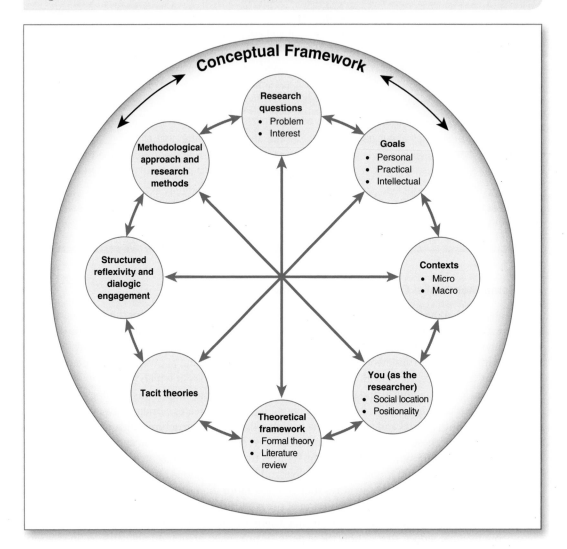

Terms and Concepts Often Used in Qualitative Research

Thought communities: Thought communities refer to the different realms or "communities" to which your research may speak or relate. These may be actual communities in a social geography sense of the term (i.e., a school community, a geographically based community), and/or they may be groups of people such as affinity groups or cultural communities. For example, if you are studying a particular aspect of teacher education, different thought communities your research may speak to or influence could include scholars of teacher education, teacher educators, and/or teachers and other educational practitioners.

Core constructs: Core constructs (sometimes called core analytic constructs) refer to the central aspects—concepts, phenomena, and topics—that guide and are at the center of a research study. These are typically articulated in your research questions; they are the key concepts, phenomena, and topic areas you seek to study and understand more fully through the research. These need to be carefully considered at the outset of and throughout research question development and research design processes since they generate the areas of theory and conceptualization that drive your research study.

expertise as you cultivate (and ultimately generate) knowledge within and possibly across thought communities already in motion (Ravitch & Riggan, 2012). In addition, Ravitch and Riggan (2012) highlight the ways that a conceptual framework supports how you locate (and perhaps argue for) your research in terms of its newness and contribution to a field or set of fields of inquiry—perhaps new contexts, questions, frames (some or all of these)—which create new terrain for you to traverse. If we stay with this metaphor, the conceptual framework is the compass, the landmarks, the navigation system, and the zoom function of your vision apparatus. You are the one to decide how these realities and tools come together in the expedition into a new topic, setting, set of theories, and so on.

The conceptual framework helps you cultivate research questions and then match the methodological aspects of the study with these questions. In this sense, the conceptual framework helps align the analytic tools and methods of a study with the focal topics and core constructs as they are embedded within the research questions. This alignment of methods and analytical tools helps to increase a study's rigor and validity (Ravitch & Riggan, 2012). The conceptual framework helps you situate the study in its theoretical, conceptual, and practical contexts; includes implications related to the study's setting and participants; situates your researcher social location/identity and positionality; and articulates how all of these aspects are related to methodological frames and processes.

A conceptual framework is the focal framework for, and process of, designing and engaging in research (Ravitch & Riggan, 2012). It details the primary or core constructs of a study and the relationships between them (Miles & Huberman, 1994; Miles, Huberman, & Saldaña, 2014). In addition, conceptual frameworks evolve and change as your understandings change: "Conceptual frameworks are simply the current version of the researcher's map of the territory being investigated. As the explorer's knowledge of the terrain improves, the map becomes correspondingly more differentiated and integrated" (Miles et al., 2014, p. 20). A conceptual framework becomes increasingly sophisticated as you become more deeply involved and your understanding of these constituent parts and the whole come together and build on and into each other. Ideally, a conceptual framework helps you to become more discerning and selective in terms of methods, grounding theories, and approaches to your research (Miles et al., 2014).

Collaboration is a key horizontal in qualitative research, and we strongly critique the enduring academic notion of the lone researcher. One issue we have with this notion of the lone researcher is that we believe that research is ideally not a solitary endeavor since genuine research reflexivity warrants multiple levels of research collaboration and engagement (L. Anderson, 2008; Denzin, 1994; Guba & Lincoln, 2005; Steedman, 1991). We believe this to be true for individual research projects as well as research team-based projects. For an individual, collaboration is vital to create the conditions necessary to challenge your assumptions and blindspots and integrate a range of perspectives into all facets of your research, and the conceptual framework is a way to seek out a range of feedback and perspectives. When working on a research team, you need ways to engender group dialogue and exchange to examine diverse perspectives. Thus, the conceptual framework can become a mechanism/strategy for aligning researchers' goals, expectations, values, and ideas about research at its outset and throughout the process (Miles et al., 2014). We have used collaborative conceptual framework development—through collective mapping and articulation processes that are grounded in the sharing of each research team member's construction of a conceptual framework memo (see Recommended Practice 2.1)—in our own research teams in ways that have contributed significantly to the design and implementation of our research. In this sense, the use of conceptual frameworks to guide a team of researchers into critical conversations about perspectives on each aspect of research design and development becomes an important consideration.

A conceptual framework can function as "a tentative *theory*" of what you are studying (Maxwell, 2013, p. 39). It is a system—defined as a functionally integrated set of elements—in the sense that it serves as an integrating mechanism that works within and across the "concepts, assumptions, expectations, beliefs and theories" that guide research (Maxwell, 2013, p. 39).

It is important to view the conceptual framework as a process rather than a product; that is, the building of your conceptual framework is a dynamic, sense-making process that happens in nonlinear stages and that takes multiple forms. The conceptual framework generates the focus of the research as it is informed and shaped by it (Ravitch & Riggan, 2012).

Multiple, intersecting components of a conceptual framework help a researcher/research team to conceptualize and articulate specific constructs and aspects central to a study as well as to examine their intersections and overlaps within the multiple contexts that shape the research. A conceptual framework includes the guiding theories or assumptions of the researcher(s), the goals and expectations of a study, and the formal and informal theories that contribute to understanding and contextualizing the focal topics of a study; furthermore, the conceptual framework allows the researcher(s) to make meaning and use of the overlaps and disjunctures within and between core constructs of your study in ways that produce deeper, more integrative understanding of the topic and contexts central to the study. A conceptual framework enables you to make an argument for the value and significance of your research in ways that reflect the intentionality of the process of uncovering the ways in which studies emerge from and contribute to the corpus of research in your substantive area (Maxwell, 2013; Miles et al., 2014; Ravitch & Riggan, 2012). In addition, a conceptual framework helps you to locate yourself in the research process, as well as to attend to various epistemological and ontological considerations and beliefs and how these shape us as researchers and therefore shape our methodological choices. Finally, a conceptual framework is constructed and continually iterates throughout your research, and it helps to refine the research simultaneously. This notion of active building and refining is central to understanding that a conceptual framework is both guiding to a study and also derived from a study.

The Roles and Uses of a Conceptual Framework

There are myriad, intersecting roles and uses of a conceptual framework, which include the following:

- Providing the overall structure in which you can articulate and examine the goals of your study and its intended audiences

- Serving as a compass for identifying the relevance and selection of, and then engaging with and integrating across, existing research literatures and theories

- Providing the idea context for you to argue for the rationale and significance of your study for/to its intended areas, fields, and/or disciplines

- Helping you to thoughtfully develop, refine, and delimit your research questions

- Providing a dynamic system for identifying and refining your understanding of your researcher positionality and engaging in a reflexive approach to your study

- Providing a methodological ecosystem for developing and continuously refining the study's methodology and research design

- Helping you to develop an appropriate framework for the collection of data and the development of instruments

- Serving as a guide and frame for the selection of data analysis frameworks and processes

- Acting as a foundation for writing and critically scrutinizing your analysis and study findings

- Helping you to see how formal theory (your theoretical framework) intersects with various aspects of your study, including research questions, methods, and goals, as well as the informal theories and working assumptions you may have about the topic

- Serving as a mechanism to consider and reflect on the significance and value of your research once it is completed, as well as to consider the next questions to be asked in the field(s) that the study speaks back into

Rather than view these aspects of the research process as linear or discrete, a conceptual framework helps the researcher recognize and assert how the aspects of research content and process influence, as they are influenced by, the guiding conceptual linkages in a given study. One way to think across these various roles and functions of a conceptual framework is to consider the notion of situating a study in its multiple, intersecting, ever-iterating contexts and concepts. This is what we mean by a methodological ecosystem, described further in the following section.

CONSTRUCTING AND DEVELOPING A CONCEPTUAL FRAMEWORK

After gaining a better understanding of the value, roles, and uses of a conceptual framework, it is important to consider how to construct and continuously develop one for/in your research. There are many ways that researchers embark on the early development of research topics, goals, and questions. The conceptual framework is a way of approaching research design/construction that focuses on what we think of as the ecology or ecosystem of a topic and the fields from which and to which they build. By ecosystem, we refer to the notion that a research study is a complex system with multiple parts that are both separate from each other and connected by and within a larger system; each tree and rock is related to each insect and blade of grass in ways that are both clear and also in need of ongoing conceptualization and explication. The concept of an ecosystem helps to reveal the interconnectedness and even interdependence of the various component parts of research studies. The conceptual framework provides the sense of interconnection and interdependence among parts of a research project. The conceptual framework is central to the construction and implementation of research. A conceptual framework embodies methodological criticality because it requires a systematic intentionality around examining and accounting for the connections between theory and research and how these are generated within and across stages of empirical work in relation to you and your goals, assumptions, social location/identity, and positionality.

The guiding sources for constructing a conceptual framework include (a) the researcher, (b) tacit theory or working conceptualizations, (c) the goals of a study, (d) study setting and context, (e) broader macro-sociopolitical contexts, and (f) formal or established theory. In the sections that follow, we discuss each of these guiding sources. These sources are not listed in priority order and are not linear. We emphasize this caveat because complexity can sometimes be reduced when complex concepts are broken into parts. We do not want to be misinterpreted as oversimplifying something as iterative, recursive, and altogether messy (in the creative albeit sometimes stressful for new researchers sense of that term) as a conceptual framework.

The Role of the Researcher in Conceptual Frameworks

The major source for constructing a conceptual framework is you, *the researcher*. As stated in Chapter One, you are the primary source of both constructing and understanding the goals and meaning of your research project. This includes your positionality, social location/identity, experiences, beliefs, prior knowledge, assumptions, ideologies, working epistemologies, biases, and your overall perspective on the world. A complex combination of all of these aspects about you as the researcher, including your identity and beliefs, shapes the goals of your study in terms of how you think about your research at the conceptual and methodological levels. Relating to the notion of the researcher as instrument (Lofland, Snow, Anderson, & Lofland, 2006; Porter, 2010), the conceptual framework helps you conceptualize, identify, engage with, and critically examine the layers of your beliefs, biases, and assumptions. Engaging in a process of identifying and reckoning with these aspects of your thoughts and feelings as a researcher, with a focus on your social identity and positionality and how these influence your belief systems, ideologies, hopes, goals, and interests, helps you to more fully and critically understand how these aspects of who you are shape what you choose to study and how you visualize, consider, and plan to study it (Alvesson & Sköldberg, 2009; Henslin, 2013; Neyrey, 1991; Tetreault, 2012; Tisdell, 2008).

Conceptualizing the researcher as a primary instrument of the research has significant implications throughout every stage of the research process in that the subjectivity, social location/identity, positionality, and meaning making of the researcher profoundly shape research processes and methods and therefore data and findings (Lofland et al., 2006; Porter, 2010). Because of the prominence of the researcher in the conceptualization, design, and conduct of qualitative research, *the positionality of the researcher is viewed as a central and vital part of the inquiry itself.* Therefore, as the researcher, you must engage in a reflexive process that helps to uncover the layers of influence on your thinking and how this shapes your study from what you choose to focus on, how you frame it theoretically, how you collect data and approach/engage with participants, and then how you interpret and analyze the data of your study and write about the findings.

Given the primary role of the researcher, you must commit to a critically reflexive process that helps you to examine the influences on your thinking about the world, the specific context(s) of your research, and what is possible in your research. This entails a commitment to identifying and engaging with each of these aspects of the research and your roles in the research (Alvesson & Sköldberg, 2009; Henslin, 2013; Neyrey, 1991; Tetreault, 2012; Tisdell, 2008). We argue that a conceptual framework provides a structure for inquiring into yourself as a major shaper of and ongoing influence on the research in all of its stages and facets. Furthermore, the conceptual framework is the engine that drives your understanding of the ways that these aspects of self and research come together to form and continuously shape research studies.

A conceptual framework helps to first identify and then clarify what you know, care about, and value as central aspects of a study and then to connect these with the various other aspects of and influences on your research (Ravitch & Riggan, 2012). Situating yourself as a researcher entails considering how your study is situated (and how and why you situate it as you do) within fields/disciplines and within your thinking of what these fields might need in terms of what many think of as "the next set of questions to be asked." By next set of questions, we mean how what is already known and written about in a field maps onto your desire to study a new dimension or aspect of that research in similar or new contexts to advance the discourse in that field. Furthermore, situating yourself and your research involves understanding how your ideologies, positionalities, and sets of relationships in the world influence your thinking broadly and specifically (Ravitch & Riggan, 2012). In terms of your working theories, it is useful to make a distinction between the formal theory aspects of your working theories and conceptualizations and those that are more organic to your thinking. These tacit theories, which are also called implicit theories, guiding assumptions, working conceptualizations, or "theories in use" (Argyris & Schön, 1978), inform the formal theories that researchers invoke and use.

The Role of Tacit Theories in a Conceptual Framework

The role of the researcher influences all aspects of a study, and another important consideration includes the ways that researchers view and make sense of the world, or what we refer to as the *researcher's tacit theories or working conceptualizations.* While we will discuss what we refer to as formal theory later in this chapter, and while your working theories and conceptualizations are certainly a part of your overall identity and social location, we want to focus on the realm of tacit theory as a way to delve into this aspect of how you, as a researcher, enter your research with your own theories of how things work. Your guiding or informal theories are important to consider before, during, and after you design and engage in your research because they form your thoughts about the research at all levels and within all approaches in both direct and indirect as well as conscious and unconscious ways that must be explicated and reckoned with—first for yourself and then for your audiences. Some of this is within the personal realm—what you believe, how it relates to your socialization into certain biases and assumptions, what perspectives you hold dear as a person, practitioner, scholar, and so on. Some of it is professional, as in the "wisdom of practice" (Shulman, 1987a, 1987b, 2004) that you have cultivated over time and that guides professional choices, decisions, and concerns. Often, these tacit theories are at the crossroads of the personal and professional aspects of who you are and what you do every day because these aspects of positionality intersect.

By tacit theories, we mean informal and even unconscious ways that we all think about, make sense of, and explain the world and the various contexts and people within it. These are the often unspoken aspects of how we view the world that can be thought of as part of our guiding epistemological and ontological beliefs. Argyris and Schön (1978) discuss the related concepts of "espoused theory" and "theory in use" to describe the difference between theories that are espoused, or consciously claimed and articulated, and the theories that are actually used and worked from within organizations, which are often implicit. We think this is a useful framework for individuals as well in the sense that each of us is guided by our beliefs and assumptions about the world, and we pull on/act from these in the decisions and actions of our everyday lives in ways that may run counter to what we purport and espouse to be our belief systems. In a process that allows us to consider the possible chasms and even collisions between the espoused and the actual theories we hold, what we at first glance and without deep reflection might say are our theories and beliefs, upon reflection, often get challenged and reshaped or developed based on a more careful and systematic charting and consideration of these views.

Based on this concept—that reflection engenders changes in your thoughts brought about by increased awareness—we think that you as a researcher should spend considerable time asking sets of questions about what tacit or implicit theories guide your interest in the topic, frame how you are thinking about the topic in a specific setting or context, and what formal theories you are drawn to and that you draw upon to help you make sense of what you are thinking and considering. Attending to these issues early on and ongoingly is enormously important to the integrity and validity of your research.

How Study Goals Influence and Inform a Conceptual Framework

In addition to making your tacit and working understandings more explicit, discerning how you think about the *goals of a study* is an important aspect of qualitative research. Most studies begin broadly with an interest or a concern of a researcher or group of researchers. What becomes important, once you decide to turn an interest or concern into an actual research study, is to map out and theoretically frame the key goals of the study.

To engage in an intentional and systematic process of developing a solid understanding of the goals and rationales for your study, we suggest a reflective inquiry process in which you ask increasingly sophisticated questions that relate directly to the goals of the research overall and at each stage of the research process. The goals of a study can stem from personal and/or professional goals, prior research, existing theory, or some combination of these influences on a researcher's thoughts, values, and interests. We recommend to our students that they think about goals in both broad and specific ways. Broadly, our students find it useful to conceptualize this in terms of their personal, practical, and intellectual goals (Maxwell, 2013). What your goals are for a study can depend on how you are coming to the research. For example, are your research questions coming out of previous research that you have conducted and want to build on? Are they emerging from questions or problems of practice that you wish to investigate to improve upon your and/or others' practice? Are they coming out of deep engagement with formal theory that you wish to engage with or push into using empirical data? The goals of your research come from multiple influences, prior and current. And so, the conceptual framework can be a mechanism for a specific process of conceptualizing and then organizing your thinking around the set of goals you have for the study and for your line of inquiry more broadly.

There are many ways to consider research goals. We suggest having a set of conversations with peer researchers and research advisers in addition to reflexive writing exercises. Specific questions to examine in writing and in conversation with "critical friends" are described in Table 2.1.

These questions may seem simple, but understanding the broad and specific goals of a study is often tricky and confusing. Even with the challenges of unearthing and engaging with these goals as you come to understand them early on and over time, it is very important to clarify and examine them carefully at the outset of a research project and continuously throughout the process.

Table 2.1   Questions for Considering the Goals of Your Study

- Am I seeking to address a specific problem or concern and, if so, as identified and framed by who?
- Am I seeking to generate a theory, test a theory, or speak to a specific theory? If so, why and in what ways?
- Am I interested in contributing to the scholarly literature and, if so, is that guided by a critique of existing theory? What is this critique?
- Do I seek to use the study to break norms of or innovate research methodology in a specific content area/field? If so, which one(s) and why?
- Do I wish to engage people in the setting in the process of identifying the problems or topics to be addressed? Why or why not? If so, how will I engage them?
- To which audience(s) do I wish this research to speak and why?
- Have I been charged with or chosen to evaluate a program and, if so, how will that guide the research process?

Conceptual Frameworks and the Role of Setting and Context

Another major piece of a conceptual framework includes the *study setting and context,* which means the actual research setting (i.e., a specific place, organization, group, community, or communities) and what we think of as the context within the setting, which means who and what aspects of that setting are central to your research. For example, while your research may be situated in the School District of Philadelphia, for example, if you are studying how students learn reading in first grade, you are going to look at a different set of core constructs and approach them using different data sources than if you are looking at issues of communication and equity between administrators and teachers.

The consideration of the overall context, combined with what we think of as the context within the context, is important because it speaks to the fact that the domain of inquiry as well as the set of stakeholders within the setting comprise an important set of factors that influence what you study and how you frame the study. Spending time considering this in some depth will help you to determine which theories and concepts you will need to consider and assemble to investigate the phenomena and stakeholders in focus in ways that help you understand the relevance of the site and setting to the overall topic, goals, and research design. The setting of a study has much to do with key concepts and constructs you will use formal theory to examine, and the conceptual framework can help you to examine these in ways that generate connections between theory building and conceptual contextualization. In Table 2.2, we suggest questions that might help you to consider these aspects and their impact on your research.

Questions like these can help you to examine the decisions you make in choosing the study setting and the participants within that setting as well as to consider the goals of the study and how these match (or do not match) the setting. Thinking about this early on can help to clarify the rationale of the study as well as help to direct the overall design and approach to specific data collection methods.

Table 2.2   Questions for Considering the Setting and Context(s) of Your Study

- How does the setting and context influence or mediate the study goals and questions?
- How does the setting and context influence or mediate the research design overall and specific methods of data collection?
- Given what the study focuses on, which stakeholder group(s) are important to include in the study and why?
- Who is left out of this list of stakeholders, and what would be the reasons not to include these individuals or groups?
- How does the context mediate the formal theories selected to help frame the study?
- Were this study conducted in a different setting, how might it be structured differently and what does that suggest about the study design?
- How are issues of power and hegemony at play in the setting and context and in my views of and assumptions about them? What impact might this have on site and participant selection?

Broader Macro-Sociopolitical Contexts and Conceptual Frameworks

In addition to the specific setting and contexts that shape your study, examining the *broader macro-sociopolitical contexts* that influence your research is of vital importance. Understanding the research setting and context and how they relate to the focus or domain of your inquiry and how the setting and focus are shaped by macro-sociopolitics is a significant aspect of engaging in critical and intentional research. What we mean by macro-sociopolitics is a combination of the broad contexts—social, historical, national, international, and global—that create the conditions that shape society and social interactions, influence the research topic, and affect the structure and conditions of the settings and the lives of the people at the center of your research (including you). Engaging in an active consideration of these forces—and the issues of equity and hegemony that they reflect—is at the core of taking an approach to research that embraces criticality since the role of broad social forces is considered central to critical research. We further discuss this in Chapter Three (and beyond).

The need to consider and include macro-sociopolitics as an aspect of your conceptual framework is important to stress, and we believe that it has (at least) two main implications. The first is the importance of examining and critically understanding the sociopolitical nature of the topic and setting of the study in ways that reflect the social, political, and structural conditions of the context of the study, which can be variably experienced by stakeholders depending on such things as role, level, and title within communities and organizations and how these relate to hierarchy. The second is that you should consider the historical moment in which your research belongs and examine how it shapes the context and setting as well as your view of and approach to the inquiry. As a researcher, you should understand the temporality that exists in a set of macro concerns and contexts that shape and mediate the setting and the research of a group or population in that setting. This is contextual as well as temporal in the sense that while there are some constants (e.g., structural oppression, hegemony), these are also shaped by the current milieu of governance, policy, economics, and guiding social constructions in society at particular moments in time (meaning that things shift over time and take on new and different meanings—race would be one good example of a concept that has changed meaning over time).[3]

The process of continued contextualization comprises what we think of as a process of "rugged contextualization," which refers to the rigorous pursuit of understanding context at these various levels as well as systematically appreciating how they inform and influence each other. A solid conceptual framework helps you to articulate and examine specific phenomena and their contexts critically and to explore and refine your sense of context at multiple levels. This means, in part, that you must learn how to complicate the concept of context even as you seek to make it explicit and understandable to those who read and engage with your work.

A major role of a conceptual framework is to contextualize research as a site of generative meaning making around the role of context, critically and intersectionally, which includes helping the researcher to examine and understand (a) the social, cultural, political, economic, and ideological milieu of the research; (b) macro-sociopolitics and their manifestation in your specific research setting, including in interactions with participants; (c) your social location/identity and positionality and its impact on design, fieldwork, and analysis; and (d) the influences of sociocultural/historical contexts of the individual(s), community, group, and/or organization in focus on the research process/product. The ability to critically consider these aspects of the research requires a commitment to intentionally identifying and then carefully examining the multiple spheres of influence on your research setting and context as well as how these affect all facets of the research. Table 2.3 details some questions to help you identify and explore these issues. We recommend addressing and considering these issues by individually reflecting and using dialogic engagement strategies.

Table 2.3   Questions to Help You Consider the Macro-Sociopolitical Contexts That Shape Your Research

- What are the contexts—local, national, global—that shape my research?
- Are there many "locals" in the sense of stakeholder groups involved in the research?
- What do I mean by "local"? What shapes my assumptions of what this means?
- How do I conceptualize culture—as static or dynamic, as variable within and across groups for example? How will/does this conceptualization influence my research?
- How do I think about issues of race, social class, gender, and sexual orientation/identity? For example, do I see them as intersectional or as binaries? Why?
- Related to the point above, how do these conceptions and biases relate to those at the focus of the study?
- What larger social and political contexts shape the setting of my research and in what ways?
- How are various forms of policy at play?
- How are politics—local, national, and global—at play?
- What social constructions guide some of the realities of the settings and participants in my study (e.g., homophobia, racism, ideas about social class mobility, ageism)?
- How might this specific moment in history influence this study, including the context and people within it? How can I consider this temporal nature of my own research?
- What are my own perspectives on and experiences with relevant social and political issues? How do these shape my research?

These are but a sampling of the questions you should ask to critically conceptualize the macro influences on the setting, context, participants, yourself, and the study. This layer of context examination is absolutely vital to designing, conducting, and achieving criticality in research.

The Role of Formal Theory in Conceptual Frameworks

An additional component of a conceptual framework includes *formal or established theory*, which we refer to as the *theoretical framework*. The conceptual framework is a "dynamic meeting place of theory and method" (Ravitch & Riggan, 2012, p. 141), and you must consider the roles that existing, or formal, theory play in the development of your research questions and the goals of your studies as well as throughout the entire process of designing and engaging in your research (Ravitch & Riggan, 2012). Formal theory—the set of established theories that are combined in relation to your ways of framing the core constructs embodied in your research questions—constitutes the theoretical framework of your study. The process we describe connotes that there must be a critical integration of theories that together construct a framework for theorizing the study. The theory integration process is guided by the ways that you, as the researcher, engage with and use specific theories and models to consider the setting, events, contexts, and phenomena you are seeing and exploring. In other words, the theoretical framework is the set of formal theories that you seek out and bring together to frame and contextualize the domain or focus of inquiry and the setting and context that shape its exploration. Formal means that these are codified through publication of a variety of sorts; it also means that such theories are recognized as accepted by (even if critiqued within) the field as a whole. An important caveat to this is that hegemony mediates, in important and often invisible or insidious ways, what formal theory gets codified and how. Hence, a critical eye toward what Sharon refers to as "epistemological domination" is vital to research that does justice to the people and contexts that are at its heart.

The theoretical framework is how you weave together or integrate existing bodies of literature—for example, adult learning theory or racial identity development models—to frame the topic, goals, design, and findings of your specific study. We distinguish a theoretical framework as different from and more focused than a literature review, which is often thought of and described as a product when it is more accurately conceptualized as the artifact of an active meaning-making and integration process. The literature review is the broader process by which you construct a specific theoretical framework, and this process is not a single moment that produces an artifact, but rather, the review of literature happens throughout the process of conducting research; it is both a guide and a by-product of research. The process of creating a theoretical framework requires that you engage at a meta-analytic level in deciding which of the theories and bodies of literature that you have reviewed are vital to the theoretical framing of your study (this often requires consultation with peers and advisers). The theoretical framework can be thought of as the way that you work out the theoretical constructions and concepts that are central to your understanding of the focal topic, the core constructs of your study, and the thought communities that are engaged in your areas of inquiry.

The theoretical framework, as we discuss in greater depth as a part of critical research design in Chapter Three, is a primary piece of the conceptual framework, and as such, it intersects with the other aspects of research development and design in ways that should be generative and active throughout the research process. Conceptual frameworks are developed and not found "ready-made" (Maxwell, 2013, p. 41). Part of this involves not considering theory immutable; it should be viewed as a point of context and reference to be engaged with (Maxwell, 2013; Ravitch & Riggan, 2012). You seek out formal theories to help understand what you are studying and why you are studying it. Furthermore, throughout the research process, you cultivate a working expertise that leads you, ideally, to speak back to theory and into theoretical debates, even challenging existing theory by exploring it within your research context. This is a very important point to consider because it sets up any research project not just as derived from or informed by formal theory but potentially as a way to explore, refine, or even refute existing theory. As well, it highlights the dynamic and fluid nature of this framework-building process. It also speaks to the importance of the researcher not assuming formal theory to be an immutable "Truth" but rather to be something that is critically considered and engaged with in ways that help frame your understandings and then to which you might speak back to with your own research findings in a variety of ways.

Researchers should ask a set of questions to understand the informal or tacit theories that we enter into our research with and their relationship with the formal theories we choose as a part of our theoretical frameworks. Table 2.4 includes some of these questions.

Table 2.4 Questions to Consider When Incorporating Formal Theory Into Your Research

- What are the formal theories from which I am drawing? Why? Where do they reside in relation to the issues into which I situate my research?
- Is the field or discipline in which I am situated (e.g., being a sociology doctoral student or a psychology master's student) limiting my ability to see multiple options for how to study this set of research questions? If so, how?
- What is the relationship between the informal kinds of knowledge and information I am drawing from and the formal theories I am using?
- Am I thinking of these in a polarized way and, if so, in what ways and why?
- What are other possible ways to consider the theories that frame my topic and setting?
- What are the possible fields or bodies of literature that can help me think beyond traditional disciplinary borders or my current understandings?
- Are there ways that the formal theories I am drawing upon reflect the hegemony of certain epistemologies, norms, and/or power structures in and beyond academia and society more broadly? How can I reckon with and address these issues in my research?
- In what ways and how might hegemonic epistemologies influence how I engage with formal theory and how I might invoke nontraditional theories to augment and perhaps even challenge my theoretical framework?

The conceptual framework, which includes your theoretical framework, is the place where you can (and we argue should) explore these kinds of questions as they shape and relate to your research. It is valuable and important to think of the intersections of informal and formal theories in ways that bring both into focus and help you understand how one leads you to the other and back again. In Chapter Three, we focus on the development of your theoretical framework as a foundational stage of research question development and the research design process.

BUILDING YOUR OWN CONCEPTUAL FRAMEWORK

Ravitch and Riggan (2012) discuss multiple strategies and practices for constructing and developing conceptual frameworks. We similarly argue that sets of reflective writings that ask a series of questions about your research are central to a conceptual framework's critical iteration. We specifically call these *structured reflexivity* and *dialogic engagement* practices. These practices, which include reflective writings and structured conversations with individuals and strategically selected groups, create opportunities for you to develop and engage with all aspects of a research process, including how you theoretically understand a topic, the methods you use, the ways you analyze data, and the variety of additional issues and concerns that help you to engage with and continually refine a conceptual framework.

Reflective practices are the key to ongoing conceptual engagement and accountability to yourself and the research. Because developing a conceptual framework is an evolving and iterative process, you should engage in these reflective practices throughout the duration of a study. The recommended practices and examples we provide are just a few ways for you to reflexively engage with your research. We present numerous additional practices and examples throughout the book to help facilitate these processes. In this chapter, we focus on the Conceptual Framework Memo (Recommended Practice 2.1) and Concept Map (Recommended Practice 2.2) as specific early practices for developing your conceptual framework. After specifically describing the practices, we include an annotated conceptual framework memo (Example 2.1) and two additional conceptual framework memos (Appendixes A and B). It is our goal that these examples help you as you construct your conceptual framework and that they do not limit or constrain the myriad possibilities for ways that you can develop, represent, and describe a conceptual framework. It is also worth reemphasizing that conceptual frameworks evolve. The examples of conceptual framework memos we present represent a snapshot of each author's thinking and processes, as conceptual frameworks are intended to iterate over time. Finally, we include an example of a conceptual framework excerpted from a dissertation (Appendix C) to help you to see the different iterations and forms a conceptual framework can take.

Recommended Practice 2.1: Conceptual Framework Memo

In our teaching of students and advising of master's theses and doctoral dissertations, we have found that memos, because they are largely seen as internal sense-making documents, allow for a kind of written engagement that frees students from the constraints of formal genres of writing. Memos can be informal or formal and can be intended solely for personal use or can serve other roles (e.g., used as communication among research team members or with various thought partners and advisers). As such, the author can truly explore ideas rather than put forth a sculpted and refined document. We therefore ask our students, in addition to writing an identity/positionality memo (see Chapter Three), to write an exploratory conceptual framework memo early in the development of their research studies. This assignment creates a structured opportunity to develop a narrative (that may include a corresponding graphical representation) of your emerging conceptual framework, receive feedback, and use it to engage with others early and continuously to develop and refine the research. A conceptual framework memo can include the following focal sections (but we argue that these can be constructed in many different ways as per your preferences in terms of how you process information):

Topics to include/consider in this memo:

1. Research topic, including context, setting, population in focus, and broad contextual framing

2. Research questions and study goals

(Continued)

(Continued)

3. Role of the following to your potential research:
   - Self (e.g., social location/identity and positionality)
   - Context(s) (e.g., institution/community, state, country, historical moment)
   - Goals (e.g., personal, practical, and intellectual—see Maxwell [2013] for a discussion of these goals)

4. Description of the relationship between your research question(s), conceptual framework, and research design choices (may also include bullet points of research design)

5. Overall methodological approach and potential research methods

6. The tacit theories that have informed your research question(s) and/or topic

7. The formal theories that guide and inform your study (theoretical framework)

8. Ways that you plan to implement structured reflexivity (individual reflection and dialogic engagement) throughout your study

Recommended Practice 2.2: Concept Map of Conceptual Framework

In this exercise, you visually map, in graphic form, the ways that your theoretical framework, research questions, researcher identity, informal theories, and so on relate and map onto each other and the methods you choose. This can be a productive activity, especially for visual learners. You then narrate this graphical representation of your work. This narrative is similar to the conceptual framework memo exercise described above but focuses on the explication of your graphic. For some researchers, the task of visually mapping their conceptual framework can seem quite daunting (we include ourselves in this category). Thus, it can make this process more productive to also verbally narrate your conceptual framework and how your theoretical framework, research questions, researcher identity, informal theories, and so on relate with and map onto each other and the chosen methods. This narrative may take the form of a short paragraph or may be longer and more closely resemble the conceptual framework memo.

In Example 2.1, Jaime Nolan does a particularly powerful job of locating herself in terms of her social location/identity and positionality in relation to the research site and participants, thoughtfully explicating the historical and current contexts at play in her research, and showing the intersection between the ideologies and theories at the heart of her study and the study methodology. She also includes a narrative about her research questions in a way that invites the reader to understand her sense-making process and discusses how the research questions are still emerging in an iterative fashion as she reads more and further develops the study focus and methodology. We also appreciate the richness of her narrative and its relationship to her graphic. Jaime has created a process-oriented memo that documents her emerging conceptual framework and locates herself and the students in context. Furthermore, she problematizes the research and the inequities it seeks to explore and work against, and she looks at the formal theory and its relationship to her own wisdom of practice and emerging conceptualizations of the research. In addition, Jaime lays out her methodological approach and begins to drill down to her research design. She does this taking a holistic, ecological approach that honors the meaning-making processes at the heart of her research.

Example 2.1: Conceptual Framework Memo and Accompanying Concept Map

An Autoethnography of Resilience:

Understanding Native Student Persistence at a Predominantly White College[4]

Jaime Nolan

October 2014

In 1774 representatives from Maryland and Virginia, as part of a treaty process, invited young men from the Six Nations to attend the college of William and Mary—founded in 1663. The tribal elders declined the offer with these words:

We know that you highly esteem the kind of learning taught in those Colleges, and that the Maintenance of our young Men, while with you, would be very expensive to you. We are convinced that you mean to do us Good by your proposal; and we thank you heartily. But you, who are wise must know that different Nations have different Conceptions of things and you will therefore not take amiss, if our ideas of this kind of Education happen not to be the same as yours. We have had some Experience of it. Several of our young People were formerly brought up at the Colleges of the Northern Provinces: they were instructed in all of your Sciences; but when they came back to us, they were bad Runners, ignorant of every means of living in the woods . . . neither fit for

(Continued)

(Continued)

Hunters, Warriors, nor Counsellors, they were totally good for nothing. We are, however, not the less oblig'd by your kind offer, tho' we decline accepting it; and, to show our grateful Sense of it, if the Gentlemen of Virginia will send us a Dozen of their sons, we will take Care of their Education, instruct them in all we know, and make Men of them.

(Palmer, Zajonc, & Scribner, 2010, p. 19)

When I arrived at South Dakota State University (SDSU) in 2011 as the university's first full-time chief diversity officer, one of the primary constituencies with whom I began working were Native students, faculty, staff, and communities. Reflecting on my experience thus far, the above quote is significant to me because the 1774 dialogue is in many ways still occurring in education institutions and in wider social discourses. Palmer's reflection on the response of the elders embodies my core operating principles as an administrator, educator, advocate, and developing researcher: Every student should have access to

an education that embraces every dimension of what it means to be human, that honors the varieties of human experience, looks at us and our world through a variety of cultural lenses, and educates our young people in ways that enable them to face the challenges of our time. (Palmer et al., 2010, p. 19)

The significance of my work lies in the disparity in educational access, opportunity, and equity available to Native communities in South Dakota, where Native Americans comprise 8.9% of the population, but only 1.4% of the students at SDSU (SDSU, 2013; U.S. Census Bureau, 2013). According to the National Indian Education Study (NIES, 2011), Native students experience culturally unresponsive and devaluing PK–12 learning environments, which reflect a history of institutionalized poverty and racism (U.S. Department of Education [DOE], 2012). The lack of Native student persistence at SDSU has continued to be a troubling issue, with the number of Native students at the university falling over the past 3 years (Integrated Postsecondary Education Data System). Yet, Native students do persist and graduate from SDSU, which indicates that resilience might underlie Native student persistence. I wish to collaboratively negotiate an understanding of the experience of Native students who persist and succeed at SDSU. The primary research question guiding my work is: *How do Native students at South Dakota State University make meaning of their educational experiences?*

Viewing the PK–12 and higher education systems as fundamentally integral to each other, I seek to understand Native students' experiences and understanding of education, community, and culture, to support our Native students through fostering their own sense of empowerment. This work will also make visible to higher education institutions the steps needed to prepare for Native students as well as students from all underrepresented communities. Subsequent questions that may guide my work include, but are not limited to:

1. How do Native students understand the roles of culture and community, and how do they negotiate cultural borderlands that impact their educational experiences?

2. How can our university foster a sense of empowerment through building sustainable university-community partnerships and instituting social justice–oriented initiatives in outreach, community building, teaching, and curriculum?

Resilience and Persistence: An Emerging Conceptual Framework

Problematized in U.S. DOE (2012) and SDSU (2013) institutional data, I will inquire into Native students' PK–16 educational experiences contextualized in the cultural borderlands they continuously negotiate. Spring (2009) characterizes the history of the education of Native peoples as de-culturalization in which the dominant culture has used education as one mechanism to destroy Native culture and assimilate indigenous peoples. Others, notably Watkins (2001) have noted similar processes specific to the education of African Americans. U.S. DOE (2012) data suggest that Native students' education continues to perpetuate de-culturalization through what Freire (1970/2000) characterized as an oppressive banking system of education, which disempowers and oppresses Native students and communities.

Theoretical Context

Theoretically, I situate my work in the literature on resilience (Hauser & Allen, 2007; Masten, 2001; Masten, Herbers, Cutuli, & Lafavor, 2008). Masten (2001) offers a counternarrative to conventional understandings of human resilience, usually portrayed as remarkable and extraordinary. Rather, resilience results "in most cases from the operation of basic human adaptational systems": "If those systems are protected and in good working order,

(Continued)

Annotation: Jaime begins this memo by situating her research topic and providing both broad and specific context on the issues at play in her work and how this relates to her emerging research topic. The goal of this memo, which was a class assignment, was to help students to examine the contextual and conceptual influences on their burgeoning research topics so that they could develop and refine their research questions. This memo was written at an early stage in the development of Jaime's dissertation research. In this beginning part of the memo, she situates the topic in her own work and in the experiences of the students whom she serves professionally. She also posits an early primary research question and set of supporting questions. These are articulated here as an outgrowth of the issues of inequity that she identifies within Native students' educational experiences.

(Continued)

development is robust even in the face of severe adversity; if these major systems are impaired, antecedent or consequent to adversity, then the risk for developmental problems is much greater, particularly if the environmental hazards are prolonged" (Masten, 2001, p. 227).

In all social contexts, including educational institutions, primary factors associated with resilience include (Hauser & Allen, 2007; Masten, 2001; Masten et al., 2008):

1. Connections to competent and caring adults in the family and community

2. Cognitive and self-regulation skills, self-reflection, self-complexity, and agency

3. Positive views of self

4. Motivation and ambition to be effective in the environment

The above factors comprise "coherent narratives" of long-term accounts of negotiating disappointments and successes filled with themes of change and connection to the past (Hauser & Allen, 2007). Theoretical contextualization of Native student academic persistence as a process of resilience is consistent with the positive deviance (PD) research framework (Pascale, Sternin, & Sternin, 2010) discussed in methods below.

An Emerging Conceptual Framework

The conceptual framework for my work is emerging from a preliminary collaborative research project on which I serve as principal investigator at SDSU as well as preliminary data from a university climate survey for which I am co-principal investigator. My co-researchers include Native students, who have persisted and thrived in an environment often overtly hostile to them, as well as faculty and staff from the American Indian Education and Cultural Center (AIECC). Based on my current work, as well as preliminary investigation of the literature on resilience, I have developed the preliminary conceptual framework shown in Figure 1.

Major concepts represented in Figure 1 include:

- Persistence: the ability to enter the university and persist through graduation

- Resilience: the process of negotiating adversity and thriving in an education setting

- Self-concept/esteem: the view of oneself along personal and academic domains

- Relationships: healthy, caring, and supportive connections

- Self-identity: the complex understanding of self

- Institutional climate: the living, learning, and work environment

Relationships. Native students have articulated that relationships with teachers, students, staff, and other members of the community are crucial to their persistence and success. The importance of relationships emerges for Native students from the central role of familial and community ties in their lives. Many of the most important relationships of which our students have spoken have been of deep educational value involving family and community members. Productive relationships are filled with empathy, awareness of self (assumptions, biases, prejudices, etc.), responsiveness (Castagno & Brayboy, 2008), and inclusiveness. According to our students, and confirmed by scholars such as Spring (2009), an impediment to developing such deep pedagogical relationships is a lack of understanding of the centrality of connection in education institutions. The arrows connecting the characteristics of relationship with the central node in the concept map are bidirectional, indicating that understanding in relationship is mutually supporting and evolving. The line connecting the relationship node to the self-concept/self-esteem node is unidirectional to illustrate the importance and impact of connection to self-concept and self-esteem.

Annotation: In this section of the memo, Jaime articulates the key theories that guide and frame her research topic and explicates the key contextual and conceptual elements of her conceptual framework and how they come together in her work and in her study of that work. She also includes her then preliminary research with her students as a way to link her emerging dissertation topic to her then current engagement in a collaborative research pilot project. Jaime does this in an early and exploratory way as a means of making sense of these concepts and how they relate to the context of the research. This is also illustrated in Figure 1.

Self-identity. Identity is complex, and in my emerging framework, foundational aspects of identity salient to this work include, but are not limited to, language, family, place, culture, class, race, educational experience, and gender. I denote self-identity and its salient factors in Figure 1 with broken lines to indicate the intersectionality and complexity of identity. Importantly, the line connecting self-identity to self-concept and self-esteem is bidirectional, which indicates the recursive and evolving nature of the relationship between self-identity and self-concept and self-esteem.

Institutional climate. Rankin and Reason (2008) conceptualize campus climate as complex social systems defined by relationships between faculty, staff, students, bureaucratic procedures, institutional missions, visions, and values, history and tradition, and larger social contexts. An institutional climate in which students thrive rather than merely survive

(Continued)

(Continued)

(Rankin & Reason, 2008) is characterized by valuing difference, inclusiveness, and responsiveness. Native students have expressed a preference for institutional spaces in which they have felt an investment in their presence and success. The arrow connecting institutional climate with self-concept and self-esteem is bidirectional, which indicates a synergy between the success of Native students and their value to educational institutions.

Figure 1 Conceptual Framework Denoting the Process of Academic Persistence of Native Students.

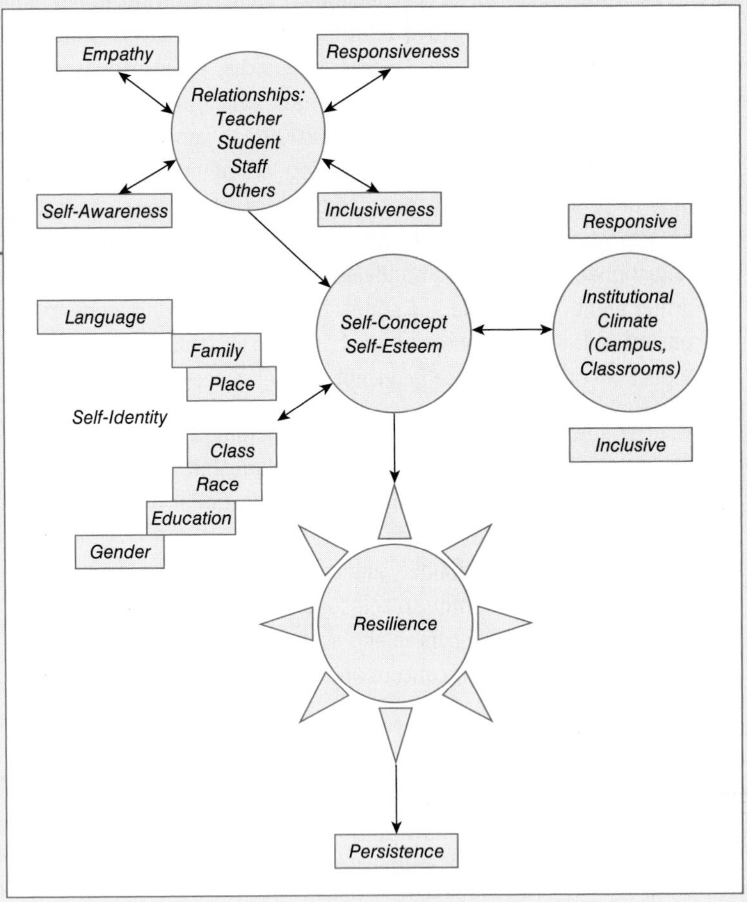

Annotation: This figure is a helpful display of Jaime's then-current thinking about the way a variety of components come together in her students' experiences and therefore in her research. This served as a beginning point upon which Jaime built her conceptual framework over time. At this stage, this figure served to help her visualize and make sense of the intersecting contexts and concepts that relate to her research questions. As she refined her research questions over time, this graphic was a useful way to consider the key constructs and theoretical frames of her study.

Self-concept and self-esteem. A Native student's self-concept and self-esteem is significantly impacted by her or his identity development, the institutional value placed on her or his presence, and the relationships she or he is able to create and maintain. The connecting arrows between all of these elements are bidirectional, which indicates that students and the institution can thrive and are integral to each other. A healthy self-concept and self-esteem promoted and valued by the institution can benefit the institution, foster Native students' sense of resilience, and contribute to their persistence in higher education.

The Role of Self and Context

October 2011: Butterfly Knowing and Learning to See in the Dark

It's fall in South Dakota. The day is azure blue and ochre and my breath catches as I breathe in the cool morning air and walk to the university car that will take me to the other side of South Dakota. I am in my first four months at SDSU and have been asked to accompany a journalism class to the Rosebud and Pine Ridge Indian reservations. This is my first trip across the state, which is much larger than I had imagined it to be. The entire state population is approximately 735,000 and has an area of more than 77,000 square miles. There is something about the vastness of this state and it being very sparsely populated that impacts my sense of place. I feel very small driving across the seemingly endless prairie and as I look up at the huge sky I feel as though my face is pressed against glass. I pause and as I look into the clouds I know that I am an outsider looking in. As we move through rolling hills and oceans of prairie grass I sense the deep and tragic beauty of a brutal history and feel a sadness deeper than any sea I know.

Delores, the professor teaching this course and leading the group, is Lakota and grew up in Pine Ridge. She has assigned her students to write stories about young people on the reservation and in each story they will write, she requires them to find the positive. She will not allow them to focus on clichéd tragedy. In spaces that seem dark and hopeless, they instead must find beauty and possibility. The students initially are righteously indignant and express frustration because they believe the truth lies in alcoholism, suicide, and abuse—in their uninterrogated white privilege. And even in the spaces where these things are part of the truth, the students do not see the

(Continued)

Annotation: This section of the conceptual framework illustrates how Jaime's personal background and personal views have influenced her conceptual framework. She begins this section with a vignette. This helps the reader to gain a different sense of the research setting and topic. Creating the vignette is also a useful (and powerful) tool for researchers to think about research from a variety of perspectives. For more on the uses of vignettes, see Chapter Eight.

(Continued)

transgenerational trauma and the role that deep poverty plays in all that is here. And, as members of the dominant culture, they certainly do not see their role in all of this. I then sense something familiar to me. In this moment, there is the presence of mestizaje, mestiza consciousness, where paradox can be held. Where there is light in the darkness.

As our time there comes to an end, I notice a quiet change beginning to take place in our students. This change, although powerful, is like a chrysalis, gossamer and fragile. Their words are not as judgmental, and they take delight in all that they have learned; what they have learned from people they had believed had nothing to offer them. There is the beginning of humility where there was arrogance. Instead of patronizing tones, there is the whisper of appreciation and respect.

The night before we are to leave, I am invited to attend a sweat with Dolores. The night is cold and clear and the sky is bright with stars. There are four others besides us who are attending the sweat. The sweat begins with a prayer from the man who is guiding the ritual. He prays to Tunkashila, which in Lakota means grandfather and/or Great Spirit. He asks that Tunkashila watch over Delores and me as we work with their young people and asks Tunkashila to guide us in our work so that the young people who come to SDSU can learn and bring their learning back to the people. Then each of us participating in the sweat is asked to offer a prayer. When it is my turn, I express my gratitude in being invited to be there and also express my thanks and acknowledge that it is an honor to be given the opportunity to work with their young people. I promise that I will do all that I can to see that their children are valued, nourished, and supported in who they are. This is something I do not take lightly. I feel the weight of it, though it is not heavy. It lives in the space between beats of my heart and whispers the truth to my bones in the quiet of every dawn.

My self. In many ways, I represent the oppressor and the oppressed, which has been a borderland (Anzaldúa, 1987) in which I have struggled throughout my life. Being a mixed-race person—Latina and White—I have experienced racial and class marginalization in the context of my own family, the product of an undesirable marriage where working class and brown skin were disdained. At many points in my life, I have struggled with the assumptions that others have placed on me. I am well educated but have experienced being unemployed and struggled against the possibility of living in

Annotation: After the opening vignette, Jaime describes her positionality and social location/identity by reflecting on past experiences that relate to the current context of her work and research as a way of situating herself within the current research topic. Jaime examines parts of her experience and identity that relate to the research topic and context as a way of framing her personal goals and experiential reasons for feeling drawn to engage in this research.

poverty. As a single mother, I have had to fight against the assumptions foisted on me and my sons about what constitutes an "intact" and "normal" family. Yet I have paradoxically moved through the world with a sense of privilege as well. I have learned that living this paradox has provided me with a lens through which I am able locate myself as a *mestiza* and position who I am in the variety of contexts in which I find myself. I have been, as Galeano (1989) writes, chopped into pieces and forced to interrogate my privilege and my assumptions in order to reassemble myself in a continuous, contextual, relational process of understanding my self, my life, the academy, and my role in it.

The context. My work with Native communities bears special meaning in a state where nearly 10% of the population is Native, where the land on which our university is built was paid for with the blood of Native peoples, where Native communities comprise the two poorest counties in the United States, where discrimination is institutionalized and taken for granted, including on the campus where I work. The narrative of this land is a narrative of tragedy, yet it is a narrative that has been silenced as an institutionalized narrative of progress—Manifest Destiny. At this historical moment, indigenous peoples throughout the world are engaged in an attempt to indigenize the academy, to reinsert silenced narratives into what the historian Ronald Takaki called the Master Narrative of American History. American history, my personal history, our individual histories, contain narratives of tragedy with which we have never fully embraced. This is my entry point, my connection, to this work, to negotiate the tragedy of oppression for both the oppressed and the oppressor.

Synthesis

My overarching research question investigates the meaning of lived experience for a group of students whose lived experience has been marginalized and deprecated for generations. These students' lived experience is unique because it occurs in the context of permeable cultural borderlands through which these students live and move. They live their histories both in the center and in the peripheries of multiple cultures. Understanding their experience as a process of resilience is consistent with a framework inclusive of concepts of self, relationships, and the institutional climate.

(Continued)

Annotation: Here Jaime engages in a synthesis of the multiple contextual factors that influence and shape her work and how they come together in her conceptualization of the research. She contextualizes the research questions as a preliminary way into her study methodology.

Annotation: In this section of the memo, Jaime overviews her research design plan. This is an important part of the conceptual framework, as all of the factors described throughout this memo impact the research design and methods choices. In this section, Jaime brings these connections to the fore both to crystallize and advance her thinking about the methodology at this early study development stage.

(Continued)

Research Design

While I will use quantitative data to initially frame and problematize the issue of the persistence of Native students in higher education, specifically at SDSU, my work will reflect a human science research design (van Manen, 1990). Epistemologically, I situate my work in the hermeneutic tradition (Gadamer, 1975; Ricoeur, 1980; van Manen, 1990) because I seek to investigate the meaning and understanding of lived experience. I also proceed from an advocacy-participatory paradigm in that my work will propose an agenda for reform with the goal of changing the lives of all participants: students, the institutions and communities in which they live, and my life as researcher (Creswell, 2007). The advocacy-participatory paradigm finds further expression in the positive deviance model based on the following premises (Pascale et al., 2010, p. 4):

1. Solutions to seemingly intractable problems already exist.

2. They have been discovered by members of the community itself.

3. These innovators (individual positive deviants) have succeeded even though they share the same constraints and barriers as others.

Positive deviance recognizes "the community's latent potential to self-organize, tap its own wisdom, and address problems long regarded with fatalistic acceptance" (Pascale et al., 2010, p. 7). The model acknowledges that once communities discover and leverage existing solutions by drawing on their own resources, "adaptive capacity extends beyond addressing the initial problem at hand" and enables "those involved to take control of their destiny and address future challenges." Positive deviance allows us to move away from and deconstruct what has been a normalizing, deficit lens through which researchers often view communities and conduct studies (Delpit, 2012; Pascale et al., 2010).

Data collection methods will include a series of phenomenological conversations (Seidman, 2006) and the collection of visual data through auto-photography (Blackbeard & Lindegger, 2007; Noland, 2006; Croghan, Griffin, Hunter, & Phoenix, 2008) and through various autoethnographic narratives. Secondary quantitative data from the U.S. DOE (2012) and institutional data from South Dakota State University (2013) specifically related to our Native students will be used to contextualize the study. I will analyze the data inductively to generate themes (Smith, Flowers, & Larkin, 2009), and I will represent my analysis as collaborative autoethnography through vignettes and visual data.

Conclusions

Considering long historical narratives of poverty, racism, and sociocultural oppression inherent in the experiences of Native students, these cultural/ educational dynamics have significant implications for the holistic well-being of our Native students at all levels of education. If universities desire to build inclusive communities, they must understand and value the experiences of students and faculty from historically marginalized and traumatized populations. The long-term value of this research lies in outreach, building the capacity for inclusive communities, and realizing social justice, which represents the larger value upon which the purpose of education rests (Carnevale & Strohl, 2013; Delpit, 2012; Nieto, 2010).

Annotation: This brief conclusion section serves as a starting point to the building of a broader rationale of the study. Here Jaime begins to position the work within its possible thought and practice communities as an outgrowth of the theoretical framework that was developed earlier in the memo.

Across this example conceptual framework memo and the examples in the appendix (Appendixes A, B, and C), we see how different researchers structure, develop, and iterate their conceptual frameworks. While the topics are different, one thing that becomes evident as we look across these narratives and illustrations of a conceptual framework is the ways in which the parts and whole of a study fit together, that is, the methodological ecosystem of a specific study. The conceptualization and construction of a conceptual framework is a valuable and generative part of the research process because it helps to bring all of the various study contexts and frames together in explicit and transparent ways that help tease out the interactions, tensions, and synergistic qualities of the parts and the whole. We believe that this process should be active, recursive, creative, and reflexive. In the next chapter, we build on the foundations offered in Chapters One and Two, as well as describe the role and process of qualitative research design and expand on our conceptualization of what criticality in qualitative research looks like.

QUESTIONS FOR REFLECTION

- What is a conceptual framework?
- What are the roles and uses of a conceptual framework?
- How does a conceptual framework guide and ground empirical research?
- How can you use a conceptual framework to build and refine your research questions and study as a whole?

- How do you construct a conceptual framework?

- What is the relationship between a conceptual and theoretical framework?

- What are the component parts of a conceptual framework, and how do they fit together?

- How do conceptual frameworks evolve?

- What are the multiple possibilities for developing and representing conceptual frameworks in narrative and graphic form?

RESOURCES FOR FURTHER READING

Conceptual Frameworks

Marshall, C., & Rossman, G. B. (2016). *Designing qualitative research* (6th ed.). Thousand Oaks, CA: Sage.

Maxwell, J. A. (2013). *Qualitative research design: An interactive approach* (3rd ed.). Thousand Oaks, CA: Sage.

Miles, M. B., Huberman, A. M., & Saldaña, J. (2014). *Qualitative data analysis: A methods sourcebook.* Thousand Oaks, CA: Sage.

Ravitch, S. M., & Riggan, M. (2012). *Reason & rigor: How conceptual frameworks guide research.* Thousand Oaks, CA: Sage.

Theoretical Frameworks

Anfara, V. A., & Mertz, N. T. (2015). *Theoretical frameworks in qualitative research* (2nd ed.). Thousand Oaks, CA: Sage.

Anfara, V. A. (2008). Theoretical frameworks. In L. Given (Ed.), *The SAGE encyclopedia of qualitative methods* (pp. 869–873). Thousand Oaks, CA: Sage.

Argyris, C., & Schön, D. A. (1978). *Organizational learning: A theory of action perspective.* Reading, MA: Addison-Wesley.

Maxwell, J. A. (2013). *Qualitative research design: An interactive approach* (3rd ed.). Thousand Oaks, CA: Sage.

Maxwell, J. A., & Mittapalli, K. (2008). Theory. In L. Given (Ed.), *The SAGE encyclopedia of qualitative methods* (pp. 876–880). Thousand Oaks, CA: Sage.

Tavallaei, M., & Abu Talib, M. (2010). A general perspective on role of theory in qualitative research. *The Journal of International Social Research, 3,* 570–577.

ONLINE RESOURCES

Sharpen your skills with SAGE edge

Visit edge.sagepub.com/ravitchandcarl for mobile-friendly chapter quizzes, eFlashcards, multimedia resources, SAGE journal articles, and more.

ENDNOTES

1. While discussed briefly in this chapter, we extensively describe the processes of dialogic engagement and structured reflexivity in Chapter Three.

2. See Maxwell (2013) for a detailed discussion of personal, practical, and intellectual goals, especially the five kinds of intellectual goals he describes.

3. For a useful discussion on race as a social construction with changing meanings over time, see Omi and Winant (1994).

4. The references in the student work throughout the book have not been included due to constraints on space.

Critical Qualitative Research Design

CHAPTER OVERVIEW AND GOALS

This chapter offers what we think of as a holistic and integrative approach to qualitative research design that supports *criticality in qualitative research*. Our primary goal in this chapter is to conceptualize qualitative research design directly in relation to the overarching and foundational goals and values of qualitative inquiry, with a particular focus on engaging the complexity of, and spheres of influence on, people's lives and experiences. Research design choices should be made with the guiding goal of seeking criticality and complexity through thorough contextualization. We describe qualitative research design as a dynamic, systematic, and engaged process of planning for depth, rigor, and the contextualization of data. It entails understanding and planning for the relationship of having a solid, detailed, intentional research design that still remains open to the kind of inductive and emergent methods required for research to be responsive and reflective of lived complexity.

We begin this chapter by defining qualitative research design. Then we detail the broad design processes, which include (a) developing study goals and rationale; (b) formulating guiding research questions; (c) developing the study's conceptual framework and linking it to research design; (d) developing a theoretical framework; (e) exploring the relationship between research design, methods choices, and writing; and (f) planning for validity and trustworthiness in/through research design. Throughout these sections, we offer suggestions and examples of activities to help guide novice and veteran researchers through the various design stages while still highlighting the recursive and iterative nature of qualitative research. This chapter culminates with a discussion of critical qualitative research design in which we specifically discuss our approach to qualitative research design.

BY THE END OF THIS CHAPTER, YOU WILL BETTER UNDERSTAND

- What qualitative research design means and entails
- The role of research questions in guiding qualitative research design
- How to develop study goals and a study rationale as part of a systematic design process
- What it means to formulate (and iterate) the guiding research questions of your study
- How to link your conceptual framework to your research design
- How to develop your theoretical framework and think about its value for your research design
- The differences between a literature review and a theoretical framework
- The role of pilot studies in qualitative research and the implications for your study design
- The value of vetting, rehearsing, and piloting data collection instruments
- The role of writing in relationship to research design and methods choices
- How structured reflexivity practices, including memos and dialogic engagement practices, can inform, guide, and complexify your research design
- How to plan for validity and trustworthiness in/through your research design
- How to critically approach research design

RESEARCH DESIGN IN QUALITATIVE RESEARCH

Qualitative research design is, most basically, the way that you, as a researcher, articulate, plan for, and set up the *doing* of your study. Research design is the overall approach to how a researcher (or research team) bridges theory and concepts with the development of research questions and the design of data collection methods and analysis for a specific study. This research design plan is based on an integration of the theories, concepts, goals, contexts, beliefs, and sets of relationships that shape a specific topic. In addition, it is grounded in and shaped by a response to the participants and contexts in which the study is carried out.

In a solid research design, theory and key guiding constructs are clear, and methods are built out of theory in ways that reflect prior learning within and across relevant fields; this theoretical examination of core concepts in the study sets the stage for a rigorous, systematic process of gathering and analyzing data.[1] Our perspective on qualitative research design is that every aspect of a study, from the early development of its guiding research questions to the selection of setting and participants, to the ways that we seek to understand the micro and macro contexts that shape all of this, is in a constant state of complex intersectionality[2] and dynamic movement.

The qualitative research design process begins at the point of interest in a topic and/or setting. From there, you engage in a process of active exploration into fields, concepts, contexts, and theories that help you to understand what you seek to know and how you seek to know it. The guiding research questions, which are cultivated through structured processes of learning, reflecting, and engaging in dialogue, are the glue between every aspect of research design. The centrality of research questions to the research is why it is vitally important to understand the core constructs of your research questions. Furthermore, the ways in which researchers need to be responsive to the phenomena and contexts of study settings means that research questions may evolve over time. This requires a mind-set that allows you to not only work hard to refine a set of research questions and a matching research design but also adopt an approach to the research that is flexible and responsive to the realities on the ground once the study begins.

As with research questions, the overall research design process is also inductive and emergent so that data collection and analysis processes can be responsive to real-time learning. This can include making changes, modifications, and/or additions to data collection methods. It can also mean that data analysis is ideally not only summative, or at the end of data collection (as is commonly the case), but is employed as a generative design tool that begins with formative analysis early on that can shape subsequent data collection and then that analysis is engaged in throughout every phase of a study. As data are collected and analyzed, aspects of the research design may change to respond to emerging learnings and contextual realities that require engaging deeply with the data as they are collected.

Data collection and analysis should not be seen as two separate phases in the research process; they are iterative and integral to all aspects of qualitative research design. The notion of the "inseparability of methods and findings" (Emerson, Fretz, & Shaw, 1995)[3]—meaning that how a study is structured and how data are collected has everything to do with the nature and quality of the data and therefore the analyses and findings that emerge from the research—underscores how integral and connected all aspects of the research process are. This connected aspect of the qualitative research process highlights the need for flexible research design. Qualitative research design often involves simultaneous processes of "collecting and analyzing data, developing and modifying theory, elaborating or refocusing the research questions, and identifying and addressing validity threats" (Maxwell, 2013, p. 2). Qualitative research design is fluid, flexible, interactive, and reflexive (Hammersley & Atkinson, 2007; Maxwell, 2013; C. Robson, 2011). Your role, as a researcher, is to connect (and reconnect) the dots between all of these intersecting parts.

OVERVIEW OF THE QUALITATIVE RESEARCH DESIGN PROCESS

In the sections that follow, we describe the processes central to qualitative research design. Qualitative data collection is iterative and often inductive, so it is important that there is a structured design approach that can help the study to achieve rigor and validity. It is the interplay between structure and flexibility that helps those who engage in qualitative research achieve validity through the collection and analysis of a quality data set that truly matches the goals, contexts, and realities that shape any given research project. A key aspect of this flexible approach to research design is understanding the range and variation of methods choices—and how they can be used creatively and responsively—so that they are employed in ways that help achieve (and even clarify) the goals of a specific study. What follows is a description of the key aspects of qualitative research design:

- Developing study goals and rationale

- Formulating (and iterating) research questions

- Developing conceptual frameworks in research design

- The focused development of a theoretical framework

- Exploring the relationship between research design, methods choices, and writing

- Planning for validity and trustworthiness in/through research design

Understanding each of these aspects of qualitative research design will be helpful to your study planning and writing processes. It is important to note that while we are laying the design process out in what may read like phases, qualitative research design, at its core, is not a linear process. We try to highlight the ways that research design evolves and changes through the recommended practices and examples in the chapter and the appendix.

Developing Study Goals and Rationale

Most studies begin broadly with an interest or a concern of a researcher or group of researchers. What becomes important, as a research study begins, is to explore and frame out the key goals of the study. As mentioned briefly in Chapter Two, Maxwell (2013) discusses the *personal, practical, and intellectual goals* of qualitative studies and suggests that you examine your understandings of study goals through these three broad conceptual categories. There are multiple goals for empirical studies, including *exploratory goals* (seeking to understand something about which little is known in an exploratory rather than conclusive way), *descriptive goals* (seeking to describe a phenomenon or experience), *relational goals* (trying to see connections between multiple variables and sets of ideas), and *causal or explanatory goals* (trying to explain causal processes within various phenomena) (Hart, 2001). It is vital for researchers to begin by critically considering the goals and motivations inspiring your research. Doing so requires, among other things, paying focused attention to the values and assumptions underlying the goals for your research and the beliefs that guide your study.

To engage in an intentional and systematic process of developing a solid understanding of the goals of your research study, we suggest a reflective inquiry process that has built into it a structured dialogic engagement around an exploration of the goals and what we think of as "the goals behind the goals," or the next layer of what is motivating your research study. By *next layer,* we mean that after the main goals of a study, there are usually other, less obvious, goals that need to be explored so that they become transparent and useful to thinking about how they shape your study. We have found that asking yourself (and being asked by others) a set of strategic questions can guide an exploration of the goals of a developing study. Examples of these strategic questions are detailed in Table 3.1.

These kinds of reflexive questions help you to refine your sense of the purposes of your study, which in turn helps you to conceptualize the parameters and content relevant to the study and begin to consider its possible uses, audiences, rationale, and significance.

Table 3.1   Questions to Consider When Developing a Research Study and Study Rationale

- What do I intend/hope to accomplish by conducting this study? For example, do I intend to improve something? Explain something? Understand sets of relationships between phenomena?
- What do I hope to understand or explain?
- Is the goal to evaluate effectiveness? If so, toward what goals and how are these defined?
- What guides the main goals of the study (e.g., wanting to refine a theory, improve a program, goals with/for the setting and participants, inform my practice)?
- What guides the next layer of goals (e.g., specific career goals, interests in a topic, desire to understand something better)?
- Who are my intended audiences for this study and why? What do I hope to share with these audiences and why?
- What do I seek to learn, know, understand, and/or explain?
- Why do I/we need to know, understand, or explain this?
- Who is the "we"? A specific field, subfield, or discipline? Multiple or intersecting fields or disciplines? Which ones and why?
- Why is this study worth conducting?
- What will this study contribute to the field(s) (e.g., filling in a gap in terms of focal context, theory development, practical implications)?
- What can/will this study contribute to theory, research, and/or practice?

Related to understanding your goals as a researcher is the development of the rationale of the study. A *rationale* is the reason or argument for why a study matters and why the approach is appropriate to the study. Rationales can range from improving your practice and the practice of colleagues (as in practitioner research), contributing to formal theory (e.g., where there may be a gap in or lack of research in an area), understanding existing research in a new context or with a new population, and/or contributing to the methodological literature and approach to an existing corpus of research in a specific area or field. Thinking about and answering the questions in Table 3.1 can aid in this process. Considering these kinds of questions is central to developing empirical studies, and it is important to understand that these rationales and goals will also lead you to conduct different types of research, guiding your many choices—from the theories used to frame the study to the selection of various methods to the actual research questions as well as designs chosen and implemented.

There are many strategies for engaging in a structured inquiry process and through it an exploration of research goals and the overall rationale of a study. These strategies can include the writing of various kinds of memos, structured dialogic engagement processes, and reflective journaling. Across these strategies, creating the conditions and structures for regular dialogic engagement with a range of interlocutors is an absolutely vital and necessary part of refining your understanding of the goals and rationales for the research. We describe each of these strategies in the subsequent sections.

Terms and Concepts Often Used in Qualitative Research

Fieldwork: In qualitative research, fieldwork entails the process of collecting data in a natural setting. This means in a setting in which the phenomenon would naturally occur (e.g., a neighborhood, organization, institution, or workplace). The term *fieldwork,* which comes from the ethnographic tradition of participant observation, is often used in qualitative research to refer to the located process of data collection.

Memos on Study Goals and Rationale

Memos are important tools in qualitative research and tend to be written about a variety of different topics throughout the phases of a qualitative study. Memos are a way to capture and process, over time, your ongoing ideas and discoveries, challenges associated with fieldwork and design, and analytic sense-making. Depending on your research questions, memos can also become data sources for a study. There is no "wrong" way of writing memos, as their goal is to foster meaning making and serve as a chronicle of emerging learning and thinking. Memos tend to be informal and can be written in a variety of styles, including prose, bullet points, and/or outline form; they can include poetry, drawings, or other supporting imagery. The goals of memos are to help generate and clarify your thinking as well as to capture the development of your thinking, as a kind of phenomenological note taking that captures the meaning making of the researcher in real time and then provides data to refer back and consider the refinement of your thinking over time (Maxwell, 2013; Nakkula & Ravitch, 1998). While we find writing memos to be a useful and generative exercise, both when we write and share them in our independent research and when we share them within our research teams, they may not serve the same role or fulfill the same purposes for every researcher. We suggest other activities and a variety of memo topics throughout the chapter (and the book) to encourage and foster engagement through writing or what we term *structured reflexivity.*

Recommended Practice 3.1: Researcher Identity/Positionality Memo

The purpose of a researcher identity/positionality memo (Maxwell, 2013)[4] is to provide a structure, at an early stage in the research development process, to facilitate a focused written reflection on your researcher identity, including social location, positionality, and how external and internal aspects of your experiences and identity affect and shape your meaning-making processes and influence your research.

We recommend that researchers with all levels of experience write this kind of memo and that you engage with this memo and add to or revise it over the course of a given study. In addition, we encourage researchers to write a new memo with each new research project since one goal of the memo is to connect aspects of your identity to the research topic and phases of the research process itself. For example, this memo is a required assignment in the doctoral-level methods courses we teach, and it is assigned before the students walk too far down the path of their independent research so that there is opportunity to challenge foundational assumptions and the relationship of who they are to the proposed study. Then the students are asked to reflect back on that memo once engaging in fieldwork so that they can further reflect on the influence of their positionality in the context of their interactions with study participants. We also recommend this memo to high school students with whom we conduct youth participatory action research (YPAR) as a way to help them understand the nonneutrality of research and to locate themselves within their justice-oriented research projects. Students of all ages and levels of research acumen find the memo to be both valuable and generative (and routinely describe the process of writing it as "more challenging" than expected).

Our students and colleagues share that they find writing this memo not only vital to their own critical understandings of themselves and their identities but also invaluable to clarifying their understandings of the topic and design process. Students also report that they revisit this memo throughout the research process to help illuminate their thinking and monitor any biases they described in the memo. Some choose to write subsequent identity memos as aspects of their identities emerge as relevant to their inquiries. (Example 3.1 is a second researcher identity/positionality memo.) We encourage our students (and you) to share these memos with a range of thought partners in ways that help you to hear constructively critical feedback on your biases as they relate to your positionality and research.

Topics to consider exploring in a researcher identity/positionality memo include the following:

- Positionality (relationship of self and roles to study topic, setting, and/or goals)

- Social identity/location (e.g., social class, race, culture, ethnicity, sexual orientation/identity, and other aspects of your external or internal identity)

- Interest in the research topic and setting

- Reasons and goals motivating the research

- Assumptions that shape the research topic and research questions

- Assumptions about the setting and participants and what shapes these

- Biases and implicit theories and the potential implications/influences of these for your research

- Guiding ideologies, beliefs, and political commitments that shape your research

(Continued)

(Continued)

- Intended audiences and reasons for wanting to address/engage them

- Other aspects of your identity and/or positionality in relationship to the research

- If engaging in team research, looking at the demographics and other features of the team in relation to your own identity and the research topic, context, and process

The above list is primarily meant to generate ideas. We are hesitant to include a list at all because we do not want to limit the possibilities; however, we provide one to help guide possible directions for this process. Still, we want to underscore that there is no wrong way of composing this memo, and, furthermore, that you can write and share multiple memos throughout the research process that relate to these topics since this kind of reflexive writing is not meant to only happen at the outset of your research.

We suggest that you consider sharing these memos with trusted colleagues and friends who can help you consider these issues productively through engaging in focused dialogue about the connections between self and the research at hand. We have also found that the sharing of these memos on research teams can be a vital source of thoughtful sharing that can generate powerful learning and exchange and help to situate the research team as a community of practice or inquiry group.[5]

In Examples 3.1 and 3.2, we include two different examples of researcher identity/positionality memos from past and present doctoral students at different stages in the research process. In Appendixes D and E, we provide two additional examples so that you see the range of ways that students approach these memos.

Annotation: In this memo, Susan reflects on her personal and professional experiences with the gendered nature of leadership at two different points in her research trajectory. Note that this is the second researcher identity/positionality memo that she wrote in relation to her study. Susan decided to structure this memo in response to a guiding question, which can be a generative way to approach a research identity/positionality memo.

Example 3.1: Researcher Identity/Positionality Memo

Susan Feibelman

Researcher Identity Memo 2

October 14, 2012

Why am I interested in the gendered nature of school leadership?

TAKE ONE (excerpted from my dissertation proposal): My interest in this topic is rooted in my personal experience as a school leader. Grounded in my first-hand experience with mentor-protégé relationships in both public and independent school settings, as well as the beneficial peer-to-peer mentoring relationships I have with other women leaders. In recent years, these conversations have acquired a frankness that reveals a growing impatience with the androcentric nature of independent school leadership and the fomenting of a "new boys club" that ensures the patriarchy's longevity (Baumgartner & Schneider, 2010). Repeatedly women's (White and of color) personal narratives describe the systematic regularity with which they are passed over for influential leadership roles. Our mutual interrogation of the context in which this practice unfolds has spawned a "mental itch" (Booth, Colomb, & Williams, 2003, p. 40) that is reinforced by a compelling body of research, which describes a similar trajectory for women leaders in public school. Fletcher (1999) refers to this phenomenon as "the story behind the story."

> All research—the particular question it finds important to ask, the point of view from which the question is posed, the source of the data used to find answers, and, of course, the interpretation and conclusions drawn from the analysis—are surely, albeit invisibly, influenced by the standpoint of the researcher. (Fletcher, 1999, p. 7)

Fletcher's use of relational theory to frame her own thinking about leadership development in corporate settings has immediate application to various forms of leadership within a school community. The principle of relational theory argues,

> growth and development require a context of connections . . . interactions are characterized by mutual empathy and mutual empowerment where both parties recognize vulnerability as part of the human condition, approach the interaction expecting to grow from it and feel a responsibility to contribute to the growth of the other. (Fletcher, 1999, p. 31)

Relational theory could as easily be applied to discussions about the practice of teacher inquiry and its self-reflective, mutually engaging, and action-oriented ethos.

TAKE TWO: My interest in this topic began with a question I started to raise with fellow teachers and school leaders following a student council election. Once again a female student running for the position of president had delivered a thoughtful, well-developed speech only to lose the election to a male classmate, whose speech was loosely organized and disarmingly comedic. The student voters responded to the candidate's irreverent charm

(Continued)

Annotation: In the first section, Susan relates her personal experiences to relevant, preexisting literature. In the second section of this memo, Susan discusses her interests to another professional experience. It is important to note that this memo was written when Susan had already developed research questions and had located her topic within relevant literature. This is an excellent example of a researcher identity/positionality memo; however, many memos will not look like this, especially earlier on in the research process.

(Continued)

by electing him president. While names and faces would change, this gendered dynamic would be played out election, after election.

Knowing the students and their track records for working on behalf of the student body, I began to question what role gender played in the election results. I also started to look more carefully at the ways in which adults in the school community modeled gender preferences through their unspoken support of certain leadership styles over others. What then was the relationship between our students' choices and the way we as their teachers might be prone to associate leadership with certain gender traits? Is this a topic for teacher inquiry?

Example 3.2: Researcher Identity/ Positionality Memo

Personal and Professional Goals for Dissertation Study

Mustafa Abdul-Jabbar

December 12, 2012

> I see my confidence growing . . . I still think it has a way to go, but I see my confidence building . . . and I feel like I have more tools at my disposal, whether it's research or people, to be able to connect with if I'm not sure. Whereas before distributed leadership, I felt limited in that.
>
> —Rosalie (personal interview, July 27, 2011) [school principal]

As an educational leadership doctoral student at the University of Pennsylvania, one of the chief-most issues I have sought to understand has been how school leaders cultivate trust within the educational organization. As a practitioner and school administrator in southeast Texas, an issue of practice that arose for me in my work was how to operate effectively in school organizations with low levels of trust, including how to systematically develop greater trust rapport and subsequent organizational capacity amongst instructional staff in low-level trust organizations. In pursuing

Annotation: In this memo, Mustafa explores and describes where his interest in the topic of relational trust in leadership emerges from, namely, his past professional experiences.

research that addresses this problem of practice, I have sought to better understand how individuals within the school organization learn to trust one another and how they come to understand one another as trustworthy.

I have chosen to study relational trust within the context of the Penn Center for Educational Leadership's distributed leadership professional development program because the leadership paradigm inherent in the program is one that moves away from a focus on school principals, as the sole drivers of teaching and learning, toward a distributed perspective as a framework for understanding leadership. Thus, acknowledging that "school leadership has a greater influence on schools and students when it is widely distributed . . . (i.e., school teams, parents, and students)" (DeFlaminis, 2011, p. 1). I feel that this leadership paradigm is more attuned to the socioemotional and social-psychological elements that are conceptual staples in research and literature on trust, more recent literature on successful leadership (see Yukl, 2009), and more consonant with my professional experience.

Distributed leadership, in that it identifies leadership as not predicated on rank/position, at least at the conceptual level, equalizes persons in the organization. For example, a teacher may be foregrounded during a particular activity while a principal is backgrounded, if the teacher's activity is more closely aligned with the core work of the organization. Results from the 2011 pilot study into how leadership team members were conceptualizing leadership found that Archdiocese team members at the piloted school site successfully internalized this idea—that the phenomenon of leadership entails more than mere formal position or hierarchy but rather leadership practice is mutually constituted by the interaction of organizational members across various situations. It was also found that team members at the pilot school site were typically unacquainted with this definition of distributed leadership practice before their professional development sessions; thus, what they recognized as leadership practice was often broadened through their participation in the distributed leadership (DL) program.

I have chosen this program context for study because I am most interested in how teachers and administrators, operating vis-à-vis one another on more equal terms, learn to trust one another. I believe that insight into this research arena will contribute to educational leadership scholarship and, through honing my understanding of the interplay between leadership and trust in schools, help me to improve my own professional practice as a school leader.

Annotation: In this next part of the memo, Mustafa explores the reasons for wanting to study the topic of relational trust within this specific context.

We hope that Examples 3.1 and 3.2 (as well as the two excellent examples included in the appendix that we urge you to read) help you to see and appreciate the incredible range and variation of approaches to and foci within these kinds of memos as well as the multiple writing styles used by the authors. We want to also remind you, after reading these, that while they are personal sense-making documents, engaging in discussion of them with colleagues and advisers is an important part of making sense of and thinking through these various influences on your thinking and being as researchers. These dialogic processes are described in the next section.

Dialogic Engagement Practices

We refer to the idea of *dialogic engagement* as systematic processes for engaging in generative dialogue with intentionally selected interlocutors about (and throughout) the research process. Dialogic engagement is an incredibly important aspect of the research process (Bakhtin, 1981; 1984; Chilisa, 2012; Freire 1970/2000; Lillis, 2003; Rule, 2011; Tanggaard, 2009). While you may constantly interact with others about your research in unstructured ways, we recommend that dialogic engagement practices be intentionally structured into the design process (and the formal research design) at different stages. For example, after conducting a few early interviews, you may meet with a partner, adviser, and/or a group of peers to think through the interview instrument, share excerpts of the data to determine if the instrument is helping to answer the research questions, think about how you are responding to participants, consider how your potential biases may (or may not) be reflected in the ways you ask questions of participants, and discuss how you respond to interviewees (Tanggaard, 2009). The important point here is to build these processes into the stages of design to create accountability to multiple interpretations and to the ongoing, rigorous challenging of your biases and assumptions. While these discussions should occur ongoingly, it is important to have structured processes that highlight the iterative and recursive nature of research built into the design. Recommended Practices 3.2 and 3.3 offer suggestions for structuring such engagement. We suggest that you decide if these sessions will be recorded and/or notes will be taken by you and/or others engaged with you. The systematic recording of these conversations has proven an important aspect (for our students and in our own work) of ongoing reflection that becomes a part of the research design and process.

Recommended Practice 3.2: Structured Sets of Conversations

These conversations are intended to facilitate deliberate and structured engagement with colleagues and peers around specific aspects of the research process. This works best if a group meets regularly and is familiar with the research topics and specific research questions of all members.

Process: Each researcher creates three to four key points or questions to discuss with at least two peers. The group divides the time they have to meet equally between all group members (it is essential to assign a time keeper if multiple researchers share at a given meeting). Each researcher can structure her time based on what she hopes to get out of the conversation and articulates her hopes and goals for the session to her peers prior to or during the session. While the conversation may evolve in unintended directions, it is important to prepare for the meeting so as to focus the conversation in ways that will most benefit the researchers.

Possible questions to consider at an initial meeting:

- What are the reasons for interest in this study?

- What are the reasons for interest in this context?

- What are the various goals for this study in terms of learning and professional engagement?

- What are some motivations for wanting to do this study?

- To what audiences am I gearing the study and why?

- How does all of this potentially shape the study?

- How can I think critically about my blind spots and assumptions and how they may influence/ shape the research?

- What theories might help me begin to delve deeper into this topic and why?

Possible questions to consider in ongoing meetings:

- What is shaping the study conceptually and theoretically and why?

- What assumptions continue to shape the research?

- How does the research design reflect these assumptions?

- Do I have the research design plan I need to achieve validity? In what ways might I improve on the rigor of the study?

- Have the study goals shifted or changed over time, and if so, why and how?

- What am I learning through vetting, rehearsing, and piloting my instruments?

- What do the early data suggest? For design refinements/changes? For the use of theory? For my working conceptualizations?

- How might others interpret these data and why?

The group should determine the topic or focus of subsequent meetings in advance, and members should prepare for these meetings, sharing documents such as memos, and data excerpts, archival

(Continued)

(Continued)

documents for peer review and discussion. For example, the group might decide to discuss researcher bias at a meeting, and members would come prepared to discuss this topic with all or parts of their memos or research journal entries that engage this topic.

In our own research group experiences, these groups can provide access to substantive data analysis thought partners since by the time of analysis, all members will be intimate with each study's research process and goals. So later in the life of the projects, we strongly urge a focus on data analysis, including the sharing and vetting of coding categories and analytic themes as is useful at various stages. The goal of these sessions is to talk through issues and questions in real time throughout the research cycle and to structure multiple subjectivities into the research process.

Recommended Practice 3.3: Paired Question and Reflection Exercise

This is a process, engaged in with a partner, to generate focused researcher reflection around key areas of importance in the study. These areas could be related to a variety of topics and conducted at different points in the research process. Examples of areas to consider include formative data analysis, instrument refinement, how the research questions are (or are not) being answered, participant representation, and researcher bias.

Process: In this paired exercise, each researcher should develop two to three key questions about her research. The partner will ask the researcher these questions and take notes about how the researcher answers the questions (while also noting her own questions that arise as she listens). The partner will then share her notes with the researcher, and the pair will discuss what stood out about the answers to the questions. With the partner (and afterward individually), the researcher will reflect on her answers to the questions and consider how they align with how she believed she would answer these questions, what underlying assumptions arose during the answers, how she portrayed participants, the overall goals of the study, and so on.

We recommend that researchers write a brief memo after this process to reflect on and document the learning and any unanswered questions. These memos could be shared with the partner one more time for additional thoughts. Furthermore, it is beneficial if the pair works together over the course of the project and engages in this process at multiple points in the research process. Our students share that they find this useful throughout the process and specifically during summative data analysis.

Reflective Journaling

It is a common (and suggested) practice for qualitative researchers to keep a research journal, and we believe that this provides an ongoing, structured opportunity for you to develop a research habit that can serve to deepen your thinking about the research process by creating more regularity and intentionality around the process of reflection. Unlike memos, which are written at selected moments throughout the research process and focus on specific topic areas, the research journal is an ongoing, real-time chronicling of your reflections, questions, and ideas over time. Research journals are useful both for in-the-moment reflection and meaning making and for charting ideas, thoughts, emotions,[6] and concerns over time. Recommended Practice 3.4 provides some guidance about keeping a research journal.

Recommended Practice 3.4: Research Journal

We recommend keeping a research journal that records (at least weekly) your thoughts, questions, struggles, ideas, and experiences with the processes of learning about and engaging in various aspects of research—from design through writing up the report. The main purposes of the research journal include that writing over time allows for the

- support and fostering of ongoing self-reflection;

- development and reinforcement of intentionality in research and good research habits;

- structured opportunity to develop and reflect on questions and ideas about research;

- researcher to keep valuable references that can be incorporated in future research;

- researcher to reflect on your thoughts, feelings, and practices in real time;

- formulation of ideas for action or changes in practice;

- documentation of your evolving frameworks for thinking about issues (including major turning points in your thinking); and

- development of meaningful questions for dialogic engagement activities.

Research journals can take a variety of forms and tend to be relatively informal. We recommend adopting any style or format that works for you; this means that entry length and structure will vary depending on the happenings of any given day or week. The research journal can also be an important source of data in qualitative research, depending on its relevance to the topic and

(Continued)

(Continued)

research questions. A research journal can be kept on your computer, in a notebook, on a smartphone, in a Prezi, or other visual format. Make sure to keep any electronic files password protected and to securely store notebooks to protect participants' confidentiality (some of our students use pseudonyms in their journals and memos). The goal of the journal is to find a format that allows for ongoing reflective writing throughout the process.

We argue that the research journal is vital throughout the research process, and it is important to begin the research process using this forum to document emerging thoughts that will shape the goals and rationale of your study and help you to build an argument for its significance.

Formulating (and Iterating) Research Questions

In the section above, we describe the process of moving from an interest or concern that you have been thinking about to the development of a focal research topic that is grounded in a solid rationale and a developing sense of the import of a study. Within this process of discovery comes the development and then refinement of your guiding research questions. Well-chosen research questions are vital to a research study and, in fact, are the center of research design. To collect the kinds of data you need to answer your research questions, you must intentionally map your research methods onto your research questions.

The development of cogent and researchable questions happens in many ways; central among them are engagement with existing theory and empirical studies in the fields related to your study and dialogic engagement with experts and peers who can help you think in focused ways about the goals and assumptions that frame and underlie your questions and study. Part of our argument for dialogic engagement in the process of question development is that the ways that you formulate research questions depend, to a significant degree, on how you conceptualize a topic or problem. This, as we discuss in Chapter Two, is informed in significant ways by the development of your conceptual framework. Since lived experience is complex and multifaceted, the research questions must be broken down into specific core constructs to be studied. We make a strong argument that you must carefully define each core construct because they are the building blocks not only of your research questions but also of the study itself. A conceptual framework helps the researcher to develop a cogent rationale for how you conceptualize the problem; identify the key influences, contexts, and factors; define these core constructs; and develop your working theories for what you think may be happening (Ravitch & Riggan, 2012).

For many researchers, especially those who do not have prior familiarity with a qualitative paradigm, this notion that research questions can be modified or changed, even once data collection has begun, is surprising and even, for some, causes concern about the rigor of qualitative research. It is the fact that research questions can be refined as we learn more about complex phenomenon and the theories that seek to explain them that helps a research study achieve rigor in a qualitative paradigm (Golafshani, 2003; Maxwell, 2013; Ravitch & Riggan, 2012; C. Robson, 2011). But, there must be certain conditions that allow this to be the case, which include (1) an intentionality in the process of developing and refining research questions, (2) a chronicling of the reasons for and influences on the key aspects of and refinements to research questions, (3) a vetting of suggested changes from multiple perspectives, and (4) early and ongoing data and theory analysis that informs changes to research questions.

There are multiple forms of research mapping—charting central goals, ideas, concepts, and processes in graphic form in a way that helps you to see core constructs and relationships between them—that can help you to make connections between central concepts, theories, and contextual aspects of the research. Recommended Practices 3.5 through 3.7 propose multiple kinds of research mapping exercises and share examples to suggest how they help to facilitate the research question development and iteration process.

Recommended Practice 3.5: Mapping of Goals, Topic, and Research Questions

Goal mapping: At the outset of a research study, map out the goals of your study visually (and in narrative form) as a way to explore each goal and chart their connections (and possibly see disjunctures) with the developing research questions. Then prioritize and cluster these goals in ways that allow you to see connections between them and that help lead toward the transition from research goals to research questions.

Topic and research question mapping: Mapping your research topic and questions is a very important stage in the development of your study. Begin by articulating the broad topic you are interested in for the study and then distill that into one to two more specific topics. Then, take these key concepts and transform them into early draft research questions that focus on the major constructs or concepts at the heart of the specified topic. This requires a careful attention to each word within the proposed topic and research questions as well as the implied or imagined relationships between them. We often ask students to prepare these individually and then have them vet them with a small group of peers who can ask them questions as they articulate the research questions, thereby helping them to get feedback on the meanings contained within each word of the questions. In our courses, we often workshop this in a fishbowl format so that students can watch each other engage in this process and then benefit from understanding the workings of the process as well as idea and process sharing for their research specifically.

Recommended Practice 3.6: Connecting Research Questions With Methods

In this approach to aligning your research methods onto your research questions, which is a crucial step in all qualitative research, you take each research question (and subquestion) and map your research methods onto it in two ways:

1. The first way is to map the specific data collection methods that you will use to attain the information required to answer the research questions. For example, if you wish to understand how doctors implement a procedure based on what they learned in a professional development experience, you would not only want to interview them, but you would need to observe them in their daily work settings to triangulate the data. You may also choose to interview their colleagues and/or patients to see what they note about how the physicians implement their learning. You might also consider putting together focus groups to initiate "groupthink" and would certainly want to see artifacts of the professional development initiative as well as of the organization and even the individual for context.

2. The second way is to map specific instrument questions onto each research question so that you are sure that your data collection instruments will in fact garner the data you will need to respond to your research questions.

See Table 5.7, Figure 5.1, and Table 10.3 for templates that you might use for this exercise. We have our students fill these out prior to class and bring them in for discussion in pairs.

Recommended Practice 3.7: Theoretical Framework Charting

To chart/map your theoretical framework, you turn to the formal theories that guide your research and represent them thematically in relation to your research questions in ways that help you to see how you are using theory to frame the research questions and perhaps the context and setting that surround them. We view this as theoretical framework building and argue that seeing the bodies of literature and guiding theories that frame the study, laid out visually, can help students to see connections and overlaps as well as tensions and disjunctures. Increasingly, our students use computer programs such as Prezi and MindMapper to engage this process.

Memos and Dialogic Engagement Practices to Support Research Question Development and Refinement

As described above, memos are useful tools for engaging in and even facilitating key aspects of the research process. In the development and refinement of research questions, we suggest that you write memos to consider what understandings and information you are seeking to gain and how these relate to each of your research questions. In Recommended Practices 3.8 and 3.9, we offer topics and prompts for memos that can assist in the development and refinement of research questions.

Recommended Practice 3.8: Memo on Core Constructs in Research Questions

This memo includes defining each of the core constructs in your research questions. For example, if you are studying professors' perceptions about the effectiveness of a civic engagement curriculum for engendering a social justice orientation in college students, you should clearly articulate what each of these constructs—that is, perceptions, criteria for judging effectiveness, civic engagement, social justice orientation—means and how you are defining them so that you will be able to understand how to approach them analytically and in terms of the research design and specific data collection methods that you would employ. In addition, you would want to consider which teachers (e.g., is it a specific group of college students using the curriculum? Are you interested in engaging with specific groups or subgroups of college students and why?). This process is intended to help you scrutinize each component part of your research questions and requires you to be precise and clear in the wording and phrasing of the questions since the entire research design will be built onto these core constructs.

Recommended Practice 3.9: Memo on Goals of Each Research Question

This memo clearly describes the goals of each of the research questions. Being clear on the goals of the research questions will help you to ensure that you are collecting the data necessary to answer them. We recommend charting out each research question and mapping goals underneath that question using bullet points to try to consider a range and perhaps even a typology of goals. This can help you to articulate each of the study goals as they relate directly to each aspect of all of the research questions.

To refine and develop research questions, memos that address all of the topics described in the Recommended Practices 3.8 and 3.9 (core constructs, goals, and the knowledge/ information sought by specific research questions) will help you to develop, scrutinize, and revise your research questions. (See Appendix F for an example of a memo about refining the research questions.) Research question development is an important process, since, as we have noted earlier, the research questions are central to the entire research process. In Recommended Practice 3.10, we outline additional ways to build on these more individualized approaches through engaging with others in dialogue about your topic and questions.

Recommended Practice 3.10: Dialogic Engagement Practices for Research Questions

Dialogic engagement exercises are a way to engage in vital conversations about your topic and questions with people who can help to challenge and support your thinking throughout the research process. We encourage you to participate in *structured sets of conversations* and the *paired question and reflection exercise* (defined in Recommended Practice 3.2 and Recommended Practice 3.3) with peers and advisers who can help you to critically explore and challenge yourself around the following topics:

Core constructs in research questions

- What are the core constructs of your study?
- How are you conceptualizing and defining each one?
- How do others define these?
- What are other possible ways of viewing and approaching the definitions of these concepts?
- What theories help to elucidate these constructs and why?
- What constructs might be missing?

Assumptions underlying research questions

- What assumptions are embedded in your research questions?
- What is shaping these assumptions?
- What might help to challenge these assumptions?
- Do others share these assumptions? Why or why not?

- What are other circulating assumptions in the research in this area?

- What data would you need to answer your research questions?

- Are there causal relationships implied in the research questions, and what implications might this have for the research design?

- What additional information do you need to answer or conceptualize your research questions?

Conceptual Framework in Research Design

While we discussed the integrative and critical role of a conceptual framework as well as described how to develop one at length in Chapter Two, we include a brief statement about it here as well since it is such an integral and vital aspect of research design that we do not want it to be forgotten in this chapter. A conceptual framework provides a specific rationale for who and what will be focal to a study; informs your choice of an overall design approach, including site and participant selection and the entire design of a study, the designation of units of analysis in the study, and the definition of the core constructs and theoretical concepts; and places you within the research in terms of your social location/identity and positionality and its relationship to the study goals and setting. It assists you in considering design integration in terms of how the research questions and design are informed by and defined in terms of the conceptual framework. Here, we focus on one key component of research design and of the conceptual framework itself—the theoretical framework—and discuss the relationship of the process of engaging in a generative review of literature to the development of your theoretical framework and the study more broadly and its design specifically.

The Development of a Theoretical Framework

There are three key aspects that scholars of qualitative research emphasize about the use of formal theory in research design. First, the role of theory is central to developing and iterating qualitative studies in formative, ongoing, and summative ways. This occurs formatively as you develop and refine aspects of your research, ongoingly as you implement your design, and ultimately when you analyze your data in relationship to existing theory. Second, theory helps to situate a study within ongoing conversations and existing theories and findings in relevant fields. Third, theory helps to add dimension and layers of understanding about a given phenomenon and the context in which it resides because it helps to deepen and extend our understandings of these concepts by conceptualizing their construction and meaning. The roles of the theoretical framework—in focusing a study, exploring and explicating relevant meanings and ideas, situating a study in its related fields (and situating those fields in relation to each other around a specific topic or issue), and helping to reveal the strengths and challenges of a study—help us to understand that theory plays not just a theoretical or conceptual role but also a methodological one. As Table 3.2 highlights, theory guides all aspects of a research study.

Table 3.2   The Roles of Theoretical Frameworks in Qualitative Research

The roles of theoretical frameworks in qualitative research include

- Influencing and shaping research throughout an entire study including in the
  - Formulation and development of research topics and goals
  - Development and refinement of research questions
  - Construction and evolution of research design
  - Development and enactment of data collection methods and processes
  - Development and enactment of analytical methods and processes
  - Framing the findings of a study and the written report
- Situating a study within preexisting literature and thought communities and the conversations happening within and across fields and subfields
- Conceptualizing different aspects and dimensions of a topic to contribute to a study's depth and complexity

We define a theoretical framework as the ways that a researcher integrates and situates the *formal theories* that contextualize and guide a study.[7] The theoretical framework of a study is developed through a multiphased process. This process often begins with a literature search, in which you seek out and map the key bodies of literature that frame a study. The next step is engaging in a formal literature review, which entails critically integrating[8] literature through writing. The term *literature review* implies that it is a discrete process. As described in Chapter Two, a review and integration of literature occurs throughout the entire research process.

Writing a literature review is a vital part of the process of your sense-making since it requires that you analyze and synthesize concepts from the literature (Hart, 2001). The process of engaging in this critical review of literature in the fields that contextualize your study has several interrelated processes and goals, including (a) tracing the etiology or history of the specific fields and topics related to the focal topics of your study; (b) cultivating expertise in ongoing and recent knowledge in these fields; (c) identifying the key theories, factors, and influences on the phenomenon and contexts to be studied; (d) gaining new and possibly innovative perspectives on how to conceptualize your research topic and guiding research questions; (e) learning the specific vocabulary and concepts in the fields that frame your study; and (f) identifying the range of methodologies employed to study related and even overlapping topics (Hart, 2001).

Early on and ongoingly, this process of engaging in the review of literature helps you to develop your argument for the goals, rationale, and significance of the study. This process is vital to understanding how to map the field and specific subject areas in terms of the questions and problems that have been addressed (or not), the key theories, concepts, and ideas. It is also crucial to conceptualizing and explaining the topic, as well as the major issues and debates in the field(s) since it is important to understand how various disciplines/fields frame the problem (Hart, 2001). We include specific questions that can help guide this discovery process in Table 3.3.

Table 3.3   Questions That Can Guide the Literature Review Process

- What are the key bodies of literature, specific works, and theorists in the areas/topics that relate to my research topic and research questions?
- What are the major theoretical debates and conversations already happening in the field(s) around this topic?
- How has this topic been studied before in terms of research design and methods?
- What methodological and theoretical assumptions have informed the way the topic has been defined and studied?
- How are these different studies and fields of study related, and do they intersect? If so, how?
- Have assumptions about the issue or context changed over time? If so, how?
- Is my topic framed or studied differently within and across fields and, if so, in what ways and why?

In order for a literature review to support your research, you should examine and articulate the aspects of the literature in an integrative and critical way, make central connections, and ask the kinds of questions described in Table 3.3. The synthesis, integration, and methodological understanding of various literatures come together to create the theoretical framework of the study. An important part of a theoretical framework often involves making an argument. You are the one who makes decisions about how to situate various literatures in relation to each other, and this is a part of how you make an argument in a theoretical framework. Your argument may also include how you frame the existing literature about a topic in relation to how it will be framed in your study. Furthermore, your argument may involve making a case (or justification) for your study.

It is important to note that the terms *theoretical framework* and *literature review* are *not* synonymous, although they are overlapping. A literature review is a process that helps you cultivate the theoretical framework for your study but is broader in scope than the theoretical framework since it includes all of the goals listed above and contextualizes the setting and context of the research as well as the topic and research questions. To transition from a broad literature review to a specific theoretical framework requires a careful, intentional, winnowing process in which you focus on that which is central to your study. This requires a meta-analytic and dialogic process in which you focus on the essential aspects of the formal theories that frame your core constructs in context. A useful analogy that Sharon shares in class is that Michelangelo was noted as saying that the extra marble left over from his famous statue of Moses was "that which is not Moses." This means that he carved away that which was not a part of his vision and understanding of the core of Moses as he wished to represent him. We think of the extra marble as "that which is not your theoretical framework." We encourage you to seek counsel as you do this, since it requires a kind of theoretical sophistication that can be daunting to a novice researcher.[9]

Writing memos as you develop your theoretical framework can be quite useful. We suggest that you address the memo topics described in Recommended Practices 3.11 and 3.12 as you work to develop your theoretical framework. We encourage researchers to share the theoretical framework and implicit theory memos with other individuals to receive structured feedback. In addition to composing memos, we recommended engaging in many of the specific dialogic engagement activities detailed earlier in this chapter, including Recommended Practices 3.2 and 3.3. We also suggest that you frequently discuss your burgeoning (and developed) theoretical framework with peers, colleagues, mentors, and friends. The process of developing a theoretical framework, like a conceptual framework, is iterative, and engaging in discussion, debate, reflection, and analysis with others and through writing is an important and generative part of the process.

Recommended Practice 3.11: Theoretical Framework Memo

This memo is intended to help you to develop your theoretical framework at an early stage and ongoingly at multiple points in your study. A theoretical framework memo might use some or all of the following questions as guides:

- Which theories am I using to frame the study topic and context?
- From which fields/disciplines do they hail? What bodies of research do they belong to?
- Why these fields and disciplines or bodies of literature? Why not others?
- How do these various framings intersect or relate to each other and to the research questions and setting?
- How am I using/engaging with these theories specifically and why?
- What are the benefits of these theories?
- What are the challenges of these theories?
- What assumptions underlie these theories and my choice to use them in my research?
- How do they cohere as a framework for my study? Why or why not? What can this tell me about my topic and setting?
- What argument am I making as I situate these theories?

Recommended Practice 3.12: Implicit Theory Memo

In this memo, you will consider the informal or working theories and beliefs that you bring to the research as a way to consider these influences on your research broadly and on your choice of formal theories specifically. Some of these may stem from earlier research and/or your professional practice; others may pertain more to your implicit or working conceptualizations as described in

Chapter Two. To engage in a process of reflection on these ideas and explore how they shape and guide the ways that you choose to engage with theory and in the broader conceptual framework, you might address/describe the following:

- What informal theories influence my choice of specific theories? My overall conceptual framework? Why? In what ways?

- Do these informal theories come from my practice? My study of various topics? Where else?

- How do they relate to my choice of theories?

- How am I using/engaging with these theories?

- What are the benefits of these theories?

- What, if any, are the challenges of these theories?

- Describe the relationship between the "formal" and "informal" theories in the study. How do these relate?

- What do I need to understand in relation to my working propositions and theories?

- What relationship(s) do I see between my theoretical framework and my larger conceptual framework? What does this help me understand about the various aspects of my study?

Research Design, Methods Choices, and Writing

Once you have established the main goals of a study, refined the guiding research questions, framed the key theories and methods used to study the topic, and made an argument for the rationale and significance of the study to the related fields, it is time to design the study methodology by *mapping your data collection and analysis methods onto your research questions directly*. There are many strategies for engaging in this process, including concept mapping and the charting of how each method maps onto each research question (see Recommended Practice 3.6) since your research questions are at the heart of your research design, both conceptually and pragmatically (Maxwell, 2013). In this section, we specifically discuss site and participant selection, the iterative and recursive nature of qualitative research design, pilot studies, and forms of vetting and piloting data collection instruments that can help you to determine appropriate and generative design choices.

Site and Participant Selection

Before you consider, or as you consider, the data collection methods for a given study, you determine the study site(s) and participants; this process is referred to as *site and participant selection* and includes the articulation of specific selection criteria for the site and the participants within that site. Selection criteria—the criteria you use to select a setting and participants—must be clearly defined. This includes the identification and clearly articulated rationale for and justification of all choices for the site(s) you choose and whom you include or exclude in the study design. Along with this comes a discussion of any challenges and limitations that go with these choices. These issues should be articulated with transparency and a clear sense of the possible limitations of these choices as well as their benefits. In addition, in the selection of participants, the researcher must pay careful attention to issues of representation in multiple senses of that term (Mantzoukas, 2004). Representation has to do with many aspects of participants' social identities, experiences, realities, and roles as they relate to the study context and study topic. For example, it may be role, experience, or positionality and/or social identities that shape what representation and sampling (another term for participant selection) look like for different studies. This will depend on the nature of the context and the research questions and is related to the process of mapping methods onto questions. We discuss participant and site selection as well specific sampling methods and their rationales in depth in Chapter Four.

Piloting

Piloting is a central aspect of designing and refining research studies and instruments. Piloting can take a variety of forms, including testing instruments, examining and noting bias, refining research questions, generating contextual information, and assessing research approaches and methods (Sampson, 2004). Piloting is commonly associated with the testing of data collection instruments in order to develop and refine them, which we discuss below. The term *piloting* is also used in reference to conducting a small-scale version of a study as a strategic prelude to conducting the larger study (Polit, Beck, & Hungler, 2001). This can mean structuring and conducting an exploratory pilot study that generates a next set of questions for a fuller study. It can also mean conducting a small-scale version of a study that will employ the same questions but with a refined research design that is rescoped based on data from the pilot study.

Piloting is a powerful tool for data-based design improvements. These improvements can include "develop[ing] an understanding of the concepts and theories held by the people you are studying—a potential source of theory" (Maxwell, 2013, p. 67). This development of theory through piloting is valuable since the data that emerge from pilot studies are based in participants' language and meaning making and therefore provide a valuable source of understanding about the meanings and perspectives of those who are a part of your research (Maxwell, 2013).

Piloting can also serve to help us understand ourselves as researchers and our research techniques by helping us hone interview skills and work on modes of interpersonal engagement, including how we frame and approach our studies with participants (Marshall & Rossman, 2016). It is important to keep in mind that in qualitative research, the purpose of piloting is to refine the research design and methods, which includes instruments as well as research questions (Creswell, 2013; Morse, Barrett, Mayan, Olson, & Spiers, 2002). After conducting a pilot study and/or piloting instruments, certain aspects of the research design may change slightly or significantly. There are times when piloting results in a researcher or research team shifting away from a particular kind of study or topic altogether. This is an important strength of piloting that can ultimately save time and energy. The many important reasons for and values of conducting pilot studies are presented in Table 3.4.

Table 3.4   Reasons for and Values of Conducting Pilot Studies

- Develop and refine research instruments
- Determine if instruments generate the necessary data to answer research questions
- Determine if and how a larger study should be conducted
- Determine any potential problems for conducting a larger study
- Develop research questions
- Develop a research design plan
- Explore the sampling strategy and rationale behind participant selection and recruitment strategies
- Train members of the research team on the appropriate methods
- Collect and analyze preliminary data and use learnings to inform subsequent study
- Determine if problems may arise in the collection and analysis of data
- Determine if the analysis methods are appropriate for the data
- Determine the resources needed to conduct the study
- Demonstrate to others (e.g., funding sources, review committee) that the researcher is competent
- Demonstrate to others (e.g., funding sources, review committee) that the study is worthwhile

Source: Adapted from van Teijlingen and Hundley (2001, p. 2). Retrieved from http://sru.soc.surrey.ac.uk/SRU35.pdf

We have seen the incredible benefits of piloting firsthand in our own research and that of our colleagues and students. Piloting cannot ensure the success of a qualitative study, but it surely contributes to its rigor and the quality of data, among other benefits. It is important to note, though, that piloting must be conducted thoughtfully and with attention to the rigor of the piloting process itself. Piloting should be conducted in thoughtful, intentional, and careful ways to make it reliable and useful (Sampson, 2004; Kim, 2011; van Teijlingen, 2002; van Teijlingen & Hundley, 2001). Furthermore, it must be documented with great care and detail in order to be useful as you move forward in your research. As you conduct the full study, you will include information about how you structured your piloting processes as a way to discuss the study design and validity.

In terms of the use of piloting as it refers to data collection instruments, we want to note that we differentiate between piloting instruments, vetting instruments, and rehearsing instruments, although these are related processes that build on each other. In the next subsections, we discuss vetting, rehearsing, and piloting instruments so that you can appreciate the differences, uses, connections, and values of each process. Ideally, you will engage in all three aspects of developing and refining your instruments since together they promote the rigor and validity of your study.

Vetting Instruments

Vetting instruments entails sharing multiple drafts of your data collection instruments with knowledgeable others who can give you critical feedback. These knowledgeable others can include classmates, peers, mentors, colleagues, experts in the field, scholars, community members, potential participants, and so on. This as an iterative process wherein a first draft would be shared with multiple readers (approximately two to three) and revised based on the careful integration of feedback, and then a second (and perhaps even third) draft is shared with these thought partners again until the instruments have been fully challenged, discussed, and improved upon based directly on this focused feedback. We do this ourselves, and we require this of our students because having your instruments vetted and discussed in relation to the goals of your study and fit with your research questions, as well as having them challenged and revised systematically, is incredibly valuable to you as a researcher. It can help uncover that which is missing or problematic, underlying assumptions and biases, issues with clarity, wording, flow, scope, and so on. Harkening back to Emerson et al.'s (1995) notion of the "inseparably of methods and findings," the quality of your data collection instruments has everything to do with the validity of your study and the quality of your data and therefore directly shapes what you are able to do with your data in your analysis and writing.

Rehearsing Instruments

Rehearsing instruments, which we view as quite valuable, is a different process from piloting instruments in at least two major ways. First, rehearsing your instrument(s) involves rehearsing (practicing and testing out) an instrument with a friend or colleague—rather than the target population—to check for flow and clarity of wording, sequencing and content of questions, and relationship of the number of questions to your ability to include follow-up questions and probes for individual meaning and terminology. Rehearsing instruments means that you engage in a mock interview so that you can rehearse the instrument and practice your interviewing style as it relates to the specific study and set of questions; it helps you to become more comfortable with the interview questions and process in a way that cannot happen until you try out the instrument in real time with real people who can react, question, and give feedback during and after the rehearsal. The fact that the person you are rehearsing the instruments(s) with is not analogous to the population of your study is a major difference in the process and has implications for the degree to which you will adjust your instrument(s) since you must remain aware that this population is different than the real intended participants. The second major difference between rehearsing and piloting instruments is that rehearsing an instrument means practicing your instrument (i.e., conducting a mock interview) but does not include collecting and analyzing data from that process. What we advise our students to do is to take notes, have the people you are interviewing take brief notes if possible, and then discuss the interview instrument and process once the rehearsal is over (you can also audio-record these to listen to later, but you need not transcribe them). When you do this in multiple rounds with different people, you are able to begin to see patterns in the interview experience and interviewees' feedback. Rehearsing instruments multiple times also means that you become increasingly comfortable with the instrument itself as well as with the opening remarks in which you discuss, with participants, the informed consent process, the structure and timing of the interviews, confidentiality, and the voluntary nature of the interview.

Piloting Instruments

When you formally engage in *piloting instruments,* you use your instruments with individuals who meet the sampling criteria for your study and collect and then analyze the piloted data. For example, when piloting an interview, the goal is to refine your instrument. Thus, you need to audio-record (with permission) and transcribe that interview and analyze the data in order to refine your instrument. To be clear, piloting instruments creates data that are used to drive data-based changes to your study. These data may or may not be used as a part of the formal data set depending on how significantly you change or revise the instruments as a result of piloting them.

When piloting interviews, for example, you engage in full-length interviews and approach all aspects of the interview as you would in a postpilot interview. As you collect and analyze the data from these interviews, the goal is to examine the data in relation to your research questions to see if you are getting the kinds of data you need to be able to answer your research questions. This may require, if possible, multiple pilot interviews to have as much data as possible. As you analyze the pilot interview data, the goal is to look for patterns in the responses. For example, you might ask the following:

- Were there similar responses to or confusion about specific questions?

- Were there particular questions that needed clarification across the pilot interviews?

- Was there any issue with the flow or sequencing of questions?

- Did you ask the right and enough (or too many) follow-up questions?

- What else emerges as you look across participant responses?

These analyses, as well as the content analyses of the responses themselves, will suggest any additional changes, revisions, additions, or deletions of interview questions you may need to make.

One important note across these three methods of instrument development and refinement is that you should carefully document how you structured and approached these processes. You should include in-depth descriptions of the specific techniques you used in your research design section that include enough detail on the actual processes as well as a focused discussion of how they influenced specific changes and revisions to your instruments and possibly other aspects of your research design. In Table 3.5, we provide a broad overview of the instrument development process, including vetting, rehearsing, and piloting processes.

The instrument piloting process, which includes vetting and rehearsing your instruments, is vital for achieving rigor and for the validity of your study. Again, we encourage our students to be sure to document (in memos and/or a research journal) the various kinds of piloting they engage in and how, specifically, it shaped or changed their instruments and broader research design.

Writing

Writing is an integral part of the research design process. Writing highlights and helps you to make sense of the iterative and recursive nature of qualitative research. We strongly believe that various forms of research writing—including memo writing, drafts of research questions and overview designs, research journal entries, fieldnotes, informed

Table 3.5   Steps for Vetting, Rehearsing, and Piloting Instruments

---

VETTING INSTRUMENTS

To vet your instruments, engage in the following processes:

1. Develop a solid working draft of your data collection instrument(s) and vet instruments by two to three thought partners, advisers, and/or knowledgeable others. To do so, either send instruments to them via email and/or, ideally, meet to vet and discuss their impressions of the instruments together.

2. Integrate their suggestions and feedback into a revised version of the instrument(s).

3. Repeat this process one or two more times until the instruments have benefited from multiple perspectives and iterations.

REHEARSING INSTRUMENTS

To rehearse your instruments, engage in the following processes:

1. Engage in rehearsing the instrument with two to three thought partners, advisers, or knowledgeable others (can be the same ones as in the vetting stage or new ones), taking careful notes throughout the process. If possible, you might audio-record these sessions (with permission). This means walking through the interview (or other data source) as if this were a real interview (but perhaps moving along more quickly than in piloting instruments).

2. Just after the interview (or other data source) is completed, debrief with these thought partners by asking them to share what they noticed about the instrument overall, specific questions, and the flow of the interview. Ask for suggestions on structure and wording of questions as well as sequencing. Also ask if and how your interview style might need refinement. As well, ask for anything problematic or missing from the instrument.

3. Record and take notes during each of these discussions if possible.

4. Once each conversation happens, look for patterns in the feedback across thought partners. For example, were there similar responses to or confusion about specific questions? Were there particular questions that needed clarification across the interviews? Was there any issue with the flow or sequencing of questions? Did you ask the right and enough (or too many) follow-up questions?

5. Use these responses to revise the instruments.

PILOTING INSTRUMENTS

After engaging in the vetting and rehearsing of instruments, pilot them by engaging in the following processes:

1. Take the vetted, postrehearsal, and revised instruments and pilot test them in their entirety with the population to be included in your study (if that is not possible, try to find a few individuals who are as close to the demographics of the participant group). These pilot tests should be structured and take the full amount of time that the real sessions would take.

2. Audio-record (with permission) and transcribe these interviews/focus groups to analyze them and also take notes during the interview if possible and not distracting to the participants.

3. Look for patterns in the responses. For example, were there similar responses to or confusion about specific questions? Were there particular questions that needed clarification across the pilot interviews? Was there any issue with the flow or sequencing of questions? Did you ask the right and enough (or too many) follow-up questions? What else emerges as you look across responses?

4. Use responses and your (and perhaps others') analysis to revise, reorder, add, and/or delete questions.

5. Look to see if these changes influence other changes to your overall research design.

6. Test the revised version if possible before starting actual data collection.

Note: We focused on interviews here, but this can apply to focus group instruments as well. While observation instruments may not be rehearsed, they can be vetted as well as used in pilot studies. Pilot testing questionnaires requires that you have people fill out the questionnaires either on paper or online as they would in the real context and with the same directions as the actual questionnaire. You then examine responses and ask respondents for feedback on structure, flow, content, directions, and so on.

consent forms—throughout the research design and overall research process are generative and vital to reflexive research. We specifically discuss the role and structure of formal writing in qualitative research, including the final research report, in Chapter Nine, and we discuss the writing practices and approaches involved in writing a research proposal in Chapter Ten.

Recommended Practices 3.13 and 3.14 are additional ways for you to engage in research design through structured writing activities in the form of memos to help you refine, develop, and justify your emerging research designs. Example 3.3 is one example of how the research design memos described in Recommended Practice 3.13 can be structured. In addition to the dialogic engagement activities described above such as *structured sets of conversations* (Recommended Practice 3.2) and *paired question and reflection exercise* (Recommended Practice 3.3), we describe a group inquiry process in Recommended Practice 3.15 that can be generative to the writing process. Sharing your writing with peers and colleagues in both formal and informal ways is important to the overall research design process as well as to the specific written products themselves.

Recommended Practice 3.13: Critical Research Design Memo

The goal of this memo is to systematically reflect on your emerging research topic and research questions (and the goals and concepts that shape them) and relate these to the plan for your data collection methods and processes. An important aspect of this memo is to pay attention to the ways that your research engages criticality in its approach to understanding context, including the impact of macro-sociopolitics on the setting and participants, as well as to setting up a research design that seeks this and other kinds of complexity and contextualization through a rigorous

(Continued)

(Continued)

process of reflexive engagement and methods consideration. (Example 3.3 demonstrates how a critical research design memo could be structured.)

Topics to consider exploring in a critical research design memo include the following:

- Iterated research questions

  o How/if the questions have changed and why

  o Describe the goals of refined research questions and how they differ from the earlier questions

- Site selection criteria and rationale questions/concerns/ideas

- Participant selection criteria and rationale questions/concerns/ideas

- Synopsis of intended methods

  o How/if these have changed and why

  o Goals of each method and rationale for each method

Recommended Practice 3.14: The "Two-Pager" Research Design Memo

The goal of this brief memo is to have you distill into a succinct document what is usually an emerging set of ideas about your topic, research questions, guiding theories, methods, and timeline for a proposed study. This memo should be used at the moment when you are seriously considering a new topic for an empirical study, often for a master's thesis or dissertation. Writing a two-page memo that distills your *topic* broadly, your possible *research questions,* the *bodies of literature* that will frame the exploration of the topic, and your possible *research design* helps you to push through ideas quickly rather than getting bogged down in a long, weighty narrative. When we advise our students to engage in this process, part of the value is to achieve a kind of emergent clarity about the research, and this happens not only through writing this memo (in part given that it is brief relative to a full proposal), but even more so through using the "two-pager" to vet the proposed study by a number of peers, mentors, and advisers who can help refine each part and the whole. Students share that this process is invaluable to the development of their thinking.

The sections of this memo are as follows:

- Topic and Setting

- Possible Research Questions

- Goals of the Study

- Bodies of Literature to Frame and Guide the Study

- Methods/Research Design Overview

- Timeline

- Questions for Your Reader(s) (these questions can also be noted throughout by using the comment function of track changes)

This memo should be kept as close to two pages as possible. There are multiple reasons for this, including that it helps you to succinctly overview your proposed study. In addition, when you are vetting this memo by multiple people, having a short and clearly developed memo can help facilitate this process.

Example 3.3: Critical Research Design Memo

Critical Research Design Memo

Brandi P. Jones

January 20, 2014

The iterative process of writing memos offers the opportunity to be reflective and thoughtful about how I approach my research study. The process of writing memos is therapeutic and restorative in ways that give me permission to breathe and enjoy the journey of research. I write in the power and memory of Sadie T. M. Alexander, the first African American woman to earn a PhD in the United States. "Don't let anything stop you. There will be times when you'll be disappointed, but you can't stop. Make yourself the very best that you can make of what you are. The very best" (S. Alexander).

Research Questions

Early in this process, my focus was on the professional identity development of Black doctoral students. While professional identity is a component of my research study, my goal is to give voice to the experiences of Black doctoral students at predominantly White institutions. Specifically, the racialized experiences defined as "encounters with faculty, peers, and institutional structures (policies, organizational norms, etc.) that reinforce racial stereotypes, engender

(Continued)

(Continued)

feelings of racial subordination and othering, and disproportionately validate and privilege members of some racial groups at the expense of others" (Harper, personal communication, December 21, 2013). My refined research questions are the following:

1. What are the racialized experiences of Black doctoral students in engineering programs?
2. How do racialized experiences affect Black doctoral students' sense of intellectual belonging in engineering fields?
3. How do racialized experiences affect Black students' postdoctoral career plans?

Literature

To examine this phenomenon, I will explore the following buckets of literature:

A. The persistent underrepresentation of people of color at various junctures in the STEM educational pipeline and subsequently in STEM fields

B. Factors and conditions that strengthen and undermine intellectual sense of belonging among graduate students in their academic disciplines/fields

C. Factors and conditions that strengthen and undermine the development and sustainability of science identities

D. Factors and conditions that compel students of color to depart the sciences

E. Racialized experiences of graduate students of color at predominantly White universities

   1. Onlyness: "The psychoemotional burden of having to strategically navigate a racially politicized space occupied by few peers, role models, and guardians from one's same racial or ethnic group" (Harper, 2013, p. 189).
   2. Stereotypes and Racial Microaggressions
   3. Racial Trauma: "Severe cases of the emotional, physical, psychological discomfort, and pain resulting from experiences with racism" (Truong & Museus, 2012, p. 228).
   4. Socialization Experiences

Annotation: After describing her research questions, Brandi briefly overviews the bodies of literature that she uses to situate her study. Articulating this at an early stage is a very important practice because you can circulate this memo to your adviser and/or peers to get feedback about the areas in which you are situating your study.

I will use existing research to guide my research design. From my initial scan of the literature, it appears that much of the research around students of color in STEM fields is focused on precollege and undergraduate populations. Furthermore, research on professional identity lacks any in-depth analysis of race. Prior research can help justify the need for my study, clarify terms, and identify possible methods (Maxwell, 2013).

Data Sources

I will collect data via questionnaires and interviews with Black doctoral students in engineering fields at predominantly White institutions. My thoughts are to select a subset of questionnaire respondents to participate in interviews. Interviews will be conducted according to the three stages into which Tinto (1993) divided the doctoral degree process: transition and adjustment, or the student experience during the first year of study; attainment of candidacy, or the period between the first year and the time a student attains candidacy; and completion of the dissertation, or the time between candidacy and the student's defense. I will also gather data from documents and resources from the National Association of Minority Engineering Program Advocates, National Action Council for Minorities in Engineering, National Science Foundation, and American Association for the Advancement of Science.

Questions About Data Sources

(1) *Do I have to narrow site selection to particular campuses or can I survey and interview doctoral students from a variety of predominantly White institutions?*

(2) *Should the participants in my sample have similar backgrounds in terms of undergraduate institutions, K–12 preparation, parental influences, and so on?*

(3) *How big does my sample need to be?*

Possible Approaches

I'm not so clear on which approach I will use. There are a few that seem applicable to my research study (I need lots of help here). Here are my thoughts thus far . . . .

(Continued)

Annotation: Throughout this memo, Brandi includes questions to her research adviser. In this way, this memo not only documents Brandi's reflexive thoughts, processes, intentions, and questions but also serves as a tool for seeking feedback from her adviser at a formative stage of study development.

(Continued)

Narrative research methods will be particularly useful given the importance of experiences and stories of doctoral students in this study. Narrative stories of a small, select group can help shed light on the racialized experiences of doctoral students at predominantly White universities. After the stories are collected, an analysis can be made in themes such as onlyness, stereotypes, racial microaggressions, and racial trauma (Harper, 2013; Truong & Museus, 2012). Given the various stages of the doctoral process, turning points or specific tensions will likely surface from the stories (Creswell, 2013). Autoethnography, a type of narrative, contains personal stories as well as the larger contextual meaning for the stories (Creswell, 2013). This can be useful as stories are told against the backdrop of institutional structures and organizational norms that limit opportunities and undermine intellectual sense of belonging of doctoral students. Restorying can be used to create a framework to link the ideas (Creswell, 2013). This framework can be useful, as not much of this research has been done on doctoral students in STEM fields.

Phenomenology would be ideal in this study because it "describes the common meaning for several individuals of their lived experiences of a concept or a phenomenon" (Creswell, 2013, chap. 4, sect. 2). I would use this approach to understand doctoral students' racialized experiences. Given my connection to this topic, it will be important to "bracket" my experiences in order to gain a fresh perspective (Creswell, 2013).

Once data are collected from doctoral students through surveys and interviews, a theory can be generated. The focus of a grounded theory approach would be on the stages of the doctoral degree process, as defined by Tinto (1993). In this study, a theory might explain the racialized experiences and sense of intellectual belonging of doctoral students at various stages of doctoral study. This can help inform developmental, retention, and success efforts of doctoral training. A grounded theory approach makes sense because there are not theories to explain the experiences of African American doctoral students in STEM disciplines.

I will use a tool, such as Lincoln and Guba's (1986) framework, to "ensure credibility, transferability, dependability, and confirmability of findings" (as cited in Truong & Museus, 2012, p. 236). Once interviews are conducted and responses are coded, follow-up interviews will be used to check accuracy of themes or theories. Given my connection to the topic and study, I will identify a colleague to be a peer debriefer to review the study.

Memos will be used throughout the research study to articulate researcher bias. Writing memos has proved to be very helpful thus far!

Questions About Approaches

Can more than one approach be used?

Pilot Study

Given that I am not an engineering doctoral student, a pilot study will be critical for me to get a sense of "concepts that are drawn from the language of participants" (Maxwell, 2013, p. 67). I will use a pilot study to gain insight on how to stage questions (I need lots of help here). Given my relationship to the topic, I have to take great care in making sure that I don't ask "leading questions" (Ravitch, personal communication, January 16, 2014). I will gain perspective on interview questions from scholars, colleagues, and recent engineering PhD alumni. My pilot study will include the following:

(1) Send survey and interview questions to colleagues and scholars for input.

(2) Survey students in various stages of the doctoral process.

(3) Conduct focus groups with students to fine-tune questions and test interview protocol.

Questions About the Pilot Study

(1) Do I have to use Black doctoral students in engineering disciplines in the pilot study? Or can I use doctoral students of other backgrounds in STEM fields?

(2) How large does my sample have to be?

(3) Can you recommend a resource that is particularly useful for developing interview and survey questions, interview protocol, and so on?

A Note About Researcher's Relationship to Study

In qualitative research methods, the relationship between the researcher and the researched is key (Creswell, 2013; Maxwell, 2013). Given my position at Princeton, doctoral students may be both eager and reluctant to share their

(Continued)

Annotation: In this section, Brandi describes and explores her positionality and assumptions as they relate to her research goals and topic. Clearly identifying these is important in the design phase so that she can monitor and reflect on these influences throughout her study.

(Continued)

experiences with me. There are some that will see this study as a way to improve the doctoral experience for Black students, and others might be hesitant to speak in front of a college administrator. This study will not only contribute to the research on African American doctoral students in STEM fields but also inform my work at a predominantly White research institution. Maxwell (2013) noted the importance of personal experience as a source of motivation (p. 24). My many interactions with doctoral students have sparked my research questions. As I pursue this research topic, I make a number of assumptions about the development of Black doctoral students at predominantly White institutions based on my wisdom of practice. I've spent countless hours listening to stories and experiences of students. I need to be aware of the "purposes and assumptions" that I bring to this relationship and study (Maxwell, 2013, p. 93).

Recommended Practice 3.15: Group Inquiry Processes

As we suggested at the beginning of the chapter, it is useful to form a group, often called an inquiry group, in which the members are familiar with each other's research. Groups should generally be between three and five people so there is a range of perspectives but also enough time for everyone to review and engage with each other's work. Each member of the group will compose the two-page research design memo detailed in Recommended Practice 3.14 and share it with the group for comments, questions, and feedback. We suggest at least 1 hour (even better would be 90 minutes) per person so that the group can thoroughly and thoughtfully engage in each part of the memo. Ideally, these conservations will be recorded, and someone will take notes for the member whose work is being discussed. Depending on time constraints, one person can share his or her memo at each group meeting.

Validity and Trustworthiness in/Through Research Design

Validity, also referred to as trustworthiness, is a key component of qualitative research design. Achieving rigor in your research leads to the kind of methodological validity necessary for studies to be considered solid. There are many aspects of validity that we focus on in Chapter Six. Here, we provide an overview to the content of that chapter because it is important that the notion of validity is discussed in relationship to critically approaching research design.

In qualitative research, the *choices and sequencing of methods* is vital to the validity of a study. It is essential that each research method is carefully mapped onto the research questions so that the kind of information (data) you need to respond to your research questions is available to you for analysis. (See Table 5.7, Figure 5.1, and Table 10.3 for templates that can help you to ensure that your methods align with your research questions.) The careful selection of specific data collection methods such as interviews, focus groups, observation/fieldnotes, questionnaires, and document review are vitally important, as is the sequencing of these methods in relation to each other and mapped across the timeline of a study. The use of multiple data collection methods is referred to as *triangulation,* which, in research design, most basically means the strategic juxtaposition of multiple data sources to achieve greater rigor and validity in a study. We discuss triangulation in depth in Chapter Six.

One important aspect to consider in research design is what methods and conditions will help you to get the most valid or trustworthy data possible. Researchers should continually ask, "How can I get the most contextualized and complicated picture possible of this group, context, and/or phenomenon?" There is no single answer to this question, as it depends on the research question(s), context, setting, and, of course, on the researcher. Multiple data sources may help to achieve rigor, but having multiple data sources does not ensure rigor. You must think about how those data sources are situated in relationship to each other, which is directly related to what we call the *strategic sequencing of methods.* This means, in part, that the order of methods and the order in which participants are interviewed or surveyed must be justified, and you should think about the rationale behind all of these kinds of methods choices in addition to the discrete data collection methods. Thus, part of qualitative research design and the notion of seeking complexity and validity is about the methods you select, the repetition or frequency with which you choose to employ them, and how they are sequenced in relation to each other. In this regard, intentional research design is an important component of validity. Rigor is then directly related to strategic research design in that it requires a careful, systematic attention to the selection of your approach at various stages and levels of the design and implementation process (Gaskell & Bauer, 2000; Golafshani, 2003; Kvale, 1996; Ravitch & Riggan, 2012).

The development of an iterative and yet structured research design, as we have stated previously, is central to rigor in that qualitative designs must be emergent and responsive to capture the complexity of the phenomenon studied. This means that there is what Hammersley and Atkinson (2007) refer to as "fidelity" to participants' perspectives rather than to specific methods or hypotheses, as is often the case in more traditional, positivist research. In addition, clarity in defining core constructs and terms is vital to the design and implementation of a rigorous study, both in the sense of theoretical rigor and in the sense of methodological clarity. Precision and clarity in defining core constructs are necessary for rigor to be possible. Also essential to validity is the articulation of a clear rationale for each design choice with a focus on data collection and analytic approaches (e.g., narrative, phenomenology, grounded theory). In solid qualitative research design, every design choice needs a cited rationale grounded in the empirical literature. Furthermore, engaging in a critical approach to instrument development (e.g., formative analysis and the refinement of a reflexive and embodied approach to data collection and analysis) is crucial to achieving validity.

As we discuss in detail in Chapter Six, research design must focus on achieving rigor. A key to achieving validity is unveiling and reflecting on researcher biases and challenging your interpretations (e.g., reflexivity, as defined in Chapter One). There is a range of strategies that you can employ to achieve validity, or at least to mitigate threats to validity. These include (but are not limited to)

- Triangulation (i.e., multiple sources, investigators, methods)

- Participant validation strategies (members checks)

- Long-term immersion in the research setting

- Soliciting feedback from insiders and outsiders (e.g., peer debriefers/inquiry group/ critical friends)

- Participatory or collaborative dialogic engagement (involving participants in some or all aspects of the research)

- Audit trail/outside auditor(s)

- Thick description (i.e., does the reader draw the same conclusions)

- Situating your study in relationship to theory, larger contexts, and other research

- Attention to and inclusion of disconfirming evidence (also referred to as negative cases or discrepant data)

We outline these strategies here as an overview and define and offer exercises that can help you to develop and engage with these strategies in Chapter Six.

We have discussed multiple aspects of research design, including developing study goals and your study's rationale; formulating and revising research questions; the role of conceptual frameworks in research design; the development of a theoretical framework; the importance of research design, methods choices, and writing; and the role of validity and trustworthiness in/through research design. Throughout our discussion of these aspects of qualitative research design, we attempted to demonstrate how the entire process is iterative, connected, and recursive. The methodological and design choices of a study cannot be separated from researcher identity and positionality or theoretical and conceptual understandings. As we close this chapter, we bring our discussion of critical research design full circle as we describe specifically what makes research critical and rigorous and therefore valid and trustworthy. We continue this discussion throughout the book and tie these ideas to ethical research in Chapter Eleven.

CONCEPTUALIZING CRITICAL QUALITATIVE RESEARCH DESIGN

Our understanding of bringing together the conceptual, theoretical, and methodological aspects of qualitative research is exemplified in what we consider *critical qualitative research design*. As discussed in Chapter One, criticality in qualitative research also includes the adoption of a critical methodological approach to your research, which seeks out and creates the conditions that allow us, as researchers, to see, engage with, and make meaning of the complexity of people's lives, society, and the social, economic, political, and historical forces that shape them. Rigorous qualitative research aligns ontology, epistemology, and methodology in complex ways that support rigor through a careful acknowledgment and engagement with these forces as they shape our thinking, the contexts of the research, and the research process and findings (Creswell, 2013).

In Chapter One, we argue that an approach to qualitative research that seeks and facilitates criticality weds the conceptual and theoretical with the methodological and looks holistically at how theory, process, context, and researcher intersect within qualitative work. In this section, we describe how the specific methodological processes that we detail throughout this chapter become what we term *critical qualitative research design*. We begin by describing what we mean by the term *critical* since it shapes how we think of criticality in important ways. In our search for a concise conceptualization of this concept, we found the Centre for Critical Qualitative Health Research's (n.d.) definition of critical research:

> The term *"critical"* refers to the capacity to inquire "against the grain": to question the conceptual and theoretical bases of knowledge and method, to ask questions that go beyond prevailing assumptions and understandings, and to acknowledge the role of power and social position. . . . The notion includes self-critique, a critical posture vis-a-vis qualitative inquiry itself.

This conceptualization of what constitutes critical helps us to begin to map what we view as a useful framing of critical qualitative research design in that it focuses on pushing into hegemonic norms, including knowledge norms, as well as examining how these concepts and habits of mind are developed in everyday social life and organizations. Willis et al. (2008) discuss the notion of a "critically conscious researcher" who challenges the confines and barriers to social change and transformation, inequity, structural inequality, and democracy in ways that resist reproducing ideas, values, and assumptions of groups that are privileged and dominant. They argue that this could be considered a "methodology of resistance" (Willis et al., 2008, p. 13). We agree with this framing and further assert that it is not simply the stance of a researcher that creates the conditions for a methodology of resistance but that it also requires a research design that pushes against hegemony and convention at every stage and within every layer of the research.

Critical qualitative research design encompasses an intentionality and a criticality about doing justice to participants' experiences (and society more broadly) through a methodology that is person centered, humanizing, and attentive to "critical" or social equity issues and that seeks to push against the kinds of research design and data analysis that assume the researcher is the knower or lone interpreter of reality (hers and others). While all researchers may not ascribe to a critical theory framework, we argue that criticality is primarily about doing justice to people's lived experiences and having a fidelity to exploring topics in deeply contextualized ways from an overall methodological approach that seeks divergence with internalized hegemony and oppressive norms in a variety of ways both ideological and procedural. An approach that engages critically in qualitative research is necessary to conduct ethical and rigorous—and therefore valid—research that achieves a fidelity to participants' experiences by situating them within broader macro-sociopolitics and considering how they become experienced and interpreted in local context(s).

We take an ideological approach to critical qualitative research design that is informed by what Cochran-Smith and Lytle (2009) refer to as an "inquiry stance." Methodologically, this involves recognizing and reckoning with power and power asymmetries. Research is not neutral; qualitative research must resist positivist assumptions and place a primacy on the interpretive and contextual in and of research. Thus, when we discuss the need to examine social location/identity and positionality through researcher reflexivity, we are talking about what we think of as transformational *un*learning as much as we are about learning. Unlearning the hegemony that resides within all of us can help us strive to conduct nonimpositional, respectful qualitative research that pushes against inequity and seeks authentic, antihegemonic engagement. Methodologically, this involves reframing and reclaiming qualitative rigor, which is reflected in all of the research design choices and processes described throughout this chapter and in the chapters to come. This process requires the building of a foundation of systematic reflexivity around research design as well as a careful and critically reframed attention to issues of validity.

Critical qualitative research design is about pushing against some of the normative research procedures that do not complicate and problematize the multiple contexts in which the research is happening. Critical qualitative research design is not solely about how you conceptualize your research or the theories that guide it; it involves how research maps onto and engages with issues and processes of equitable representation and the need to resist deficit orientations and asymmetrical interpersonal engagements. Critical qualitative research design involves maintaining a fidelity to research participants' experiences, which includes paying close attention to intragroup variability (i.e., the infinite diversity that exists within any social grouping) and resisting essentializing people and groups (Erickson, 2004). To maintain a fidelity to research participants' experiences in this way, primacy must be given to the authentic process of deeply understanding each person's perspective. This is achieved through designing for and engaging in methodological rigor. In qualitative research, you have to be reflexive about positional issues and transparent about the methodological ways you attempt to deal with them. Critical qualitative research design involves not only a mind-set but also a clear and substantiated articulation of the deliberate methodological choices you make. Without deliberate reflexivity and a methodology that places a primacy on the fidelity to participants' experiences as well as to conducting ethical research, qualitative researchers can easily reinscribe power asymmetries and deficit orientations despite the best of intentions. At the heart of critical qualitative research design is an acknowledgment of power (Cannella & Lincoln, 2012; Steinberg, 2012) and self-reflection or critique (Centre for Critical Qualitative Health Research, n.d.; Nakkula & Ravitch, 1998).

Critical qualitative research design necessitates criticality, reflexivity, collaboration, and rigor. It involves setting up the conditions necessary to critically think through all aspects of the research process, including how you will sample participants and the appropriate methods to collect and analyze data. To achieve a complexity in understanding people's lived experiences begins with the way you design your study, and your research questions are central to this. As a researcher, you must consider the complexity, inclusivity, and scope of your research questions. As we describe in Chapter One, reflexivity in critical qualitative research design involves the acknowledgment that the researcher is the primary instrument in the research (Lofland, Snow, Anderson, & Lofland, 2006; Porter, 2010); your biases and positionality must be taken into consideration from the beginning of your topic selection to its iteration into research questions and a research design. This includes that you acknowledge through reflexive engagement that you cannot avoid the trap of misinterpretation and misrepresentation. This idea *should* make qualitative researchers uncomfortable (in the sense that true introspection requires some level of discomfort), which we argue is important to the research process. The question becomes, What can be done methodologically to make the most accurate interpretations possible? We have structured dialogic engagement exercises as well as reflective memoing practices throughout this chapter and in the chapters to come, as we believe that these processes, particularly as they are discussed and explored in systematic ways with collaborators, including thought partners, mentors, and peers, will help researchers to think and engage in your research more critically.

We ended Chapter One with the section titled, "A Note on the Possibilities of Qualitative Research." As we conclude this chapter, we would like to restate our belief in the power and possibilities of an approach that fosters criticality in qualitative research. We encourage our readers to engage in a process of reflexivity and rigor that we believe can begin with adopting a sense of criticality in the processes of research design. Thus, critical qualitative research design necessitates that researchers not only think about but also understand the power of our research for both potential good and harm. In the next chapter, we build on our notion of critical qualitative research design and detail specific considerations and processes for putting design into practice in what is called data collection.

QUESTIONS FOR REFLECTION

- What does qualitative research design entail?

- How do research questions guide qualitative research design?

- How do you develop study goals and a study rationale?

- What does it mean to formulate (and iterate) the guiding research questions of your study?

- How is your conceptual framework related to your research design?

- What is a theoretical framework, and how can you develop one?

- How does a theoretical framework relate to your broader conceptual framework?

- What are the differences between a literature review and a theoretical framework?

- What role do pilot studies play in qualitative research design?

- How do the processes of vetting, rehearsing, and piloting instruments influence the data collection process?

- How are various forms of writing related to research design and methods choices?

- How can you use structured reflexivity practices, including memos and dialogic engagement exercises, to inform, guide, and complexify your research design?

- How can you plan for validity and trustworthiness in/through your research design?

- What does critical qualitative research design entail?

- How can you critically approach research design?

RESOURCES FOR FURTHER READING

Qualitative Research Design

Barbour, R. (2013). *Introducing qualitative research: A student's guide.* Thousand Oaks, CA: Sage.

Creswell, J. W. (2013). *Research design: Qualitative, quantitative, and mixed method approaches* (4th ed.). Thousand Oaks, CA: Sage.

Marshall, C., & Rossman, G. B. (2015). *Designing qualitative research* (6th ed.). Thousand Oaks, CA: Sage.

Maxwell, J. A. (2013). *Qualitative research design: An interactive approach* (3rd ed.). Thousand Oaks, CA: Sage.

Merriam, S. B. (2009). *Qualitative research: A guide to design and implementation.* San Francisco, CA: Jossey-Bass.

Patton, M. Q. (2015). *Qualitative research and evaluation methods: Integrating theory and practice* (4th ed.). Thousand Oaks, CA: Sage.

Literature Reviews

Fink, A. (2013). *Conducting research literature reviews: From the Internet to paper* (4th ed.). Thousand Oaks, CA: Sage.

Hart, C. (2001). *Doing a literature review: Releasing the social science imagination.* London, UK: Sage.

Lather, P. (1999). To be of use: The work of reviewing. *Review of Educational Research, 69*(1), 2–7.

Pan, M. L. (2013). *Preparing literature reviews: Qualitative and quantitative approaches* (4th ed.). Glendale, CA: Pyrczak Publishers.

Criticality in Qualitative Research Design

Fine, M. (1992). *Disruptive voices: The possibilities of feminist research.* Ann Arbor: University of Michigan Press.

Gitlin, A. D. (Ed.). (1994). *Power and method: Political activism and educational research.* New York, NY: Routledge.

Smith, L. T. (1999). *Decolonizing methodologies: Research and indigenous peoples.* London, UK: Zed Books.

Steinberg, S. R., & Cannella, G. S. (Eds.). (2012). *Critical qualitative research reader.* New York, NY: Peter Lang.

Zuberi, T., & Bonilla-Silva, E. (Eds.). (2008). *White logic, white methods: Racism and methodology.* Lanham, MD: Rowman and Littlefield.

Piloting

Sampson, H. (2004). Navigating the waves: The usefulness of a pilot in qualitative research. *Qualitative Research, 4*(3), 383–402.

Kim, Y. (2011). The pilot study in qualitative inquiry: Identifying issues and learning lessons for culturally competent research. *Qualitative Social Work, 10*(2), 190–206.

van Teijlingen, E. R. (2002). The importance of pilot studies. *Nursing Standard, 16*(40), 33–36.

van Teijlingen, E. R., & Hundley, V. (2001). The importance of pilot studies. University of Surrey social research update. Issue 35. Retrieved from http://sru.soc.surrey.ac.uk/SRU35.pdf

ONLINE RESOURCES

Sharpen your skills with SAGE edge

Visit **edge.sagepub.com/ravitchandcarl** for mobile-friendly chapter quizzes, eFlashcards, multimedia resources, SAGE journal articles, and more.

ENDNOTES

1. Grounded theory is different in this regard as the processes and timeline for data collection, data analysis, and review of literature can differ when conducting grounded theory studies. See Corbin and Strauss (1990, 2015) for a discussion of the specific methods of grounded theory.

2. Intersectionality refers to the idea that our identities are shaped by multiple factors and dimensions (Crenshaw, 1991). McCall (2005) defines intersectionality as "the relationships among multiple dimensions and modalities of social relations and subject formations" (p. 1771). Furthermore, Hill-Collins (2000) offers this definition: "The notion of intersectionality describes microlevel processes—namely, how each individual and group occupies a social position within interlocking structures of oppression described by the metaphor of intersectionality" (p. 487). For further reading, see Crenshaw (1991), McCall (2005), and Hill-Collins (2000).

3. This was originally a coined phrase by Emerson et al. in their 1995 book, *Writing Ethnographic Fieldnotes*. In the most recent 2011 edition of the book, they refer to it as "connecting methods and findings" (p. 15), but we prefer their 1995 terminology as it makes clear that how you engage in the research design and implementation process has everything to do with the nature and quality of the data you collect and therefore what you are able to "find" and argue with those data.

4. See Maxwell (2013) for additional discussion of researcher identity memos.

5. For additional reading on communities of practice, see Wenger (2000), and for more information on inquiry groups, see Cochran-Smith and Lytle (2009).

6. We want to thank Kelsey Jones for helping to enlighten us about the powerful role of examining your emotions as a part of the research process broadly and as a part of your conceptual framework specifically.

7. Theory is difficult to define, and there is not a shared conception of the role of theory in qualitative research (Anfara & Mertz, 2015; Maxwell, 2013). This can be very confusing to novice researchers. The goals and overall methodological approach of the study help researchers understand if you start with theory or arrive at it inductively. Grounded theory is a good example of this in that the role and process of theory development is directly related to data analysis, which starts as soon as the first piece of data is collected (Corbin & Strauss, 1990). Furthermore, in the grounded theory approach, the generation of theory involves a "constant comparison" in which emerging concepts are compared with the empirical evidence as a theory begins to be developed (Schwandt, 2015).

8. By critical integration of literature, we are referring to ways that the formal theories and literature "speak" to one another and the argument that you are making with the literature. Many books and articles have been written about writing literature reviews. See Hart (2001) as one reference for this.

9. Sharon's thanks to Ruthy Kaiser for this analogy, which can be used in a variety of generative ways.

Design and Reflexivity in Data Collection

CHAPTER OVERVIEW AND GOALS

The focus of this chapter and the next is on the considerations, processes, and techniques of qualitative data collection. We approach data collection from an ecological and holistic perspective that seeks rigor, validity, and criticality. Research participants are and should be seen as "experts of their own experiences" with much to teach us about their lives and experiences, and it is essential that you view their stories as contextualized and embedded in larger phenomena, experiences, and realities (Jacoby & Gonzales, 1991; van Manen, 1990). A primary goal of this chapter is to help you understand that your study will be driven by your research questions, goals, and participants in unique and customized ways that may need to be adjusted over time in light of fieldwork constraints and possibilities as well as emergent learnings from the data and entire research process. It can be daunting for a novice researcher to think about departing from a set research design, and that is why we stress the crucial roles of dialogic engagement and structured reflexivity throughout this book. In addition to describing qualitative data collection, this chapter provides concrete explanations and specific strategies to help systematize reflection and dialogue into what could otherwise be a daunting process. In the chapter that follows, we detail qualitative data collection methods and techniques.

This chapter begins by defining qualitative data collection as a series of connected processes and highlights the integral role that research design plays in these processes. We then discuss the role of reflexivity in data collection and describe how these reflexive processes also function as data; we term these data *researcher-generated data*. We close by discussing the design-related aspects of qualitative sampling approaches, including considerations for selecting research sites and participants.

BY THE END OF THIS CHAPTER YOU WILL BETTER UNDERSTAND

- How data collection is an iterative and recursive process

- The relationship between data collection and the other processes of qualitative research

- What it means that data are co-constructed rather than just collected

- The role of reflexivity in data collection

- The roles, processes, and steps for collecting researcher-generated data, including memos, research journals, research logs, contact summary forms, and researcher interviews

- Qualitative sampling procedures, including selecting the site and participant group for your study

DEFINING QUALITATIVE DATA COLLECTION AS ITERATIVE

Qualitative researchers should take an iterative (rather than chronological or linear) approach to data collection. This means that data collection is cyclical, emergent, and recursive—methods build upon, and are situated in relation to, each other in terms of sequencing, the nature of the data sought, and the ways that the methods will potentially support the development of each other and the data set as a whole. Figure 4.1 builds on Figure 1.1 and depicts how data collection methods are connected to all of the dynamic aspects of qualitative research.

Terms and Concepts Often Used in Qualitative Research

Data set: A data set refers to the data that have been compiled and organized to answer your research questions. This may include, for example, data from interviews, archival documents, and/or focus groups. The data in a data set have been organized and pulled from your entire corpus of data, which can include all of your data sources such as archival data/documents and artifacts; interview data; focus group data; observation and fieldnote data; survey and questionnaire data; emails and other online data; participant-generated data such as journals, reflective writing, professional documents, photos, videos, and other digital media; and any other forms of existing data.

The multidirectional arrows signify that qualitative data collection is not a linear process, but rather, it is a recursive and iterative one that is inductive—with processes that build upon and influence each other in real time as research unfolds both theoretically and

Figure 4.1 The Processes of Qualitative Research

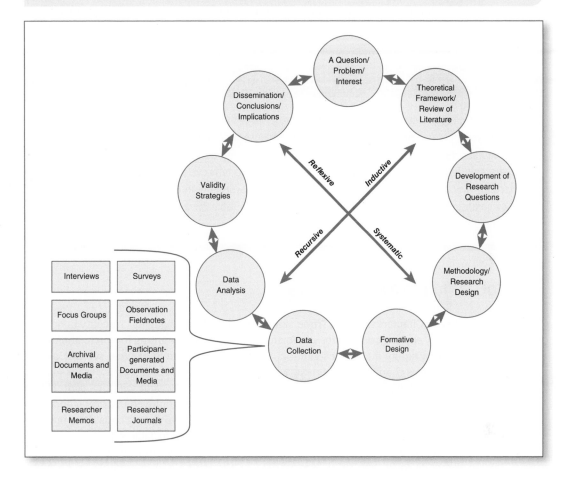

empirically—and therefore that the research is emergent in response to the learning that happens throughout the research process and especially during data collection (sometimes referred to as *fieldwork,* which is defined in Chapter Three). Viewing and approaching data collection in this more dynamic way helps you to seek complexity through intentional and strategic data collection. In this chapter and the next, we discuss data collection in a way that reflects this intentional and recursive approach.

In addition to defining data collection as a series of related, iterative processes, we want to underscore that the idea of data collection is slightly misleading. In a qualitative study, you are not merely "collecting" data. You are interacting with other individuals, and you, as the researcher and primary instrument of the study, directly impact and affect the data you "collect" (Lofland, Snow, Anderson, & Lofland, 2006; Porter, 2010). In this regard, data are generated and co-constructed rather than simply collected (Roulston, 2014). For the sake of clarity and given the terminology used widely in the field, we refer to these processes as data collection. However, recognizing that you are not simply collecting data in a unilateral way is important to conducting ethical and complex qualitative research as well as an important part of more critically approaching qualitative research.

Terms and Concepts Often Used in Qualitative Research

Design complexity: Design complexity refers to the ways that you strategically plan, design, and structure your research processes so that you can answer your research question(s) in the most complex, rigorous, and nuanced ways possible. Design complexity necessitates data triangulation and the strategic sequencing of methods as well as adopting a reflexive approach to design and data collection, which we describe throughout the book and especially in this chapter. To think about achieving design complexity, see the questions described in Table 4.1 and the questions in Table 4.2, which help you to think about reflexivity in data collection.

DATA COLLECTION AND RESEARCH DESIGN

As described in Chapters One and Three, when we think about engaging criticality in qualitative research by seeking complexity as a form of validity and accounting for micro and macro concerns about equity and representation, we think immediately of what we call design complexity. In this chapter and the next, we focus on the aspects of research design specific to the data collection process of a study or what we refer to as the *data collection plan*. This is the plan for the actual methods of data collection that will be employed in a study and their scope, timing, and sequencing. To achieve design complexity, you should consider a number of probing questions about data collection, which are described in Table 4.1.

Design complexity guides the overall data collection process and sets up the data collection methods to be as useful, appropriate to the study, and generative as possible. Revisiting the questions in Table 4.1 as you develop your data collection plan as well as begin collecting data will help you to think about issues related to collecting complex and rich data. In the next section, we discuss specific ways that qualitative researchers can strive for complexity through reflexivity and researcher-generated data sources.

REFLEXIVITY AND RESEARCHER-GENERATED DATA SOURCES

The research design does not matter unless you, as the researcher, approach the data collection process with the understanding that people are experts of their own experiences. The most well-planned and elegant research design will not be able to capture the complexity needed for qualitative research to be valid, ethical, and rigorous unless you accept the responsibility of the power you have as the researcher and mitigate that by cultivating

Table 4.1   Questions to Help Achieve Design Complexity

- Have I designed a study with enough and the appropriate data sources given what my research questions seek to examine? Given the goals of the study?
- Are the data collection methods sequenced in a way that helps me build an appropriate data set to respond to these research questions?
- Are the data collection methods structured in the appropriate ways in terms of content, scope, and sequencing?
- Have I selected the appropriate set of people who can provide the necessary information I need to answer my research questions?
- Have I engaged in multiple forms of triangulation in terms of my methods and participant selection?
- Have I selected the appropriate overall approach to getting the kinds of data that I need?
- Do I have enough and the appropriate data sources to actually provide the quality and depth of data that I need to answer my research questions with validity and confidence? What could enhance the validity of the data set?
- Given how other studies in overlapping and/or related areas are structured, is there anything I might add to my study design?
- How is theory guiding my research design and choices related to data collection?
- What assumptions underlie the study design and choices related to data collection?

and working from an inquiry stance that helps you to remain as authentic as possible to participants' experiences. Ensuring that the study methods are carried out in ways that support valid and generative data collection is in large part about your consciousness and approach as a researcher, your skills as a researcher in terms of how you engage with people, and honing those skills through a systematic process of examining yourself as the primary instrument of your research. The reflexive data generation process includes asking a series of questions, which we call reflexive data generation questions and detail in Table 4.2, about yourself as a researcher and your impact on the data and study overall. These are the kinds of questions that every researcher needs to ask throughout each stage of the research process. These questions are about you, the researcher as instrument, and paying close and systematic attention to them is important to maintaining fidelity to exploring and trying to understand the complexity of people's experiences.

In addition to these questions, a data collection plan that builds on thoughtful reflexivity and seeks design complexity is also supported by a number of the researcher-generated data collection methods that we describe. Reflexivity must be more robust and rigorous than occasional self-reflection; it can also be tracked and examined as a form of data in its own right. We refer to these kinds of reflexivity exercises as *researcher-generated data,* which are data that the researcher generates as part of the reflective process of engaging in fieldwork. In the subsequent sections, we detail data that are generated by the researcher, including memos, research journals, research logs, contact summary forms, and researcher interviews.

Table 4.2 Reflexive Data Generation Questions

- How do I present myself? The research topic and goals? What informs these choices?
- What is my communication style?
- What influences the choices I make around communication with participants within and beyond the interviews and other forms of data collection (e.g., focus groups, email exchanges)?
- What influences the kinds of communication I value from and with research participants?
- What are the kinds of knowledge or information I tend to value and gravitate toward more than other kinds (e.g., verbal vs. nonverbal, written vs. spoken, topics that capture my attention)? And what might I be missing as a result?
- Do I listen carefully? How might I improve my listening skills?
- Do I have strategies for paying careful attention and watching/tracking how my own subjectivities and biases play out in my research?
- How are these strategies working, and how might I improve upon them over time?
- Do I impose—either explicitly or implicitly—my opinions or value judgments during data collection and broader interactions with participants and, if so, in what ways?
- What assumptions lie underneath my data collection instruments (e.g., the content and wording of questions, their sequence)? The ways that I implement them?
- Am I seeing connections between my research questions and my instrument questions as responses come in?
- Am I probing for context and specifics adequately? How can I improve upon this?
- Do I cut people off or talk over them? Are there patterns related to when I seem to do this?
- Do I allow for generative pauses and silences or am I anxious around silence?
- Am I uncomfortable with people sharing private things or things that I may find problematic, uncomfortable, or even offensive? If so, do I show that during interactions?
- Am I overly confident in my appeal or my interviewing or focus group facilitation skills? If so, in what ways? How can I improve upon them?
- How does my social location and positionality influence data collection and broader interactions with participants? What are some specific examples of this?

Fieldwork and Data Collection Memos

Memos may include, among other things, your observations and reflections about various aspects of your study, including interactions with participants, data collection instruments, your skills as a researcher, and ways that you think you are influencing the data. Fieldwork and data collection memos have a variety of purposes, including making note of and considering important occurrences and key events (sometimes called critical incident memos—see Appendix G for an example of a critical incident memo), discussing decisions you need to make, and/or documenting how your research is shifting, changing, and developing over time. *Memos composed throughout the study can also be used as data.* In addition to serving as data points, data collection memos serve as connective tissue for data collection and analysis processes as well as between researchers' fieldwork reflections, ongoing data interpretations, and emergent analyses that inform future fieldwork (Nakkula & Ravitch, 1998; Ravitch & Riggan, 2012).

Fieldwork and data collection memos should be a regular and systematic aspect of the research process. A researcher or research team should engage in memo writing throughout any given study to chronicle fieldwork and data collection issues as well as to focus on emerging themes in the data. Data collection memos can be used to refine methods and to capture meaning making in process as researchers make sense of fieldwork and all incoming data (Emerson, Fretz, & Shaw, 2011; Maxwell, 2013). These memos should also include careful documentation of the following:

- Broad fieldwork and specific data collection reflections

- Researchers' experiences and interpretations as systematic "check-ins" about where each and all researchers stand in relation to the data and process

- General impressions about the space, environment, and the participants (looking across fieldnotes, contact summary forms, and/or notes on instruments)

- Researcher reflections on researcher social location and positionality and their impact on data collection

Since we view memos as a vital source of reflexivity and discovery, discussion of them is found in almost every chapter of this book. We describe two specific data collection memos in this chapter and redirect your attention to the memos described in Chapters Two and Three. We specifically recommend revisiting the Researcher Identity/Positionality Memo (Recommended Practice 3.1). This memo can be revisited and rewritten many times throughout any given research project and can be shared with critical friends and/or research teams as a way to push for critical dialogue around identity and positionality. Recommended Practice 4.1 is a fieldwork and data collection memo, and Example 4.1 is an example of a fieldwork and data collection memo written by Susan Feibelman during her dissertation process. In addition, see Appendixes G and H for additional memos that demonstrate the different structures researcher-generated data collection memos can take as well as the different roles they can serve. Just as a reminder, while memos are used for researcher sense-making, they can be shared with advisers and thought partners and can be excerpted in final reports.

Recommended Practice 4.1: Fieldwork and Data Collection Memos

The intention of fieldwork and data collection memos is for you to reflect, at various stages throughout your research, on your research project and the process of fieldwork broadly and data collection specifically over time. You might, for example, use a fieldwork memo as a place to record, reflect upon, and work through questions you have about your research process and/or changes you have had to make in your research design as a result of contextual realities or resource constraints. You might use data collection memos to note specific questions, concerns, and learnings about the data collection methods and the emerging data in real time to examine the relationship between the

(Continued)

(Continued)

methods and the data, consider methods changes, and examine incoming data in ways that can guide formative changes in the data collection methods.

Consider writing these memos not only as a way to commit to reflection through writing for yourself but also as a way to engage with others who can read the memos, ask questions, and share concerns and advice. Students working on research reports, master's theses, or dissertations also use these as a way to ask their advisers to help them think through and strategize emergent issues during fieldwork.

Potential considerations, topics, and questions:

- General or specific reflections about fieldwork experiences, including interactions with participants, observations about the site and participants, the ways that expectations are set and maintained or revised, and so on.

- General or specific reflections and thoughts about data collection methods and processes.

- How/if the data being generated answer the guiding research questions. Include excerpts of data and discuss these.

- How do your data collection methods and processes support design complexity?

- What changes may need to be made to data collection instruments, methods, and/or techniques to better answer the guiding research questions. Justify any changes with evidence from the data.

- Exploring and reflecting upon the data collection process. How, if at all, does your timeline need to be adjusted? How, if at all, has your research design changed?

- Your impressions about the space, environment, and/or the participants. How have these impressions changed over time? Pull in excerpts of data from across your fieldnotes, contact summary forms, and/or notes on protocols to help explore and explain your impressions.

- Describe your identity and/or positionality, its impact on the data, and how this may have changed over time.

Annotation: Note that Susan includes her research questions at the top of this memo. Including the research questions on a variety of your documents, including memos, is very important as it helps to keep the focus on what your study sets out to address as well as provides this important information to other readers of your memo.

Example 4.1: Fieldwork/Data Collection Memo

Susan Feibelman

April 13, 2011

Research Questions

How are women mentored to take up leadership roles in independent schools?

What networks of support exist to advance the development of women who are interested in pursuing leadership roles in independent schools?

Do female educators who are interested in taking up leadership roles in independent schools seek out mentoring relationships and/or networks of support?

How do the social identifiers of race and age inform the mentoring experience of aspiring women leaders in independent schools?

Vignette

Over a four-and-a-half week period in the spring of 2011 for my pilot study that is examining the ways in which women are encouraged to take on leadership roles in independent school settings, I completed structured interviews with seven women who are senior-level administrators in schools affiliated with the New York State Association of Independent Schools. Four of the study's participants are women of color, and three are White women. Each of the women has been a teacher and school leader in one or more independent schools prior to stepping into her current position. One woman is an assistant head of school and one is an assistant head of school and upper school head, three women are upper school division heads, one is a middle school division head, and one is a lower school division head. For three of the participants, this is their first year in their current position, while the others have been in their present role for 3 or more years.

Prior to beginning these interviews, I attended the National Association of Independent School's annual conference in Washington, D.C., where I participated in two conference sessions that addressed the role of women leaders in independent schools—"Women in Leadership: Risks and Rewards" and "Women Leading in the 21st Century." Both sessions attracted a nearly women-only audience, although the conference itself appeared to have at least an equal number of men and women participants. While neither session offered more than a superficial examination of the gendered nature of independent school leadership, as I turned around and watched women from the audience ask questions or make comments, I was struck by the emotions that swept across their faces. This was an exceptionally stunning observation. One, because I often have the sense that if I start to tell my own leadership stories, my face will reveal feelings of vulnerability that I work doggedly to hide. So I will stop myself from talking openly about my experiences, through an internal dialogue that says, "Oh no! Don't go there, you can't be that person in this conversation." Yet it was this same vulnerable affect that would appear again and again in my interview with accomplished school leaders. In the midst of a one-on-one conversation,

(Continued)

Annotation: The length of this sort of vignette will vary, but it serves to provide both faithful detail and insight into fieldwork in action.

(Continued)

almost from out of nowhere, an emotional vulnerability would wash across the speaker's face as she told a particular story. It was with this awareness I approached my interview with Ginger Truslow.

Ginger, a White woman in her early 40s, is the assistant head of school/ upper school division head of a kindergarten through Grade 12 independent girls school of approximately 500 students on the Upper Eastside. Before stepping into the school's foyer, I crossed the street, walking past an art museum and into a tony neighborhood "specialty foods" store to buy a cup of coffee. This side trip offered me a chance to orient to the surroundings of this affluent New York City enclave before stepping into the school.

Upon entering the school's formal lobby, I signed in with the "hall master" who was seated behind a desk and dressed in a coat and tie. He promptly phoned Ginger to inform her of my arrival. The foyer was filled with younger girls who had finished their school day and were preparing to leave. A handful of stylishly dressed women waited in the lobby for a child and I took a seat across from the head of school's office and waited for several minutes before Ginger arrived, a bit breathless from a meeting, and, as we ascended a monumental staircase, apologized that we had to take several flights of stairs to her office.

Ginger's office is tucked away on the third floor, is small and cozy, with two comfortable chairs and sparse decorations beyond a collection of books on shelf above her desk. Over the next 50 minutes, we would fill the space with a lively dialogue about the ways in which women who are interested in pursuing leadership roles in independent school settings seek out mentors and forge networks of support.

> I think that to me, the question is how and where we continue to create sort of these, if you will, sort of support groups or mentor groups for each other, where we can really trust and rely, we're all in competition with each other on some levels, I mean the schools not us personally, but also where can we find within each other mentors, because most of the mentors I have talked about are people who I've worked with, and I think that has been tremendous. I think what's also really informed my sense of leadership and mentorship is having moved through several different schools. I mean I'm not really interested in doing that anymore, but how we continue to find our own voice in our own schools, that perhaps is informed by our colleagues and our friends in other schools.

Annotation: Note the researcher's description of the context of the school's neighborhood setting.

Where can we do that on a sort of fairly regular basis and where will that, where can we find an effective means and also an effective group with which to do that, where we can suspend judgment and sort of develop that level of trust? You knew I went out to dinner with, oh my god Jen from Calhoun, and her names gone right out of my head . . . from Berkeley Carol . . .

Interviewer: Suzanne.

Respondent: Suzanne, I'm so sorry, and Gail Allen . . . who, Gail and I, our paths have crossed in very odd ways over the years and it was just so nice to sit down together and talk about what was happening at school and to see where we sort of had some of the same challenges, but we have to move beyond that oh . . . you know the upper school faculty are like this or this is like this and this is like this, and like bring a problem to the table and challenge each other, or maybe get to the point where you talk about something where we feel like we failed miserably or we're not being successful in communicating something.

I think that kind of collegiality is something that we as women in independent schools really have to nurture amongst ourselves.

I went to . . . NCGS has a joint program with Simmons College every other year, about leadership, and leadership in education and in some ways that was like that. You know you had to have all these surveys filled out from people you work with and then you'd get sort of all the answers and you'd look at them and you'd think wow I'm really lousy at this or I'm really good at this.

You know when you had to be with a partner you had never met before and talk about your results and I thought but there was something about it that was so easy. I think in some ways because we were so foreign to each other, the women I was partnered with were from Canada, I thought this is as safe as safe can be. You know but then we just developed this kinship . . . it's about developing a kinship with one another, and I think in some ways that's sort of what mentoring is.

Maybe that's the combination of mentoring and friendship, is the idea of kinship. Through that whole process we really had to sit with each

(Continued)

other and ask questions and challenge each other and open up to each other and it didn't feel super touchy feely in a way that was kind of a turnoff but it felt very authentic in other ways but then that group was suspended I mean right, we finished our program and we all went away and we don't talk to each other anymore. So where can we sort of create that network I think is one of the questions for me within the city, because I think that for all of us it would be tremendously helpful.

Interviewer: I agree.

Respondent: So I don't know, I've done a lot of blabbing but. . . .

Interviewer: No it's good, what's really hard I realize in trying to learn how to do this work, of asking the question and then gathering thoughts from so many different people is . . . I hear patterns right, and I want to say yes other folk are saying that they get a lot out of peer mentoring and so how does one sustain that. If there, and I believe there is a gendered quality to leadership, men are much better at creating networks of support than women are according to the literature. Men are good at asking for what they need, and women are less agile, and that there is . . . I think there's an emotional piece to it.

At NAIS this past February, there were two different sessions on women in leadership which were very different than what I have seen at NAIS before; there have been sessions about teaching girls, mentoring girls into leadership roles, but this was the first really looking at so how do you and I learn to be leaders, how do we learn to be better leaders.

So the audience was 99% female, so my take-away is it is a question that holds a meeting for women and not for men, so that's interesting, right? If you really want to think about what is supporting diversity in independent schools involve, because it does involve gender as much as it involves race and ethnicity and culture and sexual orientation and I could keep going.

Annotation: Note the researcher's engagement in written reflexivity as she seeks to explore and understand her own attitudes and behaviors, expressed in the context of this interview.

What really struck me is as I turned around and watched women who I knew and didn't know, ask questions or make comments, there was an emotionality on faces that for me was stunning. One because I always like if I start talking about my own story, I feel vulnerable, that there's a look I can feel happening on face that is like oh no don't go there, let's like . . . can't be that person in this conversation.

So I really appreciate what you said about the rawness in the room that morning and to just tell you that.

Respondent: Absolutely, and I think how we make it okay for that, because I actually, you know I think that there is definitely a place at school and there should be place at our leadership table for emotion, and I think that includes allowing people to feel elated, defeated, angry, tearful, exacerbated . . . whatever, but that's not how you conduct yourself professionally right, and I guess my question would be . . . why not? Like isn't that part of being human? So what part of our humanity are we expected to check at the door, you know we always ask that question about kids who come to our schools, say particularly a child of color who comes from a poor household, and you always ask the question what in the world does that kid check at the door when she walks through this door . . . you know.

So you live in Brownsville and you come to 75th between Park and Madison every day, like are you kidding me. You know and what part of herself does she leave behind, but I think we should actually ask that question about everybody, and I do think there needs to be boundaries between your personal life . . . like I don't want to hear about everyone's personal problems and I don't feel like work is a place to work out my personal issues . . . whatever they may be in the moment, but I do feel like you can't so cleanly separate those two.

Who you are as a person . . . and I think this is something that I am keenly aware of and the need to work on, because I think that it's something that I am often very good at hiding to great detriment to myself and those I work with, is who I am as a school person and who I am as a person, and I think you . . . I don't know why, I haven't come up with an answer but I've been asking this question now for years, and I don't have an answer without a question, but I know that I have that tendency, and I actually don't think that's good, because you don't . . . I'm not actually my best self when I'm that way. I'm not actually showing people what I intend to show them or who I intend to be, but why do we get stuck at that place?

In the above memo, you can see that Susan is making sense of her fieldwork and the data collection process itself in ways that are both reflective and formative to her study. This process of reflecting on the data collection process and the early data itself enhanced her thinking about the research and emerging data. It was also used to engage with her dissertation chair in discussing preliminary adjustments to the data collection plan. Engaging in the memos we recommend, or others like them, helps our students (and us) to chart our emerging understandings of the issues and realities at play in our fieldwork, our processes of data collection, and the role and processes of observation and fieldnotes for each new study and even new phases of existing studies.

Fieldwork Research Journal

It is a common practice for qualitative researchers to keep a research journal through-out the life span of a study. The research journal is a place to record your thoughts, questions, struggles, ideas, excitements, and experiences with the process of learning about and engaging in various aspects of research. The main purposes of the journal are to (a) help you develop good research habits related to actively engaging in researcher reflexivity; (b) provide a structured opportunity to develop and reflect on your questions and ideas about your research; (c) keep and reflect on valuable refer-ences and concepts, which you can incorporate into the research study (and future research); (d) reflect on your own thoughts and practices throughout different points in the research process; (e) formulate ideas for action or changes to your research approach; and (f) develop meaningful questions for discussions with peers, research team members, and advisers.

Journal entries are intended to provide you with an opportunity to engage in focused thinking about your perspective on concepts and experiences related to your research. Over time, these entries allow you to make deeper connections between current and past readings, ongoing self-reflection, your process of becoming a researcher, your evolving frameworks for thinking about issues (including major turning points in your own thinking), and your perspective on discussions with thought partners. The format for research journals tends to be informal; the goal is the quality of reflection. Length of entries can vary significantly. The research journal is an important data source in qualitative research and is a place, like memos, that can be messy and used for real-time sense-making. Different than memos, the journal is ongoing and flowing, and it can be viewed as a kind of phenomenological exploration into the research process itself (Nakkula & Ravitch, 1998). Of course, data entries can be shared with thought partners and advisers at any time, and this can be quite generative. They can also be excerpted, like memos, in final reports.

Research Log

In addition to a fieldwork research journal, we recommend that researchers keep track of any and all changes, additions, or modifications to the data collection plan and the overall research design, including data collection and analysis methods and processes in a *research log*. This includes changes to data collection instruments and the development of your analytical approach and methods. A research log is a way to chronicle and keep track of changes and revisions to your research methods and processes in one central location. The format of the research log can vary, and we suggest that it be kept as simple and straight-forward as possible. For example, changes, revisions, and/or additions could be noted in a simple list with bullet points; dates should be included for each entry. The goal of the log is to have a central place where you keep track of research choices and processes to account for them and their impact on the research. This is important because of the connection between methods and findings (Emerson et al., 2011) and the way that the quality of the data collected also affects analysis and findings.

Contact Summary Forms

Contact summary forms can be a vital and synthesizing part of the data collection process. Contact summary forms are used to cull and synthesize data, including participants' verbatim statements from interviews and focus groups (Miles, Huberman, & Saldana, 2014). These forms are designed to provide an efficient way to distill key aspects of your data and to manage data without losing key information. These documents can also provide opportunities for a researcher to chronicle reflections and insights that emerge during and just after each interview and focus group to make them available for further analysis for each individual researcher as well as a research team as a collective (Miles et al., 2014). Interviewers may wish to complete a contact summary form after conducting each interview and focus group or after a set of them is conducted (e.g., with a large number of interviews, you could choose to do them after every five interviews). These summarizing forms allow a researcher (or a research team) to note patterns, preliminary codes, potential issues, and so on within and across participants. They also provide researchers with field-based analysis as the data are collected, allowing a researcher or research team to more quickly get a sense of what is happening throughout the research process since these are shared and discussed over time. Miles et al. (2014) have excellent examples of contact summary forms, and we have included two of their templates in Appendixes I and J.

Researcher Interviews

At times it can be useful for you, as the researcher, to be interviewed at one or more stages of the research process. This can be a generative approach to the research process as it can elicit a meta-analysis that might otherwise be lost. There are no specific guidelines for this as these interviews can be process and/or topic specific, and they can be open-ended or semi-structured. The goal is to generate insight and reflection on the study topic and processes as well as to experience being interviewed, either using a protocol that resembles or is totally different from the ones used with your research participants. Our students often tell us that just the process of being interviewed itself is a helpful reminder of how important careful attention to the interview process is and how vulnerable you can feel while being interviewed. We recommend that these interviews are always audio-recorded and sometimes that they are transcribed for careful analysis. If these interviews are not transcribed, the recordings should be listened to so that midcourse adjustments to fieldwork strategies and methods can be cultivated. Researcher interviews can also be process oriented rather than topic specific, meaning that you could choose to be interviewed on how some aspect of the research process is going as it unfolds (or retrospectively at the end).

SAMPLING: SITE AND PARTICIPANT SELECTION

Sampling refers to the decisions you make in relation to from where and from whom you will gather the data you need to answer your research questions (Maxwell, 2013). Sampling decisions tend to be related to the individuals you will include in your study, but you may also make sampling decisions related to "groups, events, places, and points in or periods of time" (Guest, Namey, & Mitchell, 2013, p. 41). To achieve an intentional, well-reasoned sampling strategy, you will need to carefully consider your research questions and whose points of view are needed to be able to gather the data necessary to respond to your research questions with integrity and the appropriate range of relevant perspectives. These considerations lead you to make choices about site (setting) and participant selection and become the basis of your sampling strategy.

In qualitative research, these decisions are sometimes referred to as sampling but often referred to as *site and participant selection* since they include the selection of both setting and participants. Because you cannot study everything, you need to clearly define and establish criteria for the choices you will make related to the site and participant selection. In the following sections, we provide issues for you to consider as you develop your approach to both site selection and then the selection of participants within the site(s) you choose.

Site Selection

There are many considerations in the process of selecting your research setting or what some refer to as the "fieldwork site." To put it simply, research must happen somewhere, and that somewhere is important to consider directly in relation to the goals of your study and the research questions that guide it. This means that, as a researcher, you need to think carefully about why you are choosing or would choose a specific site or set of sites for your study's topic and goals. Some aspects of this selection may be scholarly in nature (e.g., wanting to engage with specific scholarly communities or theorists), and some are pragmatic (e.g., access, relationships with insiders who can provide approval). Flick (2007, p. 31) states that your site selection and sampling process consists of several steps:

1. Selecting a site (or type of site) that matches your research goals

2. Identifying situations in this site that are relevant for your issue/topic

3. Selecting concrete situations in which your issues can become visible

4. Identifying other types of situations by which your issues/topics are influenced

Considering these issues of location as you conceive your overall research design can be an important stage of study development. We find, both in our own research and in the research of our colleagues and students, that site selection can be a focal aspect of refining your overall research design in the sense that as you are working to develop and refine your research questions, considering the location can help focus the inquiry since it brings to life how the goals of the study meet the place and people in which you envision carrying out the study. Likewise, the site selection is informed by the study goals. We recommend that decisions related to site selection be an active part of dialogic engagement exercises that help you refine your topic, goals, and ultimately your research questions and study design.

To explore if your study has multiple sites or a single site, as well as some of the criteria of site selection for a given study, you must consider asking a number of site selection questions:

- Is there one specific or necessary location for your study and why?

- Do you want or need to look across multiple sites to explore a phenomenon, experience, or set of perspectives? What is the argument for this given your study focus and goals?

- What are the benefits or challenges of particular settings?

- Has the location been overly studied? (Angrosino, 2007)

- Will your presence burden the local population? (Angrosino, 2007)

- Are there ways in which studying the same research questions in different contexts would help meet the research goals?

- Are there settings within the setting, and what do those look like? How can you approach this in terms of sampling?

These are only a sample of questions that you should ask as you decide on the site(s) of your study. In general, you need to include individuals or groups (and possibly subgroups) who can provide the specific information that your research questions seek to answer. Sometimes, these individuals or groups are site specific such as at a specific organization, institution, or community. Some studies involve multiple sites, and others are not bound to a particular site or research setting (e.g., interviewing self-employed design architects across urban regions). For some, the selection of site is predetermined as when a practitioner engages in research in her own setting or a community decides to engage in participatory research within its own boundaries or when you wish to do an ethnographic study in a specific setting (see Hammersley & Atkinson [2007] for more about selecting a research setting for ethnographic research and the many considerations that go along with this). For others, the site is selected to explore specific phenomena, and therefore there might be a range of settings that could meet the needs of the study. This in and of itself requires thought and conversation. As you develop your goals for a qualitative research study, site selection often happens in tandem with research question development and refinement.

> ## Terms and Concepts Often Used in Qualitative Research
>
> **Purposeful sampling:** Purposeful sampling, which is sometimes referred to as purposive sampling, is the primary sampling method employed in qualitative research. It entails that individuals are *purposefully* chosen to participate in a research study for specific reasons that stem from the core constructs and contexts of the research questions. These reasons can include that individuals may have had a certain experience, have knowledge about a phenomenon, live or work in a particular place, or some other specified reason related to your research questions.

Participant Selection

Before selecting the specific data collection methods for a study, you need to consider the people who should be included in the study's participant group (note that we do not use the term *subjects* as we find this a quite problematic term given its meaning and the layers of power that the term reinscribes). The selection of the participant group requires a clear understanding of the goals of the research question(s) in relation to the context and populations at the center of the inquiry. In terms of the selection of participants, the primary questions to consider are the following:

- Whom do you need to include, and for what reasons and purposes do you include them, given the goals of the study, the specific research questions, and contexts that guide the study?

- Which individuals, types of individuals, and groups are particularly knowledgeable about that which I seek to learn in this study?

- Are there specific experiences, roles, perspectives, occupations, and/or sets of relationships that I seek to explore, and who can help me to explore those?

Some qualitative researchers critique using the term *sampling* in qualitative research as it stems from the goal of having a representative sample in quantitative research (Maxwell, 2013). Quantitative studies strive to employ what is called random probability sampling so that they can generalize their findings, which is not a goal of qualitative research. Instead, qualitative research, through **purposeful sampling**, provides context-rich and detailed accounts of specific populations and locations. Thus, random probability sampling is not frequently used. Rather, qualitative researchers tend to deliberately select individuals because of their unique ability to answer the study's research questions. This is often called *strategic, purposive,* or *purposeful sampling.* Purposeful sampling means that individuals are purposefully chosen to participate in the research for specifics reasons, including that they have had a certain experience, have knowledge of a specific phenomenon, reside in a specific location, or some other reason. Purposeful sampling allows you to deliberately select individuals and/or research settings that will help you to get the information needed to answer your research questions. This is the primary sampling approach used in qualitative research (Coyne, 1997; Patton, 2015). Multiple strategies can be used to achieve purposeful sampling, as detailed by Patton (2015) in Table 4.3. In addition, see Patton (2015) and Miles et al. (2014) for further discussion of the different sampling techniques used in qualitative research.

Table 4.3 Purposeful Sampling Strategies

Purposeful Sampling Strategy	Purpose: Definition/Explanation	Purposeful Examples
A. Single significant case	One in-depth case ($n = 1$) that provides rich and deep understanding of the subject and breakthrough insights and/or has distinct, stand-out importance.	Single-case examples
1. Index case	The first documented case to manifest a phenomenon; in epidemiology, the first person exhibiting a condition, syndrome, or cure; an index case often becomes the "classic" case in the literature on the phenomenon.	Understand the discovery and impact of HeLa cells, the first grown in culture that do not die, which enabled major medical breakthroughs on a number of frontiers. Document and understand the impacts of the discovery on the family of Henrietta Lacks, the source of the original cells.
2. Exemplar of a phenomenon of interest	Examining an issue in depth and over time through a single case that manifests the important major dimensions of the issue and that is accessible for intense longitudinal study.	Understand and illuminate how an elderly man and his family coped with and adapted to his severe memory loss over time and the various medical, behavioral, and psychological interventions attempted over a span of 15 years.
3. Self-study: making oneself the case	Examining one's own experience of a phenomenon of interest.	Carl Jung recorded and analyzed his own dreams as a basis for research on the nature and meaning of dreams.
4. High-impact case	A case studied and documented in depth because of the impacts it illuminates and its significance to a field, problem, or society; a high-visibility case.	Extract lessons from a case study of the successful campaign to overturn the juvenile death penalty leading up to the Supreme Court's *Roper v. Simmons* case.
5. Teaching case	An in-depth case study that offers such deep insights into a phenomenon that it serves as a source of substantial illumination about the issues documented in the case and is written to use in teaching based on the case method.	Generate a high-quality case study to teach how the Robert Wood Johnson Foundation's 20-year investment in end-of-life grant making illustrates strategic philanthropy and leads to the development of a new field.

(Continued)

Table 4.3 (Continued)

Purposeful Sampling Strategy	Purpose: Definition/Explanation	Purposeful Examples
6. Critical case (or crucial case)	The weight of evidence from a single critical case permits logical generalization and maximum application of information to other, highly similar cases, because if it's true of this one case, it's likely to be true of all other cases in that category: If it works here, it will work anywhere, or if it doesn't work here, it won't work anywhere.	Document the generalizing value of one critical case when Australian medical researcher Barry Marshall injected a bacterium, *Helicobacter pylori*, into himself to demonstrate the physiological cause of peptic ulcers and to discredit the psychosomatic theory of ulcers dominant at the time. His carefully documented study of his own body's reactions constituted a critical medical case.
B. Comparison-focused sampling	Select cases to compare and contrast to learn about the factors that explain similarities and differences.	Comparison-focused examples.
7. Outlier sampling	Cases on the tails of a distribution that would have little or no visibility in a statistical analysis, but outlier cases can reveal a great deal about intense manifestations of the phenomenon of interest.	In program evaluation, comparing successes (high outcomes) with failures (low outcomes and dropouts); studying high achievers, excellent organizations, and/or outstanding communities; studying disasters manifesting great ineffectiveness.
8. Intensity sampling	Information-rich cases that manifest the phenomenon intensely but not extremely.	Good students/poor students; above average/below average.
9. Positive-deviance comparisons	Comparing individuals or communities that have discovered solutions to problems (positive deviants) with those peer individuals or communities where the problem endures.	In the rural area of a developing country, most children were malnourished, but a few communities had healthy children: "positive deviates." Understanding the differences led to scalable solutions.
10. Matched comparisons	Studying and comparing cases that differ significantly on some dimension of interest to understand what factors explain the difference.	Comparing "great" companies with "good" companies; comparing select public, private, and charter schools; comparing resilient with nonresilient youth.
11. Criterion-based case selection	Based on an important criterion, all cases that meet the criterion are studied, implicitly (or explicitly) comparing the criterion cases with those that do not manifest the criterion. This includes *critical incident sampling*.	Quality assurance programs: Deaths in foster care, hospitals, or prisons, trigger an in-depth investigation into cause of death. Accidents or errors in hospitals trigger case studies. All airline crashes are investigated.

Purposeful Sampling Strategy	Purpose: Definition/Explanation	Purposeful Examples
12. Continuum or dosage sampling	Select cases along a continuum of interest to deepen understanding of the nature and implications of different levels or positions along the continuum.	Continuum: five stages of women's ways of knowing. Dosage comparisons for levels of program participation: high dosage—participated in all sessions; medium dosage—participated in more than half the sessions; low dosage—participated in fewer than half the sessions.
C. Group characteristics sampling	Select cases to create a specific information-rich group that can reveal and illuminate important group patterns.	Group characteristics sampling examples.
13. Maximum variation (heterogeneity) sampling	Purposefully picking a wide range of cases to get variation on dimensions of interest; two purposes: (1) to document diversity and (2) to identify important common patterns that are common across the diversity (cut through the noise of variation) on dimensions of interest.	Case studies of 20 leaders from different kinds of organizations with different background characteristics: First, document the diversity, then identify any common leadership traits or patterns.
14. Homogeneous sampling	Select cases that are very similar to study the characteristics they have in common.	Study the subpopulation of female immigrants in a job training program to evaluate the program from their unique perspective as a subgroup. Focus groups typically bring similar people together.
15. Typical cases	Select and study several cases that are average to understand, illustrate, and/ or highlight what is typical, normal, and average.	Sample and study average participants who complete a program for students who dropped out of secondary school. "Average" students are those who competed in an "average" amount of time with "average" educational outcomes.
16. Key informants, key knowledgeables, and reputational sampling	Identify people with great knowledge and/or influence (by reputation) who can shed light on the inquiry issues.	Interview long-time employees about the history of an organization or long-time residents about the history of a community.
17. Complete target population	Interview and/or observe everyone within a unique group of interest.	Following up all 19 marathon runners who lost limbs in the 2013 Boston Marathon terrorist bombing.

(Continued)

Table 4.3 (Continued)

Purposeful Sampling Strategy	Purpose: Definition/Explanation	Purposeful Examples
18. Quota sampling	A predetermined number of cases are selected to fill important categories of cases in a larger population. Quota sampling ensures that certain categories are included in a study regardless of their size and distribution in the population. Quota sampling can be flexible, a beginning point to frame the initial design and to facilitate entering the field rather than a fixed and rigid sample size; the size and composition of the sample can be adjusted based on what is learned as the inquiry deepens.	A statewide evaluation contract specifies that the study will conduct 12 community case studies with three cases from each of four regions. Having equal cases from each region is a political consideration. Janice Morse (2000) advocates quota sampling as the foundation for "organic" qualitative inquiry that grows as the inquiry unfolds and deepens.
19. Purposeful random sample	Adds credibility to a qualitative study when those who will use the findings have a strong preference for random selection, even for small samples; it can be perceived to reduce bias; purposeful random sampling is especially appropriate when the potential number of cases within a purposeful category is more than what can be studied with the available time and resources.	Resources are available to do in-depth case studies of 20 village programs in an area of hundreds of villages. Even though the sample size is too small to be considered representative or generalizable, random selection will avoid controversy about potential selection bias.
20. Time-location sample	Interview everyone present at a particular location during a particular time period.	Study everyone who works out at a fitness center every day after work (or every day before work).
D. Theory-focused and concept sampling	Select cases for study that are exemplars of the concept or construct that is the focus of inquiry to illuminate the theoretical ideas of interest.	Concept or theory-focused sampling examples.
21. Deductive theoretical sampling; operational construct sampling	Find case manifestations of a theoretical construct of interest so as to examine and elaborate the construct and its variations and implications. Theoretical constructs are based in, are derived from, and contribute to scholarly literature. This involves deepening or verifying theory.	Cases of resilience, posttraumatic stress, or other psychological constructs; case studies of organizational culture or social capital in communities as sociological constructs; studying ecological diversity in natural ecosystems.

Purposeful Sampling Strategy	Purpose: Definition/Explanation	Purposeful Examples
22. Inductive grounded and emergent theory sampling	Open-ended fieldwork reveals concepts that become the basis for subsequent sampling. "Participants are selected according to the descriptive needs of the emerging concepts and theory" (Morse, 2010, p. 235). Grounded theoretical sampling becomes more selective as the emerging theory focuses the inquiry. Additional cases are added to support constant comparison as a theory-sharpening analysis process.	In *Awareness of Dying* (Glaser & Strauss, 1965), the inquiry began by open-ended observations between nurses and patients in diverse settings; emergent concepts directed further sampling and comparisons. Grounded theory inquiries into social justice, which "often focuses on people who experience horrendous coercion and oppression" (Charmaz, 2011, p. 371).
23. Realist sampling	"Theory always precedes data collection in a scientific realist sampling strategy" (Emmel, 2013, p. 95). Explanation and interpretation in a realist sampling strategy tests and refines theory. Sampling choices seek out examples of mechanisms in action, or inaction toward being able to say something explanatory about their causal powers. Sampling . . . is both pre-specified and emergent, it is driven forward through an engagement with what is already known about that which is being investigated and ideas catalyzed through engagement with empirical accounts. (Emmel, 2013, p. 85).	A realist researcher evaluating a program aimed at helping poor families transition out of poverty would begin by identifying ideas about how the program was intended to help these families, essentially identifying hypothesized causal mechanisms within the context of the program. This conceptualization would direct the sampling. "The realist sample is always bootstrapped with theory" (Emmel, 2013, p. 83). See an example of studying perceptions of health in a Mumbai slum (Emmel, 2013, pp. 103–105).
24. Causal pathway case sampling	A *pathway case* provides "uniquely penetrating insight into causal mechanisms" (Gerring, 2007, p. 122). "Maximize leverage over the causal hypothesis" (George & Bennett, 2005, p. 172).	Case study of how leadership-initiated reform leads to increased democratization, to trace the causal pathway from leadership to reform to democratization (Gerring, 2007, p. 123).
25. Sensitizing concept exemplars sampling	Select information-rich cases that illuminate sensitizing concepts: terms, labels, and phrases that are given meaning by the people who use them *in a particular context.*	Empowerment, leadership, inclusivity, sustainable development, complexity, and innovation are sensitizing concepts.
26. Principles-focused sampling	Principles provide guidance and direction for working with people in need or trying to bring about change. Principles, unlike rules, involve judgment and have to be adapted to the context and situation. Principles-focused sampling identifies cases that illuminate the nature, implementation, outcomes, and implications of the principles.	Several agencies serving homeless youth posited a shared set of principles. Case studies of youth and staff focused on generating and analyzing evidence of the meaningfulness and effectiveness of these principles.

(Continued)

133

Table 4.3 (Continued)

Purposeful Sampling Strategy	Purpose: Definition/Explanation	Purposeful Examples
27. Complex, dynamic systems case selection: ripple effect sampling	Selecting cases where complex dynamic processes can be tracked, studied, and documented over time; such studies are inherently highly emergent. The unit of analysis is a complex phenomenon.	Tracking and documenting complex dynamic phenomena, like vicious and virtuous circles, ripple effects, adoption and adaption of innovations, viral-like spreading phenomena, crowd-sourcing dynamics, network dynamics, and so on.
E. Instrumental-use multiple-case sampling	Select multiple cases of a phenomenon for the purpose of generating generalizable findings that can be used to inform changes in practices, programs, and policies.	Instrumental-use multiple-case sampling examples.
28. Utilization-focused sampling	Select a set of cases concerning a problem or issue where sufficient depth and detail in specific cases will support rigorously identifying key factors that can credibly inform future decision making.	Detailed, systematic case studies of railroad switching operation fatalities identified five changed practices and 10 hazards to be corrected that would save lives.
29. Systematic qualitative evaluation reviews	Selecting diverse evaluation studies already completed on a program or policy and synthesizing findings across those separate and diverse evaluations to reach conclusions about *what is effective*; systematic reviews serve a meta-evaluation function (Purposeful Sampling Strategy 37 is a synthesis of qualitative *research* studies).	The Cochrane Collaboration Qualitative & Implementation Methods Group supports the synthesis of qualitative evidence and the integration of qualitative evidence with other evaluation evidence (mixed methods) in Cochrane intervention reviews on the effectiveness of health interventions.
F. Sequential and emergence-driven sampling strategies during fieldwork	Build the sample during fieldwork. One case leads to another, in sequence, as the inquiry unfolds. Follow leads and new directions that emerge during the study.	Sequential and emergence sampling examples.
30. Snowball or chain sampling	Start with one or a few relevant and information-rich interviewees and then ask them for additional relevant contacts, others who can provide different and/or confirming perspectives. Create a chain of interviewees based on people who know people who know people who would be good sources given the focus of inquiry. Researcher does the recruiting.	Seek to understand the effective relationships between agricultural extension agents and small farmers in Burkina Faso. Begin with a few farmers referred by extension agents, and then build the sample by getting referrals from the first farmers interviewed.

Purposeful Sampling Strategy	Purpose: Definition/Explanation	Purposeful Examples
31. Respondent-driven sampling, network sampling, and link-tracing sampling	A network-based strategy in which a small number of initial participants from the target population ("seeds") are studied and asked to recruit up to three new contacts in their network; initial interviewees do the recruiting, usually for compensation; used to find hard-to-access research participants because of the rarity of the phenomenon of interest and/or the sensitivity of the topic being studied.	Identify sex workers living with HIV to document and understand their coping and mutual support strategies. Trust and confidentiality are critical to obtaining a sample among rare and hard-to-reach people involved in or affected by illegal or stigmatized behaviors.
32. Emergent phenomenon or emergent subgroup sampling	Selecting a sample after the study is under way when important subgroups self-organize, or critical issues arise that affect some and not others, so those affected become the target sample.	Understand the different experiences and outcomes for two subgroups that naturally emerge in an early childhood parent education program. Women participants with and without spouses connect and cohere together. These two subgroups become an emergent comparison sample not anticipated in advance.
33. Opportunity sampling	During fieldwork, the opportunity arises to interview someone or observe an activity, neither of which could have been planned in advance.	During a site visit to a youth homeless shelter, a former staff member who helped open the shelter happens to have come for lunch and a visit with current staff friends. This is an opportunity to learn about the history of the shelter.
34. Saturation or redundancy sampling	Analyzing patterns as fieldwork proceeds and continuing to add to the sample until nothing new is being learned (especially with snowball or response-driven sampling).	After 18 interviews with female sex workers about their knowledge of and approaches to sexually transmitted diseases, the same things were being said by each new interviewee, so no new cases were added.
G. Analytically focused sampling	Cases are selected to support and deepen qualitative analysis and interpretation of patterns and themes. This is a form of emergent sampling at the analysis stage.	Analytically focused examples.

(Continued)

135

Table 4.3 (Continued)

Purposeful Sampling Strategy	Purpose: Definition/Explanation	Purposeful Examples
35. Confirming and disconfirming cases	Elaborate and deepen initial analysis by seeking additional cases (e.g., variations on or exceptions to the patterns identified in the original sample).	Case studies of nonprofit leaders showed that all those in the initial sample of 15 had spent their whole careers in the not-for-profit sector. Diversify the sample by seeking nonprofit leaders who had worked in the private sector to explore the significance of background experience on leadership approach.
36. Illumination and elaboration additions to the original sample	Deepen understanding of a finding that emerges in analysis of the original sample by adding cases that can specifically illuminate that finding.	Case studies of homeless youth ($n = 14$) turned up evidence that gay, lesbian, and transgender youth faced special problems. Three such cases were added to the sample when that pattern emerged, to strengthen the findings.
37. Qualitative research synthesis	Selecting qualitative research studies to analyze for crosscutting findings; quality criteria for which studies to include in the synthesis is a sampling issue (systematic evaluation reviews, #29 in this table, are a form of qualitative synthesis that focuses on evaluation findings).	A substantial number of separate and independent ethnographic studies of coming-of-age and initiation ceremonies have been conducted across cultures. A synthesis involves selecting studies to include in the cross-study analysis.
38. Sampling politically important cases	Attract attention to a study (or avoid attracting controversy) by adding or omitting politically important cases.	A study of adult literacy programs across the state purposefully includes a program located in the district of the chair of the legislative committee that oversees those programs.
H. Mixed, stratified, and nested sampling strategies	Meet multiple inquiry interests and needs; deepen focus, triangulation for increased relevance and credibility.	Mixed, stratified, and nested examples.
39. Combined or stratified purposeful sampling strategies	Begin with one sampling strategy, and then add a second to further focus the sample; for example, (a) begin with outliers, then do snowball sampling or (b) begin with key informants to construct a maximum-variation sample.	Evaluating the implementation of the Paris Declaration Principles for Development Aid, key informants were sampled and interviewed to then identify exemplary cases of implementation for in-depth study (outliers).

Purposeful Sampling Strategy	Purpose: Definition/Explanation	Purposeful Examples
44. Mixed probability and purposeful samples	Five examples of mixed sampling strategies: 1. *Stratified mixed methods*. Use statistical distribution to stratify for purposeful sampling (e.g., identify outliers, typical cases, or subgroups of interest). 2. *Sequential mixed methods*. Select cases from a probability sample for greater in-depth inquiry to illuminate and validate what the numbers mean. 3. *Parallel mixed methods*. Simultaneously do a survey of a probability sample for representativeness and generalizability and, at the same time, in-depth case studies purposefully chosen to provide depth of interpretation of what the survey results mean. 4. *Triangulated mixed methods*. Compare probability and purposeful samples, studied independently, with the triangulate and examine the consistency of findings with different methods and sampling strategies. 5. *Validity-focused mixed methods*. Do fieldwork (observations and interviews) to determine the validity of select statistical data (e.g., whether the procedures for gathering statistics have been followed rigorously), or find out if the control and treatment groups in an experimental design have complied with the design specifications.	A random, representative sample of people with disabilities was surveyed to determine their priority needs. A small number of respondents in different categories of need were then interviewed to understand in depth their situations and priorities and make the statistical findings more personal through stories. The qualitative sample was also used to validate the accuracy and meaningfulness of the survey responses.

Source: From Patton (2015, pp. 266–271).

Note: Sources cited in the table can be found in Patton (2015).

> ## Terms and Concepts Often Used in Qualitative Research
>
> **Unit of analysis:** In qualitative research, your unit of analysis refers to the primary focus of your research study and is most often reflected in the core constructs in your research questions. The unit of analysis is the major entity that you are analyzing in your study. It is the "who" and/ or the "what" at the center of your study. Units of analysis can be focused on people (e.g., individuals, groups), perspectives (e.g., individuals who share a common experience), structure (e.g., projects, programs), geographical units (e.g., city, neighborhood, state), artifacts (e.g., books, films, photos, newspapers), and/or social interactions (e.g., marriages, births) (Patton, 2015).

Inherent in decisions about site and participant selection is a discussion of your **unit of analysis.** Units of analysis can be focused on people, structure (e.g., projects, programs), perspectives (e.g., individuals who share a common experience), geography (e.g., cities, neighborhoods), activity, and/or time (Patton, 2015). Determining your unit of analysis is an important part of how you will select your participants (or sample). Another consideration related to participant selection is *sample size*. There are no set rules in qualitative research when it comes to having a certain number of participants: "Sample size depends on what you want to know, the purpose of the inquiry, what's at stake, what will be useful, what will have credibility, and what can be done with available time and resources" (Patton, 2002, p. 244). It is not the goal of purposeful sampling and qualitative research to generalize; thus, the sample size becomes less important than in quantitative research. The goal is to rigorously, ethically, and thoroughly answer your research questions to achieve a complex and multiperspectival understanding. Despite not generalizing to the entire population, research from qualitative studies can help to make important decisions and suggest applications to a broader population. However, qualitative findings must be considered contextually. As with all design decisions, it is the role of the researcher to acknowledge limitations and potential weaknesses of the study, including sampling decisions such as size and strategies.

Regardless of the sampling strategies you employ, it is essential to have a clear, reasoned, and explicated rationale for why you selected individuals or groups to be a part of your study. The important thing to note is that these decisions require considerable thought, exploration, and planning in order to pursue sampling with intention and to achieve validity with regard to representation. Given the notion of "the inseparability of methods and findings" (Emerson et al., 1995), the choice of the participants is clearly a central aspect of how and what you can and will learn in your research. As discussed in Chapter Three, you need to consider local, contextual, and macro-sociopolitical factors in these selections with an attention to power and politics in both the specific environment and the broader society that shapes it.

Given the vital importance of your participant pool, it is essential that as a researcher, you spend considerable time exploring and discussing with others the various benefits and challenges in choosing the participants for your study. In addition to your own exploration and engagement in dialogue about these issues, you must make your decisions transparent so that readers of your work have enough information and context to determine if your decisions make sense and how they shape your analysis and findings.

We suggest that, among other strategies, you engage in multiple conversations with peers and mentors to get a critical perspective on issues of inclusion and representation in your participant group. We also suggest careful self-reflection on how you are viewing the possibilities of participants to push into your assumptions, biases, and blind spots. Among other approaches, writing a Site and Participant Selection Memo, as described in Recommended Practice 4.2, helps many researchers gain clarity on these choices and what influences them. Sharing such a memo with critical friends and advisers is often a vital step in this learning process since it invites critique and discussion about these issues of representation that are at the heart of qualitative research. Example 4.2 is a memo about site and participant selection written early on by one of our former students, Mustafa Abdul-Jabbar, during his dissertation process. See Appendix K for an additional example of a site and participant selection memo.

What Mustafa's memo makes clear is that these choices—of site and participant selection—are ones that must be carefully considered and that require thought, reflection, and

Recommended Practice 4.2: Site and Participant Selection Memo

The goal of this memo is to clearly define the criteria by which you will use (or used) to determine a specific setting and select the participants for a study. Also, in this memo, you should explain how the methods chosen in these realms align with the research questions and goals of the study.

Potential topics to discuss include the following:

- How the proposed methods map onto your research questions

- Site selection criteria (identify and challenge them, present a rationale for each criterion)

- Participant selection criteria (identify and challenge them, present a rationale for each criterion)

- Identification, justification, and limitations of/for all methodological choices

- Issues of/concerns about representation for site, participants, and researcher(s) in relation to the study topic, goals, and setting

- Your positionality/social location and its role in informing biases and assumptions for all or some of the above choices

Annotation: Including the research questions in every memo helps to keep the focus on what your study sets out to address.

Example 4.2: Site and Participant Selection Memo

Mustafa Abdul-Jabbar

December 23, 2012

Research Questions

(1) How do leadership practices of a distributed leadership (DL) team affect relational trust between members of that team?

(2) How is relational trust related to leadership team members' perception of self-efficacy?

Archdiocese of Philadelphia

My current progress toward my dissertation proposal includes recent site and participation selection for research. I will begin this reflective memo in discussion around participant selection and then move on to discuss specific site selection(s). There have been two program implementations of distributed leadership and professional development training and support run by the Penn Center for Educational Leadership (PCEL). In the past year and a half that I have been volunteering at the Penn Center for Educational Leadership, I have spent that time vacillating between studying the older implementation of DL (i.e., Annenberg DL program with the Philadelphia school district) or the more current program implementation (i.e., Archdiocese of Philadelphia DL program). I have recently decided to conduct my research study within the context of the Archdiocesan implementation of DL.

Conception of the Distributed Leadership Program

As I stated in an earlier memo, the DL program was initially conceived to enhance effects of school leadership on student learning, through engaging classroom teachers to aid in instructional leadership processes, ensuring a wider distribution of leadership influence throughout a school (DeFlaminis, 2011, p. 1). The program intent was to carefully select (*recruitment*) teams of administrators and teacher-leaders, to develop their individual capacities for leadership (*professional development*), simultaneous with the design of action plans they collaboratively implement in their schools that was supported by ongoing mentorship (*coaching*), in order to improve institutional capacity and enhance student learning and achievement (CPRE, 2010, p. 4).

Unit of Analysis/Participant Selection

Based on the program structure articulated above, my unit of analysis will necessarily involve either the DL teams themselves (in their entirety) or select members of the DL teams. With 19 participating Archdiocese schools, each with four to eight members per team, I retain a swath of participant options. In further explication, after informal interview with John DeFlaminis— developer of the program structure and logic model—I observed that the program logic relies on utilizing teams within schools as vehicles for *distributing* leadership. Thus, I feel it is important to begin with the *team* itself as the unit of analysis because it is the program's unit of implementation. To the extent the program is effective, it should be evident in the activity and inter-activity of the team. At this juncture, the remaining challenge is to narrow my choices to specific team sites.

Program Evaluation

I am currently engaged in a program evaluation of the DL implementation with the Consortium for Policy Research in Education (CPRE). This involvement grew out of an independent study held with Professor Jon Supovitz during the fall 2012 semester and has continued through the spring 2013 semester. This project will go into the summer as well. My involvement with the evaluation has been beneficial because it provides a unique balcony view (Heifetz & Linksy, 2002) of the DL implementation assisting in meta-analysis and observation of the program itself and my role in the program. The evaluation work also allows me ready access to quantitative data (i.e., via surveys and program statistics) that I otherwise would not have had. It is within the context of the program evaluation that I have decided on the means for selecting specific research sites.

Site Selection

During the course of the CPRE program evaluation, part of the quantitative data analysis involved the creation of Likert-type scales for measuring such variables as relational trust, efficacy, perception of school issues, and so on. One of the variables measured by CPRE entailed the degree to which team members felt they had greater or lesser influence across a variety of dimensions of school leadership activity and practice. Because this particular measure was complementary to the DL framework as conceptualized by

(Continued)

Annotation: Note how this researcher describes the process of refining decisions about site and participant location and what decisions still remain.

(Continued)

Spillane (2006, 2009; see also Spillane, Halverson, & Diamond, 2004) and further corroborated through informal interview with the program developer and director of its implementation, John DeFlaminis, I decided to obtain the list of sites that scored highest on the "Influence" scale.

On a scale of zero to four, only six school sites scored 3.5 or higher. Of the six schools, only one of the schools was a school that was affected by mergings/closings and thus will be left out of the research study because it is being reconstituted/closed by the Archdiocese. Fortunately, of the remaining five schools, I have already piloted one conference research project with one of the schools and am currently piloting another study this semester with a second school within the group of five.

Next Steps

I still need to decide if two school sites/teams are sufficient or perhaps three or even all five. Depending on my methodological choices for framing this research, I may indeed choose all five teams or participants across the five teams. For example, I am currently weighing the pros and cons of either a case study framework for conducting my research or a practitioner inquiry framework. Under a case study framework, I would have to work with fewer sites but would be able to capture more of an entire team's activity and interaction. However, with a practitioner inquiry methodology, I could work with all five school sites (or at least four) and could conduct focus groups and other cross-site activities that could yield a very different research product. I confess that I'm still waiting until the end of this class, until we cover more on the benefits/challenges of case study methodology, before I make my decision. ☺

Annotation: Practical concerns of access often guide site and participant selection.

dialogue. Memos such as this one help you note and then consider issues at the heart of these choices. In addition, they create the basis for constructive conversations with peers and advisers about how to make these design choices. Finally, these memos help you to remember the choices you made throughout the research process since these often get forgotten over time. Researcher-generated data, such as memos, are an important way to record and preserve an understanding of the reasons for these structural aspects of your study.

QUESTIONS FOR REFLECTION

- How are data collection processes iterative and recursive?

- How is data collection related to the other processes of qualitative research?

- What does it mean that data are co-constructed rather than just collected?

- How is the researcher integral to data collection processes?

- What does reflexivity look like during data collection processes?

- What are the roles, processes, and steps for collecting researcher-generated data, including memos, research journals, research logs, contact summary forms, and researcher interviews?

- What key aspects should you consider when selecting the site and participant group for your study?

RESOURCES FOR FURTHER READING

Sampling, Site, and Participant Selection

Angrosino, M. (2007). *Doing ethnographic and observational research: The SAGE Qualitative Research Kit.* Thousand Oaks, CA: Sage.

Flick, U. (2007). *Designing qualitative research: The SAGE Qualitative Research Kit.* Thousand Oaks, CA: Sage.

Guest, G., Namey, E. E., & Mitchell, M. L. (2013). *Collecting qualitative data: A field manual for applied research.* Thousand Oaks, CA: Sage.

Miles, M. B., Huberman, A. M., & Saldaña, J. (2014). *Qualitative data analysis: A methods sourcebook* (3rd ed.). Thousand Oaks, CA: Sage.

Patton, M. Q. (2015). *Qualitative research and evaluation methods* (4th ed.). Thousand Oaks, CA: Sage.

Memos and Research Journals

Emerson, R. M., Fretz, R. I., & Shaw, L. L. (2011). *Writing ethnographic fieldnotes* (2nd ed.). Chicago, IL: University of Chicago Press.

Nakkula, M. J., & Ravitch, S. M. (1998). *Matters of interpretation: Reciprocal transformation in therapeutic and developmental relationships with youth.* San Francisco, CA: Jossey-Bass.

Maxwell, J. A. (2013). *Qualitative research design: An interactive approach* (3rd ed.). Thousand Oaks, CA: Sage.

Miles, M. B., Huberman, A. M., & Saldaña, J. (2014). *Qualitative data analysis: A methods sourcebook* (3rd ed.). Thousand Oaks, CA: Sage.

ONLINE RESOURCES

Sharpen your skills with SAGE edge

Visit edge.sagepub.com/ravitchandcarl for mobile-friendly chapter quizzes, eFlashcards, multimedia resources, SAGE journal articles, and more.

CHAPTER FIVE

Methods of Data Collection

CHAPTER OVERVIEW AND GOALS

Building on our discussion in the preceding chapter of the role of design and reflexivity in the qualitative data collection process, this chapter discusses the main methods by which qualitative researchers collect and generate data. These methods include interviews, observation and fieldnotes, focus groups, a review of documents and archival data (which can include online sources), questionnaires, and participatory methods of data collection. For each of these methods, we describe the structure and process of each method as well as its strengths and challenges and detail practical considerations and tips for implementing the method. We also discuss how to develop research instruments for these data collection methods. We conclude the chapter by revisiting the central role of research design and discussing the importance of ensuring that methods of data collection align with research questions so that the research generates the appropriate data to answer your guiding research questions. Throughout this chapter, we address issues of data collection related to the use of online data collection methods, engagement with social media as a form and part of data collection, and technology-mediated data collection considerations.

While qualitative data collection should be intentional, rigorous, and systematic, it should not be guided by overly rigid rules and procedures. This is a large part of what we mean by an approach to qualitative research that seeks and facilitates criticality—an approach that creates the conditions for you to holistically understand and convey the most contextualized picture of the people and phenomena in focus possible, maintaining a fidelity to the complexity of participants and their experiences, which can mean that data collection methods are dynamic and emergent (rather than fixed). Thus, while we describe specific methods of data collection in this chapter, we want to be clear that these methods must be chosen directly in relation to study specifics rather than being prescribed.

BY THE END OF THIS CHAPTER YOU WILL BETTER UNDERSTAND

- A range of methods and techniques for generating qualitative data, including interviews, observation and fieldnotes, focus groups, questionnaires, documents and archival data, and participatory methods

- The key values and characteristics of these different methods and techniques for generating qualitative data

- The considerations for and strengths and challenges of each of these qualitative data collection methods

- How to critically examine the data you generate and the ways that you, as the researcher, influence these data

- The roles and uses of technology and social media in data collection

- The centrality of research questions to research design and data collection

- Ways to make sure that your methods align with your research questions

INTERVIEWS

> The skilled questioner and attentive listener know how to enter into another's experience. (From Halcolm's Epistemological Parables, as cited in Patton, 2015, p. 421)

> An interview is a social interaction with the interviewer and interviewee sharing in constructing a story and its meanings; both are participants in the meaning-making process. (Holstein & Gubrium, 1995, p. 8)

As a mainstay of qualitative data collection, interviews are at the center of many qualitative studies since they provide deep, rich, individualized, and contextualized data that are centrally important to qualitative research. The primary goals of qualitative interviews are to gain focused insight into individuals' lived experiences; understand how participants make sense of and construct reality in relation to the phenomenon, events, engagement, or experience in focus; and explore how individuals' experiences and perspectives relate to other study participants and perhaps prior research on similar topics. In qualitative research, it is centrally important to delve into each person's experience and to relate those to other participants' experiences so that you can come to understand a fuller range of perspectives and experiences about a particular topic or phenomenon. For example, you might consider what is shared and what is unique as well as what accounts for these similarities, complexities, and differences. In that sense, you are interested in the individual himself or herself as well as understanding the range and variation of experiences and perspectives within a group of one sort or another (e.g., cultural group, affinity group, employee group, family, community). Qualitative interviews do not typically seek uniformity in questioning but, rather, pursue what Sharon has termed *customized replication*—that is, while they share key questions, interviews seek the customization of each conversation through individualizing follow-up questions and probes for specifics within each interview. This means that while similar questions are asked across study participants since they are vital to collecting data that can answer the research questions, they are not all asked the exact same questions in the same order; this requires that the interviewer create individualized follow-up questions and contextualizing probes both prior to and during the interview.[1]

If planned for and approached well, the interview becomes a forum and process by which you can explore people's perspectives to achieve fuller development of information within and across individuals and groups while keeping similar lines of questioning that help you to look within and across experiences in ways that help decipher meaning, experience, similarity, and difference. In this way, qualitative interviews seek range and variation in people's meaning-making processes, experiences, and points of view toward being able to understand and then communicate about relationships between these complicated realities and viewpoints (Brinkmann & Kvale, 2015; Fontana & Prokos, 2007; Weiss, 1994). Interviews, when conducted in this manner, help to achieve the complexity in thought and method that is central to not only what we consider criticality in qualitative research but also to conducting valid and rigorous qualitative studies.

While the use of interviews in qualitative research is typically expected, it is important to understand why you would choose to engage in interviews as a part of your study design. Weiss (1994) highlights the key reasons for choosing to conduct qualitative interviews as a major data source for your study, which include that, given the guiding research questions and goals of the study, you seek to

- Develop full, detailed, and contextualized descriptions of experiences and perspectives
- Understand and integrate multiple individual perspectives
- Describe processes and experiences in depth
- Develop holistic descriptions of perspectives, realities, experiences, and phenomena
- Learn how participants interpret events and experiences
- Bridge intersubjectivity between researcher and study participant

The reasons to select qualitative interviews are central to and reflect the naturalistic and interpretive values of qualitative research. There are various approaches to and methods of conducting qualitative interviews, which we explore in the remainder of this section.

Key Characteristics and Values of Qualitative Interviews

There are a number of *key characteristics and values of qualitative interviews* that we believe are helpful for you to keep in mind as you consider the overall research design and the design of specific data collection methods of any study. These characteristics and values include that qualitative interviews are relational, contextual/contextualized, nonevaluative, person centered, temporal, partial, subjective, and nonneutral (to be clear, we argue that these characteristics also apply to other qualitative data collection methods such as focus groups, participant observation, and so on.). We describe each of these characteristics below and provide important considerations related to them in Table 5.1.

- *Relational:* Interviews constitute a relationship, however brief. What we mean by this is that trust and reciprocity are vital to and at the center of healthy relationships. Thus, these values need to be centralized throughout the interview process—including in the recruitment of participants, engagement in setting up the interviews, the actual interview(s), and any follow-up, including additional data collection and participant validation techniques (discussed in depth in Chapter Six). Trust and reciprocity are contextual, and they might mean different things to different people in different interview scenarios and contexts. Interviews should be built on a concept of mutual engagement and reciprocal transformation rather than power asymmetry or interrogation of any sort (Gilligan, 1996; Gilligan, Spencer, Weinberg, & Bertsch, 2003; Josselson, 2013; Nakkula & Ravitch, 1998).

- *Contextual/contextualized:* Every interview is conducted within multiple, intersecting contexts, including the setting and people involved as well as the broader macro-sociopolitical contexts that shape them. A skilled, reflective interviewer understands that the interview happens in a complex ecosystem of someone's life and that the job of the interviewer is to understand the responses (as data) in individualized and contextualized ways. This also entails seeking to understand how these various contexts and forces shape the individual's experiences and perspectives in relation to the study's goals. As we often say to our students, "Context is everything." General questions lead to general responses, which are almost always less useful than probing for specifics. Specific and contextual follow-up questions and probes are often necessary to get a more complex construction of someone's experience as well as an understanding of the concepts and terminology used by each and every participant.

- *Nonevaluative:* In qualitative research, the goal is to understand what participants think, feel, and experience rather than to judge or evaluate their accounts or them as people. This means that interviewers try to understand the perspectives, experiences, thoughts, feelings, and/or ideas of those interviewed, not evaluating or judging these, rather exploring, learning about, and seeking to understand them. This is important to understand given that researcher bias, if unexamined, can shut down open communication in interviews as judgmental or evaluative responses are felt by those being interviewed even if in subtle ways (such as affirming certain kinds of responses over others, making involuntary positive or negative facial expressions, and the like). For qualitative interviews to be useful and generate quality data, you, as the interviewer, must make every effort to identify and reckon with your biases and judgments and not impose these through the use of evaluative language and nonverbal communication.

- *Person centered:* The person engaging in the interview (the participant or what some refer to as the *interviewee,* although we see this language as too passive and therefore less respectful than *participant*) must be attended to and engaged with as the focus of the interview and as an expert of his or her own experience. The fidelity needs to be to the participant's experience of the interview rather than to the data collection instrument (protocol) or even the research questions. In this sense, the interviewer centralizes the relational quality of the interview and the experiences of the person being interviewed in ways that show careful attention to the individual as well as respect for the opinions, feelings, and ideas expressed. This means, for example, that if an interview seems to cause a participant distress, you should discontinue engaging in questioning, as the experience of the participant is considerably more important than the data.

- *Temporal:* We suggest that researchers work to understand what we think of as the temporal quality of interviews (and research more broadly). By this we mean that a particular moment in time—historical, institutional, social, and/or personal time—can have great impact on what is shared or not shared in an interview. This entails understanding that the temporal nature of someone's developmental trajectory, the number of years in a job, the current political milieu and social climate, a specific moment in the life of an institution, and/or the time of day of an interview can have great influence on what is shared, how it is shared, and how it is conceptualized and articulated. It is very important to understand the temporal quality of qualitative interviews since doing so will help you gain a clearer sense of context and fluidity in experience and meaning making.

- *Partial:* Interviews are snapshots of a moment in time. Peg McAdam, an incredible teaching assistant Sharon had the privilege of learning with while at the Graduate School of Education at Harvard (who is an expert gardener and researcher), related data collection to an analogy of a garden in that a gardener does not begin planting, even on a seemingly empty field, with nothing. A gardener always starts with something in the sense that the soil is rich in history, in nutrients, in all that has brought it to its current state, and that this history—and all that it carries with it—has implications for its current state, its possibilities, as well as what happens next in terms of learning and knowing how to properly tend to it. We think of this analogy quite often in terms of interviewing and data collection: An interviewer meets a participant within his or her already happening life, with all of its existing qualities and properties that have been derived from years of growth and development. The interview takes place in a preexisting ecology in ways that the researcher must be cognizant of in terms of both how to approach and how to analyze the data that come from the interview. This entails understanding the complexity and partiality of any single interaction (or even multiple interactions). We caution our students not to be overly confident that they get the whole of other people's experiences from what is usually a quite brief interaction, however powerful and enlightening it may be.

- *Subjective:* Life is subjective—in the sense that human beings are ever-engaged in their interpretation of life, and multiple subjectivities operate within any given experience and context—and therefore so too are interviews. Interviews are not, by definition, neutral or "objective," and they seek to understand people's positions, views, experiences, and particular subjectivities in a number of important ways that are about understanding individual subjectivity. This understanding of the ways that qualitative researchers frame research—as inquiry that attempts to uncover, understand, and engage with multiple subjectivities within individuals' natural contexts—relates also to the subjectivity of the researcher as well as intersubjectivity, or how the researcher and interview participants' subjectivities layer and develop in relation to each other. Qualitative interviews seek not only this kind of individualized learning but also an understanding of the subjective quality of the interview process, interactions, and data from all angles.

- *Nonneutral:* Interviews are far from neutral, no matter what the topic. This means that there are layers of bias, assumption, and politics (macro and micro) and other kinds of influences on the interview and even the interview relationship and interaction itself. The nonneutrality of interviews is important to keep in mind; otherwise, you might think you are getting the "objective truth" of someone's reality when in fact the data are mediated by an endless number of factors and dynamics, many of which are found in the interview interaction itself. While we may not always know and in fact often do not have an understanding of these influences, an awareness that they exist creates more room for contextualizing the data and even at times asking questions in ways that can lead to unearthing and engaging with some of these spheres of influence on people's perspectives and experiences.

These qualities and values of qualitative interviews are important to consider as they can help you to think about and actively consider the many variables that shape research broadly and interviews specifically. It is especially important to think critically about these values and characteristics as you construct **research instruments** (including interview protocols), which is described in the next section. We provide specific questions in Table 5.1 to help you process these characteristics when developing and conducting interviews.

Terms and Concepts Often Used in Qualitative Research

Research instruments: In qualitative research, research instruments refer to the tools that you develop and use to collect the study data. These data collection instruments, which can also be called protocols or guides, include the questions, prompts, and/or procedures that guide data collection. For example, an interview instrument will have a sequenced list of specific questions or prompts that researchers will ask participants during interviews. Other examples of instruments include observation templates or protocols. These are often used on research teams to make sure that observations conducted by multiple researchers capture similar information. The extent to which instruments, guides, and protocols are structured depends on the overall approach that guides the methods as well as the guiding research questions.

Table 5.1   Considerations for Developing and Conducting Interviews

Characteristic	Considerations
Relational	• How am I describing the research in terms of topic, goals, and process?
	• How am I presenting myself and engaging with interview participants?
	• How are expectations being set?
	• What promise(s), if any, am I making to participants?
	• Have I discussed with participants how their data will be used?
	• What assumptions are playing out as I interview participants? How might this affect the relational ethos of interviews?
	• How am I dealing with power asymmetries, including the interview relationship and its possible implications?
	• Have I established trust and a comfortable interview environment? How would/do I know? How can I explore this question throughout the research process?
	• How is the participants' engagement in my study mutually beneficial?
Contextual	• What contexts (micro and macro) are relevant to the study and research question(s)? How do I get at those in relation to specific interview questions?
	• Have I asked questions that help to explore and understand these contexts?
	• Have I probed for specifics that are contextualized in place and time?
	• How are participants describing these contexts? What do they assume I know as they describe these context(s)? What do I assume I know as they describe them?
	• Are there aspects of the context or culture(s) of the setting that are implicit? If so, how can they be explicated?
Nonevaluative	• In what ways am I biased about participants' responses?
	• How may I be evaluating or judging various choices, perspectives, opinions, and so on both consciously and unconsciously?
	• What kind of communication, verbal and nonverbal (e.g., head nods, affirming statements, words/phrases), did/do I make during and after a participant answers a question? Are there patterns to those?
	• How might this communication have affected how the participant experienced the interview and/or future responses to questions?
	• How am I accounting for and examining my presence in the interviews?
	• How/am I practicing being nonevaluative in interviews?
Person centered	• Am I respecting the participants by clearly reviewing the informed consent[2] form and giving participants adequate time to review it and ask questions?
	• Am I being respectful of the participants by making sure that I have permission for the interview to be audio-recorded *before* I turn on the recorder? Am I clearly communicating that it is the choice of the participant as to whether the interview is recorded?

(Continued)

Table 5.1   (Continued)

Characteristic	Considerations
	• Am I clearly communicating that participants can refuse to answer any question and can stop participating in the interview at any time?
	• Am I paying close attention to the participant and respectfully listening and engaging?
	• Am I placing a primacy on collecting the data over the participants' experiences and, if so, in what ways?
Temporal	• What specific questions can I ask to understand if there is a temporal quality to some of the participants' responses (i.e., to get at change over time, to see if past, present, and future are different or aligned)?
	• Have I accurately captured the context of the setting and participants' experiences as the participants describe them? Am I clear on how this context may have changed over time?
	• What accounts for some of these changes in context and experience? Are there local influences? Macro-sociopolitical influences?
	• How might these forces (historical and present) influence the ways that the participants respond to specific questions?
	• Are there specific individual developmental forces at play here (e.g., age, identity development, years of experience in a setting or position)? If so, how can I understand that from the participant's perspective?
Partial	• How can I develop my study design to seek as much complexity and depth as possible? How do the interviews (and other methods) fit into this larger design?
	• How can I structure the study instruments to be as probing for depth and context as possible?
	• How can I engage with participants in data collection that helps them understand how the study fits into their larger experiences and perspectives?
	• How am I acknowledging, as I begin formative data analysis, that the interview is a snapshot of a person's lived experience?
	• What other kinds of data collection would help me to deepen and contextualize my emerging understandings from the interviews?
Subjective	• How is this process shaped by my subjectivities and assumptions? Which ones specifically and how?
	• What can I do to engage in thinking about these subjectivities and assumptions in focused ways?
	• How are the participants' renderings of their experiences and perspectives subjective?
	• How can I probe for as much detail as possible into their interpretive processes and assumptions?
	• How am I accounting for the ways that my subjective experience and the subjective experience of the participants come together?
	• How can I use a range of reflexivity exercises to help me engage with my subjectivities? What are my goals in that?

Characteristic	Considerations
Nonneutral	• How am I accounting for the inherent power asymmetry present in the interview (because I am the one asking questions)?
	• What can I do to address issues of bias, power, and their impact on the interview participants and our interactions?
	• What larger sociopolitical forces might be manifesting themselves in my micro-interactions with participants? How can I address this?
	• How am I representing the multiple layers of bias (mine and the participants') present in the interview?
	• Is the wording of particular questions playing out in ways that are problematic for individual participants? If so, what might account for this? How can I address it?

Constructing Qualitative Interviews

The central values of qualitative interviews described thus far are foundational to how interview instruments (also called interview protocols or guides) are structured, since the ways that researchers structure and sequence interview questions have a direct impact on what can be learned from them. To this end, Patton (2015, pp. 444–445) notes six kinds of interview questions that seek to understand these various realms of people's experiences and interpretive processes:

1. Experience and Behavior Questions—These questions focus on what a person has done, will do, or is currently doing.

2. Opinion and Values Questions—These questions are designed to explore what a person thinks and believes about a topic, experience, phenomenon, or event and the value she or he places on it.

3. Feeling Questions—These questions explore what people feel and their emotional experiences (rather than or in relation to what they think).

4. Knowledge Questions—These questions seek facts and information that someone knows about a topic, phenomenon, event, or specific context.

5. Sensory Questions—These questions probe for what people have experienced at the sensory levels as in what they see, hear, touch, taste, or smell.

6. Background/Demographic Questions—These questions focus on people's social location, identities, and positionalities as they conceptualize and describe them.

It is important to understand the range of types of interview questions, the goals associated with each kind of question, and the differences among them. These kinds of interview questions are certainly not mutually exclusive and in fact are most often used in different varieties of combinations. See Appendix L for sample interview protocols organized around research questions. Interviews tend to be structured, semi-structured, or unstructured. Semi-structured interviews are the most common in qualitative research, and we describe each of these interview approaches in this section.

- *Structured:* Structured interviews, also referred to as survey interviews (Weiss, 1994), are based on fixed-item, precategorized responses that ask identical questions to each individual. The use of probes and follow-up questions is limited to those on the instrument so as to achieve uniformity across interviews. This permits comparisons among subgroups, and the responses to fixed-item interviews can serve as data that can drive statistical models.

- *Semi-structured:* In semi-structured interviews, the researcher uses the interview instrument to organize and guide the interview but can also include specific, tailored follow-up questions within and across interviews. In this approach, the interview instrument includes the specific questions to be asked of all respondents, but the order of questions and the wording of specific questions and subquestions follow a unique and customized conversational path with each participant. Probing and follow-up questions may be suggested on the interview instrument, and they are used as needed in the interview.

- *Unstructured:* Unstructured interviews, also called open-ended interviews, adopt a process that allows for interviews to be completely inductive and tailored to each participant's experiences. These interviews do not follow a prespecified guide. Researchers may identify areas of research interest or topics that they want to address, but they do not necessarily use an instrument with specific questions.

Developing Interviewing Skills

The quality of the information obtained during an interview is largely dependent on the interviewer. (Patton, 2015, p. 427)

No matter which kinds of interviews you are engaged in, becoming a skilled interviewer takes practice and requires considerable feedback and reflection. Here, we offer ideas to help you develop your interviewing skills.

- Audio- or video-record several interviews that you conduct and watch them with peers and advisers to engage in a discussion of the process and ways to refine and improve your approach.

- Observe experienced interviewers engaging in a set of interviews and debrief with them about their strategies. Alternatives to this could include collaboratively reviewing their transcripts and/or a video of their interviews if the in-person observation is not possible.

- In a group, take turns interviewing each other while the other group members observe. This is often called "fishbowl" interviews. The observers will provide feedback and ask questions in the group format. You can also assign roles to group members when interviewing. For example, one person could be a participant who tends to answer interview questions with one- or two-word answers, and another could be a participant who is overly talkative but not on-point. Role-playing these interviews can help everyone involved hone and refine their skills.

Table 5.2 Advice for Before, During, and After Interviews

Before the interview begins, consider the following:

- Select a setting that is quiet and free from distractions. Factors to consider when selecting a location for the interview might include noise, traffic and hustle and bustle, and proximity to others whom the participant might not appreciate hearing or seeing the interview.

- Make sure to clearly explain the purpose of the interview. This is incredibly important and should be considered carefully so as to avoid being leading (or misleading).

- Review the informed consent form with participants. Sometimes this is emailed in advance to participants, and sometimes it is given as the participant and interviewer sit down together; either way, before the interview begins, allow ample time for participants to review the consent form (or to go over it with them verbally) and ask questions. This also includes informing participants that they are free to withdraw participation in the interview at any time as well as to refuse to answer any question. (See Chapter Eleven for a detailed discussion of informed consent and other ethical considerations.)

- Request consent to record the interview. Do this after all general questions have been answered and do not turn on (or take out) your recorder until after you have consent to record the interview. (The way you ask for an interview to be recorded can affect how participants answer. Sometimes, it can be better to allow participants to check a box stating their preference so that they are less swayed to answer yes than when asked verbally.)

- Address terms of confidentiality. Be careful not to confuse confidentiality with anonymity since it is highly difficult to ensure true anonymity unless data are aggregated. (Refer to Chapter Eleven for specific considerations related to confidentiality and anonymity.) Be sure to carefully explain who will have access to the recording and transcriptions and under what circumstances.

- Explain the format, structure, and process of the interview. Explain the type of interview you are conducting in terms of the structure of questions, the process, the timing, and your expectations and role. We often find it a good practice to confirm with participants how much time they have allotted for the interview.

- Discuss what happens once the interview ends. Does the study require more of the participant's time and in what ways? When? Also be sure to set up clear (and realistic) expectations in terms of follow-up contact.

- Make sure that the participant feels comfortable physically and emotionally.

- Ask if the participant has any questions about any of the above before you get started with the interview.

- If you have consent, make sure the recorder is turned on and recording. (You should test the recorder before the interview and peek at it at least once during the interview. It is a good practice to always have spare batteries and a backup plan.)

(Continued)

155

Table 5.2 (Continued)

During the interview, consider the following:

- Provide clear information on the process and timing of the interview.

- Try to engage participants in the interview as soon as possible (rather than talking at them).

- Before asking about controversial matters (such as feelings and conclusions), first ask about more neutral kinds of things as you establish a sense of each other and (hopefully) a level of comfort and trust.

- Attentively and actively listen to the participant. Weiss (1994) provides examples of good and bad interview transcripts. In the good interviews, the interviewer speaks little. Remember that the goal is to listen.

- Ask follow-up questions. Follow-up questions should be guided by your research questions as well as the participant's responses. You might ask for specific examples of something a participant describes or for a participant to clarify something.

- Ask participants to define and explain their understanding of key terms and concepts they use.

- Use the participant's own language to ask new and follow-up questions.

- Remember that context is everything and try to seek out contextualized responses.

- Make eye contact throughout the interview and be sure to show that you are engaged.

- Take notes unobtrusively. Taking notes can help you to remember to ask specific follow-up questions. However, it is *very* important that you remain present and attentive (eye contact is key) in the interview and take notes as unobtrusively as possible.

- Remember that the pacing and structure of questions should reflect that you are listening carefully.

- Share your appreciation for the participant's time and thoughtful responses in verbal and/or nonverbal ways throughout the interview.

- If the participant seems distressed or anxious, be sure to check in about this (and pause or discontinue the interview if needed).

- At one or more moments during the interview, give a time check so the person has a sense of how much time has passed and how much is left.

- Leave time for open comments, reflections, and questions at the end of the interview.

After the interview, consider the following:

- Adequately thank the participant and appreciate her or his time and responses.

- Write notes on a copy of the instrument with the participant's name (real or pseudonym) and the date on it so that you can clarify or follow up with any questions and key ideas.

- Write down any observations you made during the interview. For example, was the person particularly emotive (in a positive or negative way) at specific times and, if so, when? Were there any surprises during the interview?

- If appropriate and possible, follow up by email or with a handwritten note of thanks.

- Engage in interviews with invited observers, including peers who can then debrief with you and offer feedback. Make sure to get permission from the participants before inviting any observers to an interview.

- If having someone else physically present at your interview is not possible, having peers and advisers review transcripts and offer feedback can also be quite informative. It can be useful and generative to do this in a group format.

As previously mentioned, it is important that the interviewer appear (and be) nonjudgmental. This can be difficult to achieve in situations in which your views and belief systems are different from those of the person you are interviewing. It is also important to note that you can still be biased when your beliefs are similar to participants, and this bias may sometimes be harder to see. You must be attuned to and conscious of both verbal and nonverbal messages that you give off during each interview. This means that you must be cognizant of your own body language and verbal expressions as well as be flexible when pursuing certain lines of questioning, including rephrasing, probing, and/or backing off as necessary. Above all, the interviewer has to be a good listener and genuinely care about the experience of the research participants. It is also important to note that interviews are ideally audio-recorded and transcribed since analysis of various sorts requires that there is a text to code and annotate. We discuss the process of transcribing interviews in Chapter Six and analysis in Chapters Seven and Eight. Helpful advice and suggestions for preparing for and conducting interviews as well as developing interview instruments are presented in Tables 5.2 and 5.3.

There is great range and variation in the approach and structure of qualitative interviews. You should be aware of these differences and ultimately use an approach that best allows you to answer your research questions and accomplish the goals of your study. For further reading in this area, see the reading resources at the end of this chapter.

Technology and Interviews

As technology and social media increasingly mediate all facets of society, the field of qualitative research, like most fields, is finding itself needing to adapt to new ways of approaching methodology, which of course includes interviews. Many researchers have begun to consider new forms and forums for interviewing, including "electronic or virtual interviewing" or "online or Internet-based interviewing," respectively, that is, interviews that happen without the physical presence of the interviewer and interviewee (Fontana & Prokos, 2007; O'Connor, Madge, Shaw, & Wellens, 2008). O'Connor et al. (2008) discuss two main types of online interviewing—namely, synchronous interviews in which a written and/or oral exchange takes place in real time via online conferencing, chat room, or VOIP (Voiceover Internet Protocol) technologies such as Skype or Google Hangout and asynchronous interviews in which interviews are conducted via email, a listserv, or an online bulletin board environment. While a deep treatment of this topic is beyond the scope of this text, it is worthy of mention and brief discussion of the advantages and disadvantages of this technology in interviews specifically and qualitative studies more broadly, and at the end of the chapter, we direct you to several resources that engage this topic in depth.

Some advantages to technology-mediated interviews include potential access to a greater number of people given geographical and other access and mobility issues,

Table 5.3   Tips for Developing Interview Instruments (Protocols)

- Provide clear information on the process and timing of the interview in an inviting, clear, and transparent overview description.
- Frame the interview with respect for the person's time and expertise and set this as the primary tone and value of the process.
- Make sure questions are not worded in ways that are leading.
- Include mostly open-ended questions so that the process is inductive. (This depends on the goals and selected interview approach.)
- Include clear follow-up questions that probe for specifics and examples whenever possible. These follow-up questions should be taken into consideration when determining approximately how long the interview will take.
- Be sure questions are simply stated, avoiding jargon and confusing terms.
- Avoid overly complex or webbed questions in which you are actually asking more than one question. These kinds of questions are often referred to as double-barreled questions.
- As you develop the instrument, remember the economy of time and the relationship between time and the quality/depth of responses.
- Include reminders in the instrument to probe for implicit assumptions and conceptualizations whenever possible.
- Take an inquiry/discovery stance in questions (and demeanor).
- Pilot and adjust instruments. (Refer to Chapter Three for deeper discussion of the role of piloting and how to pilot instruments.)
- Include a final open-ended question that allows participants to provide any other information they prefer to add and/or their impressions of the interview.

potentially lower cost and less inconvenience to participants given that there is no travel, and a potential for a more rapid study given that interviews can be scheduled back to back because no travel is involved. Some negatives and challenges include a loss of being together in the same place, which can mean a lack of careful reading of nonverbal expressions; less engagement and attention; difficulty establishing the relationship and a sense of safety, trust, and comfort; and possible access issues given lack of access to technology in many parts of the world. Procedural issues, technical issues, and technology issues also must be considered. As O'Connor et al. (2008) state,

> Researchers who have used online interviews, whether synchronous or asynchronous, report many differences between online and onsite interviews. In particular, issues relating to interview design, the building of rapport, the virtual venue and research ethics present the online researcher with challenges . . . [including] . . . online recruitment, representativeness, interview conduct and design, respondent identity verification, building rapport and research ethics. (p. 276)

These kinds of interviews also highlight issues with the nature of online interaction, issues of informed consent, and adjustments needed to make online interviews of all sorts work. Pros, cons, and trade-offs to these choices need to be carefully considered and discussed with those who can help you to make informed choices. The use of various interview forms ultimately depends on the study goals and research questions as well as the resources of the researcher. Online interviews present advantages and challenges and may or may not be suitable for certain research projects (O'Connor et al., 2008). As you consider these methods and processes for your own research, please be sure to carefully consider their benefits as well as their limitations.

Interview Transcripts

Sharon often says to students, "Observations without fieldnotes are mere memories, not data." Similarly, we stress the importance of real-time recording and transcription of interviews (and focus groups, which are also referred to as group interviews) since we consider transcripts as more reliable than interview notes. Interview notes, careful and detailed as they may be, are not real-time accounts of the interview since they are mediated by memory, which can be imprecise and self-serving. Thus, interview transcripts are of central concern as a hallmark of qualitative data; they are most often viewed as necessary to rigorous data collection and analysis. As Kvale (2007) states,

> Transcribing the interviews from an oral to a written mode structures the interview conversations in a form amenable to closer analysis, and is in itself an initial analysis . . . The amount and form of transcribing depends on such factors as the nature of the material and the purpose of the investigation, the time and money available, and—not to be forgotten—the availability of a reliable and patient typist. (p. 94)

What becomes clear here is that transcripts are not only important to data collection, as they are the way that interviews produce real-time data. They are also useful for valid and rigorous data analysis, and we discuss these analytical considerations and issues in detail in Chapter Six. Without transcripts, it is difficult to engage in intensive, iterative data analysis. Even listening to recordings lacks a kind of deep interaction with the text of the interview. It is important to note, however, that there are times when the appropriateness or necessity of a specific interview or focus group scenario makes it impossible to record an interview or requires the researcher to turn a recorder off. In these instances, we suggest that a researcher take careful notes during the interview and then attend to the development of these notes as close to the interview session as possible (much like turning in the field jottings into fuller fieldnotes).

Turning the spoken word into the written word, as Kvale (2007) and others state, is a vital aspect of qualitative data collection. Many considerations need to be taken into account so that transcripts are valid and reliable. Transcription, to be clear, is not simply transactional; it transforms the spoken into the written and in that sense is an interpretive act, which we discuss in more detail in Chapter Eight. We share some of the key considerations in Table 5.4, refer you to further reading resources at the end of the chapter, and build on this discussion in the next chapter.

Based on our experiences and those of our colleagues and students, we strongly suggest that you create an intentional, efficient, and user-friendly indexing system for your

Table 5.4   Considerations for Transforming Recorded Data Into Transcripts

- What technology will you use to record interviews (e.g., digital recorder, video recorder, voice recognition program)?
- What transcription procedures will you follow? This includes choices about how to move from recording to text, which involves a series of technical and interpretational issues (e.g., whether transcripts are verbatim, preserving speech patterns such as "um" and "like").
- How will you deal with reliability and validity issues (including transcript quality and structural choices about transcripts)?
- How will files be securely sent to transcription services if used (e.g., does the transcriber have a secure process for receiving/sending files)?
- What is your timeline for transcribing interviews? (Note: A 1-hour interview takes about 5 hours to transcribe verbatim.)
- If using voice recognition or transcription services (e.g., Rev.com), how will you review recordings against transcripts to check for inevitable errors?

transcripts. This includes assigning a number to each transcript, page numbers on each page of every transcript, and numbers on each line of every page. This will enable you to create a system to identify where specific passages and quotes are as you engage in multiple readings and codings of the data. We describe this more fully in Chapter Six when we discuss validity and in Chapters Seven and Eight as we discuss data analysis.

OBSERVATION AND FIELDNOTES

Observation and fieldnotes is an important qualitative method because it allows for the researcher to see and record firsthand the activities in which research participants are engaged in the context(s) in which these activities happen. Observation can often be used as a method of data triangulation—defined in the next chapter and meaning here the use of multiple data sources to achieve a range of contextual data—since research shows that the validity of self-reporting (such as in interviews and focus groups) comes into question; therefore, researcher observation in real time is often used to validate information garnered from focus groups, interviews, and questionnaires (Emerson et al., 2011; Maxwell, 2013).

Qualitative observational research explores and describes the mediating contexts on behavior, attitudes, beliefs, and interactions, including organizational, relational, and cultural knowledge(s). Observation employs interpretive and naturalistic approaches to understanding people and activities in their multiple and intersecting contexts, including aspects of social identity and positionality. Furthermore, observational research enquires into individual and/ or group behavior and interaction as a method to find the meanings and significance of these behaviors in context; it attempts to identify, explore, and explain complex social structures and contexts as they shape the behaviors and experiences of those who participate in them. It is very important to note that *observation without fieldnotes does not constitute data;* it is only through the recording of activity through various stages in the fieldnote-writing process (described in the next section) that the observation becomes data.

There are a number of positive aspects and benefits of observation and fieldnotes as a data collection method:

- Observation and fieldnotes can be flexible and exploratory.

- Observational research findings, when combined with other data collection methods, can serve to enhance validity because they are considered triangulated data.

- Observation allows researchers to observe people in their natural settings so that understandings are contextualized.

- Reviewing your fieldnotes over time can allow for the building of insight into the roles, contexts, and realities of the research setting and people within it over time.

- Reviewing your real-time jottings and transforming them into fieldnotes facilitates reflective engagement (and ideally dialogic engagement at strategic moments) with the observational process.

In addition to the strengths of this method, there are also potentially challenging aspects of observation and fieldnotes, which include the following:

- There can be issues with reliability and transferability since observational findings often reflect a unique population or context and therefore cannot be transferred to other individuals, groups, or contexts. Furthermore, the sample of individuals may not be representative of the population and/or the behaviors observed are not representative of the individual.[3]

- Since observation is interpretive, and therefore inferential, fieldnotes can reflect researcher biases and assumptions in ways that then become codified as unbiased data.

- Issues of what is called **reactivity**—the ways that the researcher affects the behaviors and processes observed in the research as a result of being there—can impact the observations of the participants in a setting. Reactivity is a concern in direct observation since individuals may shift or change their actions rather than being natural and authentic, yet the researcher all too often records the activity as if it is natural.

- The activities and interactions observed can be decontextualized if observations are short term and/or timing is incongruent or inconsistent in a variety of ways.

- Observation and fieldnotes are highly inferential in that they can, if they remain unchallenged, cause challenges to the data set since inference can be easily sedimented into coding and analysis as data and therefore influence emerging interpretations in ways not grounded in participants' realities.

Despite these challenges, observation and fieldnotes is a widely used and respected method in qualitative research. Furthermore, it is also important to note that these challenges are not necessarily unique to this method. As with all data collection methods, it is important to understand their benefits and limitations as well as the multitude of choices you can make around how the methods are structured. The ways that fieldnotes can be structured are described in the next section.

Terms and Concepts Often Used in Qualitative Research

Reactivity: Reactivity refers to ways that the presence of a researcher affects the research setting and the behavior of the participants. This most commonly refers to when participants change behaviors as a result of being observed but can also refer to when interview participants change responses due to perceived reactions from the researcher. This reaction to researcher presence impacts the data, thereby affecting the validity of the study.

Fieldnotes

Depending on your approach, the fieldnotes generated by observation can be descriptive, inferential, or evaluative, and fieldnotes are often a combination of two or more of these approaches. *Descriptive* observational fieldnotes do not require the researcher to make inferences; rather, the idea is that the researcher observes and simply describes what has been observed as neutrally as possible. This can at times be confusing and even misleading since often what seems objective is in fact inference, and it is important to pay attention to your interpretive filters. *Inferential* observational fieldnotes require that the researcher understand that she is making inferences—interpretations and assumptions beyond the data—about what is observed and the underlying motives, affect, or emotions of the events and behaviors observed. *Evaluative* observational fieldnotes mean that the researcher is consciously making an inference and a judgment about the nature and motives of the behavior or events observed. It is important to understand these various kinds of fieldnotes and to understand the goals, roles, and differences between them. Broadly, fieldnotes often include descriptive as well as inferential data. While it is important to acknowledge the differences between these, the lines are often blurry, especially between description and inference. This makes systematic and structured reflexive engagement with data even more important.

Again, it is vital to remember that without recording observation through writing, there are no data. This is why we consider observation and fieldnotes as one method. It is essential to record your observations through the careful and systematic process of writing fieldnotes. Writing fieldnotes is an absolutely vital part of the observational research process, and it is important to understand that there are many approaches to fieldnote writing with varying reasons and processes that relate to each choice. In their widely used book *Writing Ethnographic Fieldnotes*, Emerson et al. (2011) argue that writing fieldnotes requires that a researcher work over time to develop a number of specific, applied skills that include the following:

1) Moving from theory (or problem statement) to what you need to observe

2) Understanding your theoretical construction of the focus and guiding questions

3) Learning to discipline how you see, hear, and take notes in the setting

4) Capturing social interactions in words (how to observe and write about "order" or sequences of action)

5) Learning how to write an analysis: developing a sense of style or becoming conscious of stylistic and representational choices

6) Seeking the **emic** (in contrast to **etic**) perspectives and language of insiders in the setting

As Emerson et al. (2011) make clear, there is a sequential process of writing fieldnotes, which goes from in-the-field "jottings," which are contemporaneously written notes that you take (either in a notepad or on a computer depending on the context) while at the research site. These jottings are turned into more coherent written accounts of what is observed, which is the process of turning them from jottings into fieldnotes, and happens after the researcher leaves the field but ideally close to the time of the observation. The "real-time jottings" are an essential grounding and resource for writing the fuller fieldnotes, which should be written shortly after being in the field so that they are closer to the time and memory of actual observations and are therefore more reliable.

Terms and Concepts Often Used in Qualitative Research

Emic: Emic refers to culturally and contextually embedded conceptualizations and descriptions of beliefs, behaviors, and ways of being that are meaningful and resonant to individuals and/or groups. In qualitative research, emic refers to being inductive in your approach so that rather than imposing your own concepts and terms, you create the conditions necessary for participants to offer and articulate their own conceptualizations in language that is organic to them as they describe an account, event, or phenomenon.

Etic: Etic refers to descriptions of beliefs, behaviors, or ways of being or belief that are attributed by an outside observer (the researcher) that are not culturally embedded or necessarily organically conceptualized, articulated, or contextualized. In qualitative research, etic means that the language used to describe an account is not organic to the participants. We discuss the concepts of emic and etic in relationship to validity in Chapter Six and data analysis in Chapter Eight.

As previously noted, Emerson et al.'s (1995) notion of "the inseparability of methods and findings" is important to consider across the research process and with specific regard to observational fieldnotes. As they argue, separating the real-time experience of the observation and jottings that happen in the field from the written fieldnotes about the observation, as if they are objective and separate, confuses the meaning and goals of field "data" because it treats these data as "objective information" rather than as interpretive and specific to the observer and his or her subjectivities. Situating the observational fieldnotes within the subjective interpretation of the researcher and yet seeking to keep the notes as close to the events as possible is the goal of this process. Writing quality fieldnotes is a complex process rife with many choices, and we strongly recommend reading Emerson et al. (2011) on this since their book is the seminal and most comprehensive work in this area to date.

Participant Observation

Participant observation, which is the primary method used in ethnography, is a form of observation in which you, as the researcher, establish a presence or role in the setting. Your role in the setting is negotiated and renegotiated ongoingly through gatekeepers as well as with other participants. Participant observation involves varying degrees of immersion in a research setting as well as varying degrees of participation and observation. The extent to which you are immersed in a setting and the degree to which you participate in daily activities will depend on a variety of factors, including the goals of your study, resources, and access. For example, the extent to which you can be immersed and participate in a setting may be limited in certain populations such as when studying individuals in prison (Lichtman, 2014). As with all qualitative research, participant observation acknowledges that the researcher is the primary research instrument. Furthermore, "unlike passive observation where there is a minimum interaction between the researcher and the object of study, participant observation means establishing rapport and learning to act so that people go about their business as usual when you show up" (Bernard & Ryan, 2010, p. 41). While the extent of participation in research settings can vary, reflexivity is central to participant observation. All research involves aspects of participant observation as "we act in the social world and yet are able to reflect upon ourselves and our actions as objects in that world" (Hammersley & Atkinson, 2007, p. 18).

The hallmarks of participant observation include that the researcher

- seeks out observable data related to the focal topic based on existing theory and/or methodology in systematic ways;

- carefully records action and interaction in context, including that which might be unnoticeable without focused and intentional observation;

- engages in the process of "making the familiar strange and the strange familiar" (van Hardenberg, n.d.), meaning that you need to move between insider-outsider sensibilities; and

- ongoingly monitors observations and includes evidence of personal bias or prejudice.

Writing fieldnotes while conducting participant observation can sometimes be challenging. Researchers using participant observation should think about "*what* to write down, *how* to write it down, and *when* to write it down" (Hammersley & Atkinson, 2007, p. 142). These determinations are guided by the goals of the study, the research questions, your style and preferences, and the extent to which you are immersed and participating in a setting. There is no "correct" way to write fieldnotes, and many texts and researchers have specific preferences and approaches (e.g., creating columns for observation and in-the-moment inferences or using [brackets] for inferences). What we argue is most important, as described above, is that fieldnotes are composed as close to the actual observation as possible.

Researchers, no matter the approach to observation and fieldnotes that you take, can engage in structured exercises that help you to understand both the skill of disciplined and systematic observation as well as the art of writing useful fieldnotes. In both learning about the subjectivity and ways to focus observation and in understanding the range of technical choices in the writing of fieldnotes, Recommended Practice 5.1 is a fruitful exercise for novice researchers' understandings of the process and the need to hone your skills of observation and related data generation through the construction of fieldnotes. In addition, the memo described in Recommended Practice 5.2 will help you to help you engage with and examine your data from observations and fieldnotes.

Recommended Practice 5.1: Observation and Fieldnotes Exercise

This exercise is intended to highlight multiple aspects of the data collection method of observation and fieldnotes. The experiential component of the exercise is at the heart of the learning. Our students find this exercise very insightful in the ways that it brings out what individuals focus and do not focus on as well as ways that individuals structure their fieldnotes. These can be easily adapted for a class setting, research team, or an inquiry group.

Process:

1. In groups of three to four, pick a place where you can easily observe people (can be a coffee shop, restaurant, library, or other public space). Without talking to your partners, spend 30 minutes taking notes about what you observe/hear.

2. After the exercise, turn your individual in-the-field jottings into fieldnotes and email these fieldnotes to all of the group members.

3. After emailing each other the extended fieldnotes, review your group members' notes, carefully reading these to get ready for the group discussion outlined in Step Four.

4. Using your own and your group members' fieldnotes, engage in a discussion of the following:

   - How did your accounts overlap and/or differ?

   - How did your approaches to observation and notetaking differ (e.g., style, focus)?

(Continued)

(Continued)

- What does this exercise teach you about observation?

- What does this exercise teach you about fieldnotes?

- What other specific questions or ideas did this experience raise about the research process?

Take notes about these points as you discuss with your group.

5. Each group will informally report to the larger group (class/research team/other inquiry group) about their learnings and any questions/comments/observations that arise from this experience. This whole-group discussion should generate engaged conversation about learnings and lessons arising within and across the groups.

Recommended Practice 5.2: Observation and Fieldnotes Memo

Choose two to three entries of observation fieldnotes that you have written and write a memo that responds to each of the following questions:

1. What do my fieldnotes focus on? Exclude? What might this teach me about my own subjectivities and biases as a researcher?

2. What stylistic and structural choices am I making in my fieldnotes and why? Are there other choices I should consider given the goals and roles of observation in my study design?

3. What are the goals of engaging in observation and fieldnote writing given the goals and guiding questions of the study?

4. What additional forms of data might I need to answer my guiding research question(s)?

5. Do the data in my fieldnotes challenge and/or support other data sources? How and why or why not?

6. How might my observations help me to formulate or refine a research question that I can systematically answer through engaging in further data collection and analysis?

7. How might I challenge my subjective interpretations of what I am observing?

8. How does my status as an insider and/or outsider in the setting influence this process? What does that teach me about the research process?

A final note to consider about observation and fieldnotes is the role of technology and how it mediates this process. One consideration is whether and how the use of a laptop, smartphone, or other computer device makes sense in a specific context in terms of if it is distracting or potentially problematic. For example, in Sharon's fieldwork in Nicaragua, the use of a laptop in some contexts (i.e., in specific rural community contexts) for field- notes might seem strange and out of context and therefore problematic, whereas in others (for example, in Managua, the major city and in the schools and organizational contexts in which the research takes place), it would be a comfortable medium. Furthermore, in Nicole's work in secondary schools, laptops can be distracting to both students and teach- ers. Despite the convenience, the use of technology can influence the research setting in positive and negative ways that must be weighed against its benefits. Another consider- ation is if you want to use audio or video recording instead of or in addition to written fieldnotes, which then creates the question of what you will *do* with those recordings and how you will manage what might quickly become an enormous quantity of recordings. The anthropologist and ethnographer, Philippe Bourgois, suggested to Nicole that she dic- tate her fieldnotes into a digital recorder to save time and make sure that the fieldnotes are recorded as close to the observation as possible. These recordings are then transcribed. As with all methods choices, you should consider the limitations and benefits of these tech- nologies as they relate to your research questions as well as the contexts in which your fieldwork is carried out. We refer you, at the end of this chapter, to additional resources that can help you in this realm.

FOCUS GROUPS

Focus groups, which are sometimes referred to as group interviews, provide an opportunity to collect data that may not arise in an individual interview because of the group dynamics and interactions. Focus groups create the conditions to observe (and even facilitate) com- munication between research participants as a means to generate data in a group forum. Focus groups are particularly suited for studies that seek to understand questions related to social interaction processes (Morgan, 2010). Furthermore, studies that explore attitudes, opinions, and experiences in specific groups and contexts are also well suited to the use of focus groups. Focus groups can be used to examine how knowledge and ideas develop, are constructed, and operate within a given contextual, cultural, or group context. The kinds of data that focus groups tend to generate can be related to content (e.g., the topic or subject of the group discussion) and/or process (e.g., the group dynamics and interactions) (Millward, 2012; Morgan, 2010). The research questions and the goals of a study determine if you focus on the specific content, processes, or both.

Focus groups can create the context for generating *groupthink,* which is when an indi- vidual introduces a topic and the rest of the group focuses on this topic and ultimately generates a group understanding. Depending on the goals of your study and your research questions, groupthink can be useful or problematic (or some combination). Positively, the group can build on and create ideas as a collective that yields robust and relationally grounded data. Groupthink is a common aspect of focus groups, and if you are focusing on individuals' unique and specific experiences, focus groups may not be the right method.

Focus groups also allow participants to comment in relation to each other's thoughts, experiences, and responses to specific questions and therefore generate emergent topics for group inquiry and discussion that go beyond responses to the questions on the instrument. Focus groups intentionally structure group interaction as part of the method of data collection. Participants are encouraged to engage with each other by building on each other's responses, asking each other questions, exchanging stories, and/or commenting on each other's experiences and perspectives. Group engagement is useful in studies in which the researcher seeks to have participants explore issues of importance to them, in their own terminology, and generating their own questions (Stewart, Shamdasani, & Rook, 2006). Focus groups can vary from the very structured and facilitator led to more participant run to somewhere in between (Lichtman, 2014). See Table 5.5 on page 169 for suggestions related to conducting focus groups.

While one-on-one interviews provide a depth of information on each individual in a study, focus groups foster thinking and understanding among individuals and subgroups. There are some important advantages of focus groups and ways that they can be used to achieve a number of goals, which we describe below.

- Focus groups place a primacy on understanding that which is shared and unique across participants' attitudes, experiences, priorities, uses of language, and frameworks of understanding.

- They encourage participants to co-generate and explore questions and develop analyses of common experiences.

- Focus groups encourage a variety of kinds of communication from participants, including relational modes of participation that build off of others' responses.

- They help participants to identify and explore group norms and culturally embedded values.

- Focus groups can serve to provide a form of quality control since participants can provide a kind of checks and balances to each another that can mediate false or extreme views (Patton, 2015).

- They can encourage open dialogue about potentially taboo or embarrassing topics and allow for the expression of challenges or criticism. (There is also the belief that the opposite can be true.)

- Focus groups facilitate the expression of ideas and experiences that might be left underdeveloped in an interview with a single participant.

- They can engage participants who cannot read or write (as compared to written questionnaires).

- Focus groups encourage participation from those who may be reluctant to be interviewed on their own but engage in the discussion generated by other group members.

- Focus groups can be used at various points to inform study design, design of contextually appropriate methods, and design of instruments (Barbour, 2007).

Table 5.5 Suggestions for Focus Groups

- Ideally, we argue that focus groups should range in size from four to six participants to ensure ample time for in-depth discussion of focal topics as well as the inclusion of all participants' perspectives across questions and topics. Sometimes researchers use focus groups to "beef up" participant numbers, and we do not recommend this. We urge you to consider the quality of data that you will be able to get given the economy of time to questions and people in a given group.

- Think carefully about your recruitment strategies and the individuals comprising the groups so that they meet the intended sampling goals and set the right context for shared disclosure and engagement.

- Have a notetaker present who can also be a peer debriefer. This is helpful since facilitating focus groups requires considerable focus and energy, and leading the group and taking notes simultaneously can be difficult. In addition, being able to debrief afterward and having an alternate perspective (or two, or three . . .) will push your interpretations and help with improving your facilitation style.

- Set clear ground rules and narrate the timing and process to participants so there is understanding among the group. For example, make clear that you want to hear from everyone, that people should respect each other's opinions but encourage generative and respectful disagreement.

- Try to engage participants in the focus group as soon as possible and create the conditions for everyone to participate equally.

- Before asking about controversial matters, first ask about more neutral kinds of things as you establish a sense of each other and a level of comfort in the group.

- Think about technical aspects of focus group data collection, including what you will need to identify speakers in a transcript (for example, having people say their names when they speak or keeping an order of names in notes) and how you will record in a group setting.

- As with all data collection, reflect individually and/or discuss with others to consider emergent understandings of themes and to refine the instruments for future focus groups.

- Think carefully about issues of confidentiality in focus groups since people hear each other's responses in real time and can identify those responses in written reports.

- Clearly communicate issues of confidentiality to participants. This is especially important with focus groups since you cannot control how other members in the group will treat what participants share.

- Consider the setting of a focus group in terms of comfort and implications for individual members and the collective.

- Engage in thought experiments to anticipate problematic group dynamics and troubleshoot ahead of time in terms of recruitment, sampling, and facilitation style as well as boundary and tone setting.

- One great way to practice focus group facilitation is to work with a group of peers in a fishbowl format to view and offer suggestions to each other related to facilitation style and so on.

In addition to the potential advantages of focus groups detailed above, there are also potential limitations, including times when participants may be reluctant to share their experiences and perspectives in a group format. When conducting a focus group, researchers must be mindful of group or organizational politics and how the structure and grouping of focus group participants can affect the process and therefore the data. For example, grouping employees with their bosses or other members of groups that experience tension could create additional problems and should be avoided unless there are compelling reasons to do so and all parties are willing to participate. Another potential challenge of focus groups is related to issues of confidentiality between group members. You cannot promise confidentiality when using focus groups since each member of the group must rely on not only your discretion but also the discretion of other group members, and as a researcher you are not able to make those assurances. Furthermore, topics that are considered taboo or difficult to discuss can pose specific issues in focus groups, including that those whose experiences or perspectives prove to be outside the norm or unpopular in a group may not feel comfortable speaking out or sharing their thoughts and experiences within the group format. In addition, there are issues brought about by the fact that focus groups are a great method for eliciting and identifying major themes but are less useful for uncovering and engaging with more subtle or hard to get at differences. This can make the data from focus groups less specific and therefore potentially less useful in answering certain kinds of research questions. One additional challenge is that focus groups rarely take place in people's natural environments, and so there can be distance from the contexts of their lives and the specific experiences in focus. This can be the case with interviews as well, but it is much more likely with focus groups given logistical considerations (Patton, 2015).

Another consideration in using focus groups is that while groupthink can be useful, it also means that responses are not independent, and certain opinions and beliefs may be left out or informed by other responses as they build from them in conversational style. This must be kept in mind when analyzing focus group data (and when facilitating the group). How focus groups are facilitated is quite important, and if your ability to set up and structure the group well and attend to group dynamics is weak, the data will suffer. You need to hone specific focus group skills, which overlap with interviewing skills but also account for managing group dynamics, making room for people to fully articulate their thoughts, and creating the conditions necessary for people to share their perspectives and experiences in ways that provide quality data. Finally, the number of participants in a focus group and the number of questions asked directly relate to the quality of data that focus groups can generate. We urge our students to carefully consider the goals of focus groups and make sure that they will generate data with requisite depth given the specific goals of a study. This can mean having fewer participants in each focus group, asking fewer questions, facilitating group exchange in explicit ways, setting clear ground rules for participation, and encouraging turn taking. Table 5.5 details additional suggestions for using focus groups.

In addition to the suggestions in Table 5.5, we suggest that you to return to Table 5.1 and 5.2 for interviewing tips that will help you to consider additional aspects of focus group planning, facilitation, and follow-up. In addition, the emergence of online focus groups creates issues for facilitation that in some ways mirror those discussed previously regarding individual interviews but that also relate to the group dynamic in the virtual space. For example, using Adobe Connect (a virtual conferencing platform) means that the facilitator can see each person but the participants do not necessarily see each other. Furthermore, people can write during these sessions and see each other's written comments, but the synchronicity must be tracked between that which is spoken and written. All of these features of various online conferencing technologies have implications for data collection that must be seriously considered. See the texts suggested at the end of the chapter for additional resources related to focus groups, specifically Patton (2015, p. 476). It is important to note that the same considerations for interview transcripts apply to focus group transcripts as well.

DOCUMENTS AND ARCHIVAL DATA

Organizations and groups are "data-rich environments" in which much data already exist that can (and should be) be tapped into (Robson, 2011; Schein, 2004). The review of existing, relevant, and contextual documents is an essential component of the data collection and analysis process (Patton, 2015). Documents that already exist in a group or organization (or other context) are called naturally occurring documents—that is, they exist without the involvement, facilitation, or instigation of the researcher. These existing documents are often an important source of context and history that can help us, as researchers, understand the complexities of what we study better by providing a form of data triangulation to first-person accounts. In this sense, the kinds of data that already exist (often referred to as *archival data* or existing documents) serve to supplement the forms of data that the study precipitates through collection with study participants.

As Wolcott (1994) states, "Everything has the potential to be data, but nothing *becomes* data without the intervention of a researcher, who takes note—and often makes note—of some things to the exclusion of others" (pp. 3–4). A researcher or research team should spend considerable time identifying, collecting, organizing, reviewing, and analyzing all relevant documents that contextualize and relate to the phenomenon under study and the context more broadly. The review and triangulation of these documents in relation to primary empirical data is essential to understanding the context in which the research happens. There are a number of ways to think about this aspect of data collection since it includes many possible data sources, including existing documents and data sources of great variety, including online data sources (e.g., websites, blogs, listserv communications, public domain sites) and participant-generated data (e.g., photos, films, reflective writing, blogs).

Bogdan and Biklen (2006) divide archival data into three main categories that we find helpful: personal documents, official documents, and popular culture documents. *Personal documents* include data sources that are developed by individuals such as letters, emails, scrapbooks, notes and lists, past writing of a variety of sorts, reflective writing, and blogs. These data are generated by the individual and are therefore useful as a window into their individual thought processes broadly and specifically in relation to the topic in focus. *Official documents* include those kinds of documents that are developed, produced, or disseminated by institutions (e.g., businesses, schools, or agencies) that can be used for internal and/or external audiences and uses. These kinds of data sources can include websites, mission statements, job descriptions, handbooks, memos, meeting minutes, press releases, training materials, brochures, and so on. These documents are helpful in understanding the formal or organizational context of the study and provide context for the focal topic. *Popular culture documents,* also called "publicly accessed documents," can include blogs, films, books, magazines, photographs, and other popular and publicly accessed materials. These kinds of documents provide context and can also be used to explore people's meaning-making processes in relation to publicly consumed materials, images, and messages.

It is also important to note the role of technology and social media in terms of what constitutes preexisting or archival data. Lee, Fielding, and Blank (2008) discuss the influences of the Internet on what they refer to as "data capture using the Internet" and using the "internet as an archival resource" (pp. 8–14). The kinds of data that exist on the Internet should be considered in relation to your data collection plans, participants, and the contexts in which the data collection occurs (Lee et al., 2008). Furthermore, there is a range and variation of media materials used in qualitative research (see Brinkmann, 2012, for a discussion of these), and qualitative researchers must consider multiple aspects related to how social media affect research, participants, and researchers. Social media are a form of dynamic data that must be attended to in terms of how media are used and in terms of how participants engage with data that exist in social media (Brinkmann, 2012). Considering the range of ways that technology, online engagement, and social media influence what is knowable about the people and places in studies is an important concern.

Many ethical and validity issues must be considered throughout the process of deciding upon relevant and ethical data using the Internet. We discuss these issues in Chapter Eleven and refer you to texts that can help you consider these issues in the resources for further reading section at the end of this chapter and in Chapter Eleven.

A SURVEY APPROACH AND QUESTIONNAIRES

A survey approach to research is often used to gather information about individuals' attitudes, beliefs, and behaviors. It is important to clarify that surveys are a design choice rather than a specific method (Mathers, Fox, & Hunn, 2009). Survey methods can include interviews (both in person and over the phone) and administering questionnaires to study participants that ask them to respond to questions that can be completed in person, online, over the phone, or by mail. Survey interviews are primarily used in quantitative research because they tend to be highly structured and therefore not as conducive to the goals of qualitative interviewing (Mathers et al., 2009).

Questionnaires are what people tend to think of when they think of surveys. Questionnaires can be a useful data source within a larger data collection plan for a variety of reasons that relate to triangulation of methods. To this end, there are advantages to questionnaires, which include the following:

- They can be an efficient way to collect data from a range of people across locations.

- Responses can be easier to compile and analyze than other forms of data.

- Significant amounts of information can be collected from a large number of people in a short period of time.

- They are relatively cost and resource effective.

- Individuals can remain anonymous.

- They can be carried out by the researcher or by any number of people with limited effects on their validity and reliability.

- The results of questionnaires can usually be quickly and easily quantified by either a researcher or through the use of a software package.

While the above are possible advantages to using questionnaires, there are also possible disadvantages of questionnaires, including the following:

- Responses provide only a limited amount of information without explanation and contextualization.

- They work best when their questions (or items) are objective (e.g., one's age) rather than subjective (e.g., one's feelings about an event, changes in perspective over time) (Patten, 2001).

- They do not tend to generate rich or contextualized data, and therefore responses can be hard to analyze.

- They can be inaccurate because people tend to give socially acceptable responses even when the questionnaire is anonymous (Patten, 2001).

- There is no way to know if a respondent is being truthful.

- It can be difficult to tell how much thought has gone into responses, which can affect accuracy.

- People may read and understand questions differently and therefore reply based on their own interpretation of the question (and there is no mechanism to know).

- There is a level of researcher imposition in the design of questionnaires, which means that there is much that researchers are not able to learn.

- They can restrict access if disseminated via Internet since that requires a networked computer.

- They require literacy and therefore might marginalize those who are not literate.

Items on questionnaires can be or forced-choice or open-ended (Fink, 2006). Forced-choice questions require participants to select an answer from those specified (including Likert-scale questions); these are typically not used in qualitative research. Open-ended questions can take multiple forms such as short-answer questions, journals, or essays. Many design decisions must be taken into consideration when using a survey approach, including the sample size and issues of representation; the content, type, wording, and order of questions; the logistics of administration; and how the data will be analyzed (Mathers et al., 2009). These considerations cannot be adequately covered in this text. For more information on the construction and administration of questionnaires, see the reading resources at the end of this chapter. In Table 5.6, we include tips for designing effective questionnaires.

Remember that technology may, and increasingly does, play a role in the dissemination and collection of questionnaires (e.g., using SurveyMonkey or some other online survey

Table 5.6   Tips for Effective Questionnaire Design

- Carefully consider your study goals and align the questionnaire items directly with them.
- Use each guiding research question to create clusters of questions.
- Consider the forum/structure/timing of the questionnaire and create questions that "fit" with this forum.
- Write a brief but substantive introduction to the questionnaire that addresses the purpose and guides the participants.
- Provide clear instructions that are visually differentiated (through use of capital letters, bold, or underlined type).
- Think about the formatting and appearance of the questionnaire so it is clear and easy to navigate.
- Consider questions from the participants' points of view so they are clearly stated and relevant, subquestions align with the main question, and the overall process is not too burdensome.
- Anticipate a range of possible answers to troubleshoot and refine your questionnaire.
- Think about the order, flow, and sequence of the questions so they are logical.
- Possibly vary the types of questions to create texture (e.g., a strategic combination of open-ended, closed-ended, Likert-scale questions).
- Think about how the data will be processed and analyzed as a way to consider the structure and length of the questionnaire and its various subsections.
- Pilot questionnaires and refine based on feedback.
- Consider whether and why you would want a questionnaire to be anonymous or not and plan accordingly (particularly if it is part of a triangulated data collection plan).

tool to construct and disseminate questionnaires) and that this comes with both challenges and benefits, including access to networked computers, perceptions of confidentiality, facility with online methods, email saturation, and related issues. As well, programs (like SurveyMonkey) can help batch and analyze data in ways that can be useful since they provide customizable questionnaires as well as back-end programs that include analytical tools, sample selection, and data representation tools. These tools should not take the place of rigorous, qualitative data analysis, which we discuss in Chapters Seven and Eight.

PARTICIPATORY METHODS OF DATA COLLECTION

In addition to the more common kinds of qualitative data collection methods discussed above, there are forms of data collection in which researchers co-construct and co-generate data with participants. In participatory approaches to research, participants are also researchers and are from within the community or group at the center of a study (with some possible outside accompaniment); they may define the topics or problems to be addressed, design their own studies, and engage in collaborative data collection and analysis toward answering their guiding questions and goals in ways that emerge from within their own communities. (We say "may" because there is a continuum of involvement within and across various participatory approaches.) We review four of the most commonly used participatory data collection methods since, while they are less common among our students, understanding their value and uses may influence you to adapt aspects of these methods to your own studies. These data collection methods stem from the field of participatory action research (PAR), which is a specific methodological approach to research with its own history, literature, values, goals, and commitments. In a participatory action research approach, data are generated by participants—individuals and communities—with a goal of creating the structures and conditions for people to tell their own stories in ways unmediated by researcher influence, presence, or engagement.

There are many participatory techniques used by community or group members to explore and document their own contexts, communities, networks, and experiences. While these methods are beyond the scope of the book, we mention several here so that you know that there is a range of additional data collection methods that can be further explored. At the end of the chapter, we recommend further reading on participatory and community-centered methods of data collection. The data collection methods overviewed here include (a) photovoice; (b) community-based oral testimony; (c) social network, community, and institutional mapping; and (d) transect walks. Sharon tends to use aspects of each of these approaches across her research projects, both her participatory research and her broader research, because they provide a different kind of window into the contexts of the research as well as increased insight into participant experiences and perspectives.

Photovoice

Photovoice refers to the collaborative taking, interpretation, and discussion of photographs as contextualized data that participants generate (Wang & Burris, 1997). This method of data collection is derived and developed out of a concern for authentic representation(s) of the perspectives and experiences of marginalized populations (including those who are nonliterate). Participants are provided with cameras; asked to chronicle daily experiences, contexts, and/or events; and then engage in group discussions about the images and their contexts, using the photos to narrate their ideas and their worlds. The guiding belief is that the choices of foci and frames for the photos are valuable, and then as people speak (and hear) about the meaning of the visual imagery, the narration of these choices requires the sharing of feelings, experiences, expertise, knowledge, and the like in ways that are more contextualized and embedded within daily experience than traditional data collection methods. As explained by Wang and Pies (2004),

> Photovoice is not intended to be a methodology in which an entire body of visual data is exhaustively analysed in the social scientific sense. As a participatory methodology, photovoice requires a new framework and paradigm in which participants drive the analysis—from the selection of their own photographs that they feel are most important, or simply like best, to the "decoding" or descriptive interpretation of the images. (pp. 100–101)

This method can be adapted in a variety of ways for a multitude of contexts and can include video recordings as well (for example, Sharon often gives handheld video recorders to people in the communities in which she engages in research as well as digital cameras). We have seen the photovoice process, and the photos and brief videos themselves add much to data sets and collaborative research processes. We suggest that you refer to the reading resources at the end of this chapter for more on this data collection method, including discussions of issues of ethics related to taking photos or video of others, issues of consent, and challenges of the method.

Community-Based Oral Testimony

Community-based oral testimony is embedded within community contexts since it is an ecological approach to data generation and collection that focuses on people's narration of their lived realities in a relational context. This approach relies on extended, largely unstructured conversations with individuals and groups through which a carefully selected range of participants tell their own stories in their own words and convey their understandings of how their lives, communities, and national contexts shape their experiences in relation to the focus of the research (e.g., if a study is focused on transitional justice, participants might tell stories of their experiences with past injustices of a variety of sorts). One aspect of this method is that people engage in these processes in their natural settings and without the imposition of more formal interview structures and processes. Another is that it is an expressly relational approach to individual and group narration of experiences and perspectives.

This method is viewed as a way to push against more traditional data collection methods that many in the PAR world argue is shaped and constrained by power dynamics between researchers and "subjects." The belief is that within these more traditional power dynamics, participants' answers to questions are shaped, constrained, and even artificially determined by researcher imposition. Some even argue that this can happen when outside researchers "collect" oral testimonies rather than them staying within the community itself. One way to address this is to ask participants to "interview" each other so that the conversations can be shaped by mutual dialogue rather than by top-down, impositional agendas and expectations about what researchers value or want to hear. In addition to its focus on creating the conditions for people to share their own stories in more organic and shared ways than a traditional interview, another goal of this method is to be able to engage with a truly representative sample of people from various parts, roles, identities, and relationships within a given community and in relation to the study topic and goals (Golden Institute, 2011). As a researcher, you still need to consider issues of reactivity since your presence and the presence of recording devices can have a significant influence on what is shared (or not shared).

Social Network, Community, and Institutional Mapping

Mapping—of social networks, communities, and institutions—is a useful and inexpensive tool for gathering community and institutional data from and with insiders that can be descriptive or diagnostic in nature. Mapping exercises are useful for collecting contextual and baseline data that can be used in myriad ways, including to help create a foundation for community engagement and ownership of planning by including a range of perspectives, communities, and different groups and subgroups within communities (Sadler & McCabe, 2002). As Warren (2004) states, "Maps are more than pieces of paper. They are stories, conversations, lives and songs lived out in a place and are inseparable from the political and cultural contexts in which they are used" (cited in Rambaldi, 2005 p. 5). In this sense, engaging with community members as they map their networks (social, familial, professional, communal), their communities (resources, power centers, structures), and their institutions (organizational mapping) is a powerful and direct way to gain significant insight into how various stakeholders construct their understandings of their communities.

Corbett and IFAD Consultative Group (2009), in their thorough review of participatory mapping processes, describe the important uses of these kinds of mapping exercises as data collection. They state that such maps provide community depictions of important features and locations. They assert that participatory maps include

> depictions of natural physical features and resources and socio-cultural features known by the community. Participatory mapping is multidisciplinary. What makes it significantly different from traditional cartography and map-making is the process by which the maps are created and the uses to which they are subsequently put. Participatory mapping focuses on providing the skills and expertise. For community members to create the maps themselves, to represent the spatial knowledge of community members and to ensure that community members determine the ownership of the maps and how and to whom to communicate the information that the maps provide. The participatory mapping process can influence the internal dynamics of a community. This process can contribute to building community cohesion, help stimulate community members to engage in land-related decision-making, raise awareness about pressing land-related issues and ultimately contribute to empowering local communities and their members. (p. 4)

In addition to helping researchers to understand the terrain of the research in a community's current state, mapping might have already occurred for institutional and organizational description and development purposes ongoingly and therefore could offer archival data. For example, when Sharon begins any PAR work within communities, an early question she asks is if any kinds of formal or informal mapping have taken place and if there are artifacts of those processes that individuals and community members might share with and explain to her. Mapping discussions should be recorded whenever possible and are considered data and a part of the formal data collection process.

Transect Walks

Transect walks are common in participatory approaches to research and can be used across a variety of contexts and research approaches. Transect walks are useful for introducing a researcher or research team to a community because local community members serve as guides and orient the researcher or team to the geographic area by walking them to various places that they deem relevant and important and narrating to the researcher(s) as they guide the walk. These "walks can be either casual, random or systematic based on a check-list if items to be probed" (Mukherjee, 2002, p. 140). As Mukherjee describes them,

> Such walking exercises of outsiders with others can be undertaken with the objective of gathering some general idea about the locality or a specific area or location within an area. Or, alternatively, such a walk can be systematic and broadly structured with a specific purpose in view, for example, studying a local utilisation pattern of a localised natural resource, for example, a forest or studying the social aspects of a locality such as the type of housing conditions in an area. Such a "walk" can precede or follow "verbal" and/or "visual" interactions. For example, a focused walk can be undertaken to a site based on participatory mapping exercise of a locality. (p. 139)

During or after these walks, which are structured to reach any number of selected points in a community or area, a transect diagram is often created, and data from the walk can be triangulated with local maps and other kinds of local community information (Dudwick, Kuehnast, Nyhan Jones, & Woolcock, 2006). In addition to the diagram, transect walks help researchers to understand multiple perspectives as communities are described in real time through local stakeholders. This data collection process allows for active questions, responses, and the sharing of the setting in real time in ways that are quite generative and useful. These walks can be recorded through the use of audio or video recording as well as fieldnotes and photos. Researchers often debrief these walks with community members and write memos that reflect learnings and questions that emerge from this process. Transect walks can be a one-time event or can happen at various points and with various individuals and groups over time to seek out multiple perspectives and changes over time. They can be used within a range of contexts.

The participatory approaches to data collection described in this section are included to be helpful to your understanding that additional, non-researcher-directed methods of data collection and collaborative analysis exist. At the end of this chapter, we provide additional resources in this area, including texts that offer more detail on participatory action research broadly, the specific data collection methods included here, and many additional participatory data collection methods.

REVISITING DESIGN

As Chapter Three highlights, your research questions are central to your study. As you think about the data collection methods that you will use, you must map these methods onto your research questions. Maxwell (2013) suggests creating a matrix that helps you to link your methods with each research question. Similarly to Maxwell, we suggest that you answer the questions presented in Table 5.7 and represented again in Figure 5.1 to consider how your methods are (or are not, which is vital to discern) mapping onto your research questions. You may answer these questions in the form of a matrix, in narrative form, or in graphical form as in Figure 5.1. The matrix in Table 5.7 can be helpful because

Table 5.7   Considerations to Ensure Methods Align With Research Questions

Research Questions	Core Constructs	Study Goals	Site and Participant Selection	Design and Methods	Rationale	Instruments
What am I studying?	What terms, phrases, concepts, and ideas in my research questions need to be defined?	What do I hope to accomplish with the research? What principles and/or ideals do I deem important?	What is my unit of analysis? Where and from whom will I get the data I need? Why?	How will I get the data I need to answer my research questions?	How will these methods with this population answer my questions? Why is this valuable?	How will I specifically design my data collection instruments to answer my research questions?

we suggest that you include a different row for each research question to ensure that you address each research question individually.

As Figure 5.1 illustrates, your research questions are at the center of your study. Recognizing this and the iterative nature of qualitative research is especially important to keep in mind during data collection. Furthermore, the dynamic way that the processes illustrated in Figure 5.1 evolve underscores the notion that data are co-constructed rather than merely collected. Each data collection method has strengths and challenges, and there is no generally "right" or "best" method. The methods you employ and how you employ them are driven by the research questions and the recursive relationship between these questions and engagement with participants as well as structured reflexivity processes as research questions can (and often do) evolve (in a nonlinear sense of that term) as a study progresses.

Our discussion of critical research design in Chapter Three entails that researchers maintain fidelity to research participants' experiences and pay close attention to intragroup variability (defined in Chapter Three as the diversity within groups). To do this, primacy must be placed on the authentic process of working to understand people's perspectives in contextualized ways; this is achieved through methodological rigor throughout the entire study and especially during the data collection processes. The questions and considerations we present and the structured reflectivity and dialogic engagement practices presented in this chapter as well as in Chapters Two, Three, and Four are a part of achieving this rigor. In addition, to achieve qualitative rigor, you should consider not only each of these data collection methods in complex and strategic ways but also how they can productively overlap and complement each other and how they should be scoped and sequenced in relation to each other. By keeping the research questions at the center of your study, you allow for methods to emerge inductively as data are collected and analyzed. We discuss the dynamic relationship between data collection and data analysis in Chapters Seven and Eight.

Figure 5.1 Considerations to Ensure Methods Align With Research Questions

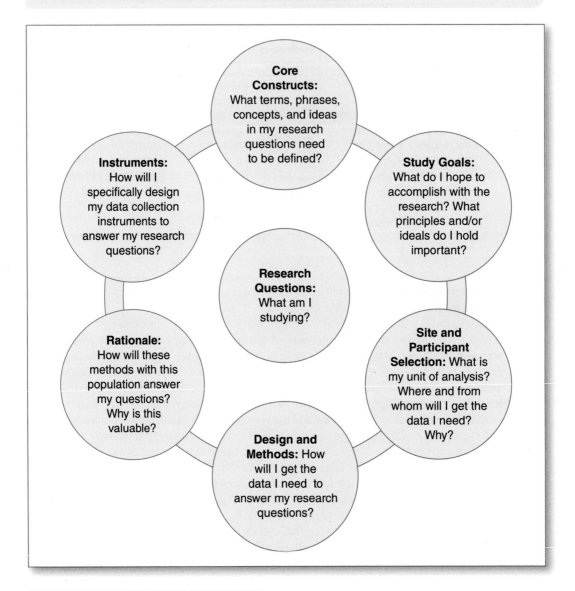

QUESTIONS FOR REFLECTION

- What are the key values and characteristics of the different methods and techniques for generating qualitative data, including interviews, observation and fieldnotes, focus groups, questionnaires, documents and archival data, and participatory methods?

- What are the considerations for and strengths and challenges of each of these qualitative data collection methods?

- How can you critically examine the data you generate and the ways that you, as the researcher, influence these data?

- What are the considerations, strengths and challenges, and the roles and uses of technology and social media in data collection?

- What does it mean that research questions are central to research design and data collection?

- What can you do to ensure that your methods align with your research questions?

In the next chapter, we describe how achieving qualitative rigor is directly related to issues of validity. We explore strategic sequencing of the data collection methods described in that chapter as well as broader strategies and methods that can be used to address validity and trustworthiness as you collect and analyze data.

RESOURCES FOR FURTHER READING

Data Collection

Glesne, C. (2016). *Becoming qualitative researchers: An introduction* (5th ed.). Boston, MA: Pearson, Allyn & Bacon.

Guest, G., Namey, E. E., & Mitchell, M. L. (2013). *Collecting qualitative data: A field manual for applied research*. Thousand Oaks, CA: Sage.

Miles, M. B., Huberman, A. M., & Saldaña, J. (2014). *Qualitative data analysis: A methods sourcebook* (3rd ed.). Thousand Oaks, CA: Sage.

Patton, M. Q. (2015). *Qualitative research and evaluation methods* (4th ed.). Thousand Oaks, CA: Sage.

Sapsford, R., & Jupp, V. (Eds.). (2006). *Data collection and analysis* (2nd ed.). London, UK: Sage.

Interviews

Brinkmann, S., & Kvale, S. (2015). *InterViews: Learning the craft of qualitative research interviewing* (3rd ed.). Thousand Oaks, CA: Sage.

Chilisa, B. (2012). *Indigenous research methodologies*. Thousand Oaks, CA: Sage.

Gubrium, J. F., Holstein, J., Marvasti, A. B., & McKinney, K. D. (Eds.). (2012). *The SAGE handbook of interview research: The complexity of the craft*. Thousand Oaks, CA: Sage.

Josselson, R. (2013). *Interviewing for qualitative inquiry: A relational approach*. New York, NY: Guilford.

Kvale, S. (2007). *Doing interviews: The SAGE Qualitative Research Kit*. Thousand Oaks, CA: Sage.

Patton, M. Q. (2015). *Qualitative research and evaluation methods* (4th ed). Thousand Oaks, CA: Sage.

Rubin, H. J., & Rubin, I. S. (2012). *Qualitative interviewing: The art of hearing data* (3rd ed.). Thousand Oaks, CA: Sage.

Tanggaard, L. (2009). The research interview as a dialogical context for the production of social life and personal narratives. *Qualitative Inquiry, 15,* 1498–1515.

Weiss, R. S. (1994). *Learning from strangers.* New York, NY: Free Press.

Observation and Fieldnotes

Angrosino, M. (2007). *Doing ethnographic and observational research: The SAGE Qualitative Research Kit.* Thousand Oaks, CA: Sage.

Bernard, H. R., & Ryan, G. W. (2010). *Analyzing qualitative data: Systematic approaches.* Thousand Oaks, CA: Sage.

Emerson, R. M., Fretz, R. I., & Shaw, L. L. (2011). *Writing ethnographic fieldnotes* (2nd ed.). Chicago, IL: University of Chicago Press.

Guest, G., Namey, E. E., & Mitchell, M. L. (2013). *Collecting qualitative data: A field manual for applied research.* Thousand Oaks, CA: Sage.

Focus Groups

Barbour, R. (2007). *Doing focus groups: The SAGE Qualitative Research Kit.* Thousand Oaks, CA: Sage.

Guest, G., Namey, E. E., & Mitchell, M. L. (2013). *Collecting qualitative data: A field manual for applied research.* Thousand Oaks, CA: Sage.

Krueger, R. A., & Casey, M. A. (2015). *Focus groups: A practical guide for applied research.* (5th ed.). Thousand Oaks, CA: Sage.

Morgan, D. L. (1997). *Focus groups as qualitative research* (2nd ed.). Thousand Oaks, CA: Sage.

Patton, M. Q. (2015). *Qualitative research and evaluation methods* (4th ed.). Thousand Oaks, CA: Sage.

Stewart, D. W., Shamdasani, P. N., & Rook, D. W. (2007). *Focus groups: Theory and practice* (2nd ed.). Thousand Oaks, CA: Sage.

Document Review

Bogdan, R. C., & Biklen, S. K. (2006). *Qualitative research for education: An introduction to theories and methods* (5th ed.). Boston, MA: Pearson.

Brinkmann, S. (2012). *Qualitative inquiry in everyday life: Working with everyday life materials.* Thousand Oaks, CA: Sage.

Lee, R. M., Fielding, N., & Blank, G. (2008). The Internet as a research medium: An editorial introduction to *The SAGE handbook of online research methods.* In N. Fielding, R. M. Lee, & G. Blank (Eds.). *The SAGE handbook of online research methods* (pp. 3–20). London, UK: Sage.

Rapley, T. (2007). *Doing conversation, discourse and document analysis: The SAGE Qualitative Research Kit.* Thousand Oaks, CA: Sage.

Questionnaires

Fink, A. (2006). *How to conduct surveys: A step-by-step guide* (3rd ed.). Thousand Oaks, CA: Sage.

Mathers, N., Fox, N., & Hunn, A. (2009). Surveys and questionnaires. The NIHR RDS for the East Midlands/Yorkshire & the Humber. http://www.rds-yh.nihr.ac.uk/wp-content/uploads/2013/05/12_Surveys_and_Questionnaires_Revision_2009.pdf

Patten, M. L. (2001). *Questionnaire research: A practical guide* (2nd ed.). Los Angeles, CA: Pyrczak Publishing.

Participatory Methods of Data Collection

Chevalier, J. M., & Buckles, D. J. (2013). *Participatory action research: Theory and methods for engaged inquiry.* New York, NY: Routledge.

Cooke, B., & Kothari, U. (Eds.). (2001). *Participation: The new tyranny?* London, UK: Zed Books.

Chilisa, B. (2012). *Indigenous research methodologies.* Thousand Oaks, CA: Sage.

Dudwick, N., Kuehnast, K., Nyhan Jones, K., & Woolcock, M. (2006). *Analyzing social capital in context: A guide to using qualitative methods and data.* Washington, DC: The International Bank for Reconstruction and Development/The World Bank.

Kindon, S., Pain, R., & Kesby, M. (Eds.). (2007). *Participatory action research approaches and methods: Connecting people, participation and place.* London, UK: Routledge.

Mukherjee, N. (2002). *Participatory learning and action: With 100 field methods* (Studies in Rural Participation 4). New Delhi, India: Concept Publishing Company.

Tolman, D. L., & Brydon-Miller, M. (Eds.). (2001). *From subjects to subjectivities: A handbook of interpretive and participatory methods.* New York: New York University Press.

Internet-Based Data Collection

Denissen, J. J. A., Neumann, L., & van Zalk, M. (2010). How the Internet is changing the implementation of traditional research methods, people's daily lives, and the way in which developmental scientists conduct research. *International Journal of Behavioral Development, 34,* 564–575.

Fielding, N., Lee, R. M., & Blank, G. (2008). *The SAGE handbook of online research methods.* Thousand Oaks, CA: Sage.

Paulus, T. M., Lester, J. N., & Dempster, P. G. (2014). *Digital tools for qualitative research.* Thousand Oaks, CA: Sage.

Salmons, J. (2014). *Qualitative online interviews: Strategies, design, and skills.* Thousand Oaks, CA: Sage.

Salmons, J. (Ed.). (2011). *Cases in online interview research.* Thousand Oaks, CA: Sage.

ONLINE RESOURCES

Sharpen your skills with SAGE edge

Visit **edge.sagepub.com/ravitchandcarl** for mobile-friendly chapter quizzes, eFlashcards, multimedia resources, SAGE journal articles, and more.

ENDNOTES

1. Some qualitative studies follow structured interview instruments (protocols), but this tends not to be the norm.

2. We discuss and complicate the notion of informed consent in Chapter Eleven.

3. It is important to keep in mind that representative samples are neither the norm nor the goal of qualitative research.

CHAPTER SIX

Validity: Processes, Strategies, and Considerations

CHAPTER OVERVIEW AND GOALS

This chapter presents the concept and related processes of validity (also referred to as trustworthiness) as more than a set of procedures or techniques but rather as considerably different from the quantitative paradigm from which the concept of validity was originally adapted. Validity is an active methodological process, a central value of qualitative research, and a research goal. Validity needs to be attended to from the research design phase through data collection to analysis and writing up your findings.

We begin this chapter by providing historical and present contexts for the concepts of validity and trustworthiness, defining the terms, discussing debates about their place in qualitative research, and describing the different criteria for engaging in and producing studies that are considered valid and trustworthy. Next, we specifically detail qualitative methods and techniques to achieve validity (we refer to validity throughout the chapter and book for reasons that we explain here). We conclude the chapter with a discussion about rigor and validity in qualitative research.

BY THE END OF THIS CHAPTER YOU WILL BETTER UNDERSTAND

- What it means and entails to conduct valid qualitative research
- The various terms and concepts that relate to processes of validity
- Different processes and considerations related to conducting valid qualitative studies
- Specific strategies and techniques you can design and implement to help work toward validity, including triangulation, participant validation, the strategic sequencing of methods, thick description, dialogic engagement, multiple coding, and structured reflexivity practices
- The relationship between qualitative research design and validity
- The various debates about validity and trustworthiness in the field of qualitative research

Engaging critically in qualitative research includes the recognition that validity should not be approached as a checklist of "technical fixes" (Barbour, 2001). Rather, it should be understood as a central, complicated, and challenging goal that should be met with and shaped by a conceptual framing of what constitutes rigor and validity given the complexity of qualitative research and a set of systematic, methodological processes that seek to reach that goal (with an understanding that it is imperfect). We view validity as a value, goal, and complex set of intersecting processes, which is why the concept is discussed throughout this book and not solely in this chapter. At the same time, we believe that setting up and conducting rigorous and valid qualitative studies requires specific knowledge and skills and therefore devote this chapter to a focused discussion of these concepts and processes. We share in this chapter, and indeed the whole book, the caveat that researchers can and must work toward validity but that this is a complex concept and an ultimately elusive objective given the layeredness of qualitative research and the complexity of the lives and contexts it seeks to understand and represent.

OVERVIEW OF VALIDITY AND TRUSTWORTHINESS IN QUALITATIVE RESEARCH

The concept of validity is often confusing for novice researchers, which is in part because there are many terms for the notion such as "authenticity, goodness, verisimilitude, adequacy, trustworthiness, plausibility, validity, validation, and credibility" (Creswell & Miller, 2000, p. 124). In addition to the multiple terms that are seemingly used interchangeably, some qualitative researchers reject the concept of validity, asserting that it is not compatible with qualitative research (e.g., Kvale, 1995; Lather, 1993; Wolcott, 1990) and that it is borrowed from quantitative research and therefore based on epistemological frames incongruent with qualitative values. When qualitative research was garnering acceptability as it developed as a field in the early 1970s, the positivist research paradigm was still regarded as the "gold standard," and the standards of quantitative research were used to develop qualitative research constructs. In the 1980s, Lincoln and Guba, two seminal scholars in qualitative research, placed concerns about the standards and criteria used to judge the quality of qualitative research at the forefront of debate in the field (e.g., Guba, 1981; Guba & Lincoln, 1981b, 1989; Lincoln, 1995; Lincoln & Guba, 1985). They attempted to define appropriate criteria that better accorded with a naturalistic inquiry/constructivist paradigm rather than within a positivist one, specifically examining the differences between the two paradigms on the nature of reality, the nature of the inquirer-"subject" relationship, and the nature of "truth statements." Guba and others set in motion what has become a complex and diffuse set of debates not just about the terms *validity* and *trustworthiness* but, more robustly, about these issues broadly and their place within debates about qualitative rigor, validity, and value.

Validity, in qualitative research, refers to the ways that researchers can affirm that their findings are faithful to participants' experiences. Put another way, validity refers to the quality and rigor of a study. A discomfort with the positivist origins of the concept of validity and its emphasis on objective truth has led some methods scholars to use the term *trustworthiness* as an alternative, while others use the terms interchangeably or use other terms such as *quality* or *rigor.* As described above, the debate is not merely a semantic one, but rather it pushes into questions about epistemological traditions. While the field has not settled on a shared understanding or term, the terms *validity* and *trustworthiness* are most commonly used and evoke the importance of ensuring credibility and rigor in qualitative research. We do not want to get caught up with the specific word that one uses—*validity* or *trustworthiness*—to describe the processes and approaches that qualitative researchers use to assess the rigor of qualitative studies. Whichever terms you use, the concept of developing valid and trustworthy studies is paramount in qualitative research. Throughout this book, we use the term *validity* since it remains the prominent word in the field and since we believe it has been reframed and reclaimed within the qualitative tradition, although we acknowledge the problems with this term. Furthermore, we frame validity in ways that place a primacy not simply on the specific validity concepts or procedures used to attempt to achieve it but also on doing justice to the complexity of research participants' experiences and thoroughly contextualizing their lives, perspectives, and experiences in ways that help to present the most complex and therefore valid renderings possible.

Given the growing prominence of qualitative research within and across fields, and given that it is increasingly scrutinized in relation to proposals for funding and policy mandates, issues of validity are more central than ever (Barbour, 2001). Here we highlight two distinct approaches to issues of validity—"transactional validity" and "transformational validity" (Cho & Trent, 2006). Transactional validity includes techniques and attempts to achieve a "higher level of accuracy and consensus by means of revisiting facts, feelings, experiences, and values or beliefs collected and interpreted" (Cho & Trent, 2006, p. 321). This approach to validity differs from transformational validity, which is "a more radical approach [that] challenges the very notion of validity" (Cho & Trent, 2006, p. 320). Transformational validity is an "emancipatory process leading toward social change" that "involves a deeper, self-reflective, empathetic understanding of the researcher while working with the researched" (Cho & Trent, 2006, pp. 321–322). The transformational approach to validity, and qualitative research more broadly, is informed by a critical social theory perspective in which the validity of the research is understood and assessed by the action it generates (Cho & Trent, 2006). These approaches to validity do not need to be dichotomous as "transformational approaches seeking ameliorative change can and should be combined, when deemed relevant by the researcher(s) and/or participants, with more traditional trustworthiness-like criteria" (Cho & Trent, 2006, p. 333).[1]

Regardless of the approach used, validity in qualitative research can never be fully ensured; it is both a process and a goal (Cho & Trent, 2006). Having a valid study cannot be achieved merely through the use of specific, technical strategies (Barbour, 2001; Cho & Trent, 2006; Maxwell, 2013). However, there are methods that researchers use to help increase the rigor, and thus validity, of qualitative research studies, and we describe these throughout this chapter.

ASSESSING VALIDITY AND TRUSTWORTHINESS

Validity is an approach to achieving complexity through systematic ways of implementing and assessing a study's rigor. As discussed previously, qualitative research demonstrates a fidelity to participants' experiences rather than specific methods (Hammersley & Atkinson, 2007); this is equally true for the concept of qualitative validity. In contrast with quantitative researchers, "qualitative researchers use a lens not based on scores, instruments, or research designs but a lens established using the views of people who conduct, participate in, or read and review a study" (Creswell & Miller, 2000, p. 125). The different lenses that shape the validity work of qualitative researchers include the lens of the researcher, of the research participants, and of other individuals external to the study. In addition to these lenses, specific criteria for assessing validity differ for qualitative researchers depending on the qualitative paradigm to which they subscribe (Creswell & Miller, 2000).[2]

Validity Criteria: Credibility, Transferability, Dependability, and Confirmability

Qualitative researchers should adhere to a set of different standards or criteria than quantitative researchers to assess validity or trustworthiness given the differences in values between the paradigms; these standards include credibility, transferability, dependability, and confirmability (Guba, 1981). While arguing for different qualitative standards, Guba (1981) juxtaposes these criteria onto the respective quantitative notions of internal validity, external validity, reliability, and objectivity.[3] These standards may be inadequate to assess rigor in many qualitative studies (Toma, 2011), and rigor can be assessed in many ways and need not parallel quantitative standards. We say this to emphasize that qualitative researchers should develop validity approaches that align with the research questions, goals, and contexts of their studies. Despite this, in this section, we define the commonly accepted concepts for assessing rigor in qualitative research, including credibility, transferability, dependability, and confirmability, because they help researchers to conceptualize, engage with, and plan for various aspects of validity.

Credibility is the researcher's ability to take into account all of the complexities that present themselves in a study and to deal with patterns that are not easily explained (Guba, 1981). This is akin to the quantitative notion of internal validity (Guba, 1981; Lincoln & Guba, 1985; Miles et al., 2014). Internal validity entails that "the researcher can draw meaningful inferences from instruments that measure what they intend to measure" (Toma, 2011, p. 269). In other words, in qualitative research, internal validity, or credibility, is directly related to research design and the researcher's instruments and data. The attempts to establish credibility are achieved by structuring a study to seek and attend to complexity throughout a recursive research design process, and the notion of credibility is a good example of the concept of "the inseparability of methods and findings" (Emerson et al., 1995). Credibility is an important part of critical research design. While there is not—and should not be—a checklist that can be applied for achieving validity, we present a set of questions that may be helpful to consider when thinking about the credibility of a study in Table 6.1. Qualitative researchers attempt to establish credibility by implementing the validity strategies of triangulation, member checking (what we think of and describe as participant validation), presenting thick description, discussing negative cases, having prolonged engagement in the field, using peer debriefers, and/or having an external auditor (Toma, 2011). We discuss these strategies in detail in the section that follows.

Transferability, which is juxtaposed with external validity (Guba, 1981; Lincoln & Guba, 1985) or generalizability (Toma, 2011), entails that qualitative research is bound contextually. The goal of qualitative research is not to produce true statements that can be generalized to other people or settings but rather to develop descriptive, context-relevant statements (Guba, 1981). In this regard, transferability is the way in which qualitative studies can be applicable, or transferable, to broader contexts while still maintaining their context-specific richness. Lincoln and Guba (1985) pose an important question that helps us to understand the concept of transferability and the parallel notion of external validity: "How can one determine the degree to which the findings of an inquiry may have applicability in other contexts or with other respondents?" (p. 218). Because a primacy is placed on fidelity to participants' experiences in qualitative research, it is important to understand that the goal of qualitative research is *not* to produce findings that can be directly applied to other settings and contexts. However, qualitative research can certainly be transferable to other contexts. Methods for achieving transferability include having detailed descriptions of the data themselves as well the context (also called thick description) so that readers/research audiences can make comparisons to other contexts based on as much information as possible (Guba, 1981). This allows the audiences of the research (e.g., readers, other researchers, stakeholders, participants) to transfer aspects of a study design and findings by taking into consideration different contextual factors instead of attempting to replicate the design and findings.

Dependability refers to the stability of the data. Dependability is similar to the quantitative concept of reliability (Guba, 1981; Lincoln & Guba, 1985). Qualitative research studies are considered dependable by being what Miles et al. (2014) describe as consistent and stable over time. Dependability entails that you have a reasoned argument for how you are collecting the data, and the data are consistent with your argument. In addition, this notion means that data are dependable in the sense that they are answering your research question(s). This entails using appropriate methods (and making an argument for why the methods you use are appropriate) to answer the core constructs and concepts of your study. The methods for achieving dependability are the triangulation and sequencing of methods and creating a well-articulated rationale for these choices to confirm that you have created the appropriate data collection plan given your research questions. As with the other validity constructs, a solid research design is key to achieving dependability.

Confirmability, which is often described as the qualitative equivalent of the quantitative concept of objectivity, takes into account the idea that *qualitative researchers do not claim to be objective* (Guba, 1981). Instead, qualitative researchers seek to have confirmable data and "relative neutrality and reasonable freedom from unacknowledged researcher biases—at the minimum, explicitness about the inevitable biases that exist" (Miles et al., 2014, p. 311). In other words, building on a foundational premise of qualitative research that the world is a subjective place, qualitative researchers do not seek objectivity; however, your findings should be able to be confirmed. Thus, one goal of confirmability is to acknowledge and explore the ways that our biases and prejudices map onto our interpretations of data and to mediate those to the fullest extent possible through structured reflexivity processes (such as the ones described throughout this book). Methods to achieve confirmability include implementing triangulation strategies, researcher reflexivity processes, and external audits (Guba, 1981). As described in Chapter Three, researcher positionality and bias are important aspects of qualitative research that must be scrutinized, problematized, and complicated. Because the researcher is viewed as a primary instrument in qualitative research (Lofland, Snow, Anderson, & Lofland, 2006; Porter, 2010), the researcher must challenge herself and be challenged by others in systematic and ongoing ways throughout all stages of the research. And this must be concertized within the research design itself.

Table 6.1 includes questions that are intended to help you begin to think about the validity constructs described here (as well as the additional constructs described in this chapter) in an ongoing and practical manner that should guide you from design through implementation and data analysis. We encourage revisiting your responses to these questions multiple times throughout a study in the form of memos or dialogic engagement exercises (see Chapter Three for a detailed discussion about the uses of memos and dialogic engagement in qualitative research).

Types of Validity

To address five main ways of approaching validity in your research, we turn our attention to the work of Maxwell (1992), who aptly describes five categories that he considers generative to understanding qualitative validity: descriptive validity, interpretive validity, theoretical validity, generalizability, and evaluative validity. We briefly describe these categories here since we view them as central to understanding the range and variation of concerns in the realm of validity.[4]

Descriptive validity refers to the factual accuracy of the data, and the other four categories described are dependent on this aspect. Important factors to consider related to Maxwell's category of descriptive validity include the recording and transcribing of interviews (e.g., Who recorded interview transcripts? Who transcribed them? Are there errors or omissions?) and the taking of fieldnotes (e.g., When—as in how close to the actual fieldwork—were jottings turned into fieldnotes? How close to the actual events are these recorded observations? Were there multiple observers, and do they each have fieldnotes?). Descriptive validity might sound quite transactional, but when you consider "the inseparability of methods and findings," it is extremely important that your data are accurate (Emerson et al., 1995). Descriptive validity underscores the way that validity is achieved throughout the entire research process because your analysis will be inherently flawed if your data are imprecise or incomplete.

Interpretive validity is the match between the meaning attributed to participants' behaviors and the actual participants' perspectives. This includes the accuracy of your analysis vis-à-vis the lived experience of the participants in your study. Methods for achieving this are closely related to data collection and analysis. For example, how you interpret someone in the moment of an interview and the follow-up questions you ask is one aspect of interpretive validity. In addition to interpretive decisions made during data collection, interpretive validity is affected by how you analyze the entire corpus of data. Methods for achieving interpretive validity include trying to use the words and concepts of the people studied; this is referred to as *emic* accounts in qualitative research. The contrasting concept to emic is *etic,* which entails that the language used to describe an account is not organic to the participants, meaning that it is introduced by the researcher. Descriptive validity can refer to both etic and emic terms while interpretive validity is concerned with emic notions. The concepts of emic and etic have important implications for validity and are vital to your ability to attend to contextualization and complexity in your data. It is foundational to think about people articulating their own experiences in qualitative research because everything else is interpretation. Revisiting the ever-important notion of the inseparability of methods and findings, interpretive validity is necessary to attend to in data collection, analysis, and representations.

Theoretical validity is the ability of a study to explain the phenomena studied, including its main concepts and the relationships between them. This concept explores the relationship between your empirical study, other empirical studies, and other theories that may or may not be empirically based. Theoretical validity is about the ability to have your data speak to existing theory and/or to have existing theories inform your data. Methods for achieving theoretical validity include ensuring that an applicable theory is provided and that it explains the data.

As described above, *generalizability* is not a recognized goal of qualitative research. However, Maxwell (1992) addresses this oft-asked question by pursuing the question of if the findings would exist with different data. Maxwell describes internal and external generalizability as related to the quantitative concepts of internal and external validity. For example, he asserts that internal generalizability, which he states is more useful to qualitative researchers, is whether generalizability exists within the community studied. This entails determining how individuals from the same community agree and understand their experiences. This does not mean that researchers are looking to generalize but rather ask questions about people's experiences vis-à-vis each other's experiences. We caution against using the term *generalizability* and find Maxwell's use of it somewhat problematic because a significant number of qualitative research studies are not looking for people's experiences to coalesce. In addition, as Maxwell also discusses, the sampling used in qualitative research is normally purposeful and not representative because generalizability is not the goal. As with all validity measures, researchers must determine appropriate and rigorous ways of assessing our studies.

Evaluative validity is whether the researcher is able to describe and understand data without being evaluative or judgmental. Evaluative validity, like generalizability, is not as central to qualitative research as descriptive, interpretive, and theoretical validity because many qualitative researchers do not attempt to evaluate what we are studying (Maxwell, 1992). Methods for achieving evaluative validity include paying careful attention to the words one uses given that language both reflects and generates meaning. For example, look at adjectives, descriptions of people's behaviors, and ways you frame things. Think about words and phrases such as "well-being," "healthy," "normal," or "risky." Some researchers might consider the use of such terms neutral; however, we argue that language is not neutral and, therefore, that qualitative researchers need to be very thoughtful and deliberate about our choice of words and must be able to explain why we make the choices we do.

Table 6.1   Reflexive Validity Questions

Validity Construct	Questions to Consider
Credibility	• What are ways to create a research design that not only seeks complexity but also attends to the real-life complexities that exist in any group, setting, community, and so on? • How do my site selection criteria and sampling strategies contribute to an authentic rendering of the context? • How do my methods align with my guiding research questions? • Have I designed my study so that the data set is rich and has multiple contributing data sources that complement each other? • How do I understand and engage with patterns that I am seeing in the data? • How will I interpret and make sense of my data so that my assumptions and biases are challenged? • What is the role/what are the roles of the research participants themselves in shaping the research and challenging my interpretations? • How am I going to connect the puzzle pieces?
Transferability	• How am I describing the contextual factors that shape and mediate my study? • Am I providing enough contextual data and framing for outsiders to be able to fully contextualize the study's findings? • How will I/do I interpret and make sense of my data in ways that are contextually embedded and authentic? • How do I describe the setting, participants, and specifics of the setting? Is there enough thick description? • Have I made the contextual relevance and embeddedness clear enough in my analysis and write-up?
Dependability	• Why and how did I choose my research methods? • Why are these the most appropriate methods to answer my research questions? • How do my methods map onto my research questions? • Have I designed the study in a way that assertively seeks rigor? How so? What might this be improved upon? • What might be challenged about my study design, data collection process, and/or analytic process? How can I proactively address these critiques and concerns? • Have I vetted/challenged my research design with critical friends and advisers to know what other methodological possibilities and limitations to the design may exist?
Confirmability	• Do I have my own agenda, and am I imposing that upon the data, thereby influencing the findings? If so, in what ways? How can I interrupt that? • Would someone else have similar conclusions/interpretations? How might they differ? • Whom can I engage within the research process itself to challenge my thinking? • At which points throughout the study should I seek out thought partners around issues related to my subjectivity and positionality?

Validity Construct	Questions to Consider
Descriptive validity	• Who is transcribing the data? Why? • What choices am I making related to transcription (e.g., verbatim transcription, pauses noted, punctuation intact, time intervals noted)? • How am I ensuring the accuracy of my data? • How are fieldnotes taken in terms of timing, structure, focus, and so on? • How might I get feedback on these aspects of my research design and the ensuing data set?
Interpretive validity	• What kind of language am I using? Emic and/or etic? What is the rationale for this? • How can I track the emic and etic language used so that it helps inform my data analysis? • How would the participants in the study feel and think about the language I am using? • Am I asking participants about the resonance and/or dissonance they may feel upon reading my account of their perspectives/experiences?
Theoretical validity	• What theories does my study speak to, build on, engage, and/or challenge and in what ways? • How did I select and determine these theories? Can I clearly explain my rationale? • Were the theories selected before data collection? How do I explain my rationale? • Might there be additional theories that I should consider during and after data collection? • With whom might I explore the role of theoretical framing in my study?
Generalizability	• How would members of the community I am studying interpret my findings? How can I try to explore this? • Are the data contextualized adequately so findings are connected to the context(s) that shape them? If not, how can I address this? • Am I using appropriately contextualized language so that claims do not seem sweeping or generalized in a problematic sense? • How can I resist claiming generalizability and yet still discuss the potential relevance of my findings for other contexts and studies?
Evaluative validity	• How would the participants in my study feel about the language I am using? How can I explore this? • Do I hold judgments that are invisible to me, and if so, how can I engage and challenge them? • How are these judgments influencing the data and my conclusions? • Are there judgments I hold that I am aware of but not engaging with critically, and if so, what are they and how can I challenge myself and get others to challenge me on these? • Who might I ask to challenge my judgments and how they influence the research? At which points throughout the process should I address these issues?

Each qualitative approach has additional validity measures that we will not go into because they are outside the scope of this book. However, we encourage readers to consult these depending on the qualitative approach(es) you choose to implement. We have included texts at the end of the chapter that can support further exploration of this topic as it pertains to specific studies.

Terms and Concepts Often Used in Qualitative Research

Thick description: Thick description is the way in which qualitative researchers describe a research setting in writing; the goal is for a researcher to accurately and thoroughly describe the important contextual factors. Thick description is an important aspect in increasing the complexity of your research by thoroughly and clearly describing the study's context, participants, and related experiences so as to produce complex interpretations and findings that allow audiences to make more contextualized meaning of your research. Thick description connotes a depth of contextual detail, usually garnered through multiple data sources, including observation and fieldnotes; it allows readers to have enough information and a depth of context so that they can picture the setting in their minds and form their own opinions about the quality of your research and your interpretations.

SPECIFIC STRATEGIES AND PROCESSES FOR ACHIEVING VALIDITY

Critically approaching validity involves recognizing that transactionally or procedurally employing validity strategies will not generate rigor. Technical strategies can assist in developing a rigorous study, but they alone do not guarantee rigor (Barbour, 2001). Merely implementing these strategies will not produce valid studies; they must be situated and understood in relationship to the inherently local and complex nature of qualitative research. Technical strategies should be combined with systematic understandings throughout research design and analysis.

While employing a specific strategy does not ensure a study is valid, there are many methods that qualitative researchers can use to help foster and support validity. In this chapter, we describe the specific methods of triangulation, participant validation (member checks), the strategic sequencing of methods, **thick description**, dialogic engagement (others refer to this as peer debriefers, critical friends, or critical inquiry groups), multiple coding, structured reflexivity practices, and mixed-method research. These can be used together, at various times, and should be thought of as options that can work independently or interdependently, depending on the specific research questions, goals, approach, methods, and the local and contextual factors of your research.

Triangulation

Triangulation is a set of processes that researchers use to enhance the validity of a study. It is commonly thought of as having different sources or methods challenge and/or confirm a point or set of interpretations. Broadly, triangulation involves "taking different perspectives" (Flick, 2007, p. 41)[5] or "examining a conclusion (assertion, claim, etc.) from more than one vantage point" (Schwandt, 2015, p. 307). Specifically, triangulation entails seeking "convergence among multiple and different sources of information to form themes or categories in a study" (Creswell & Miller, 2000, p. 126). These sources of information "can involve the use of multiple data sources, multiple investigators, multiple theoretical perspectives, or multiple methods" (Schwandt, 2015, p. 307). It is important to note that the goal of triangulation is not always to seek convergence. Researchers can learn much when perspectives differ (and we would argue, as discussed in more depth in Chapter Eleven, have an ethical responsibility to create the conditions for being challenged). The important point is that researchers should understand the crucial need to seek out and engage with multiple perspectives to answer our research questions.

There are multiple forms of and approaches to triangulation, and the processes of triangulation can include methodological triangulation, data triangulation, investigator triangulation, theoretical triangulation (Denzin, 1970/2009), and perspectival triangulation. Below we describe each of these forms of triangulation.

- *Methodological triangulation:* The triangulation of data collection methods includes within and between methods. Within-methods triangulation means that researchers use one method (e.g., interviews) but different strategies associated with that method (e.g., having different comparison groups in the interviews). Denzin (1970/2009) considers between-methods triangulation to be more robust as this includes using different methods such as observations and fieldnotes as well as in-depth interviews in ways that are generative to the overall data set.

- *Data triangulation:* The triangulation of data sources is similar to purposeful sampling in that "researchers explicitly search for as many different data sources as possible which bear upon the events under analysis" (Denzin, 1970/2009, p. 301). According to Denzin (1970/2009), data sources can be triangulated according to time (i.e., data collected at different times of the day), space (i.e., data collected at different places), and/or person (i.e., data collected from/about different people). This allows researchers to collect data using different sampling strategies and to examine data at varying times and places as well as with different individuals.

- *Investigator triangulation:* Investigator triangulation entails that there are multiple researchers involved in a given study. As discussed throughout the book, we encourage collaborative research because it has the potential to produce more complex data given the generative exchange between researchers. However, when tasks are simply delegated on research teams, rather than shared in an environment of professional exchange, this may not be useful from a triangulation perspective, and additional problems can arise. Ideally, this type of triangulation involves all researchers participating in and actively communicating about the same processes so as to increase validity and reduce bias (Denzin, 1970/2009).

- *Theoretical triangulation:* Theoretical triangulation is the inclusion of a range of theories to frame the study topic in context and is, actually, rarely achieved. However, Denzin (1970/2009) considers this an important aspect of validity. The triangulation of theory will help prevent researchers from coming up with atheoretical findings and selecting data to suit particular theories as well as encouraging researchers to broaden the relevance of studies by considering different theories.

• *Perspectival triangulation:* This form of triangulation is related to sampling/participant selection and is part of *data triangulation,* discussed in the second bullet above. This requires the intentional, systematic inclusion of a range of participant perspectives (e.g., people with various roles, occupations, or relationships within a group or setting as it pertains to the topic and goals of a study) to seek range, nuance, complexity, disagreement, and generative tensions in perspectives and in the data set as a whole.

The goal of using triangulation strategies is to consider if you have enough data and the right kinds of data to provide you with a quality and depth of information to answer your research questions confidently; it is about making sure you have engaged in a systematic and rigorous pursuit of trustworthy data collection that will allow for the most authentic and stable interpretations to emerge.

Despite its growing visibility in research studies, there are critiques of triangulation, which center on not viewing it as easy or as a panacea as it is sometimes presented. It is a complex process that, like all validity strategies, does not ensure quality and rigor. While triangulation may theoretically sound easy to implement, it can actually be difficult in part because "data collected using different methods come in different forms and defy direct comparison" (Barbour, 2001, p. 1117). Furthermore, "the production of similar findings from different methods merely provides corroboration or reassurance; the absence of similar findings does not, however, provide grounds for refutation" (Barbour, 2001, p. 1117). While we consider triangulation to be a crucial strategy (or set of strategies) to establish validity, it is important to acknowledge that engaging in triangulation does not necessarily make a study valid. The assumption underlying triangulation is that there is one reality instead of the possibility of multiple realities and perspectives. As stressed in the previous chapters, your research questions guide the methodological choices you make. A study that truly triangulates in some or all of the ways described above is also tentative and even critical of whether and how such triangulation contributes to the validity of the study even as you engage heartily in the design and implementation of triangulation (Blaikie, 1991). It is important to note that there is triangulation in data collection (as described in this section) and that there is triangulation in data analysis as well (sometimes referred to as analytic triangulation), which we discuss in more detail in Chapter Seven. We find that while students often design their studies well for triangulation in data collection, they do not engage in rigorous analytic triangulation, which is vital to the goals of these various forms of and approaches to triangulation and validity.

A final consideration related to these cautions about triangulation is the alternative concept of *crystallization* (Richardson, 1991). In this regard, rather than conceiving of three points of convergence in triangulation, a crystal offers multiple dimensions and perspectives. (See Richardson [1991, 1997] for more on the concept of crystallization.) Crystallization offers a less bounded concept in and a helpful metaphor that underscores the layeredness of the processes involved. While we do not use the term *crystallization,* our approach to triangulation is similarly flexible. Triangulation is a misnomer given that there are often more than three points of data or process points when incorporating triangulation.

Participant Validation Strategies (Also Known as "Member Checks")

Member checks is a term commonly used for how a researcher will "check in" with participants in his or her study. This is also called respondent validation (Barbour, 2001). We prefer the term *participant validation* since it connotes what we describe here as a more process-oriented and person-centered approach to challenging interpretations by creating the conditions for study participants to speak into and about a study. Member checks are often discussed as a validity measure to establish credibility, and Lincoln and Guba (1985) consider them the most important validity measure used to establish credibility, stating that "if the investigator is to be able to purport that his or her reconstructions are recognizable to audience members as adequate representations of their own (and multiple) realities, it is essential that they be given the opportunity to react to them" (Lincoln & Guba, 1985, p. 314).

Terms and Concepts Often Used in Qualitative Research

Participant validation strategies (member checks): Participant validation strategies, commonly referred to as *member checks,* are processes by which researchers "check in" with participants about different aspects of the research to see how they think and feel about various aspects of the research process and the parts of the data set that pertain to them. These strategies can be technical, including having participants verify the accuracy of statements and/or transcripts. These strategies can also include more relational approaches to engaging with participants to elicit their thoughts and responses to your interpretations and analytical concepts in more in-depth ways at various points throughout the research process. A goal of participation validation processes is to create the conditions that help you to explore and ascertain if you are or are not understanding participants' responses, how you are understanding them, and to be challenged on your data collection processes and your interpretations of the data.

Participant validation strategies, which can and should be engaged at a variety of stages throughout the research process, include some form of connecting with or "checking in" with the participants in a study to assess (and challenge) the researcher's interpretations and the accuracy of your analysis. There are different ways to engage in participant validation, including checking in to determine accuracy, engaging in sustained dialogue over time, and reflexively involving participants in collaborative processes (Cho & Trent, 2006, pp. 334–335). The way you involve participants will vary depending on the approach, goals, and research questions.

Potential questions to ask participants (depending on the nature and context of the study and the participants) include the following:

- Does this transcript reflect and resonate with your perspective? How might it differ and why?

- Is there anything that this transcript does not capture?

- Is there anything problematic in the interview and/or the transcript?

- What do you think about how I am interpreting your words and your story? Those of others in the study? Any specific concerns I can address related to this?

- Does my interpretation/description resonate with you (in what ways yes or no)?

- Is my interpretation/description authentic for you (in what ways yes or no)?

- What have I misunderstood?

- Is there anything that I am missing?

- Are there specific areas you would like to clarify? Add to?

- Is there anything you would like to suggest I consider in my analysis?

- Is this how you would categorize this idea/concept/comment?

- Do these codes make sense/resonate with you? Do these code definitions resonate? Why or why not? What can I understand you better as I code and analyze your words and the data set more broadly?

- Do my descriptions feel appropriate and accurate? If no, can you tell me more about whatever it is that I have not described well?

- Do these findings feel resonant to you? How might they feel more accurate and fitting?

- Are there assumptions and/or biases that you see underneath anything I have written or said that you feel I should challenge?

Terms and Concepts Often Used in Qualitative Research

Codes: In qualitative research, codes are tags or labels that researchers use (through a process called *coding*) to organize data into manageable units or chunks so that you can find, group, and thematically cluster various pieces of data as they relate to your research questions, findings, constructs, and/or themes across the data set. All data can be coded, including transcripts, fieldnotes, archival data, photographs, videos, research memos, and research journals. Once a researcher develops codes through specific processes of reading and organizing the data, codes are then defined succinctly (usually in a phrase or brief sentence). We discuss codes and coding processes in depth in Chapter Eight.

Participant validation can occur informally during an interview (or focus group) and/or can be structured more formally as a follow-up interview or a meeting (or set of conversations) to discuss emerging analytical constructs, codes, analytical themes, and findings with participants. In addition to being both structured/formal and unstructured/informal, the process of participant validation ideally occurs continuously (Guba, 1981;

Lincoln & Guba, 1985). There are many approaches to participant validation as well, although they tend to be talked about as a standard check-in about/in relation to an interview transcript. We suggest that researchers think creatively about a range of possible approaches to participant validation, including the standard interview transcript review, as well as more in-depth follow-up conversations or interviews that probe for participants' perspectives on a number of aspects related to the study's data and analysis. These more in-depth conversations or interviews are themselves a form of substantive data generation (as long as they are recorded and transcribed).

Participant validation, as noted above, can also inform data analysis as participants are invited to engage in reflection and critique of codes, emerging theories about the data, analytic findings, and so on. But there is an important caveat: For participant validation processes to be valuable requires that the researcher actively engage with and respond to the participants' critiques, amendments, interpretations, and suggested additions to the data set. This sometimes requires considerable reflection and even additional data collection and/or changes in the analysis of your data. These participant validation processes are ideally considered a part of the data set, and as such, they must be attended to and recorded systematically in the same ways as all other forms of data. We see all too often that students will include "member checks" in their study timeline at the end of a study when it is clear that there is no time to address what they learn from this process methodologically. This is a problem as it renders this step as less useful if useful at all (as well as potentially disrespectful to participants).

As stated above, participant validation strategies should be employed at multiple points in a study. When striving for complexity and fidelity to participants' experiences, it is imperative that the data that emerge from participant validation strategies inform the study in ongoing and substantive ways, which necessitates being intentional about when they are timed throughout the process. This means that participant validation techniques must be planned for and scheduled within the research design and its corresponding timeline. (We strongly urge researchers to be very careful about "time creep" with participants. This means not taking more of their time than requested and requires telling participants upfront all time commitments and processes in which you will ask them to engage and seeing those as a covenant not to be broken.) Furthermore, just conducting "member checks" does not ensure that your study is "valid" or "accurate." As we discussed at the beginning of this chapter, achieving validity and rigor does not entail checking things off of a list.

Participant validation is not a perfect process, and the power asymmetries involved in all aspects of research, including participant validation, should be acknowledged and reckoned with. For example, it is important to consider how a researcher affects the value of participant validation processes, not just when they are conducted. You need to strive to set up a dynamic in which a participant would feel comfortable challenging your interpretations and engaging in critiquing and the sharing of alternate perspectives. This includes that the researcher must be clear from the beginning that you are only an interpreter and therefore do not have interpretive authority over the narrative that is wholly unchecked (Nakkula & Ravitch, 1998). Furthermore, you should think about how to negotiate between potentially contrasting perspectives (between researcher and participants) (Barbour, 2001).

Despite these potential challenges, participant validation is an important process that should, when appropriate to the research design, be used to strengthen the rigor and validity of qualitative research. As with all aspects of research, it is critically important to respect participants' time and understand that some participants may not be willing to grant researchers additional time for such activities. In addition, issues of access may make participant validation processes difficult if not impossible. In Chapter Eleven, we address these issues in relationship to conducting ethical and respectful studies. Researchers should proactively consider issues that may arise and how they inform the participant validation choices in research design.

Strategic Sequencing of Methods

What we refer to as the *strategic sequencing of methods* is directly related to having a robust research design, as discussed in Chapter Three. This is an important consideration when attempting to achieve validity because of the variety of methods you can use to answer your research question(s), and the order in which these methods occur can affect both the nature and quality of data you are able to collect. Depending on the study's goals and research questions, researchers should pay careful attention to what we think of as *within- and between-methods sequencing*. For example, researchers using narrative or phenomenologi-cal research may need to conduct multiple, lengthy interviews to get the depth of exchange that these approaches require. The ways that interviews are structured, including sequen-tially in terms of the rest of the data collection plan, are important to the quality of the data and the validity of the study. For *within-methods sequencing,* you should consider how questions flow and follow each other in your data collection instruments so that you collect data in as contextualized and emic a way as possible (i.e., how are questions sequenced, are they clustered by theme, etc.). This involves a strategic building of information from the participant's perspective, so it requires thought about how conversation happens and how engagement is most natural and authentic. This also requires improvisation, since each conversation is different and unique. Examples of *between-methods sequencing* include how interview questions may be informed by observations that can help you to make the inter-view questions more contextually specific. For example, town hall meetings may be informed by ways that focus groups can help bring to the surface certain dynamics and tensions in groups. In addition, focus groups might be used to get a sense of the range and variation of viewpoints across a group for further in-depth interviewing. There are multiple ways to sequence methods; the important part is to remember that each of these choices should have a clear rationale and that those rationales should be carefully articulated in your data collection plan.

Depending on your research questions, framework, participants, and access, how you sequence the methods of your study will affect the quality of your data. As we have stated above, we equate validity with complexity. Thus, sequencing is about building the com-plexity of a study through a strategic collection of data that facilitates generative tensions in the data set, not necessarily having more data or following prescriptive checklists. An important aspect of criticality in research design and the notion of seeking complexity, validity, and trustworthiness is about the methods employed, the frequency with which they are used, and how they are sequenced in relation to each other. Conducting a rigorous and valid study that attends to complexity begins with a solid research design.

Thick Description

Thick description is an important aspect in increasing the complexity of your research by thoroughly and clearly describing the study's contextual factors, participants, and experiences so as to produce complex interpretations and findings, which in turn allows audiences to make more contextualized meaning of your findings. The strategy of thick description is proposed to enhance transferability (Guba, 1981) and credibility (Creswell, 2013). Understanding the concept of thick description can be confusing (Ponterotto, 2006; Schwandt, 2015). Many definitions contrast the concept of thick description with thin description, as Ponterotto (2006) notes, but this does not tell us what thick description actually is. Denzin (2001) offers this definition: "Thick description creates the conditions for thick interpretation. . . . Thick interpretation gives meaning to the descriptions and interpretations given in the events that have been recorded" (p. 117). Denzin summarizes the characteristics of thick description by stating that it builds on multiple and triangulated methods, is contextual and historical, captures individuals' experiences and meanings in a situation, "allows the reader to experience vicariously the essential features of the experiences that are described," and does not "gloss what is being described" (p. 117). Thick description involves researchers "describing and interpreting observed social action (or behavior) within its particular context," as well as providing clear descriptions of that context so that readers can understand participants' thoughts and feelings (Ponterotto, 2006, p. 543).

It is the role of the researcher to provide thick description so that thick interpretation can be made and then presented to readers (Ponterotto, 2006). Thick interpretation, which builds on thick description, "attempts to unravel the multiple meanings that are present in any interactional experience" and should be "meaningful to the persons studied" (Denzin, 2001, p. 117). Thus, thick description leads to thick interpretation and ultimately thick meaning of the study's findings for researchers, participants, and research audiences (Ponterotto, 2006). Thick description is often considered a strategy for ethnographers; however, we, like Ponterotto (2006), believe that it can be employed in all types of qualitative research. What thick description looks like in a study varies depending on the specific methods used as well as the study's goals and research questions. For example, in an ethnography, this may involve the use of extensive fieldnotes that describe the setting and participants in depth. In interview-based studies, thick description may involve contextualizing participants' responses so that readers can understand contextual factors in which quotes are presented and discussed. Context is incredibly important in qualitative research, and thick description is one way to help contextualize your data so that readers can determine the validity of your findings. For more on this concept, see Ponterotto (2006) and Denzin (2001).

Dialogic Engagement (Also Known as Peer Debriefers, Critical Friends, and Critical Inquiry Groups)

We discussed dialogic engagement at length in Chapter Three, and we revisit it here as a validity strategy that other texts may refer to as peer debriefers, critical friends, or critical inquiry groups. The goal of sharing your research with others is to create the conditions in which others (and yourself) can challenge your interpretations of the research process and data at all stages throughout the research process. In Chapter Three, we suggested specific activities for doing this, what we term *dialogic engagement practices,* and we recommend that you refer back to these practices throughout your study and not just during the initial research design phase. We consider dialogic engagement a crucial aspect of validity and believe that it is necessary to engage with, in structured ways, a range of individuals (e.g., stakeholders, colleagues, peers, mentors, friends, participants) who can challenge you on your assumptions, biases, and interpretations in a variety of ways and at many stages throughout a research study. Ideally, you might have a combination of people who would be considered insiders and/or outsiders[6] to a research setting. When engaging critically with others about your research, important questions to consider specifically around interpretation include the following:

- How are you interpreting:
  - The setting? The participants? The overall context?
  - Your research questions and the goals of your study?
  - The processes of framing the study and making your access/entry?
  - Your social identity and positionality and they relate with and shape the study design and process?
  - Participant engagement and the development of the research relationships?
- What shapes and mediates these interpretations?
- How are these interpretations unique to you? Influencing the research?

We discuss specific questions of interpretation related to data collection in Chapters Four and Five and to data analysis in Chapters Seven and Eight. However, we want to underscore the role of dialogic engagement experiences throughout all aspects of the research process—not just research design or data analysis—as crucial to fostering complex and rigorous, and therefore valid, qualitative studies. Of the validity strategies we describe in this chapter, we consider participant validation strategies and dialogic engagement exercises crucial to conducting not only rigorous but also ethical studies. Furthermore, these are necessarily intersecting constructs in the realm of qualitative research. We discuss the ethics of qualitative research more in Chapter Eleven but want to state clearly here that it is unethical, especially as a lone researcher (but in any research), to not have your assumptions and interpretations challenged at multiple points throughout the research process. Without such questioning, assumptions and interpretations go unchecked and can negatively affect the research in a variety of ways that may be seen or may be invisible to the researcher(s). Again, the notion of the "inseparability of methods and findings" (Emerson et al., 1995) helps us to understand that there are implications for our research to not engaging reflexively and collaboratively.

Strengths and Challenges of Dialogic Engagement

Interrupting normative and hegemonic thinking is no easy task, and this is why we encourage researchers to structure dialogic engagement as interruptive experiences that include opportunities for people to push against the grain of their socialization so as to recognize their biases (hooks, 1994). When you structure these dialogic experiences, think strategically about the people with whom you choose to have these experiences. Are you selecting them because of convenience and friendship, their relationship to the context, their expertise, and/or their ability to challenge you? Friends who will challenge you (often called "critical friends" in the qualitative literature) are important because you can feel safe enough to allow them to see your biases and prejudices (easier said than done) and to help you uncover your working assumptions and their influence in/on your research. However, friends may not serve that role for all researchers. It is very important that you think about and strategically design encounters in which you will have your research and your thinking truly challenged. Having the interruptive experiences that hooks (1994) discusses is often uncomfortable. However, these experiences are necessary not only for validity but also ethically. If you are comfortable throughout your research, you are probably missing something(s) that would help you to be a more ethical and tuned-in researcher. Raising our thresholds for discomfort and anxiety is a critically important part of reflexivity in research since facilitating change in our deeply held beliefs and biases often requires discomfort and even pain (Ravitch, 1998). As poet Khalil Gibran (1923) states, "Your pain is the breaking of the shell that encloses your understanding" (p. 52). While it may often feel more like discomfort than pain, some form of disequilibrium is necessary for true self-reflection to be integrated into your raison d'être.

Since we are all limited by our own subjectivities and biases, engaging with others is the *only* way we can truly challenge ourselves (Nakkula & Ravitch, 1998). For dialogic engagement to help achieve this, the selection of critical friends is of paramount importance in the sense that you must choose people who will actively challenge you and who will do so in ways that are constructively critical. This kind of engagement requires trust and a sense of confidentiality—that the people with whom you engage will honor that you are learning about yourself and making yourself vulnerable and possibly sharing things that show some of your more unflattering biases and assumptions. These kinds of relational considerations should be viewed and approached as central in the selection of your research thought partners. Given the prominence we place on the need for and value of dialogic engagement, as well as the broader concept and processes of collaboration (formal as well as informal) with any number of thought partners (e.g., peers, faculty advisers, mentors, research participants) throughout the research process, we want to describe some of the benefits and challenges of collaboration in qualitative research. While on its face, collaboration might seem wholly positive, you should problematize certain aspects of collaboration to understand how it does or does not positively influence validity. A more critical approach to collaboration includes being aware of its limitations and aspects that researchers should be cautious of as well as recognizing its benefits. Below we detail considerations and examples of concerns related to collaboration and dialogic engagement.

- Considerations should be made in relation to how power and authority influence the ways in which a researcher takes in feedback from advisers, peers, inquiry group members, and research participants. It can be difficult to see how our relationships and contexts of interaction affect how we understand feedback and negotiate between our perspectives and those of our thought partners. For example, when an adviser offers advice, it is usually weighted more than a peer. But perhaps it is at times the peer who is offering insights that we should place a primacy on?

- Researchers should pay attention to issues of interpretation and how various interpretations are considered and weighted, especially when there are differences or contradictions. For example, there may be times when a researcher's analyses result in things participants or critical friends find difficult to hear or believe to be incorrect—but that the researcher believes are supported by data/existing literature and theory.

- It can be confusing to know how to attribute and make sense of tense communications or issues that arise in collaborations. For example, if someone is continually challenged in ways that are emotionally difficult, is it defensiveness and/or perhaps an unhealthy alliance? This example speaks to how a researcher should negotiate and make good judgments about what is and is not actually a productive collaboration.

- There can be multiple kinds of issues when research participants challenge the researcher's processes, goals, and/or interpretations. On one hand, how can you pay careful attention to what might be useful in participants' suggestions when they challenge you? On the other hand, how can you keep a fidelity to your interpretations if you believe the participants might be influenced to retract or change what you believe to be a "truer" representation of their perspective?

- Considerations of how collaboration influences confidentiality are crucial. For example, sharing transcripts is one way to get multiple perspectives on your data and your interpretations of the data but presents challenges to matters of confidentiality.

This is just a sample of the kinds of issues that can emerge from engaging in collaborative dialogue and processes. However, we do not intend this to disincentivize collaboration. Rather, we hope it will help you to plan for, approach, and engage in collaboration in more constructively critical ways. Just as none of the other validity strategies in this chapter can ensure a study's validity, collaboration is not a panacea and must be complicated and problematized.

Multiple Coding (Also Known as Interrater Reliability)

The concept of "interrater reliability" or what we, Barbour (2001), and others refer to as "multiple coding" can serve as an additional validity measure especially for research teams, although as Barbour points out, it is not without its limitations. *Multiple coding* involves when a researcher, set of critical friends, and/or research team creates a set of codes and code definitions, and then each team member codes and analyzes a specific, agreed-upon number of transcripts (or other data sources). The goal of this process is to see how people are coding the data, specifically looking to see if interpretations overlap, intersect, and/or are divergent and in what ways. This concept is discussed in more depth in our discussion of coding and analysis in Chapter Eight; however, we mention it here as a validity strategy and to serve as another example of how validity is related to all aspects of the research process.

Multiple coding is one way that researchers approach the issue of the subjectivity of interpretation in an effort to mediate that at the analysis and coding level. Lone researchers can also achieve this by hiring independent research consultants or having individuals volunteer to help them assess how they are coding their data. Barbour (2001), in a critique of what she deems a too-superficial approach to improving rigor in qualitative research, describes how what is important about multiple coding is not "the degree of concordance between researchers" but "its capacity to furnish alternative interpretations and thereby to act as the 'devil's advocate'" (p. 1116). As with all validity strategies, multiple coding should not be seen as an easy solution to the concerns of interpretation and bias.

Structured Reflexivity Processes in Validity

We describe a multitude of structured reflexivity practices in depth in Chapter Three, including specific memos, dialogic engagement practices, research journals, mapping strategies, and so on. We encourage readers to consider such practices as validity strategies. Systematically and critically engaging with our biases, interpretations, processes, and reflections can help us to produce more complex and ethical research. These processes and planned experiences are integral to establishing the validity of a study, and that is why it is crucial to systematically structure these throughout your research process. Structured reflexivity processes are vitally important to conducting rigorous and valid studies and directly relate to the concept we discussed previously of the *researcher as instrument* (Lofland et al., 2006; Porter, 2010). If we, as researchers, do not actively and critically monitor and challenge our biases and positionality, the complexity and rigor of our studies, no matter how theoretically robust the design, will be undermined. In addition to critically and systematically examining your biases, you should be aware of and engage with other aspects of you as a researcher, including your research skills. Related to this, we present questions to consider in Table 6.2.

These are just a sample of possible questions, and you can engage with them in a variety of ways, including through the use of memos and dialogic engagement practices. Understanding the myriad ways that we, as researchers, affect our data is incredibly important not just as an aspect of validity but to the quality of an entire research study. All of the validity strategies will not produce "good" research without an acknowledgment of and systematic processes to interrogate the multitude of ways that researchers impact their research. This requires an openness to critical self-reflection, to approaching our research as what Hidalgo (1993), like Sanday (1976), talks about as being "introspective ethnographers"—that is, researchers who seek to understand the culturally embedded meanings and constructs of various people and places with an eye that not only looks outward with an inquisitive gaze but also that necessarily turns inward, in tandem, to critically explore and increasingly understand the self as the sense maker, interpreter, and situated other, to interrogate the cultural meanings that influence us, shape our beliefs and assumptions and biases and therefore all facets of our research.[7]

Table 6.2   Questions to Consider Related to Research Skills

- What is my communication style and why (i.e., what forces have shaped it and how might it differ from other people's)?
- How can/do various aspects of my communication style affect how I engage in data collection?
- What are some ways that my cultural and personal communication style might influence the research? The interviews specifically?
- What are the kinds of knowledge or information I tend to value and gravitate toward more than others and why? How might this influence what I notice or engage with more or less in the research?
- Do I tend to listen carefully? How might I get feedback on this and assess this during interviews?
- Am I present and engaged enough to ask good follow-up questions? How can I improve upon this?
- Do I assume too much understanding/familiarity with participants and why?
- Do I assume too much difference/distance from participants and why?
- How do I engage in interviews? Am I affirming? Do I share too much? Am I too distant? Do I seem judgmental? What shapes these choices and behaviors?
- What biases do I bring to early and ongoing data analysis? How do these shape the analysis process?

As discussed at length in Chapter Three, *memos* are useful tools for designing rigorous research, and having a robust research design is the first step in achieving validity. In addition to the memos we suggest throughout the book so far, we specifically detail a memo related to validity in Recommended Practice 6.1.

Recommended Practice 6.1: Validity/Trustworthiness Research Design Memo

The goal of this memo is to encourage you to systematically consider issues of validity at various points throughout your study. We recommend composing multiple iterations of this memo to continue to monitor validity as you progress through your study.

Potential aspects to consider include the following:

- Describe the research questions and goals.
- Discuss how and if you are addressing the following validity strategies:
  - Triangulation
  - Participant validation
  - Strategic sequencing of methods

  o Thick description

  o Dialogic engagement strategies

  o Multiple coding

  o Structured reflexivity practices

  o Mixed-methods research

- Articulate how each of the above validity strategies maps onto and informs your research questions and goals.

- Describe and explain any lingering questions you have regarding threats to the validity and rigor of your study.

- If any major changes have been made to your study, discuss these in relationship to threats to validity and/or validity strategies.

As an example of this kind of exploration of validity in a study, we include as Example 6.1, an excerpt from a dissertation written by Susan Bickerstaff.

Example 6.1: Validity Excerpt From a Dissertation Proposal

Excerpt From Susan Bickerstaff's Dissertation Proposal, *Challenging "Dropout": Why and How High School Dropouts Return to School*

Qualitative methodologists (e.g., Creswell, 1998; Denzin & Lincoln, 2003) long attempted, in many cases unsuccessfully, to divorce qualitative design from quantitative standards of rigor (i.e., generalizability, validity, and objectivity). In his review of the literature on standards in qualitative research, Toma (2006) identifies four alternative components of a rigorous qualitative research design: credibility, transferability, dependability, and confirmability. I have included several data validation techniques into the proposed research design in an effort to address these four components. "Credibility," writes Toma, "is established if participants agree with the constructions and interpretations of the researcher"

Annotation: Because this is an excerpt from a dissertation proposal, this example is more formal than other validity memos may be. As with all memos, they can vary in terms of formality depending on the intended audience and purpose.

(Continued)

(Continued)

(p. 413). Participants in research projects have traditionally been thought of as "subjects," people who can be studied but whose voices are rarely heard. During the final phase of my project, member checks and follow-up interviews will allow me to check my assumptions and enlist participants into a meaningful collaborative relationship in which my inferences are offered for feedback and critique. Johnson (2007) found that the analytic memos she gave to her participants as part of the member checks were too text-heavy and complex, so she distilled the memos into poems written in the voices of her participants. Her participants then responded to these poems, clarifying and correcting as necessary. During fourth-round interviews with participants, I may also employ a method known as Interpersonal Process Recall (Carspecken, 1996, p. 163). IPR, originally developed as a therapeutic approach, involves playing a tape of an interview or focus group back to the participant(s) and allowing them to pause the tape to comment on and interpret the data. In addition to serving as a form of member check, IPR offers a rich additional source of data that demonstrate how students' narratives change over time. If discrepancies between my interpretation and my participants' interpretation of the data cannot be resolved, both interpretations will be presented with commentary in the final analysis.

The second component of rigorous qualitative design is transferability. Toma (2006) likens this characteristic to generalizability in quantitative research. He suggests that researchers who connect their methods of collection and analysis to theory offer opportunities for readers to connect the findings to a larger body of work and therefore see the import of the study beyond the local context. I hope that by positioning this study within the literature on poststructural narratives, possible selves, and youth literacies, my work can speak to the issues for a wide range of students returning to education after leaving high school.

Dependability, Toma (2006) argues, can be achieved through a number of design features. He suggests that to be dependable "findings must go beyond a snapshot" (p. 416). Inherent to my theoretical framework is the assumption that narratives are constructed differently depending on the time and place of their construction. By following the focal students over 12 months, my hope is that I can identify which narrative features remain stable over time and which change as the students progress through the program. Triangulation of multiple data sources offers another assurance that a researcher's analysis is dependable and valid. Student interviews, focus groups, observations, and documents offer me a wide variety of data sources, which can be checked against one another. Careful attention to and analysis of discrepant data, in the final report, will lend both credibility and dependability to my project.

Confirmability "is the concept that the data can be confirmed by some-one other than the researcher" (Toma, 2006, p. 417). Beginning in Phase Three, peer debriefing will become a regular component of my data analysis. I will work with two fellow doctoral students, sharing sections of my low-level codes, reconstructive horizon analysis, and high-level codes, I can solicit feedback on the extent to which my inferences can be validated by the data. Carspecken's (1996) method of reconstructive horizon analysis in itself offers a source of confirmability in that it forces the researcher to record each step she makes between literal-level coding and high inference coding.

Finally, my ongoing reflexivity, both during data collection and analysis, will be essential to the "trustworthiness" (Lincoln & Guba, 2003) of my research design. I enter this site an outsider on many levels. I bring with me assumptions about the experiences of these returning students. My position as a "visitor" in the program, as a researcher, and potentially as a tutor, will affect the ways in which the students narrate their stories to me and the ways in which I interpret those stories. As I remain in the site over time and develop a relationship with each participant, those effects will diminish but never disappear. While the experiences and voices of the students are of primary importance, my own experiences in the site cannot be obscured.

To do so would imply a positivist orientation in which the findings I report are truth rather than a representation of student experiences. My criticalist stance demands I work to call attention to inequalities, injustices, and power imbalances. Part of that work must be a recognition of and admission of my own position of power. As I embark upon this project, I endeavor to represent the students and their stories as transparently and as responsibly as possible.

> Annotation: This example highlights how the researcher's position to the research topic and setting cannot be ignored. This is important for reasons related to validity as well as related to ethical considerations.

Mixed-Methods Research

Mixed-methods research designs, which strategically combine aspects of qualitative and quantitative methods, can be an additional way to seek qualitative rigor and validity *depending on the research questions, goals, and arguments you are trying to make.*[8] While there are researchers who believe that qualitative and quantitative methods are distinct and even incompatible, others believe it is possible for skilled researchers to use both. This is increasingly the case over time as researchers understand that while positivist and interpretivist paradigms have different assumptions about the world and the social processes and experiences within it and use different methods in their research, you should map methods onto research questions, and certain questions require both qualitative and quantitative data. Most researchers have come to understand that these methods are not exclusive.

To be clear, the way that employing a mixed-methods design can help establish validity depends on the study's research questions and goals. For example, you may use quantitative data as additional points of triangulation in relation to specific kinds of research questions that may require things such as pattern identification, frequency counts, and/or other kinds of causation-focused questions. However, as with all triangulation, qualitative researchers are not always looking for points of convergence or patterns. There is much to learn from data that speak to different things, and this strengthens validity because the research has the potential to become more complex.

The integration of the frameworks and methods of both qualitative and quantitative methods can allow for great insights to emerge. This is contingent on a clarity of the goals and uses of both kinds of data through a process in which you map out the research questions to align the data collection methods onto these guiding questions. The choice of quantitative or qualitative or the strategic combination of both is driven by the goals and guiding questions of each specific study. For more information about mixed-method research, see Creswell and Plano Clark (2011) and Greene, Caracelli, and Graham (1989).

RIGOR AND VALIDITY IN QUALITATIVE RESEARCH

Across the validity strategies described above, we assert that transparency in all aspects of the research process is not only necessary to conducting ethical qualitative studies but also an important aspect to achieving validity. In addition to being clear and transparent about all aspects of a study with participants (what we think of as *internal-facing transparency*), threats to validity and the presence of bias, which is present in all studies both quantitative and qualitative, should be clearly articulated in a final research project whether that is a paper, film, or other form of representation[9] (what we think of as *external-facing transparency*). In a research report, this may take the form of a limitations section, a section entitled "researcher roles and issues of validity," a section on reflexivity and positionality, and/or across several of these sections. Not all researchers include such sections in final reports, perhaps in part because they fear it will make their data and their findings appear to be questioned and therefore less valuable. However, the trustworthiness of a researcher (and therefore the study) increases when researchers transparently discuss the limitations of the study and their interpretations. Thus, in the same way that Hammersley and Atkinson (2007) refer to a "fidelity" to participants' perspectives rather than to specific methods or hypotheses (as is often the case in more traditional, positivist research), we also argue for the researcher to have a fidelity and accountability to the integrity of the research process achieved through these various kinds of approaches to research rigor.

In addition to transparency, collaboration is a central aspect of validity and rigor that we have discussed throughout this chapter and book. Collaborative processes may not be a formal aspect of the methodological approach of every study, but multiple kinds of collaboration with colleagues and peers should be an intentional and systematic part of every study. If you are conducting participatory research, it is extremely important to establish structured processes for collaboratively conceptualizing, engaging in processes of, and assessing validity as joint data collection and analysis as well as ongoing, collaborative, formative conceptualization, implementation, and assessment of the research process. Even for researchers not conducting participatory research, involving the stakeholders in "checking" your data, interpretations, and analyses can help to create more rigorous and valid studies. Related to our discussion of collaborative processes is the common, although we believe at times misguided, practice of lone qualitative research wherein there is one researcher and nothing built into the design that challenges his or her interpretive process and the relationship between reflexivity and validity. We consider collaboration as a vitally important measure to achieving validity, and this collaboration may take multiple forms depending on the specific study. When analyzing data, which we focus on in Chapters Seven and Eight, there are additional validity measures to consider that can be accomplished through collaborative involvement and/or external auditors. Of course, these can also be accomplished by lone researchers, but as we have discussed previously, the complexity (and therefore validity) increases when data and analysis are subjected to multiple interpretations.

As detailed throughout this chapter, there is a reciprocal relationship between rigor and validity, and both are central to the integrity of qualitative research. Validity is an active and iterative process of achieving research rigor that relies on you, as the researcher, to make reasoned and grounded decisions that faithfully attend to the complexity of participants' experiences.

QUESTIONS FOR REFLECTION

- What makes a qualitative research study valid/trustworthy?

- What are the primary critiques of validity in qualitative research?

- How will you address the different validity constructs and their respective considerations in your research?

- What strategies can you implement to increase the validity of your studies?

- How are these strategies important and useful as well as problematic?

- How is qualitative research design related to validity?

- How is a study's validity related to its rigor?

In the chapter that follows, we continue to address issues of validity and rigor as they apply to the process of qualitative data analysis.

RESOURCES FOR FURTHER READING

Barbour, R. S. (2001). Checklists for improving rigour in qualitative research: A case of the tail wagging the dog? *British Medical Journal, 322,* 1115–1117.

Cho, J., & Trent, A. (2006). Validity in qualitative research revisited. *Qualitative Research, 6*(3), 319–340.

Ellingson, L. L. (2009). *Engaging crystallization in qualitative research: An introduction.* Thousand Oaks, CA: Sage.

Golafshani, N. (2003). Understanding reliability and validity in qualitative research. *The Qualitative Report, 8*(4), 597–606.

Kvale, S. (1995). The social construction of validity. *Qualitative Inquiry, 1*(1), 19–40.

Marshall, C., & Rossman, G. B. (2016). *Designing qualitative research* (6th ed.). Thousand Oaks, CA: Sage.

Maxwell, J. A. (1992). Understanding and validity in qualitative research. *Harvard Educational Review, 62*(3), 279–300.

Maxwell, J. A. (2013). *Qualitative research design: An interactive approach* (3rd ed.). Thousand Oaks, CA: Sage.

Mays, N., & Pope, C. (1995). Rigour in qualitative research. *British Medical Journal, 311,* 109–112.

Mays, N., & Pope, C. (2000). Assessing quality in qualitative research. *British Medical Journal, 320,* 50–52.

Toma, J. D. (2006). Approaching rigor in applied qualitative research. In C. F. Conrad & R. C. Serlin (Eds.), *The SAGE handbook for research in education: Engaging ideas and enriching inquiry* (pp. 405–423). Thousand Oaks, CA: Sage.

ONLINE RESOURCES

Sharpen your skills with SAGE edge

Visit edge.sagepub.com/ravitchandcarl for mobile-friendly chapter quizzes, eFlashcards, multimedia resources, SAGE journal articles, and more.

ENDNOTES

1. Our thanks to Justin Jimenez for his help in understanding the etiology of the concept of validity.

2. See Guba and Lincoln (1994) for a discussion of these qualitative paradigms, which include postpositivist, constructivist, and critical.

3. Other scholars (e.g., Toma, 2011) use the term *generalizability* instead of *external validity.*

4. For a more detailed description of each category, see Maxwell (1992).

5. For a detailed history and description of the concept of triangulation, see Flick (2007).

6. As we discussed in Chapter One, we do not want to create binaries around positionality topics such as insider/outsider. However, we use the terms because they can be helpful to think about the various ways you and others are situated in relationship to your research.

7. See Erickson's (2004) "Culture in Society and in Educational Practices" for a wonderful discussion that expands and complicates static notions of culture.

8. To be clear, we do not see mixed-methods research as a criterion for validity; it is a validity strategy *only* when appropriate to the specific study itself.

9. Forms of representation are described in Chapter Nine in more detail.

CHAPTER SEVEN

An Integrative Approach to Data Analysis

CHAPTER OVERVIEW AND GOALS

Data analysis tends to be an aspect of qualitative research that is the least understood, and it is often thought of as a nebulous process (some of our students call it a "black hole of confusion"). While, as we discuss in this chapter and the next, there is no "right" way to conduct data analysis, and it is often a "messy" endeavor, it should not be nebulous.

As we describe throughout the book, being intentional and transparent in all of your research processes is incredibly important. This is especially true for data analysis. Transparency helps to establish rigor and validity, as others are able to ascertain and assess your methods and ultimately understand how you arrived at your findings. Transparently describing your analytical processes is part of critically approaching data analysis. Adopting a criticality when analyzing data, which we describe throughout the next two chapters, involves a number of actions that help you to acknowledge and address the inherent power you have as a researcher and faithfully ensure that your interpretations are as accurate as possible.

We begin this chapter with a discussion of what data analysis is and distinguish between the acts of analysis and interpretation. We then provide an overview of the different analytical approaches to qualitative data analysis and describe what it means to approach data analysis critically. This chapter also details an *integrative approach to qualitative data analysis*, which includes the formative and iterative nature of qualitative data analysis and underscores the value of developing a particular kind of intentionality around interpretation broadly and formal data analysis processes specifically. This includes working from an understanding of the interpretative power in/of data analysis and the need for an ethical and critical approach to data analysis that fully respects participants and seeks to understand and do justice to their lived experiences.

BY THE END OF THIS CHAPTER YOU WILL BETTER UNDERSTAND

- What qualitative data analysis is

- How analysis and interpretation differ and how they relate

- The range and variation of different qualitative approaches to analyzing data

- What an integrative approach to qualitative data analysis entails

- How qualitative data analysis is iterative and recursive and what this means for how you engage in data analysis

- How qualitative data analysis is formative as well as summative

- The ways in which qualitative data analysis requires data and theory triangulation

- How the power differentials within data analysis are an ethical and methodological concern

- The ways in which qualitative data analysis is a process of seeking out alternative perspectives

Terms and Concepts Often Used in Qualitative Research

Qualitative data analysis: Qualitative data analysis encompasses the processes that qualitative researchers employ to "make sense of their data." Broadly speaking, data analysis is understood to include a variety of structured processes for looking across your data set to identify and construct analytic themes and, ultimately, turn these themes into what are commonly referred to as "findings" that help you to answer your research questions. Qualitative data analysis is the intentional, systematic scrutiny of data that occurs ongoingly throughout the research processes. This analysis of data often involves the specific processes of *data organization and management, immersive engagement with data,* and *writing and representation*. These processes are discussed in depth in Chapter Eight. An integrated approach to qualitative data analysis, which we discuss in this chapter, involves understanding that qualitative data analysis is iterative, formative, and summative; should be based on data and theory triangulation; addresses issues of power; and seeks out alternative perspectives. A challenge when analyzing qualitative data is to engage with and make sense of a significant corpus of data in a process that carefully reduces the amount of data, identifies significant patterns in those data, and does so in a way that allows you to construct an analytic framework for communicating the essence of what your data reveal (Patton, 2015).

DEFINING AND CRITICALLY APPROACHING QUALITATIVE DATA ANALYSIS

Data analysis is often described as both an art and a science; it is a crucial component of qualitative research because data cannot speak for themselves (Schwandt, 2015; Willig, 2014). Qualitative data analysis is the intentional, systematic scrutiny of data at various stages and moments throughout the research process. This scrutiny often involves the specific processes of *data organization and management, immersive engagement with data,* and *writing and representation.* These processes make up our three-pronged approach to data analysis, and we describe each of them in depth in the next chapter. In this chapter, we describe what qualitative data analysis is, how analysis is related to other aspects of the research process, and how analysis is generally approached.

Data analysis relates to all aspects of qualitative research and should not be thought of as isolated at one summative moment. We consider this ongoing analysis part of the iterative and recursive nature of qualitative research and as a central aspect of validity. We discuss this aspect of analysis in more depth toward the end of this chapter, as a part of our integrated approach to qualitative data analysis. However, we highlight this here because data analysis is all too often conceptualized and discussed as occurring in separate processes or distinct stages. Part of critically approaching qualitative data analysis involves recognizing that qualitative data analysis begins as soon as the first piece of data is collected and continues throughout the entire research process. Conceiving of data analysis as iterative and recursive underscores how all aspects of qualitative research directly impact your study's findings, including your views and assumptions, how you design your study, the literature you review, the data you collect, how you analyze your data, and then how you communicate your processes and findings to an audience. Critically and ethically approaching data analysis necessitates that you think about (and articulate for your readers) how your role in the creation of data affects the arguments you ultimately make. For example, consider how instrument questions were asked and how the wording and framing of questions influenced how participants answered them (and then your analysis of those answers).

A challenge when analyzing qualitative data is to engage with and make sense of a significant corpus of data in a process that carefully reduces the amount of data, identifies significant patterns in those data, and does so in a way that allows you to construct an analytic framework for communicating the essence of what your data reveal (Patton, 2015). The traditional notion of qualitative data analysis is that researchers collect a large amount of data (although size is relative) and, upon completion of data collection, then analyze those data and write up their findings. While this is a summative process of data analysis that many qualitative researchers ultimately follow, it overlooks several important aspects and components of data analysis, including the iterative and recursive nature of the qualitative research cycle of a study and formative as well as ongoing data analysis. These processes are vital to holistic and rigorous qualitative research studies. This cyclical nature of qualitative data analysis commences at the beginning of data collection and continues throughout as the process of collecting data transitions into summative data analysis once all of your data are collected. This iterative and recursive nature of qualitative data analysis is part of what we mean by critically approaching data analysis. We detail additional aspects of a more critical approach to qualitative data analysis in Table 7.1.

Table 7.1   Considerations for Critically Approaching Qualitative Data Analysis

Critically approaching qualitative data analysis involves

- Acknowledging the iterative, recursive, and ongoing nature of data analysis
- Paying careful attention to participant and researcher language, context(s), and perspectives
- Understanding the role of and engaging in formative analysis in the further development of data collection instruments and to understand emergent themes so that they can be attended to within the data collection process
- Understanding the relationship between the various data sources/data collection processes and the nature, content, and scope of the data set
- Engaging in triangulated—in terms of data and theory—data analysis processes
- Seeking out and engaging with disconfirming evidence in generative ways
- Engaging with thought partners in your investigation into your data and the various influences on data interpretation with attention to researcher identity, positionality, and assumptions
- Attending to issues of **interpretive authority** in systematic ways in an effort to resist its imposition to the extent possible

Terms and Concepts Often Used in Qualitative Research

Interpretive authority: Interpretive authority in qualitative research refers to the power of the researcher to be the interpreter and translator of people's lived experiences and perspectives. This becomes especially problematic when researchers believe that they have the "true" or "correct" version of someone's story. There is not a single truth or reality in qualitative research, and out of respect to participants and their lived experiences, it is important to acknowledge that there is inherent power in all forms of research. Thus, qualitative researchers should systematically acknowledge and reckon with this power and attempt to resist it as much as possible both through their own individualized reflexive engagement and through creating the conditions and processes of dialogic engagement in which assumptions, biases, and interpretations are rigorously challenged. We discuss this throughout this chapter, present specific recommendations in Table 7.3, and revisit the topic again in Chapter Nine.

We encourage you to approach data analysis in a way that is structured, yet also fluid and flexible. As we discuss throughout this and the next chapter, there are many broad and specific analytic approaches, methods, and techniques to help you analyze your data. However, the goal of data analysis should be to maintain fidelity to participants as experts of their own experiences. Understanding the connection between data collection, analysis, and findings is crucial to conducting valid, rigorous, and ethical qualitative research, and it is an important part of critically approaching qualitative research.

ANALYSIS AND INTERPRETATION

As we begin this chapter, we want to distinguish between data analysis and interpretation. *Analysis* and *interpretation* are often used synonymously; qualitative research is commonly referred to as "interpretive research," and given the interpretive nature of qualitative research, there can be a conflation of the terms *analysis* and *interpretation*. Broadly, interpretation is a way that individuals make sense of their world. Interpretation occurs all the time; it is a natural part of being human.[1] Related to qualitative research, interpretation includes how a researcher explains, understands, and/or represents study participants and their lives. While interpretation is an endemic part of qualitative data analysis, data analysis is distinct from and more specific than the broader sphere of interpretation in that it involves engaging in intentional and systematic processes of interpreting data. In this regard, qualitative data analysis refers to deliberate, systematic, and structured acts of interpreting data and then describing data in ways that reflect both process and insight.

It is important to our understanding of criticality in qualitative research that researchers acknowledge that individuals are interpreting all of the time, including when we, as researchers, are observing, engaging in interviews, and reading through data. While some texts discuss "bracketing"[2] your personal opinions, biases, and/or evaluative language (bracketing refers to formally creating visual ways of [bracketing] personal reactions to the data in real time), we believe that this is incredibly difficult to do in ways that actually achieve the desired goals. Even when we think we are being objective, there is subjectivity inherent in all acts of qualitative research (and in all aspects of life more broadly), and even as we try to explicate or parse out our subjectivity, much alludes us. Systematic efforts to become more conscious of, and track how, subjectivities are at play in research are useful (even if not wholly effective in the ways that some suggest they are), but they need to be considered carefully.

The notion of *intentionality* in data analysis is why the distinction between analysis and interpretation is important. Intentional, systematic data analysis is crucial to conducting valid and rigorous qualitative studies that present findings that uphold a fidelity to the data and therefore to people's experiences and perspectives.

Analysis, then, can be thought of as how you analyze your data set knowing that subjectivity is deeply embedded within your data as well as within how you interpret those data. Your subjectivity can be harnessed as a vibrant part of the analytic process since it is embedded in all layers and phases of the interpretive processes that constitute the various aspects of qualitative research—from the development of a topic and research questions through data collection to the final writing up of your study—and surely throughout the iterative cycle of data analysis. Understanding the relationship of broad interpretive processes to specific analytic procedures—and the role of subjectivity across these domains—is an important part of an approach to qualitative data analysis that is based in intentionality and criticality of thought and process. We thread this notion throughout the remainder of the chapter.

OVERVIEW OF QUALITATIVE DATA ANALYSIS PROCESSES

Qualitative data analysis encompasses the processes that qualitative researchers employ to "make sense of their data." Broadly speaking, data analysis is understood to include a variety of structured processes for looking across your data set to identify and construct analytic themes and, ultimately, turn these themes into what are commonly referred to as "findings" that help you to answer your research questions. Furthermore, you may use your data-based findings to contribute to the ongoing dialogue in the fields that are related to your study's domain of inquiry.

The process of qualitative data analysis has multiple, intersecting phases and is nonlinear in ways that we describe throughout this chapter. Importantly, while the methodological approach that you use (e.g., grounded theory, phenomenology, case study, narrative research, general qualitative analysis techniques) will shape your specific analytic process, there are shared features across most approaches to qualitative data analysis. These shared features include concurrently collecting and analyzing data, composing memos throughout data collection and analysis, coding data in some form, using writing as an analytical tool, developing analytical ideas and concepts, and connecting your analysis to prior theory and literature (van den Hoonaard & van den Hoonaard, 2008, p. 186). These activities tend to occur in the overlapping phases of reduction, reorganization, and representation of data (Roulston, 2014).

A goal of data reduction is to determine the "phenomena of interest," that is, the key phenomena that you seek to explore within your data corpus in relation to the questions and goals of your study (Roulston, 2014, p. 304). This can be determined by beginning with an open-coding process that narrows the data from your corpus to create a data set that you will continue to analyze. However, the specific research questions and research approach will influence the way researchers approach this phase. Data reduction may involve carefully reviewing your data to determine what is of interest to your specific research questions and topic. For example, within a phenomenological approach, "researchers reduce data by eliminating repetitive statements and data irrelevant to the phenomenon being examined" (Roulston, 2014, p. 304). Data reduction often occurs in tandem with data reorganization in which data are grouped and categorized (or coded) according to specific features, commonalities, differences, or other aspects of the data. The data reorganization phase includes comparing codes and categories to begin to determine themes and preliminary findings. We have discussed how qualitative research is iterative and recursive, but this is especially true in data analysis. As you begin to determine emerging themes, you recheck these themes with your data and search for alternative explanations. In the representation phase, researchers develop arguments based on the data and previous research and represent their data in various ways, including using quotes from transcripts to support themes, diagrams and other visual representations, and narratives of participants' experiences (Roulston, 2014).

Different researchers have different approaches to data analysis, and we describe our specific approach in the next section and in depth in the next chapter. Despite specific differences based on the approach, the processes of condensing and organizing data so that you can engage with the data as you develop interpretations are central across most approaches to qualitative research. Whether you use the terms of *reduction, reorganization, and representation,* Miles et al.'s (2014) terms of *data condensation, data display,* and *conclusion drawing/verification,* or other related terms, understanding that these processes are nonlinear and ongoing is a key feature of an integrative approach to qualitative data analysis.

While most data analysis approaches include variations on data reduction, organization, and representation (Roulston, 2014), there are important differences in the various approaches to qualitative data analysis as well as considerations and challenges that exist within and across these approaches. When you hear researchers discuss qualitative data analysis, you might hear them say that they are using grounded theory, narrative analysis, or some other specific methodological approach to analysis. However, a large number of qualitative researchers, especially novice qualitative researchers, employ what are thought of as general qualitative approaches, and we devote the bulk of this chapter and the next to more general qualitative analysis for that reason. Still, we overview the most common data analysis approaches in Table 7.2 to help you gain a broad sense of the different analytical approaches and urge you to see the resources at the end of this and other chapters to further your knowledge of specific analytic approaches.

Table 7.2 Overview of Qualitative Approaches to Data Analysis

Approach	Analytical Focus
Phenomenology	• Focuses on the experience of the participant(s) in an examination of shared experience • Analysis remains as close to data as possible
Discourse analysis	• Systematic focus on the language used • Analysis focuses on the ways that meaning is created through language
Grounded theory	• The goal is to generate theories from the data • Employs inductive processes that stay close to the data through a constant comparison method
Ethnography	• Like grounded theory, the analytical focus is not theoretically driven and employs inductive processes • The goal is to derive contextualized cultural meaning making and gain an insider perspective of a phenomenon through prolonged immersion in a setting
Action research	• Rejects the notion of expert and engages a group process of data analysis • Places primacy on participants' perspectives • The goal is to engage in analysis of data in real time to influence change
Narrative analysis	• Focuses on individuals' storied experiences • The specific focus can be on the content or the structure of the stories as narratives
Thematic analysis	• Develops themes to answer research questions • Used in multiple methodological approaches

Source: Information in the table is summarized from Willig (2014).

Despite the specific analytic approach or combined approaches, all researchers make conscious (and unconscious) decisions about the roles of the researcher, participants, data, and theory. Willig (2014) describes the need for researchers to be clear about decisions related to data:

> Every study makes assumptions about the type of knowledge it seeks to produce and it is given direction by the types of questions which it asks of the data. Every study needs to be clear about what 'status' it attributes to the data, that is to say, what it wants the data to tell the researcher about. In this sense, every qualitative study, irrespective of which specific method is used, interprets its data because the data never speak for itself. It is always processed and interrogated in order to obtain answers to particular questions, to shed light on a particular dimension of human experience and/or to clarify a particular aspect of an experience or a situation. (p. 147)

Whether you use a specific methodological approach or a general qualitative approach, it is important to understand that data do not speak for themselves. Table 7.2 is intended to provide the broad brushstrokes for each of these approaches but is therefore necessarily reductive and incomplete. Entire books are devoted to data analysis for each of these approaches, and there are many different ways to approach analysis and different analysis methods within each approach. For further reading on the different data analysis approaches, see Willig (2014), Creswell (2013), and Miles et al. (2014).

More general approaches to qualitative data analysis typically employ thematic analysis, and as such, we briefly overview thematic analysis here. Furthermore, many analytical approaches include work with themes (Braun & Clarke, 2006; Gibson & Brown, 2009). *Thematic analysis* involves noting relationships, similarities, and differences in the data. A theme reflects important concepts in the data and is often understood as "a generalized feature of a data set" (Gibson & Brown, 2009, p. 129). The generalized nature of themes is a phenomenological critique of thematic analysis because it may not include the particularities of individuals' experiences (Gibson & Brown, 2009). However, as we have asserted throughout the text, context and fidelity to participants are paramount in qualitative research. Themes do not necessarily need to reflect patterns and commonalities. Many factors influence what ultimately become themes, and central to these are your research questions (Braun & Clarke, 2006). For example, depending on the questions you ask, you can create a nuanced account of one or two themes instead of an overall depiction of your entire data set (Braun & Clarke, 2006). Furthermore, the researcher, as the primary instrument in qualitative research, actively constructs and develops themes through various processes. A key aspect of critically approaching qualitative data analysis is to make transparent the processes and activities that helped you, as the researcher, determine themes from the data. Themes do not simply emerge from data (Braun & Clarke, 2006).

In the sections that follow, we describe our overall approach to data analysis, which can be applied to a variety of analytical approaches. In the next chapter, we describe specific processes and procedures for analyzing qualitative data.

AN INTEGRATIVE APPROACH TO QUALITATIVE DATA ANALYSIS

By *integrative approach to qualitative data analysis,* we mean an approach that iteratively integrates within and across design, data collection, data analysis, and theoretical framework building throughout the various stages and activities of your research project. This requires engaging a specific kind of criticality in your approach to the entire data analysis endeavor. This includes that the researcher

- keeps a focus on the integration of formative data analysis into the research process to ongoingly shape data collection;

- integrates across the various data sources to see connections, disjunctures, and opportunities for further exploration and inquiry (sometimes referred to as analytic triangulation);

- engages with related theory in formative and inductive ways that challenge thinking and help conceptualize what is happening in the data;

- engages in reflexive and collaborative processes that challenge interpretations and analytical procedures;

- understands that the overall framing of and approach to the data analysis process is intentional and systematic as well as creative and emergent; and

- uses the ever-emerging conceptual framework as a guide and ballast for data analysis processes and emerging interpretations of the data so that integrative connections can be made between the research questions, the goals of the study, the context(s) that shape the study, the data, relevant formal theory, researcher identity and positionality, and study methodology.

Central to this kind of approach is understanding that your data inform your overall research design and ongoing data collection. Furthermore, an integrative approach to data analysis highlights the way the methods that you use are directly related to the findings you will develop. An integrative approach to data analysis involves understanding how all aspects of the research process shape the nature, scope, and content of your data set and is a vital aspect of the data analysis process. This approach to data analysis requires that you engage in the process of looking across data sources in triangulated ways that help you to get the most complex and valid picture of the phenomenon and perspectives in focus. Thus, an integrative approach to qualitative data analysis involves, or integrates, what we think of as five key framing notions for data analysis:

(1) Qualitative data analysis is an iterative and recursive process.

(2) Qualitative data analysis is formative as well as ongoing and summative.

(3) Qualitative data analysis requires data and theory triangulation.

(4) Qualitative data analysis necessitates the recognition and address of the power differentials within data analysis as ethical and methodological concerns.

(5) Qualitative data analysis is a process of seeking out alternative perspectives.

In the sections that follow, we describe each of these five key framing notions for an integrative approach qualitative data analysis. We view the combination of these framing concepts as central to engaging critically in qualitative data analysis.

Qualitative Data Analysis Is an Iterative and Recursive Process

First, *qualitative data analysis is an iterative and recursive process*. This foundational and generative concept involves understanding that data analysis begins, as an active and ongoing process, from the moment the first piece of data is collected. Furthermore, it requires an understanding that each subsequent analytic process informs and builds on the other in ways that, if paid attention to, can help you to see the layers and complexities in your data and ultimately critically inform and ground your findings. It is important to understand the ways that your methods—meaning the broad methodology as well as the various data collection methods and then each data source—are directly related to the findings since the data constitute the material on which your findings are based. Thus, understanding the relationship between the study instruments, the data they do (or do not) produce, and data analysis is an additional component of what we mean by an iterative and recursive approach to data analysis. The analytical procedures that are employed in data analysis should stem from a fidelity to people's lived experiences, meaning that all methods should seek to garner data that are individualized, contextualized, and as rich as possible. Methodologically, this comes from constructing rigorous qualitative studies beginning with your research questions. Thus, methods and findings are directly linked in that the analytical approaches you use depend on the research questions, instrument questions, and what you seek to learn with/from individuals. Engaging in data analysis that is iterative means that you harness these various data sources and processes as vital parts of a meaning-making process and do not parse them out of your analysis process and the findings to which the process leads you. Moreover, it means that you see their refinement as emergent and responsive to what you are learning in real time. It is only through ongoing data analysis that such insights and design improvements can be cultivated.

A primary goal of qualitative data analysis is to be focused on and authentic to what study participants actually say, how they say it, and from within which contexts they share particular thoughts or experiences. Related to this is the need for analysts to be transparent about the ways questions are worded and how the researcher and situation shapes responses (data). The fidelity must be to participants and the information they share; this means that researchers should try to resist the urge to assume meaning, make sweeping inferences, or overassert what can be learned from partial engagements. For example, interviews and observations are only snapshots of lives already in process and far beyond our grasp given the limits of what we know from our data collection methods. This does not mean that qualitative researchers need to be apologetic; however, it does mean that we, as researchers, need to be careful not to generalize or assume beyond the scope of our data.

Understanding the connections between data collection, analysis, and findings is crucial to conducting valid, rigorous, and ethical qualitative research. Seeing all of this as a process of constructing meaning, in ways that are messy and generative, helps you to understand that data analysis is layered and that the process of coming to your findings should be intentional even if dispersed over time and data source. The iterative nature of data analysis is directly related to how it is also formative, which we discuss in the next section.

Qualitative Data Analysis Is Formative and Summative

Second, *qualitative data analysis is formative and summative*. As described above, qualitative data analysis is often conceived of as a summative process that occurs after data have been collected. While summative data analysis is clearly an important aspect of analyzing data, it is only one aspect. Formative data analysis is the analysis of data that commences as soon as they are collected. Approaching data analysis formatively entails understanding and engaging in multiple analytic processes, including conceptualizing the ways that data analysis should drive the refinement of research questions and data collection instruments. For example, you may conduct a handful of interviews and start to look across those transcripts to see how you might adjust your instruments, how you might adjust your research design, and/or how you might tweak your research questions. Specifically in terms of instrument refinement, which is a key reason for early data analysis, after reviewing the transcripts from these interviews, you look across people's responses, and you notice that certain instrument questions are not effective. Perhaps they were worded awkwardly, did not get at the core constructs in your research questions, and/or prompted responses that were too general, lacked context, or were not individualized or specific enough to be useful. You might also notice that you are running out of time in many or most of the interviews and therefore will need to prioritize certain questions or even take some out to have time for follow-up questions and prompting for examples and explanation/unpacking of language. These kinds of adjustments require formative data analysis.

Formative analysis (also referred to as preliminary or early analysis) begins with the refinement of your methods (and sometimes the research questions) and continues throughout the entire research process as data are collected. However, the formative usefulness of analysis is not just because it provides the insights you need to tweak and refine your methods and instruments; formative data analysis also means that you are starting to get a sense of what you are learning from the data. Given its range of useful and vital roles in the qualitative process, it is important to systematize formative data analysis processes. In addition to reflexive processes such as writing memos, there are other strategies in which you can engage to formatively analyze your data, including *piloting*. Even if you are not conducting a full-scale pilot study, the practice of rehearsing and piloting your instruments as well as vetting them by a number of thought partners (these processes are described in depth in Chapter Three) are other ways that you can formatively analyze your methods and preliminary data to refine and adjust the study design and/or study instruments as necessary.

Formatively analyzing your data often leads to a refinement of the research questions and methods for data collection. Revising research questions may happen for a variety of reasons, including data from early interviews that signal that you are not getting the data you need to answer your research questions. It could be because you do not have the right research questions, or you realize that the questions are cast incorrectly. In addition, this could happen when you engage in data collection and analysis and realize that the scope of your research questions is too broad or not broad enough. The message here is that you should engage in data analysis from the beginning of your study until the very end and, throughout the process, look for all the various ways that your data can teach you about the methodology, participants, and the substantive focus of the study.

Throughout the data analysis process, we recommend paying close attention to and thinking about the following questions:

- How did I ask the various questions in my interviews/focus groups/questionnaires? Based on the data from within and across these data sources, could this be improved upon and, if so, how?

- In what ways does how I asked these questions possibly relate to or shape how participants answered them? What does that help me consider in the revision of instruments and in my early and ongoing data analysis?

- Did I ask enough/the right follow-up questions to really understand what each participant meant in contextualized ways? If not, what should I change/add/refine?

- Who introduced and placed emphasis on various central concepts and terms? How will understanding this help me further collect and engage with the data appropriately?

- How do I analytically understand my role in the creation of data so that the arguments that I make are valid and authentic?

- How do the various stages of data collection and analysis intersect and inform each other?

- How did early data analysis inform subsequent data analysis? How did that shape or inform the findings?

Data and Theory Triangulation in Qualitative Data Analysis

The third key framing notion for qualitative data analysis is the concept that *qualitative data analysis requires data and theory triangulation*. People often refer to and think of triangulation in data collection, which we argue is crucial. There is also triangulation in data analysis (sometimes referred to as analytic triangulation), which can mean both theoretical triangulation (using multiple theoretical perspectives to ground, make sense of, and challenge emerging learnings from your data) and data triangulation (juxtaposing and looking across the data sources for tensions and ways the various data challenge and/ or support emerging theories) as a part of the analytic process. We argue that these can be intertwined in important ways that strengthen your analysis and findings. As discussed in Chapter Six, triangulation involves bringing together multiple perspectives to answer your research questions (Flick, 2007). It can include the triangulation of methods, data sources, investigators, theories (Denzin, 1970/2009), and perspectives. In this chapter, we focus solely on the triangulation of data sources and theories since they are central to qualitative data analysis.

Analytical Data Triangulation

Data triangulation involves examining data sources collected at different times, places, and/or about/with different people (Denzin, 1970/2009). Here we extend this idea of data triangulation into the data analysis process, which we term *analytical data triangulation*. This means that as you analyze your data, you do so through an approach that explores and integrates across the various data sources to help round out and challenge your understanding of the participants and their perspectives/experiences as individuals, within groups and subgroups, and in relation to the data sources themselves. This requires that you create a careful data analysis plan. (See Appendix M for an example of a data analysis plan.) This plan details how you will sort and read the data in ways that juxtapose data sources and examine the data in what we think of as "analytic slices" (the image we have is of standing above a tree that was sliced open and seeing the rings from a variety of vantage points that each tell a different story of the history and experiences of the tree). One example is that Sharon has a large, hollowed, intricately carved gourd that was made by an elder artisan in Otavalo, a largely indigenous town in the Northern Imbabura Province of Ecuador, where she engaged in teaching and applied research for some time. As he gave her this gourd, the artisan told her that the carvings (see Image 7.1), which are each quite intricate depictions of scenes of everyday life in his community with narration in the local dialect, could be understood in several different ways. There are 40 small panels of carvings on the gourd, and this man told her that she could read it as 40 mini-stories, 3 mid-range stories (one for each of the three rings of panels), or one ongoing story with 40 scenes. This has always remained in Sharon's mind as an example of how you can look for different relationships in and frames for data and that the organization of the analysis affects what is learned.

To take a different example, if you are conducting a study that seeks to understand the television-watching habits of adults and their relationship to people's social identities, you would organize the data in different analytic slices to explore various relationships and possible ways of viewing and understanding the data. These may include sorting and analyzing the data by age, region, profession, levels of education, and any other way you wish to explore connections between social identity and television watching as your focal research question. It could also mean that for all participants you would look across all data sources in which they engaged (e.g., their interview transcript, television-watching log, and their parts of a focus group transcript).

Engaging with analytical data triangulation requires that you consider and organize analysis of the data in ways that help to view it from various analytic angles. These processes resist a more simple and traditional approach to data analysis, which is to review all data chronologically. Organizing the data so that you can explore the various data sources within and across participants and other organizing constructs is a vital aspect of learning your data set and then creating arguments about what is going on in the data.

Image 7.1   Representation of Alternative Frames for Analysis

Analytical Theoretical Triangulation

In Chapter Six, we define theoretical triangulation as using more than one theory to interpret and make sense of your data. This helps prevent you from developing findings that are either atheoretical and/or conform to particular theories. *Analytical theoretical triangulation* involves understanding the back and forth between theory and data as well as your own sense-making and the sense-making of your colleagues and peers with whom you discuss what you are learning from the data. This intentional interplay between data and theory can be a somewhat "messy" in that it is not a straightforward, linear, step-by-step process. Maxwell (2013) discusses some of the risks of stacking the deck too much with formal theory, but for most qualitative researchers, it is important to engage with at least enough theory from related fields to inform your study and methods but not too much so as to make conclusions before you begin the process. This can be tricky as you need to find a balance between using theory to drive and ground a study, using it to support emerging learnings, and using theory to frame and/or challenge your study findings.[3]

To think about how to analytically triangulate and complicate the way you engage with theory, it is helpful to consider how theory is used. Qualitative researchers tend to develop middle-range theories, challenge preexisting theoretical understandings, and/or apply middle-range theories to new areas (Kelle, 2014).[4] This engagement with theory can take the form of "top-down" theory, which uses deductive practices and preestablished theoretical concepts to analyze data, or "bottom-up" theory, in which concepts are inductively created from the data (Gibson & Brown, 2009). Like Gibson and Brown (2009), we do not dichotomize these concepts, as most qualitative researchers use both inductive and deductive approaches. We encourage you to think of theory as helpful in allowing you to see concepts and relationships that you might not have noticed. In this way, theory is "a means of working with data in some *particular* and *motivated* way" (Gibson & Brown, 2009, p. 11). You may hear qualitative researchers discuss having "a theory" that they are using. This can seem overwhelming for novice researchers, and students have often come to us lamenting that they do not have a "theory" that explains their data. While we assert that theoretical triangulation is crucial for developing sound findings, this does not mean that you should feel pressure to work with a preexisting theory or develop a new theory. Rather, "it is usually better to think about the ways in which very specific *theoretical components* may be used for analyzing data than to worry about the requirement of having 'a theory', which can sound rather daunting" (Gibson & Brown, 2009, p. 12). Situating your analysis within theoretical components that you speak into and help to illuminate different aspects of your data will help to create analytically complex and robust findings.

Recognizing and Addressing Power Asymmetries Within Qualitative Data Analysis

Fourth, as with all research, there is significant power involved when interpreting individuals' experiences and narratives. Critically approaching data analysis involves a *recognition and address of the power differentials within data analysis as an ethical and methodological concern.* In addition to this notion, explicit steps to resist impositional analysis are crucial. Reckoning with the power asymmetry inherent in most forms of data analysis[5] and, ultimately, the representation of other people's experiences and perspectives is a critical aspect of data analysis. While data analysis cannot be sufficiently achieved through checklists (Barbour, 2014), there are ways that you can rigorously analyze data throughout the research process to attend to issues of power and interpretive authority. We detail these strategies throughout this chapter and present specific considerations in Table 7.3.

It is an act of power to interpret someone else's reality and tell their story (Lather, 1986; Nakkula & Ravitch, 1998). You must think very carefully about this throughout the entire research process but especially in analysis and representation. We specifically discuss this notion in relationship to writing and representation in Chapter Nine. In addition to addressing interpretative power and other issues related to bias in data analysis, you must also consider how the data were collected, whose voices were left out, and the specific methods used and their relationship to the extant data. An additional consideration to keep in mind when thinking about representing other people's lives is that data are co-constructed and not simply "collected," and therefore researchers must account for their role in the creation of data, including its limitations (Roulston, 2014). It is important to be transparent about your interpretations and to acknowledge that there are assumptions underlying and shaping your interpretations and that you also may make inferences, which should be named as such. We, as qualitative researchers, can never fully mitigate impositional interpretation, but we can understand that this is a concern and a complex issue in our research and be honest and transparent in our processes and acknowledge our limitations.

There are no easy answers for to how to resist enacting your interpretative authority in problematic ways; as with all decisions, including the language you use, be transparent that the claims you are making are your interpretive claims. Describe where the codes and themes in your study came from and how they were developed, including if they were emically or etically derived (Maxwell, 2013). To resist the imposition of interpretative authority entails both transparency and integrity regarding how you use and define concepts. This makes having a clear understanding of the analytical constructs in a study, how you define them, and how study participants define them quite important. There is no way to avoid imposing your interpretive authority when making sense of other people's lives and experiences; however, we describe a few strategies in Table 7.3 to help you resist the uncritical imposition of this authority.

We will describe in more detail several of the ideas above as they relate to issues of analytic writing and representation in Chapter Nine, as we discuss writing up your themes and findings and making claims about others' perspectives and experiences in final reports.

Table 7.3   Considerations for Trying to Resist Interpretive Authority

- When possible, include as much raw data as possible (carefully selected and analyzed of course) while remembering that data do not speak for themselves.
- Actively engage in structured reflexivity processes and dialogic engagement exercises (see the practices in this chapter as well as Chapter Three for additional examples of these).
- Actively engage in participant validation (see Chapter Six) throughout your analysis processes so that your participants can give feedback on and challenge your emerging and final analyses.
- Avoid generalizations and statements devoid of context.
- Stay as close to the data as possible in your interpretations and analytic write-up (findings).
- Actively engage in validity strategies such as triangulation, participant validation, the strategic sequencing of methods, thick description, dialogic engagement, and structured reflexivity practices (see Chapter Six for specific descriptions of these).
- Acknowledge that, despite all of the validity strategies employed and your reflexive diligence as a researcher, your interpretations are just that—your interpretations. This does not mean that you have to be apologetic or that you cannot make assertions. It does, however, mean that you need to be transparent and indicate what data support the findings, how they support them, and how you arrived at the key themes and findings.
- Throughout every stage of the research, keep a log/description of your processes, including times when you compose memos, refer back to memos, check in with participants, revise instruments, engage with theory, consult with peers, and so on. These may or may not be documented in a research journal in more detail. If you do not keep a research journal, the act of documenting when you do what with a brief description of it will help you to be transparent in your processes as you return to the analysis plan in your text and revise it based on what actually happened methodologically and allow you to think about and reflect on how you are engaging with your data and research design.
- Acknowledge limitations of the study in the final report/product and describe these in relationship to findings.

Seeking Out Alternative Perspectives

Finally, *qualitative data analysis is a process of seeking out alternative perspectives.* By *alternative,* we mean perspectives that will challenge your own. One way of seeking alternative perspectives is to be engaged with other people in your research process. Dialogic engagement is an incredibly important aspect of the overall research process as well as to data analysis specifically. While we encourage this kind of collaboration throughout all stages of the research process, here we focus on its role in driving a more authentic and generative data analysis process.

Dialogic engagement should be intentionally structured into all of the data analysis processes. Furthermore, you should actively seek out people who will challenge you in a variety of ways, including specifically challenging your assumptions, biases, preconceived notions, and how each and all of these shape the ways that you think about the data and the people in your study. We urge you to find people who represent a variety of perspectives that may be linked to their years of experience within a certain milieu or position, their social location/identity and positionality in relation to the focal group in your study, their relationship to the sample or topic, their identification with a specific cultural background or region, their methodological and/or research experience, and so on. Furthermore, you may select individuals specifically because they may or may not have stated experiences related to your research and/or because you are comfortable having your assumptions and biases challenged with them. From these various vantage points, the goal is that you will be challenged in ways that will advance your own reflexivity skills and therefore your critical analytic skills as well as help you to gain additional (and alternative) perspectives about your data. This will happen only if the expectations and norms of these relationships are set to include challenging each other in ways that are honest, direct, and supportive. There are a variety of ways to engage with these thought partners. For example, you can engage with others in reading and thinking together about an individual interview transcript, looking across a number of transcripts, looking across multiple data sources for one participant, looking across fieldnotes, and/or considering the role of theory as it relates to the data. These critical conversations can happen in an open-ended way that seeks broad interpretative differences and/or can be specific as in developing preliminary codes, reading analytic memos together, and interrogating preliminary themes.

The goal of dialogic engagement is to structure into your analytic process a systematic engagement with alternative perspectives about your data and your analysis of those data early on and ongoingly. The important point here is to build these processes into the processes of analysis to create accountability to multiple interpretations and to challenge your biases and assumptions. Logistical considerations related to these meetings include deciding if these sessions will be recorded and/or notes will be taken either by you and/or by others engaged with you. The systematic recording of these conversations is often an important aspect of ongoing reflection that becomes represented in the research process as another layer or form of data. The above suggestions are geared toward an individual researcher, but research teams can and should engage in these activities as well, both within and beyond their teams.

In Chapter Three, we offer ideas about what we refer to as *structured sets of conversations,* which are a set of exercises intended to facilitate deliberate conversations with colleagues and peers about specific aspects of the research design process. This can be easily adapted for the analysis process specifically, which is what we have done in Recommended Practices 7.1 and 7.2. This process works best if a group meets regularly and is familiar with the research topics and specific research questions of all members. In our research group experiences, these groups can provide access to wonderful data analysis thought partners since all members are acquainted with each study's research process and goals. We have found the sharing and vetting of codes and their definitions to be useful at various stages. The goal of these sessions is to talk through issues and questions about your data and data analysis in real time throughout the process of analysis and to structure multiple subjectivities into the analytic process.

Recommended Practice 7.1: Structured Sets of Analytical Conversations

Prior to or during these conversations, a researcher typically shares a piece of data. This can be one full transcript or pieces of several data sources juxtaposed (e.g., several pages of a few transcripts focused on one topic or person plus supporting data that relate to those excerpts). The researcher can either share the data and simply ask, "What stands out to people?" and/or can ask one to three specific questions about the data and/or her analysis to discuss with at least two peers. The group divides the time they have to meet equally between all group members. (It is essential to assign a timekeeper if multiple researchers share at a given meeting.) Each researcher can structure her time based on what she hopes to get out of the conversation and articulates her hopes and goals for the session to her peers prior to or during the session. She can also share her emerging interpretations and see how people perceive them, if they agree, challenge them, and so on. While the conversation may evolve in unintended directions, it is important to prepare for the meeting so as to focus the conversation in ways that will most benefit your research. Possible questions to consider at an initial meeting include the following:

- What is happening here? What stands out from the data?

- How do you think about or describe what stands out?

- How does it relate to the guiding research questions?

- Does this relate with what emerged with other participants? In what ways does it overlap or depart from that?

- How do you think about the person's story in relation to other participants?

- What assumptions do I bring to the data?

- What do the early data suggest? For design refinements/changes? For the use of theory? For my working conceptualizations?

- What kind of analytic approach will help me understand and best engage with my data?

- What is missing?

Possible questions to consider in ongoing meetings include the following:

- What is shaping the study conceptually and theoretically and why? How does it relate to the data?

- What assumptions continue to shape the research and the analysis of data?

- How might others interpret these data and why?

- What emerging stories am I discerning and why? What patterns, if any, am I seeing and why?

- How does context mediate what I am seeing?

- How might I be misinterpreting the data?

- What kinds of codes am I thinking about and why?

- What is shaping my sense of how to write about these data? How to represent the realities of the participants as they conveyed them?

The group should determine the topic or focus of subsequent meetings in advance, and members should prepare for these meetings, sharing data and related documents such as memos, archival documents, and even related articles and studies if useful for peer review and discussion. For example, the group might decide to discuss researcher bias in analysis at a meeting, and all members would come prepared to discuss this topic with all or parts of their memos or research journal entries that engage this topic and/or even do some writing and exchange within the meetings.

Recommended Practice 7.2: Paired Question and Reflection Analysis Exercise

This is a process, engaged in with a partner, to generate focused researcher reflection around key areas of importance to the analysis of your data. These areas could be related to a variety of aspects of data analysis and conducted at different stages of the process. Examples of areas to consider include the following:

(Continued)

(Continued)

- Formative data analysis, including preliminary codes and interpretations

- Instrument refinement

- How the research questions are being answered (or not) by the data

- Key themes in the data

- Relationship of data to research questions and theory

- Participant representation

- Researcher bias

In this paired exercise, each researcher should develop two to three key questions about her data. The partner will ask the researcher these questions and take notes about how the researcher answers the questions. The partner will then share her notes with the researcher, and the pair will discuss what stood out about the answers to the questions. With the partner (and afterward individually), the researcher will reflect on her answers to the questions and consider how they align with how she believed she would answer these questions, what underlying assumptions arose during the answers, how she portrayed participants and layers of analysis, and so forth.

We recommend that researchers write a brief memo after this process to reflect on and document the learning and any unanswered questions. These memos could be shared with the partner one more time for any final thoughts. Furthermore, we recommend that the pair work together over the course of the project and engage in this process throughout the research. Our students share that they find this useful throughout their research and specifically at the summative analysis stage.

QUESTIONS FOR REFLECTION

- What is qualitative data analysis?

- What is the relationship between data analysis and interpretation?

- What are the key aspects of most approaches to qualitative data analysis?

- How is qualitative data analysis iterative and recursive? What does this mean for how you engage in data analysis?

- What does it mean that qualitative data analysis is formative as well as summative?

- In what ways are data and theory triangulation necessary in qualitative data analysis?

- How are the power differentials within data analysis an ethical and methodological concern?

- What does it mean to seek out alternative perspectives in qualitative data analysis?

In the next chapter, we build on what we have discussed here as an integrated approach to data analysis. We detail our three-pronged approach to data analysis and describe the specific processes and methods to analyze your data as well as provide examples and exercises to help guide you through the analytical processes.

RESOURCES FOR FURTHER READING

Bogdan, R. C., & Biklen, S. K. (2006). *Qualitative research for education: An introduction to theories and methods* (5th ed.). Boston, MA: Pearson.

Coffey, A., & Atkinson, P. (1996). *Making sense of qualitative data: Complementary research strategies*. Thousand Oaks, CA: Sage.

Creswell, J. W. (2013). *Qualitative inquiry and research design: Choosing among five approaches* (3rd ed.). Thousand Oaks, CA: Sage.

Emerson, R. M., Fretz, R. I., & Shaw, L. L. (2011). *Writing ethnographic fieldnotes* (2nd ed.). Chicago, IL: University of Chicago Press.

Marshall, C., & Rossman, G. B. (2016). *Designing qualitative research* (6th ed.). Thousand Oaks, CA: Sage.

Miles, M. B., Huberman, A. M., & Saldaña, J. (2014). *Qualitative data analysis: A methods sourcebook* (3rd ed.). Thousand Oaks, CA: Sage.

Roulston, K. (2014). Analysing interviews. In U. Flick (Ed.), *The SAGE handbook of qualitative data analysis* (pp. 297–312). London, UK: Sage.

Weiss, R. S. (1994). *Learning from strangers: The art and method of qualitative interview studies*. New York, NY: Free Press.

Willig, C. (2014). Interpretation and analysis. In U. Flick (Ed.), *The SAGE handbook of qualitative data analysis* (pp. 136–149). London, UK: Sage.

ONLINE RESOURCES

Sharpen your skills with SAGE edge

Visit **edge.sagepub.com/ravitchandcarl** for mobile-friendly chapter quizzes, eFlashcards, multimedia resources, SAGE journal articles, and more.

ENDNOTES

1. Our perspective on interpretation is informed by hermeneutics, which is a German-based strand of interpretive philosophy that frames and considers how individuals interpret and make meaning of the world in ways that are individualized and contextualized within their own backgrounds, the context(s) in and connections through which they are interpreting, and how these shift and change over time (Nakkula & Ravitch, 1998).

2. For more on the concept of bracketing, see Gearing (2004).

3. For more on this, see Maxwell and Mittapalli (2008).

4. Kelle (2014) defines grand middle-range theories by stating that you will engage with theory "derived either from grand theories (about universal social processes or structures) or from middle-range theories (about social phenomena in a limited domain). In the process of theorizing, the researchers' previous theoretical knowledge (containing concepts with limited empirical content) is integrated with members' knowledge ('lay theories') in order to construct empirically contentful categories and statements about the investigated domain" (p. 565).

5. We say most forms since participatory action research seeks to resist this through ongoing, collaborative, and democratic processes of co-constructing the entire research process, including data analysis.

CHAPTER EIGHT

Methods and Processes of Data Analysis

CHAPTER OVERVIEW AND GOALS

Building on our discussion of an integrative approach to qualitative data analysis in the previous chapter, we present a three-pronged approach to qualitative data analysis. We describe the processes, procedures, and phases of qualitative data analysis and discuss the influences and techniques necessary for a critical, rigorous, and valid process of analyzing your data. The prongs consist of *data organization and management, immersive engagement,* and *writing and representation.* The chapter details specific processes for each of these aspects of data analysis and presents ways to integrate validity considerations throughout analysis.

BY THE END OF THIS CHAPTER YOU WILL BETTER UNDERSTAND

- What a three-pronged approach to data analysis looks like and the specific strategies and methods you can use in data analysis processes

- How to manage and organize your data and what this means for data analysis

- How to develop a solid data management plan

- The analytical decisions related to transcribing data

- Processes for precoding data

- What it means to immerse yourself in your data and how to do so

- Different ways to "read" data

(Continued)

(Continued)

- What coding is and how to do it

- The different approaches to analyzing qualitative data

- The role of Computer-Assisted Qualitative Data Analysis (CAQDAS) in data analysis

- How to holistically analyze data items through connecting strategies

- Ways to structure and approach dialogic engagement exercises as you analyze your data

- Processes for developing analytic themes

- Considerations and processes for ensuring validity during data analysis

- How writing is an ongoing part of data analysis

- What you can do to work to resist imposing interpretative authority to the fullest extent possible

During data analysis, which begins as soon as data are collected, you should recursively organize and reorganize your data, engage with your data through multiple strategies, and write and rewrite as you represent your data. These processes, which we term *data organization and management, immersive engagement,* and *writing and representation,* are described throughout this chapter and summarized in Figure 8.1 on page 239.

DATA ORGANIZATION AND MANAGEMENT

Data organization and management are often overlooked as important, ongoing processes that not only support and help refine sense-making and data analysis but also are integral parts of analysis. There are many aspects and stages of data organization and management, and we specifically detail the following considerations:

- How to organize, manage, and keep track of your data

- The use of transcripts as both practical and analytical considerations

- Engaging in precoding your data

Data Management Plan

Qualitative studies tend to produce a significant amount of data. For example, a 1-hour-long interview will likely produce a 30 + -page transcript; a 2-hour-long observation session will likely produce between 8 and 15 single-spaced pages of fieldnotes (see Chapter Five for discussion of jottings and fieldnotes). Having a plan in place for how you will organize your

Figure 8.1   Three-Pronged Data Analysis Process

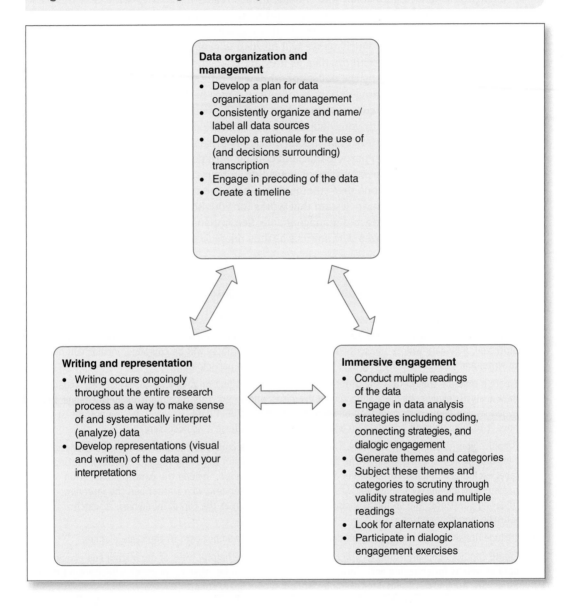

Data organization and management
- Develop a plan for data organization and management
- Consistently organize and name/label all data sources
- Develop a rationale for the use of (and decisions surrounding) transcription
- Engage in precoding of the data
- Create a timeline

Writing and representation
- Writing occurs ongoingly throughout the entire research process as a way to make sense of and systematically interpret (analyze) data
- Develop representations (visual and written) of the data and your interpretations

Immersive engagement
- Conduct multiple readings of the data
- Engage in data analysis strategies including coding, connecting strategies, and dialogic engagement
- Generate themes and categories
- Subject these themes and categories to scrutiny through validity strategies and multiple readings
- Look for alternate explanations
- Participate in dialogic engagement exercises

corpus of data as the data are collected is essential. You should be intentional throughout data collection, and it is vitally important that you not wait until the end of data collection to engage with and organize the data. Therefore, it is essential that you begin to organize and manage your data as soon as you begin collecting them, which means that you should create a *data management plan*.

Creating a specific data management plan is important for five major reasons. First, engaging in data organization and management creates familiarity with (and therefore more holistic knowledge of) your data and supports your meaning-making processes. Second, staying organized and therefore in tune with your data also helps to facilitate formative data analysis, which, as we will discuss, is vital to solid research. Third, engaging in a systematic way with the data as they are collected helps you to see if gaps in the data need to be filled in by the collection of new or different kinds of data. Fourth, it can be extremely overwhelming to have hundreds of pages of data that you must organize all at once before you can substantively begin engaging with your data, so organizing and familiarizing yourself with your data as you go helps to prevent this stressor. Fifth, as you organize your data, a variety of questions inevitably arise that you can use to collect additional data, check in with participants, and consult with peers and advisers about in order to seek alternative perspectives about your data in terms of validity, confidentiality, and any and all other questions that emerge throughout the data collection process. You will need to find an organization system that works for you; however, in Table 8.1, we provide general tips for organizing and managing data. See Appendix M for one example of a data analysis plan that organizes data sources as they relate to each specific research question.

Table 8.1   Tips for Organizing and Managing Your Data

☑ Before you begin collecting data, have a written plan for where data will be stored, how data will be protected, how data will be transcribed (and by whom), how notes will be compiled, memos you plan to write, and peer data analysis groups in which you plan to participate.

☑ Create a timeline that details the different phases of data collection and ways to formatively engage in data analysis, memos, dialogic engagement exercises, and so on. This timeline is likely to change and evolve as you begin data collection and formatively analyze your data. However, it is important to have it to keep yourself accountable and on track (and to adjust it as necessary).

☑ Organize your data files consistently. This includes having a consistent way of saving word-processing files that make retrieving them in the future easier. In addition to saving files consistently, routinely label each piece of data. For example, at the top of an interview, include the date, time, location, name of participant (or pseudonym), who conducted the interview, who transcribed the interview, lingering questions, and so forth. Include similar information at the top of all memos, research journals, and other notes.

☑ Include line and page numbers on all data and consistently number data in the same manner.

☑ Create Contact Summary Forms as you collect data. These are helpful ways of organizing and managing data as well as for engaging with data, as you may include codes, themes, or other reflections on these forms. See Appendixes I and J for examples of Contact Summary Forms.

Transcripts

In addition to having a plan and system for managing your data, you should develop a system and rationale for how your data will be transcribed. As you organize your data, you will most often need to transform data into text; this typically involves the use of *transcripts,* a representation of spoken data. Gibson and Brown (2009) provide a helpful definition of a transcript:

> When researchers speak of a transcript, they are referring to a mode of representing a piece of data that has been gathered. Data refers to material that has been collected (or *generated*) in the course of research, while transcription is the process of rendering that data into a new representational form. Through transcription, researchers *represent* or better still *re-present* the data that they have gathered. (Gibson & Brown, 2009, p. 109)

Transcripts, which are primarily textual, can be generated from audio or video recordings. It is important to keep the original audio and video files to refer to throughout the data analysis process, as the transcript is a representation of the data. Focusing on the work of the educational anthropologist, Frederick Erickson, Ravitch and Riggan (2012) discuss the nonneutrality of transcripts and frame them, as Erickson (1996) does, as an active rather than mechanical part of the analysis process. Drawing on Erickson, Ravitch and Riggan state, "Taking a theoretical approach to transcription— deliberately emphasizing aspects of what is being transcribed in order to examine them more closely—is not preparation for data analysis but analysis itself"[1] (p. 98).

Not only is the transcript a representation, but it is also a form of interpretation. We therefore believe that transcriptions should be as verbatim as possible. However, we acknowledge that calling a transcript verbatim does not make transcripts and the process of transcribing objective or entail that a researcher will ever capture all of the nuances in an audio or video recording. While acknowledging these limitations, we still assert that transcripts should be verbatim. There is an interpretive authority when you take out people's natural speech patterns such as "ums" and "ers," and we question where that stops. For example, do you start translating individuals' speech into "Standard English," which, as many argue, is a social construction? What people often refer to as "cleaning up the data" raises ideological, political, and ethical issues and should be carefully considered. Still, some scholars question the efficacy of verbatim transcriptions because they view "ums" and "ers," for example, as extraneous.

Having verbatim transcripts does not signify or ensure rigor (Barbour, 2014); however, we believe that researchers must maintain fidelity to participants' experiences, words, and genuine articulation of their experiences. When transcripts, which we also consider to be representations, are deliberately altered, the researcher is not maintaining such fidelity. If transcripts are not transcribed verbatim, this must be acknowledged and explained with a clear rationale for this choice. It is not sufficient to say that interviews were audio recorded and transcribed. You should state who transcribed them (e.g., members of the research team, external individuals, voice recognition software programs) and how they were transcribed (e.g., verbatim, in an edited way). If transcripts are "cleaned up," which we strongly advise against (and find the terminology offensive), you must describe how. In addition, when transcripts are transcribed and translated from one language to another, it is best to include the original language and the translation, as well as indicate who translated it. Anyone who has ever learned another language knows how much is lost in translation, and for this reason, the original language (and regional dialect within that language) should be provided whenever possible. Sometimes, you can use brackets or footnotes to indicate phrases or words that do not translate easily or accurately. These issues are at the center of qualitative inquiry since they demonstrate the role of interpretation in all aspects of qualitative research (Kelly & Zetzsche, 2012).

All transcripts are translations (Kress et al., 2005), and transcription is representation in that researchers are making choices about what is important (Gibson & Brown, 2009). For example, where do you add punctuation? What gestures will you write down? What gestures will you leave out? How will you indicate eye contact, tone of voice, and short and long pauses? There are many judgments that researchers make, and it is imperative that you are thoughtful, open, and transparent about your reasons behind your transcription decisions. Kowal and O'Connell (2014) note that

> all transcription is in principle *selective* and entails the inevitable risk of systematic *bias* of one kind or another. Nonetheless, this risk can be countered by making decisions on the basis of reasoned choices rather than arbitrary, non-reflective ones. Consequently, both basic and applied researchers in the social sciences must approach transcription with a very critical eye (and ear). (Kowal & O'Connell, 2014, p. 66)

In addition to explicitly describing your decisions surrounding transcription, discuss with your study participants how transcripts will be represented in the final product (Roulston, 2014). This can help you to make and justify decisions regarding transcription choices. See Table 8.2 on page 243 for practical considerations and suggestions regarding transcription.

Precoding

As you begin organizing your data and generating transcripts, you should engage in what is called precoding. This is a process that many qualitative researchers engage in before they begin coding their data. Before we describe precoding, we briefly define what coding is (with a more in-depth description to follow). Broadly speaking, coding entails ways of organizing and labeling your data that help with analysis. One purpose of coding is data organization, and coding supports analysis by allowing for the identification of patterns across multiple data points or sources, the identification of relationships within data, the establishment of common themes/elements across nonuniform data, and for looking critically across stakeholder groups for shared and divergent patterns. *Codes* are descriptive or inferential labels that are assigned to units of data (Miles et al., 2014). Codes help researchers to organize data into manageable units or chunks to engage with analytically. All data can be coded, including transcripts, fieldnotes, archival data, photographs, videos, research memos, and research journals. As van den Hoonaard and van den Hoonaard (2008) state in *The SAGE Encyclopedia of Qualitative Research Methods,*

> Any analysis of data involves some form of coding. Coding reflects both the personal analytic habits of researchers and the general principles that flow from particular qualitative research methodologies and theoretical perspectives. In its most stringent form, the analysis of data can entail line-by-line coding of text whereby researchers capture every empirical and conceptual occurrence in each line. In conversational analysis, even the duration of pauses is measured and used as data. At the other end of the continuum, some researchers adopt a more flexible approach, perhaps coding whole paragraphs or groups of sentences at a time. (p. 187)

Table 8.2 Practical Considerations Regarding Transcription

- Be sure to have page numbers and line numbers on every page and line of the whole transcript. This allows you to create a notation system for indexing parts of transcripts that you can return to later for analysis and for the use of quotes in the final report.
- Decide if certain structural choices are important, for example, time stamps, time intervals for pauses, notations for pauses, notations for nonverbal communication, and so forth.
- We believe that it is beneficial to transcribe recordings yourself because it helps you to understand the interview/focus group better and to relive and reflect on how it went in real time. But this is not always possible since transcribing is quite time-consuming. If you are having someone else transcribe for you, be aware of issues of confidentiality. For example, if you are using a transcription service, be sure that files can be sent in ways that are secure. If you are choosing a colleague or someone known to you to transcribe, be sure that this choice does not compromise confidentiality (i.e., if that person is familiar with the participants).
- If someone else is transcribing your data, be sure to be very clear with them about all of your structure and style choices.
- Be sure to locate transcription services that are vetted and reviewed for quality and turnaround time as well as price.
- Even if someone else transcribes your data, we recommend listening to the original recording as you read the transcript the first time so that you can make sure that the transcript is accurate and to fill in any "inaudibles" that the transcriber may note. (This is the notation used for when parts of an interview or focus group are indecipherable, but we find that we often can fill these in since we know the participants and/or were present at the interview and therefore listen to their statements in context.)
- After transcribing a recording or receiving a transcript from a service, look to the notes that you took during the interview/focus group to fill in anything that would be useful to interpreting the words on the page. This might include nonverbal behavior in responses to specific questions, pauses, or body language. This might also include any notes taken if there was a point during or after the interview in which the participant shared ideas or thoughts that were not recorded and therefore did not make it into the transcript and so on.
- It can also be useful to listen to the recordings, particularly if you did not transcribe the interviews yourself because listening to them can be different than reading. Some of our colleagues suggest that researchers listen to these recordings while walking or otherwise being away from your desk to help resist distraction by writing as you listen. Others suggest listening to them while reading over the transcripts so that you can add notes on anything you hear that is either different than the transcript or more about tone or other affective aspects of the interview.
- While new voice recognition software programs exist, be sure to do a trial run with these to determine the quality.
- Make sure to allow ample time for transcription of the entire data corpus and include this in your timeline.

Precoding is a process of reading, questioning, and engaging with your data (e.g., transcripts, artifacts, fieldnotes) before you formally begin the process of coding the data. Reading and familiarizing yourself with the data can help the data to become less "opaque" as preliminary codes are developed (van den Hoonaard & van den Hoonaard, 2008). Precoding, which is a form of open coding that we describe subsequently, can take many forms. If done by hand, it might include circling, color coding with markers or highlighters and/or underlining key words or phrases that stand out, writing notes or questions in the margins, jotting your first impressions, noting specific terminology, and so on. These precoding processes can also be completed with a computer by highlighting text or using the track change function in Word documents. To be clear, precoding is an important analytical process that occurs before you summatively analyze your data. It helps you to familiarize yourself with the data, generate potential codes, see if you need to revise aspects of your design, and determine what literature you need to consult. We suggest composing the memo described in Recommended Practice 8.1, or one similar to it, during and after this process.

Recommended Practice 8.1: Precoding Memo

The purpose of this memo is to capture what you learned from the processes of precoding your data. This memo can be very useful as you continue the data analysis process and to refer to if you think your analyses may be "straying" too far from the data.
   Potential points to address:

- Emerging learnings

  - What stands out?

  - What seems noteworthy?

  - How do the data relate to my guiding research questions?

- Lingering questions

  - Were some of my questions leading? If so, in what ways? How might I account for that in the data?

  - What data, if any, do I still need to collect?

  - What are the limitations of these data?

  - What other questions do I have after thoroughly reading through my data set?

- Reactivity

  - How am I, as the researcher, influencing the data?

  - Is there anything I can do to address this?

  - How do I see my presence/influence in the data?

- Ideas/thoughts about potential codes
  - What prompted these codes?
  - Are they emic or etic?
  - Are they inductive or deductive?
  - Are they related to theory? If so, in what ways?
- How do the emerging learnings map onto and/or challenge my theoretical and/or conceptual frameworks?
- What literature do I need to consult/reread?

IMMERSIVE ENGAGEMENT

Immersive engagement involves multiple ways for you to immerse yourself in, critically engage with, read, and analyze your data. The primary processes for immersive engagement with data include

- multiple data readings (first unstructured and then structured/goal-oriented readings);
- implementing data analysis strategies, including, coding, connecting, and dialogic engagement strategies; and
- generating, scrutinizing, and vetting themes.

We describe each of these immersive engagement processes in this section.

Multiple Data Readings

As data are collected and organized, it is important to consider how you will *read the data* at various points in a study. As stated earlier, you should formatively and ongoingly analyze your data as they come in, and this entails "reading" and "rereading" your data in an effort to support various analytic processes. In addition, once you have collected all of your data and begin summative data analysis, you will continue to read the entire data corpus, which refers to all of the data collected, multiple times with differing foci and goals.

We suggest that the first reading of the entire data corpus be a thorough reading uninterrupted by coding data to get the overarching context and sense of the lay of the land. We refer to this as an *unstructured reading* and strongly suggest this as a vital first step to get oriented to and immersed in the entire data corpus. An analogy Sharon uses with her students is that when she was an undergraduate, she took a world religions course and was asked to read the Old Testament in 1 week. Since she had grown up reading it weekly, as many religions do in a congregational forum, she thought she knew the themes and lessons contained within the text. However, engaging in reading the entire book at one time created the conditions by which she started to see meta-themes present through-out and specific patterns across subsets of stories and specific protagonists that were not as evident when the book was read in chunks over time. This reading of the whole text proved quite different from the compartmentalized readings of her youth and taught her a great deal about this text in terms of its macro and meta themes and even its organiza-tion and the implications of that on her interpretations. So too does a read of your entire data corpus when it is uninterrupted and therefore more comprehensive and holistic. This unstructured read can also help you to narrow your data corpus to form your data set.

While it is difficult for many (us included) to resist the urge to "be efficient" and code during this first reading, not doing so provides a focus and pace that can allow you to get through a considerable amount of the data set in a few sittings (ideally closely scheduled together). It also allows for particular kinds of connections and insights to emerge in ways that they might not were the process guided by coding. After this unstructured reading, you should do two things: (1) write an analytic memo that helps you digest and make sense of your thoughts as you read and (2) discuss this with peers, colleagues, and/or an adviser within close timing of reading it in order to share your emerging sense of the data, ask questions, express concerns, and engage with suggestions and feedback.

While we suggest an unstructured first read of the entire data corpus, there is not a "correct" way or specific order for you to read your data; however, it is important to justify and have a clear rationale for your choices (and for formal texts such as master's theses, dissertation proposals and dissertations, and other published texts to include this in the data analysis section of the report). While there are many ways to do this, it is necessary to conduct multiple readings of your data set. After reading the entire data set multiple times, with various readings focused on specific analytical goals, you may read the data again to look at them through various lenses, for example, at small groupings of partici-pants around particular concepts or ideas or around aspects of their social or professional identities and experiences that might analytically help you make sense of how who they are might influence their responses, depending on your research questions.

Suggested ways to "read" your data:

- Formative readings (see Recommended Practice 8.2 for the suggested memo topics to help you to engage with the formative nature of data analysis)

- Unstructured readings

- Chronological readings (i.e., looking at the data set by date)

- Readings by method (i.e., interviews, focus groups)

- Readings by participant (i.e., tracing one person through the data set, for example, looking at her interview and/or focus group responses, her writing, the fieldnotes in which she is involved)

- Inductive readings (i.e., reading parts or the whole data set for insights, themes, and patterns)

- Deductive readings (i.e., reading to seek out ways that the data speak to specific concepts found in prior research and/or theory)

- Readings specifically centered on the research question(s)

- Readings that explore how aspects of participants' social identities, backgrounds, and/or experiences influence their responses and inform your analysis

As you read your data, you may look for patterns, the range and variation of experiences within and among participants as a whole and/or within subgroups, and/or the specific context(s) of the setting and how that shapes the data and helps you make sense of them. Much of the qualitative literature discusses looking for patterns, and we want to reemphasize that you may or may not be looking for patterns. What you are looking for depends on your research questions and the goals of the study.

In sum, throughout the analytic process, you will conduct multiple readings of your data. The order and structure in which these readings occur is an important part of the intentionality of data analysis. It is important to articulate the reasons behind all of your decisions related to data analysis, and just as the sequencing of methods is an important validity strategy, the order in which you read and analyze your data is also important to the analyses you conduct.

Recommended Practice 8.2: Formative Data Analysis Memo

This memo serves a variety of purposes. It aids in the refinement of research questions, design, and instruments. It can also help researchers to document their initial impressions of the data, which can lead to preliminary insights and learnings. This memo is also an important part of systematizing the process of formatively analyzing your data. We recommend composing a formative data analysis memo at multiple stages throughout the data analysis process. These memos may broadly reflect on what has been learned so far or they might specifically consider one interview or one instrument. In addition, these memos may reflect on the data in relationship to your conceptual or theoretical framework. As with all memos, be sure to date the memo and include the data collected so far and/ or the specific data being analyzed. Of course, we recommend sharing these memos with thought partners so that you can discuss some of the questions and points that emerge from them.

Potential topics to address in these memos include the following:

- What am I seeing across the data at this point? (Pull excerpts from data and discuss.)

- What stands out to me? Why? (Pull excerpts from data and discuss.)

- Do I have the necessary questions on my instruments given what I seek to learn? Too many? Too few?

- Are my instrument questions worded appropriately? Sequenced well?

(Continued)

(Continued)

- How can what I am seeing in the data help me to ask better follow-up questions of participants?

- Am I getting the data I need to answer my research questions? Why or why not?

- Am I asking the right overall research questions, or do they need to be tweaked or refined?

- What do I need to read or with whom might I speak to better understand what I am learning?

- How do the data map onto my conceptual framework and/or theoretical framework?

- How does my conceptual framework and/or theoretical framework need to evolve?

- What lingering questions do I have about (the data, my instruments, the theory, etc.)?

- How does my subsequent data collection need to change?

- As I think about all of these issues, what can I improve upon? Seek consult about?

Data Analysis Strategies: Coding, Connecting, and Dialogic Engagement

As stated previously there are many ways that you can approach and analyze your data, and just as there is no "correct" way to "read" data, there is also no "correct" way to analyze data. Maxwell (2013) describes this succinctly:

> I want to emphasize that reading and thinking about your interview transcripts and observation notes, writing memos, developing coding categories and applying these to your data, analyzing narrative structure and contextual relationships, and creating matrices and other displays are *all* important forms of data analysis. (p. 105)

The data analysis strategies we specifically recommend include coding, connecting, and dialogic engagement strategies.

Coding

People often think that coding and analysis are synonymous, and it is very important to distinguish between them. Coding is a process of assigning meaning to data. A code can be a word or phrase that explains or describes what is going on in the data (Corbin & Strauss, 2015); codes can also represent analytical ideas. Coding is typically a part of qualitative data analysis, but it is only one of many aspects of qualitative data analysis. Data analysis begins when you start organizing and thinking about your data, and coding is the most common way to do this. However, coding your data does not mean you have analyzed them. Researchers differ about whether coding constitutes analysis. We consider the act of labeling or tagging data (i.e., coding) as a necessary and useful component of analysis but not as the whole of analysis itself, meaning that we believe many fetishize coding and approach it as if it is *the* act of analysis. We see it as a part of a larger set of analytic processes, both internal and external. However, to be clear, the thought and sense-making processes that go into creating, defining, and refining codes as well as generating categories are important components of data analysis. Coding organizes and "breaks [data] down into manageable segments, and identifies or names those segments" (Schwandt, 2015, p. 30). Depending on the approach and amount of interpretation involved in the coding processes, it is often considered more descriptive or explanatory (Schwandt, 2015). As we describe in the sections that follow there are different ways of and approaches to coding your data.

Approaches to Coding

Qualitative data analysis often involves both *inductive* (coming from the data) and *deductive* (coming from other sources such as theory or prior research) coding processes. While qualitative data analysis is often characterized as inductive, deductive processes can also be incorporated in your analytical approach (Schwandt, 2015). The inductive and deductive nature of qualitative data analysis highlights "the inseparability of methods and findings" (Emerson et al., 1995) since the questions that you ask and the way that you ask them directly shape the type of data you collect. For example, when analyzing data, it can be important to determine whether the participants or the researcher introduced specific terminology or concepts within an interview (or other data source) since this can help you understand the origin of a thought, word, or concept. This is related to the concepts of emic and etic. Language generates meaning, and understanding whether terms are emically or etically (or a combination of both) discussed can be helpful to understanding someone's narrative and the meaning he or she assigns to particular ideas, perspectives, or experiences.

Inductive approaches to data analysis are also called *emic, insider, bottom-up,* or *in vivo* processes. Generally, an inductive approach to coding stays as close to the data as possible. For example, *in vivo coding* uses the participants' words to label data segments instead of researcher-created words or phrases. It is important to note that some codes may be in vivo, whereas other codes may be inductive but not use the participants' language. Deciding when and how to code data is a skill that researchers develop. It is also important to note that using only in vivo codes does not necessarily confer rigor. Emic or in vivo codes should be applied when the participant's language accurately captures what was expressed. Deductive coding is also referred to as *a priori, outsider, etic,* or *top-down approaches.* These processes involve reading the data and looking for something specific. You may take what you are looking for from prior literature, prior research, or even your own informal theories or "wisdom of practice" (Shulman, 2004). Inductive and deductive codes are not mutually exclusive, and in many studies, the strategic combination happens through multiple readings for each kind of coding.

Coding Processes

As we have stated throughout this text, there is no "right" or "wrong" way to approach qualitative research. You will develop processes and methods that work for you. For example, Miles and Huberman (1994) begin with a provisional "start list" of codes prior to collecting data that come from the conceptual framework, research questions, and areas of interest. Generally, coding data involves reading for regularly occurring phrases, terms, interactions among actors, strategies and tactics, consequences, and patterns of participation (Miles et al., 2014). Below, we describe and define different coding processes.

Coding involves assigning codes, which can be words or phrases, to "chunks" of data. When coding, include enough text so that the meaning of the quotation remains clear when removed from context. *Open coding* is when you highlight sections of text or label them in some fashion. This form of coding is similar to the "first-level coding" that Miles et al. (2014) describe, which involves summarizing segments of data. Open coding can be done a number of ways. Some people do it by hand, use sticky notes, or use different color highlighters. Others may use features in word-processing documents such as highlighting or using track changes and commenting features. Some researchers use specific qualitative data analysis software programs referred to as Computer-Assisted Qualitative Data Analysis Software (CAQDAS); we discuss the uses of these programs in Feature Box 8.1. Open coding may involve multiple rounds and readings. A first round of coding may be used to determine what stands out, and then a second round may focus specifically on aspects of your research questions.

Axial coding, which is also called thematic clustering coding or "pattern coding" (Miles et al., 2014), occurs once you have established codes. It is a process of going from coding chunks of data to starting to see how these codes come together into coding categories or clusters from which you will situate sets of constructs or concepts in relation to each other to make arguments and develop findings. We tend to begin our analysis process with open coding because it lets us see the lay of the land or the forest. Our next couple of "reads" of the data might be related to research questions to help us see the trees. There are, of course, multiple ways that you can approach this depending not only on the study's goals and research questions but also on the contextual factors of the study. Appendixes R and S.4 present coding schemes and example codes to help you see two ways of approaching the coding process.

Creating what is called a *code set,* or a group of codes, is an iterative process. It is often best for novice qualitative researchers to start broadly and then work to develop more narrow and specific codes as the analysis progresses. In the early stages, codes should be primarily descriptive rather than analytic or theoretical. *Descriptive* categories tend to stay close to the text. For example, emic codes or categories tend to be descriptive or what Maxwell (2013) calls substantive, which entails that they are often generated from inductive, open coding. It is important to note that descriptive categories are not necessarily emic; they may be descriptive but still represent the researcher's understandings rather than the participants' (Maxwell & Chmiel, 2014). *Theoretical* categories, which tend to be more etic, "place the coded data into a more general or abstract framework" that stems from prior theory or research or from theory developed inductively (Maxwell, 2013, p. 108). Theoretical categories connect the coded data directly to your theoretical framework (Maxwell & Chmiel, 2014).

Multiple codes can be assigned to a particular quotation or piece of data, and codes can overlap. It is important to be judicious; everything in a transcript does not need to be coded. In addition to looking for patterns, as Miles et al. (2014) suggest, look for "gaps" in your code set. These gaps might be meaningful quotations that do not fit under any of your existing codes. What you look for when you are coding your data will depend on your research question as well as if you are conducting inductive or deductive (or other) reads

Table 8.3 Concepts to Look for When Coding Data

- Repetition in and across various data items
- Strong or emotive language
- Agreement between individuals
- Concepts that are not discussed or commented on
- Disagreement between individuals
- Mistakes and how/if they are solved

Source: Information in the table is paraphrased from Gibson and Brown (2009, p. 134).

of your data. In the Table 8.3, we summarize additional aspects that Gibson and Brown (2009) recommend looking for when coding.

Other Coding Considerations

As soon as you begin coding your data, you should begin the process of defining your codes. *Code definitions* consist of brief (ideally a short phrase and no longer than a sentence) definitions for each code so that you are clear what each one means and how they are distinct from each other. We, along with many of our students, find developing the definitions to be a clarifying process. This is especially important when multiple people are analyzing data, so that there is clarity and consistency in how codes are conceptualized and assigned, but it is also important for individual researchers to be clear on what their codes mean. Code definitions should be clear and concise. Because data analysis is an ongoing process, make sure to revisit your data set when you revise or edit codes and code definitions. This includes going back to data that you may have already coded. See Example 8.2 for an example of code descriptions and definitions.

Schwandt (2015) describes specific actions to avoid when coding, including staying only at the descriptive level, conceiving of coding as a technical act devoid of theoretical underpinnings, and thinking of codes as fixed instead of dynamic. In addition, it is important to not conflate coding with analysis. As previously stated, coding is one form of reading, interpreting, and organizing data, but there are other ways to read and sort data. The important thing in coding is to engage in ways that resist these troublesome tendencies and engage a systematic and consistent approach to the data analysis. Structuring your analytic processes to include dialogic engagement is a vital aspect of analysis since it helps to challenge and refine your working assumptions, interpretations, and ideas about what the data tell you broadly and specifically in relation to your research questions. One way to structure conversations with thought partners and advisers is to write, share, and discuss a series of analytic memos.

As you code your data, you should systematically and critically reflect on what you are learning, what you still need to learn, and how codes relate to other codes. We encourage documenting these reflections in memos. In Recommended Practice 8.3, we present topics for you to address in coding memos throughout the analysis process, and Example 8.2 is one example of a coding memo. See Appendix N for an additional example of a data analysis memo.

Recommended Practice 8.3: Coding Memos

As you develop codes, define them, refine them, and begin to develop themes, it is important to capture your analytical thinking in the form of memos. Potential memo topics to address include the following:

- How your codes categorically relate to each other. This is a particularly useful topic to address as you begin to reduce codes to themes. Describe the different codes, pull excerpts of those codes from your data, and explain how they map onto other codes.

- The coding process. Summarize, in detail, your process of coding. Describe precoding, the order in which you read your data, and how codes were developed (inductively, deductively, both), and give examples of data that map onto specific codes. Having this process detailed will help you to be transparent and articulate your process in your final report. It will also help you to think about possible limitations to your study.

- Developing and refining specific codes. This is a beneficial topic to address when defining or refining codes; it can help you to become clearer about what your codes mean and how you are using them. If on a research team, all team members should write this memo about the same code to see if and how their code conceptions align. Select a code or two that you are currently using to code your data. Describe this code, include several data excerpts that you have coded according to this code, articulate why these data should be coded in this manner, and discuss any lingering thoughts or questions related to this code. Questions include the following:

  - What does this code mean?
  - How does this code relate with and map onto my data?
  - How does this code relate with and map onto my research questions?
  - Am I using this code consistently? If not, what might that suggest?
  - What other codes are related to this code?

Example 8.1: Coding Memo

Susan Feibelman

Memo: Coding—Pilot Study

April 3, 2011

Research Questions

How are women mentored for leadership roles in independent schools?

- MENT   **Mentor**
- HOS_MENT   **Head of School as Mentor**

Annotation: Each of the codes that Susan developed was given a color that she used when coding her data. This memo illustrates one way of making sure that your codes align with your research questions.

- PER_INIT   **Personal Initiative**
- KIN   **Kinship (Friendship)**
- MOD   **Models**

What networks of support exist to support the development of women who are interested in pursuing leadership roles in independent schools?

- PROG   **Professional Development Programs**

Do women who are interested in pursuing leadership roles in independent schools seek out mentors and/or networks of support? (How do they ask or learn to ask for support and mentoring?)

- PROF_GLS   **Professional Goals**
- PER_GRW   **Personal Growth**

Do the social identifiers of race and age inform the independent school leadership roles that women take on?

- DIV_W   **Diversity Work**

Miscellaneous

- L R   **Leadership Role(s)**

Emerging Questions

Do women step into leadership roles without waiting to be invited in?

How do women learn the culture of independent schools, and how does the diversity of independent school cultures shape that learning experience?

Do women step into leadership roles before a formal mentoring relationship has emerged?

How does a woman's first year(s) in an independent school set the stage for her potential to take on leadership roles? What is the relationship between leadership and institutional culture? How do assumptions about gender roles inform this relationship?

How do the participants describe what they wanted/needed in a mentor-protégé relationship?

Can a head of school (HOS) be an effective mentor? What are the inherent risks to having one's supervisor serve as mentor?

How does proximity shape mentoring relationships?

Example 8.2: Example Code Descriptions and Definitions

Ceci M. Cardesa-Lusardi

Code Descriptions and Definitions

Theme	Code Description	Code	Code Definition
Life-altering critical event	Atrocities to self	AT1	An extremely wicked or cruel act involving physical violence or injury to the study participant.
	Atrocities to loved one	AT2	An extremely wicked or cruel act involving physical violence or injury to study participant's loved one—either a family member or close friend.
	Atrocities to acquaintance or community member	AT3	An extremely wicked or cruel act involving physical violence or injury to study participant's community member (childhood friends, acquaintances, tribesmen, or tribeswomen).
	Atrocities heard about loved one or friend	AT4	An extremely wicked or cruel act involving physical violence or injury to study participant's loved one or friend.
	Atrocities heard about in general	AT5	An extremely wicked or cruel act involving physical violence or injury done in general to Liberians and/or Liberia.
	Atrocities— witnessed of loved one	AT6	An extremely wicked or cruel act involving physical violence or injury to study participant's loved one.
	Atrocities— witnessed of acquaintance/ community member/s	AT7	An extremely wicked or cruel act involving physical violence or injury to an acquaintance and/or community member.
	Atrocities— witnessed of stranger(s)	AT8	An extremely wicked or cruel act involving physical violence or injury to someone who is or was a stranger.

Theme	Code Description	Code	Code Definition
	Atrocities—shift of daily life at the beginning of War	AT9	Study participant's sharing of "how things used to be" before the War
Life-altering critical event - forced migration	Self—within Liberia	FM1	Study participant forced to flee home and displaced within Liberia due to political conflict, war, or genocide; state authorities unable or unwilling to protect them b/c of nationality, race, religion, tribal identity, political opinion, or social group.
	Self—outside Liberia	FM2	Study participant forced to flee home and relocated outside of Liberia due to political conflict, war, or genocide; state authorities unable or unwilling to protect them b/c of nationality, race, religion, tribal identity, political opinion, or social group.
	Loved one— within Liberia	FM3	Study participant's loved one forced to flee home and displaced within Liberia due to political conflict, war, or genocide; state authorities unable or unwilling to protect them b/c of nationality, race, religion, tribal identity, political opinion, or social group.
	Loved one— outside Liberia	FM4	Study participant's loved ones forced to flee home and relocated outside of Liberia due to political conflict, war, or genocide; state authorities unable or unwilling to protect them b/c of nationality, race, religion, tribal identity, political opinion, or social group.

(Continued)

(Continued)

Theme	Code Description	Code	Code Definition
	Others—within Liberia	FM5	Study participant witnessed and/or heard about strangers forced to flee home and displaced within Liberia due to political conflict, war, or genocide; state authorities unable or unwilling to protect them b/c of nationality, race, religion, tribal identity, political opinion, or social group.
	Others—outside Liberia	FM6	Study participant witnessed and/or heard about strangers forced to flee home and relocate outside of Liberia due to political conflict, war, or genocide; state authorities unable or unwilling to protect them b/c of nationality, race, religion, tribal identity, political opinion, or social group.
Education	Education self (acquired)	ED1	Study participant acquired formal education at primary or secondary school, and/or tertiary level: postsecondary education, college, or university degree, including graduate school and/or professional school.
	Education self (desired)	ED2	Study participant desired formal education at primary or secondary school, and/or tertiary level: postsecondary education, college, or university degree, including graduate school and/or professional school.
	Education self (robbed by War)	ED3	Study participant believed to have been robbed by the war of formal education at the primary or secondary, and/or tertiary level: postsecondary education, college, or university degree, including graduate school and/ or professional school.

Theme	Code Description	Code	Code Definition
	Education children/siblings	ED4	Study participant's child/ren acquired/ing formal education at primary or secondary, and/or tertiary level: postsecondary education, college, or university degree, including graduate school and/or professional school.
	Education in relation to the "next generation" of youth	ED5	Study participant's education level in relation to the "next generation" of youth regarding formal education at the primary or elementary school, secondary or high school, and/or tertiary level: postsecondary education, college, or university degree, including graduate school and/or professional school.

Feature Box 8.1 Computer-Assisted Qualitative Data Analysis Software (CAQDAS)

As previously discussed, historically, researchers have used a variety of methods to sort, code, and analyze data. These have included a range of manual methods, including the use of colored pencils or highlighters to code data (with color codes), creating a numerical system with numbers corresponding to codes, using a specific word to indicate the codes, and making notations of codes and analytic insights in the margins of paper or electronic documents. As van den Hoonaard and van den Hoonaard (2008) state in *The SAGE Encyclopedia of Qualitative Research Methods,*

> These seemingly archaic approaches to analyzing data have had a pervasive effect on today's approaches. Many researchers still abide by some of these techniques of coding, whereas others have adapted them to computer-generated texts. (p. 187)

Computer-Assisted Qualitative Data Analysis Software (CAQDAS) can be helpful to organize and sort data, especially when working with large data sets. It is important to remember that qualitative data analysis software does the mechanical or more transactional work but not the thinking. CAQDAS does not conduct analysis (Flick, 2014; Gibbs, 2014). The software can help organize

(Continued)

(Continued)

analytical processes and help manage the data (Gibbs, 2014). Gibbs (2014) describes that the basic functions of CAQDAS include the development and revising of code lists, coding data often by highlighting text and selecting a code (or codes) to be applied, retrieving coded information (i.e., all excerpts that have the same or multiple codes), writing memos and linking these memos to particular codes or excerpts of text, searching the data (for specific text or specific codes), developing diagrams, and importing other documents, video and audio files, and images in one central location. Gibbs states that most of the software programs available will work well for coding and thematic analysis and that the choice is often determined by preference and expense. Popular programs include Atlas.ti, Dedoose, Cassandre, MAXQDA, NVivo, and Transana. In sum,

> CAQDAS software is essentially a database that holds source data, such as transcripts (including ethnographic notes), video, audio, memos, and any other documents that are available in electronic form, and then supports the annotation, coding, sorting, and other manipulations of them and keeps a record of all this activity. (Gibbs, 2014, p. 281)

As described above, coding is a part of data analysis, but the process of coding does not constitute analysis. When using CAQDAS, "coding is simply a process of attaching a name or tag to a passage of text, or an area of an image, or a section of a video or audio recording" (Gibbs, 2014, p. 283). The sense-making, processing about the codes and other categories, and generating themes are made by the analyst(s). As others have stated (e.g., Flick, 2014; Gibbs, 2014), CAQDAS does not conduct analysis; it can support researchers in analytical processes.

For researchers using numbers, CAQDAS can be especially useful as it can also provide descriptive statistics on qualitative data. Marshall and Rossman (2016) acknowledge how some qualitative researchers use numbers to count the frequency of patterns or experiences among participants. As they also point out, this use of numbers is quite different from the way that quantitative researchers use them. It is important to understand that CAQDAS does not conduct analysis in the way that statistical programs, such as SPSS, do (Flick, 2014). We caution trying to quantify qualitative data; however, we urge you to stay as close to the data as possible. For example, saying 15 of 20 participants discussed a certain topic more accurately describes your data than stating that "many participants" discussed a certain topic. Thus, using numbers—what some people call descriptive statistics—to specify and clarify statements can be useful in qualitative research, and CAQDAS can aid in this process.

A cautionary note about using software comes from our reading of a variety of researchers who critique these modes of inquiry or at least question their increasing and seemingly uncritical adoption (uncritical in the sense that some researchers do not seem to see or acknowledge how it can change the process). In fact, in some circles, these programs are controversial. On the positive side, as van den Hoonaard and van den Hoonaard (2008) state, these programs can help researchers to opt for larger samples since they can help them to engage with large amounts of data, but they also note that this does not ensure quality:

There is, however, no guarantee that the analysis is any better, because it may foreclose on the interplay among creative insights, memoing, and continuing development of codes that results from an ongoing connection with the raw data. Some argue that the software imposes a structure that may imperceptibly constrain the analysis. (pp. 187–188)

Added to this is the commentary that since some researchers see these computer processes as technical acts rather than creative analytic acts, some choose to have their research assistants or peers use the software to code data (rather than doing so themselves), so they become increasingly removed from the process. This concern is about the ways that software can create distance between the researcher and the data. We do not take a position on this because we believe there are positives and negatives to the use of CAQDAS, but we believe it is important to understand the debates, critiques, and excitements about these kinds of programs. In our own practice as researchers, if we employ software, it is to help organize and sort data. We conduct the analysis manually. We strongly advise that novice researchers spend focused time learning manual methods of data analysis even if you also use software or plan to do so in the future since this helps create familiarity with the data and processes of analysis.

It is important to note that most data analysis software programs cost money but have free trials, although these limit the amount of data you can analyze. It is also important to mention that it takes time to learn how to use the various kinds of software, and this time needs to be factored into the choice of using them and into your research timeline. We encourage you to discuss this choice with research advisers, team members, and peers to determine its fit for your study and to learn about their suggestions, cautions, and thoughts about its use in your specific study.

Connecting Strategies

An additional analytical approach involves using what Maxwell (2013) describes as connecting strategies instead of categorizing strategies, which tend to compare and contrast data and primarily include coding. *Connecting strategies* seek to develop the context of the data and not isolate excerpts of the data in the way that coding does. Connecting strategies are ways of analyzing data that "do not focus primarily on *similarities* that can be used to sort data into categories independently of context, but instead look for relationships that *connect* statements and events within a context into a coherent whole" (Maxwell, 2013, p. 113).[2] This type of analysis involves looking holistically at a transcript or fieldnote instead of sorting out portions of each into different categories. For example, connecting strategies, which are more prevalent in narrative approaches to data analysis, may focus on the structure or the significance of an interview transcript (Maxwell & Miller, 2008). Analyzing connections within a specific context often involves noting relationships that unite data into a narrative (Maxwell & Chmiel, 2014). Categorizing (coding) and connecting strategies are complementary and are often used in conjunction and should not be thought of as mutually exclusive because only using connecting strategies can prevent you from seeing alternative ways of interpreting context (Maxwell & Chmiel, 2014; Maxwell & Miller, 2008). One way to integrate them is to deliberately look for similarities and differences as well as connections when analyzing data (Maxwell & Chmiel, 2014). Maxwell and Chmiel (2014) state that the data displays of Miles and Huberman (1994), which we discuss subsequently, are one good example of this integration.

Dialogic Engagement

Another important way to immersively engage with your data is through interaction with others, which brings us to the analytical strategy of dialogic engagement. Dialogic engagement activities, defined in Chapter Three, are important ways of engaging with and analyzing data. One aspect of dialogic engagement includes collaboratively analyzing your data. Benefits of collaborative data analysis, formally defined as joint effort by two or more researchers to produce an agreed-upon interpretation of data, can include intercoder reliability, deeper local understandings as a result of incorporating multiple perspectives, greater reflexivity, and useful applied knowledge (Cornish, Gillespie, & Zittoun, 2014). As we discuss in depth in Chapter Eleven, there are also potential problems with analyzing data collaboratively. For example, if transcripts are shared with others, they should be de-identified. When discussing a research project with participants, you will need to inform them of who may look at their data and how their confidentiality will be protected. We also do not want to imply that collaboration denotes rigor. No single aspect or phase of data analysis denotes a rigorous qualitative research study. However, sharing your data with other individuals in ways that respect participants and that help you to challenge and understand your own interpretive process helps to generate richer and more complex data analysis. Despite this, not all studies, especially those that guarantee anonymity, are suited to collaborative analysis.

An important part of collaborative analysis is shared dialogue. Even when working as a lone researcher, there are many benefits to systematically structuring points across the duration of the study to generate dialogue with others about your data. Ultimately, a goal of qualitative data analysis is to faithfully represent participants' experiences in the most complex and contextualized manner possible. Subjecting your interpretations to the scrutiny of other individuals (both internal and external to the research setting) is an important part of achieving this complexity. Again, while there are cautions about collaboration, which include that not all collaborations are useful and that thought partners must be carefully considered at the selection stage and throughout, engaging with others in meaningful ways that inform your data analysis generally yields more benefits than drawbacks. In Recommended Practice 8.4, we detail an additional dialogic engagement exercise, *peer data analysis review,* as a helpful way to engage others in your data analysis processes. Practices such as these, and the ones described throughout the book—which can be adapted to specifically focus on data analysis—should be done at multiple points during the research and data analysis processes. Subjecting your interpretations to the perspectives of others is an important component of conducting rigorous and valid qualitative studies. In the next section, we describe additional strategies for scrutinizing your analyses and themes.

Recommended Practice 8.4: Peer Data Analysis Review Session

The goal of this exercise, which we do routinely with our students and colleagues, is to get outside perspectives on a section of data that you are analyzing. We recommend working in groups of three to four. Each individual will bring an excerpt of data that is approximately two single-spaced pages in length. The research question(s) and one to two sentences describing the setting/context should be included at the top of the first page.

We encourage students to select a "meaty" piece of data, one that, while brief, is interesting, confusing, and/or intriguing. The goal is to choose something that you want help thinking about. The excerpt can be from one data source (e.g., an excerpt from one interview or fieldnote entry) or can include excerpts from multiple data sources. We limit the length of the excerpt because of time constraints; however, your group can determine the length of the excerpt and duration of the session, which should be divided evenly among all individuals with a timekeeper to ensure equal time. Each group participant should determine in advance how her time will be spent and communicate that to the group. For example, you may want to have your peers do an initial read of the data, paying attention to what stands out to them. Alternatively, you may already have codes developed that you bring and share with your group to see if they would code this excerpt in a similar manner and to discuss how they might define them and other kinds of codes that they identify.

Generating, Scrutinizing, and Vetting Themes

As you progress through the iterative processes of data analysis, you will begin to develop, from your codes and how they will ultimately be combined and clustered, analytic *themes,* or categories. Themes represent important concepts in the data and are often generalizable to a data set (Gibson & Brown, 2009). Despite this, themes do not necessarily need to reflect patterns and commonalities. A variety of aspects influence what ultimately become themes, and central to these are your research questions. Themes do not simply emerge from data (Braun & Clarke, 2006). You, as the researcher and the primary instrument in your research, actively construct and develop themes. This occurs as you continue to engage with your data through the analysis strategies described in the previous section as well as through writing processes, which we describe subsequently.[3]

Analyzing your codes is one overall approach for generating themes (Braun & Clarke, 2006). You may begin to see relationships between codes, combine codes, and discard some codes. In Table 8.4, we detail what a sample process for developing themes may look like. It is important to understand that researchers generate themes in a variety of ways. We include this sample only to help you to understand the general process. Many qualitative researchers recommend using data displays to help you develop themes.[4] Data displays, as Miles et al. (2014) describe them, take the form of either matrices (rows and columns) or networks (lines and arrows) and are ways of displaying information to help draw conclusions. These can be useful when consolidating data from different sources or types of data as well as seeing relationships between data (Miles et al., 2014). See Appendixes O and P for examples of network data displays. See Appendix Q for an example data display that uses emically developed categories from a pilot study and maps them onto the research questions of a new study.

Table 8.4   A Sample Process for Developing Themes

- Read all of your coded data
- Begin to group and combine (and possibly delete) codes

  ○ As you do this, analyze your codes and develop themes
  ○ Consider overlaps, disjunctures, patterns, and what they tell you about the data
  ○ These themes may have the same names as your codes or they may be different

- Document these themes, what they mean, and what codes went into them in memos

  ○ Look to your research questions for the relationship of codes to themes
  ○ Look to theory as a framework to analyze data in thematic ways

- Re-revisit and recode your data with your themes

  ○ Determine what is missing
  ○ Generate subthemes

- Tell the story of your themes in writing

  ○ Include data that support your themes
  ○ Explain how your themes fit into the broader understanding of your data
  ○ Explain the relationship of the themes to your research questions
  ○ Consider the various ways that theory can inform the themes

- Throughout these processes, write memos to make sense of your emerging understandings and discuss these with thought partners and advisers as needed

Analysis and Validity

As you analyze your coded data to develop themes, you will refine and revise these themes. This involves going back and rereading your entire data set to determine if your themes accurately reflect your data, and this may also involve recoding sections of your data (Braun & Clarke, 2006). There are no set rules for how much a concept should be present in your data in order to become a theme. This is one reason why it is important to scrutinize themes by checking and rechecking your interpretations against your data as well as looking for alternative explanations and possible misinterpretations. A primary concern is to not "force" the data to conform to your preconceived notions, and one strategy to incorporate is to search for what is sometimes referred to as "negative cases" (Roulston, 2014). This involves looking for cases (e.g., people, instances) that do not fit a particular pattern or your current understanding of the data; people call these *disconfirming evidence, negative cases, discrepant data,* or *outliers.* We find this language somewhat problematic because it has the potential to inscribe (or reinscribe) deficit orientations or a marginalizing stance toward those who do not fit what the researcher may consider to be the norm. This seems particularly ironic in qualitative research since a core value of the paradigm is to understand each individual's unique experience in ways that are contextualized and specific as well as to understand the range and variation of experiences, perspectives, and beliefs. Before researchers consider an individual or group an "outlier," there should be focused discussion about why this term is applied and what it means in the context of the study. While looking for disconfirming evidence is an important analytical strategy, it is important to not just look for what you might call "outliers" but also to look for evidence that will challenge and complicate your findings. Furthermore, we like to consider outliers our teachers. What can we learn from them? How do they complicate and challenge our understandings and interpretations? The goal of looking for disconfirming evidence or "outliers" is to challenge your preconceived notions as well as the themes you are developing. The more you challenge, question, and look for alternative explanations, the more complex your interpretations will be. A key lesson here is to question and challenge what you (or others) view as a norm and why.

In addition to seeking alternative explanations, the validity strategies we describe in Chapter Six are important to continuously engage with throughout the research process and especially during data analysis. For example, what we refer to as *multiple coding* (Barbour, 2001)—which means multiple researchers coding the same data to look for shared understanding and to check the validity of the codes as they are developed, refined, and codified—should not be considered a one-time exercise to see how individuals agree and/or disagree about codes but should be an opportunity for research teams to develop more complex understandings of their data (Barbour, 2014). We highlight validity strategies in relationship to data analysis in Table 8.5 and encourage readers to revisit Chapter Six for a more detailed discussion of these strategies.

In addition to the validity considerations we present in the table Table 8.5, if you are not transparent in your processes and documenting how you developed themes, readers cannot fully ascertain the validity of your study. Stated differently, you must document and describe your specific process of coding and theme development and relate these explicitly to your interpretations and analysis so that readers understand how you derived meaning from the data. Related to this, Braun and Clarke (2006) detail five pitfalls to avoid when conducting thematic analysis; we summarize these as follows:

1. Failing to analyze the data by either stringing data extracts together and/or simply paraphrasing the content of extracts.

2. Making questions from an instrument the analytic themes. In this regard, no analytical work has taken place.

3. Having themes that do not make sense, overlap too much, and/or do not cohere.

4. Having data that do not support your claims and/or support other claims.

5. The theoretical framework and the analytical processes, research questions, and claims do not align.

Table 8.5   Validity Strategies and Data Analysis

Strategy	Data Analysis Considerations
Triangulation	• Have I subjected my data to methodological, data source, researcher, and theoretical triangulation? • How do my data align or converge? • How do my data differ or diverge? • How am I making sense of the points of alignment and divergence? • How do I need to revisit and challenge my interpretations based on what I am learning from my data?
Participant validation	• How have I engaged participants in my study? • How am/did I adjust my study design based on early engagement with participants? • Have I shown the participants their respective transcripts? Have I shared interpretations and analysis? What kinds of questions did I ask? How were these asked, and how did participants respond? • *Important note: Participant validation strategies should NOT happen only at the end of a study. As described in Chapter Six, they must be ongoing so that you can make use of and adjust the study/analysis based on engagement with participants. When implemented at the end of a study, the participant validation becomes more of an item to check off of a list and less of a validity strategy.*
Strategic sequencing of methods	• As I formatively analyze my data, what choices should I make about the order in which I implement methods? • How do the choices about the methods I am using/I have used affect my analysis and interpretation? • Am I considering the way that methods are connected to findings as I review the data on participants and how the data sources relate with and build on each other? • What are ways to analyze data that bring multiple data sources into productive tension with each other?
Thick description	• How am I describing my data? • Are my descriptions sufficiently detailed? • When individuals not familiar with the research context read about my data, can they accurately understand the context and setting? • Am I providing enough information about the contextual factors and participants' thoughts and experiences so that others can make their own interpretations of my data?
Dialogic engagement	• How have I engaged others in the data analysis process? What might I do differently or do to improve this? • What have I learned from these engagements? • How am I making sense of and formatively incorporating their feedback?

Strategy	Data Analysis Considerations
	• How might I more rigorously invoke and engage with other people's perspectives on my data and my analysis of these data?
	• *Note*: See Chapters Three and Six for specific dialogic engagement exercises.
Multiple coding	• In what ways did our coding align?
	• In what ways did our coding diverge from each other?
	• How am I (or are we) defining codes?
	• What stood out during and after this process?
	• What about our coding scheme needs to be revised?
	• What alternate understandings of the data exist that I may be overlooking?
Structured reflexivity processes	• How are you, as the researcher, influencing your data and analysis?
	• What are other possible interpretations?
	• See specific structured reflexivity exercises, including memos, dialogic engagement exercises, research journals, and other strategies, described throughout the book. As data analysis is formative and ongoing, these exercises should be practiced throughout the research process.
Disconfirming evidence	• How have I challenged myself to come up with alternative explanations of my interpretations?
	• What stands out as counter to my emerging themes and potential findings?
	• Am I thinking there is a norm? If so, why and what shapes my perspective on that?
	• What data do I need to engage with (or collect) to learn more about possible alternative explanations?
	• What can I learn from participants and/or experiences that are different from the patterns I developed?

Ultimately, the goal of qualitative analysis is to describe a rich, detailed, and complex account of your data that maintains fidelity to participants' ideas, perspectives, and experiences and how they communicate them. Conducting a thorough, robust, and transparent analysis is a primary way of maintaining this fidelity.

Terms and Concepts Often Used in Qualitative Research

Data saturation: In qualitative research, data saturation refers to the point at which you are no longer finding "new" themes in your data. This can be a problematic term, as there can be infinite themes or ways or looking at data. Furthermore, this concept implies that there is a reality that you can find, which is something that qualitative research pushes against. However, based on your research questions, you can reach a point of data saturation in that you are continuing to see recurring patterns and concepts in your data or have enough data to sufficiently answer your research questions.

Data Saturation

Although data analysis is an iterative process, studies ultimately must end. You may have heard researchers discuss reaching a point of data saturation. Data saturation, as defined by Saumure and Given (2008) in *The SAGE Encyclopedia of Qualitative Research Methods,* refers to the following:

> Saturation is the point in data collection when no new or relevant information emerges with respect to the newly constructed theory [that happens as the data are being collected]. Hence, a researcher looks at this as the point at which no more data need to be collected. When the theory appears to be robust, with no gaps or unexplained phenomena, saturation has been achieved and the resulting theory is more easily constructed. (p. 195)

Data saturation entails how you are able to tell that you can speak back to the core constructs in and goals of your research questions. As you engage with your data and move from primarily descriptive codes, or what Maxwell (2013) terms *substantive codes,* to more analytical or theoretical codes, you are beginning to make sense of the sorted data. What you are specifically looking for depends on your research question(s). Some qualitative researchers no longer refer to a saturation point so as to not imply that researchers can find a single "Truth" (Marshall & Rossman, 2016). It is crucial to involve others as you attempt to determine if you have reached saturation or if this concept is even relevant to your study. Even if you are a lone researcher, colleagues and participants should be consulted before you make that determination.

WRITING AND REPRESENTATION

Writing is integral to qualitative data analysis, and it involves far more than the formal process of writing a final report. Writing occurs ongoingly throughout the research process, including the writing that you do to engage with your data analytically; the writing of memos, journals, reflections about participant validation, and other validity measures; and the defining and refining of codes. Engaging with the data through multiple strategies, including various kinds of analytic writing, will help you to determine how to best craft your final product.

As van den Hoonaard and van den Hoonaard (2008) state,

> Qualitative researchers arrive at a more profound analysis of the data when they engage in writing up the data as soon as possible. These short or lengthy writing bouts often yield insights that were not readily apparent even after the coding had been completed. Indeed, researchers may find that they need to go back to the data to recode for concepts that became apparent during the initial writing up of the data. (p. 187)

Some individuals employ visual or graphic displays as a starting point for bringing their analysis together. Others process by writing and composing drafts upon drafts (we include ourselves in this category) to process our data and analysis. Others spend time drafting and revising outlines. There is no one approach to beginning to compose the final write-up. In Chapter Nine, we discuss writing up final reports as well as additional writing strategies and complicate the notions of writing and representation. Here we include writing and representation since it is a key aspect throughout the process of data analysis and is not simply a summative act.

Memos

We have discussed memos extensively throughout the book and want to emphasize their contribution to the analytical process. Memos should be used throughout qualitative data analysis for theoretical and personal reflection; theory building; considering evidence for theoretical links; describing research processes, procedures, questions, and emergent issues with detail and transparency; and documenting new and different questions. Writing memos not only captures but also facilitates thinking (Maxwell, 2013). Furthermore, looking back at your memos over time can be very helpful in data analysis, and depending on the research questions, memos are sometimes coded as data. Either way, tracing your thinking throughout a study and reading the memos of your research team colleagues can be insightful and help to document and reflect on your analytical processes as well as to move your analysis forward. Analysis is

> both the concept and the thought processes that go behind assigning meaning to data. Analysis is exploratory and gives consideration to different possible meanings in data and then keeps a record of the thought that took place before arriving at a possible meaning. The thought process is recorded in a memo. This makes analysis a dynamic and evolving process. (Corbin & Strauss, 2015, p. 58)

We have described several memos throughout this chapter as well as in other chapters. We detail an additional memo in Recommended Practice 8.5 as another way of engaging with your data by creating a vignette, which is a carefully selected moment, event, piece of data, or reflection on the research process that you write up in a short piece of text (usually one to two paragraphs) for reflection and discussion about its meaning. We encourage students to choose a moment or piece of data that feels representative or emblematic of something they see as interesting or generative in their data and/or research process.[5]

Recommended Practice 8.5: Vignette Memo

This is a specific type of memo in which you include a vignette that speaks to what you are learning and seeing in your setting. The vignette can take multiple forms, and there is no specific way you should compose it. In addition to the vignette, include a few reflective paragraphs at the end of the memo. Use the vignette as a basis for reflecting on one or two of the following topics:

(Continued)

(Continued)

- Your research design

- The relationship between questions and methods

- The roles of being a researcher

- Other central issues about which you have questions or concerns

Other Writing Considerations

The amount of raw data included in final written reports varies depending on the goals and the audience(s). For example, a 20-page journal article is quite different from a 50-page evaluation report, a 30-page master's thesis, or a 200-page dissertation. The research questions, goals, purposes, and audience(s) of the study often determine how much data are included in a final report. However, including more data does not mean that a study is "better" or "more authentic." All studies, regardless of the amount of data included, must resist the imposition of interpretative authority to the fullest extent possible. Final research products of all kinds (e.g., reports, articles, video documentaries) are edited by researchers, and because of this, qualitative researchers must acknowledge (both in our minds and in the products we generate) our power and the limitations of our inferences.

As you read through and write about the data, the notions of emic and etic,[6] described in Chapters Five and Six and discussed in relation to coding above, are not just about the articulation or conceptualization across researcher and participant (or co-researchers if engaging in a participatory approach), but the notions are also important in challenging your understanding of how things get conceptualized in the field and ultimately in representation of the data. Many qualitative texts discuss emic and etic in a linear fashion—occurring at the moment of introduction of a researcher to the setting or a concept. This may often be the case; however, it is important to understand the way notions of emic and etic can affect all aspects of a study and especially when conducting analysis. Thinking about the ways concepts and ideas are introduced and discussed in a study highlights the powerful role of language in analysis. Writing, which we describe as integral to data analysis, is also incredibly powerful because researchers have the ability to represent others' lived experiences, and we discuss this power and the many considerations that come with it in the next chapter.

QUESTIONS FOR REFLECTION

- What are the primary strategies and methods you can use to analyze your data?

- What strategies can you use to manage and organize your data?

- What decisions and implications should you consider related to transcribing data?

- What are the processes for precoding your data?

- What does it mean and look like to immerse yourself in your data?

- What is coding and what are the different approaches to coding data?

- What are some considerations when using Computer-Assisted Qualitative Data Analysis Software (CAQDAS)?

- How can you structure dialogic engagement exercises as you analyze your data, and what do these exercises help you do?

- What are the key processes for developing data-based themes?

- How do you works toward validity during data analysis?

- What is the role of writing in data analysis?

- How can you work to resist imposing interpretative authority throughout the analysis process?

RESOURCES FOR FURTHER READING

Bernard, H. R., & Ryan, G. W. (2010). *Analyzing qualitative data: Systematic approaches*. Thousand Oaks, CA: Sage.

Gibbs, G. (2007). *Analyzing qualitative data*. Thousand Oaks, CA: Sage.

Gibson, W. J., & Brown, A. (2009). *Working with qualitative data*. Thousand Oaks, CA: Sage.

Miles, M. B., Huberman, A. M., & Saldaña, J. (2014). *Qualitative data analysis: A methods sourcebook* (3rd ed.). Thousand Oaks, CA: Sage.

Patton, M. Q. (2015). *Qualitative research and evaluation methods* (4th ed.). Thousand Oaks, CA: Sage.

Rapley, T. (2007). *Doing conversation, discourse and document analysis*. Thousand Oaks, CA: Sage.

Saldaña, J. (2013). *The coding manual for qualitative researchers* (2nd ed.). London, UK: Sage.

Silverman, D. (2015). *Interpreting qualitative data* (5th ed.). Thousand Oaks, CA: Sage.

Sullivan, P. (2012). *Qualitative data analysis: Using a dialogical approach*. Thousand Oaks,, CA: Sage.

ONLINE RESOURCES

Sharpen your skills with SAGE edge

Visit **edge.sagepub.com/ravitchandcarl** for mobile-friendly chapter quizzes, eFlashcards, multimedia resources, SAGE journal articles, and more.

ENDNOTES

1. For more on this, see Chapter Five in Ravitch and Riggan (2012), which uses Frederick Erickson's (1996) work, "Going for the Zone: The Social and Cognitive Ecology of Teacher-Student Interaction in Classroom Conversations," to explore the role of one's conceptual framework in analysis and engage analytic issues such as transcription in an innovative way.

2. See Maxwell and Miller (2008) for examples of connecting strategies and Gee (2011) for a discussion of identifying connections in data.

3. Themes are outcomes of coding (Saldaña, 2013). While a code may sometimes ultimately become a theme, it is important to note that themes do not just emerge from the data (Braun & Clarke, 2006).

4. See Miles et al. (2014) for a detailed discussion and examples of data displays.

5. For more on vignettes, see Erickson (1986). For examples of the use of a vignette, see Wong (1995) and Ravitch and Tillman (2010).

6. For additional discussion of the concepts of emic and etic, see Headland (1990), Lett (1990), and Morris, Leung, Ames, and Lickel (1999).

Writing and Representing Inquiry

The Research Report

And maybe stories are just data with a soul.

—Brené Brown (2010)

How [stories] are told, who tells them, when they're told, how many stories are told, are really dependent on power.

—Chimamanda Ngozi Adichie (2009)

CHAPTER OVERVIEW AND GOALS

Throughout the book, we describe the integral nature of writing within and across all aspects of qualitative research and suggest many writing practices, including a range of memos, collaborative writing activities, and a research journal. These help structure and model a process of continuous writing as a means of active sense-making and engagement with the research. In the same way that we describe data analysis as iterative, so too is the process of writing in qualitative research. The more you engage with your data through both formal and informal writing processes, the better you are able to understand the data, think about them critically and meta-analytically, and ultimately communicate the meaning of your data to others. All of this writing leads into the writing of the final study report, which is the primary focus of this chapter.

There are many aspects to consider when formally writing up qualitative research, including how to make informed decisions about the goals, audience(s), purposes, and focus of the study; the format and structure of final reports; incorporating and representing data; voice, language, and participant portrayal; and the processes of outlining, drafting, and revising written reports. Building on previous chapters, we again underscore the power inherent in qualitative research and in the representation of study participants, their contexts, and aspects of their experiences and argue for respectful, authentic, and ethical representations of individuals and groups.

BY THE END OF THIS CHAPTER YOU WILL BETTER UNDERSTAND

- The ways in which writing is an integral part of all aspects of the research process

- The relationship of ongoing research writing to the final research report

- The importance of considering your study's purpose, goals, and audience(s), as well as the processes that can help you decide upon these aspects of your study

- Considerations related to a research report's structure and format

- The components of a final research report

- Issues related to the ways in which data are included and represented in research writing

- The balance of and lines between description, interpretation, data inclusion, and analysis in written reports

- Considerations relating to issues of voice, language, and participant portrayal in research reports

- Understanding how your interpretations of data are/should be relational, contextual, nonevaluative, person centered, temporal, partial, subjective, and nonneutral

- Ethical considerations related to writing and representing research

THE CRAFT OF WRITING: OUTLINING, DRAFTING, AND REVISING

Writing can be thought of as a set of technical skills, as an art, and/or as a craft. For most people, writing takes a great deal of time and energy, effort, and process. Each writer has a distinct style. When asked if there is a "right way to write," Sharon often shares the story with our students of the differences between her own approach to writing up her research and that of her primary writing partner in graduate school. While Sharon would write a general outline and use that to structure her report by writing into the outline section by section, putting out her thoughts in quite rough ways and then working to refine them in multiple, painstaking drafts (somewhere between 8 and 12 revision cycles), her research thought partner would not write a single word until she had worked out a detailed outline in her mind and had crafted much of the prose in her head. By the time she wrote, she only needed one to two subsequent revisions. The two women both ended up with high-quality qualitative research reports, but their writing processes and styles were strikingly different. We begin this chapter with this story to make the point that you should proactively think about your approach to writing, including outlining, drafting, and incorporating feedback into drafts. There is no right way to engage in these processes, but there are more planful approaches that should match your style and needs and that may have implications for your timeline and process. A few of these approaches include creating an outline and a detailed timeline, writing as often as you can, and engaging with others throughout the writing process. We discuss each of these in the subsections that follow.

In the same way that we argue for the utility of creating a detailed timeline for data collection, we argue that it is important and valuable to create a detailed timeline for the writing process. The timeline should include opportunities for seeking and incorporating feedback from others as well as for your own processes—from outlining to writing various sections through the entirety of the writing process. Set realistic goals and give yourself time to write while holding yourself accountable to your goals. If working on a dissertation or master's thesis, share this timeline with your chair and/or adviser to be sure that you have allowed enough time for the various stages of the research. If working on a course paper, be sure to allow enough time for inevitable drafting, revising based on feedback, and questions that will emerge as you write so that you have time to reach out to your professor and peers for advice if needed. If working in a research group, make sure that each and all group members have a say in the timeline and agree to the interactions and deliverables that a shared timeline will require.

One very important early step in the report writing process is creating an outline, which should be vetted for feedback from thought partners, co-researchers, and/or advisers and should also be included in your timeline. As you create a detailed outline for your study report, consider the relationship between your guiding research questions and the themes that you have in focus. An outline should include all sections (major and subsections) of the final report; it helps you to zoom in on the themes and findings from your data specifically since this will create the new analytic and discussion sections of the report. As you develop a detailed outline, consider the structure of the report, any page limits and requirements, and how you want to build your story and your argument. However you structure your outline, we suggest using it to directly guide the structure and content of your report. Make sure that you have a clear understanding of how much space you have for the analysis and findings, as well as the ensuing discussion, recommendations, and so on, since this may guide you to make choices about how many findings you choose, which findings are more or less in focus, and how you will structure the reporting of your data.

Once you begin writing, consider how you want to approach writing the first draft of the report and then how this will parlay into subsequent drafts. Consider your writing style, preferences, constraints on time, inclusion of various readers and revisions of drafts, and so on as you chart out this process. Many qualitative researchers emphasize the importance of allowing the time needed to adequately reduce and condense your final product "until you feel you've captured the essence and essentials of your study" (Saldaña, 2011, p. 141).[1] This includes incorporating feedback as you write and revise. To this end, Wolcott (2009) states,

> The critical task in qualitative research is not to accumulate all the data you can, but to "can" (i.e., get rid of) much of the data you accumulate. That requires constant winnowing, including decisions about data not worth entering in the

first place. The idea is to discover essences and then to reveal those essences with sufficient context, yet not become mired by trying to include everything that might possibly be described. (p. 39)

This winnowing of content that is nonessential is one mark of solid research reports since it allows for the focus to emerge. We believe that for most researchers, especially novice researchers, having other sets of eyes on your emerging report significantly helps the winnowing process. "Good" qualitative research reports, regardless of the format, genre, or style, are engaging, clear, useful, rigorous, respectful of participants, focused, insightful, and relevant (Saldaña, 2011). This requires time and the careful integration of feedback from at least a few peers and/or advisers.

There is a lot of good advice about writing,[2] but like all advice, take what is useful to you and do not feel discouraged if you do not implement the same routines and strategies as your colleagues and advisers. We offer two pieces of advice related to writing: to write often and to share and discuss your work with others as frequently as possible. While eliminating distraction and having large chunks of time to write is ideal, it may not always be possible. This is why we suggest writing as often as possible. For us, writing is a cogent form of sense-making that helps us to develop and further our ideas. It helps us engage in focused dialogue with participants, peers, and mentors about our work so that our interpretations are challenged in generative ways. Thus, when we first begin writing, we do not pay close attention to syntax and style but rather to content, with the understanding that we will produce multiple drafts. For example, we lost count after Chapter One of this manuscript reached its 20th iteration. Syntax and style matter, but in an effort to begin writing, we do not worry about these issues in our first drafts. Rather, those drafts are focused on the conceptual connections and central arguments we are making, and we build those with complete focus. We suggest that you save your various drafts as distinctly named files since you may want to refer back and use parts of earlier drafts that are changed or deleted based on later insights and feedback.

In addition to writing often, we strongly recommend regularly sharing your work with others to get feedback that informs your writing and revisions at all stages of the report development process. We have suggested dialogic engagement activities throughout this book, and we highly recommend forming a writing group in which you are familiar with each other's work, but you do not have to be researching similar topics or even using similar methods. This can be a group that you actually get together and write with or it can be a group that you meet with to share aspects of your thinking and writing over time in order to get focused feedback. See the previous chapter for who may be the best individuals to include in a writing group. It is important to have individuals who will challenge you on your biases and assumptions, and this is particularly useful at the writing stage.

Discussing your work with others is helpful in all writing processes. Specifically, speaking about your work in a structured format can help you to solidify nascent conceptualizations. Recommended Practice 9.1 describes one way that you can verbally share your work with peers in order to receive feedback. This process, which we call *mini presentations,* will help you to clarify your ideas and your thinking about how you are narrating the story of your data as well as help you to polish and refine presentation skills.

Recommended Practice 9.1: Mini Presentations

In your research or writing group (or with a group of peers), practice giving a 3- to 5-minute "presentation" of your study that focuses on its primary research questions, goals, and findings. The group should determine the time limit ahead of time and strictly adhere to this. We recommend not allowing more than 5 minutes. This forces you to be succinct and to practice distilling the key aspects of your study in a limited amount of time. This exercise can be great practice for presenting your work, which can help you to clarify and solidify your ideas. Furthermore, practicing in front of peers can allow you to "safely" determine gaps in your argument and areas for improvement in your presentation content and style. Your group may develop a specific protocol for providing feedback on the presentations, or each group member may come prepared with specific questions that she wants the rest of the group to consider in regards to her presentation and/or topic. We recommend delivering these mini presentations multiple times as you write up your study and, if possible, assigning someone in the group to take notes so that the presenter can focus on the conversation and have a second set of notes to compare to any notes he may have also taken. These sessions can also be recorded so that you can return to the full conversation if need be.

Note: This whole exercise can be done in a course context as well with either voluntary or assigned pairs or triads who can work together over the course of a semester both in and out of class time.

WRITING A FINAL REPORT: BUILDING ON, DEEPENING, AND CODIFYING ANALYSIS

> Reporting is not separate from thinking or from analysis. Rather, it is analysis. (Miles et al., 2014, p. 325)

This chapter builds on the previous chapter about data analysis because writing is an integral part of analytical processes. Writing is the way to take analytic insights and transform them into narratives. As Gibson and Brown (2009) state, "Analysis is not analysis until it is written down" (p. 194), which means that without writing up your analysis, through the development and articulation of themes and findings that become your key arguments in response to your research questions, it stays an internal thought process rather than a piece of research that others can engage with and learn from. And while we argue throughout the book that writing is an integral part of sense-making and should *not* occur only at the end of a study (since, among other reasons, that would be incredibly overwhelming for you as the researcher), the writing that comes at the end of a study can be cathartic and deserves additional attention and explanation.

The writing process, which is a part of the three-pronged approach to data analysis discussed in the previous chapter, continues and even deepens as you begin to formally "write up" your research. In the subsections that follow, we discuss aspects that are important to consider in qualitative research reports.

Goals, Audience(s), and Purposes

The goals, audience(s), and purposes of your study directly influence writing decisions, including the format, style, and content of your written report. Being clear on what you want to accomplish and whom you intend to read the report will help you to be systematic and purposeful in the choices you make throughout a study and specifically at the writing stage.

The audience(s) you envision speaking to with your written work, which is an outgrowth of the study's purpose(s), is a central consideration in how you structure the report. How you structure and write reports depends on the readers you envision and conversations you want to enter into within specific fields or disciplines. For example, writing for a policy audience with readers who will likely base policy and funding decisions on the findings of your report would lead you to a different structure and tone than writing for an academic audience that is primarily interested in theory generation and methodology. Another example is that writing for an audience of practitioners in your field could mean that the report might be more implementation oriented since the audience may be most interested in using the findings to inform practice or structure programs and therefore that it is possible that less time is spent on the formal literature review. (We want to be careful about overgeneralizing with these examples since there are not necessarily clear boundaries that separate that which is scholarly with that which is practice oriented. For example, some practitioner research is practitioner scholarship (Ravitch & Lytle, 2015), which would mean that there is the same level of review of formal theory and methodology and that it is not necessarily or exclusively intended to be guiding to practice or program development. It depends on which practitioner audiences you are speaking to/with.) The expectations of these audiences may be different, and this shapes your approach to writing and the structure and tone you take up in the final study report. Likewise, writing a dissertation, master's thesis, or empirical study for a research course means that you have what we think of as "audience near," meaning your dissertation committee, thesis adviser(s), or professor, who often hail from different fields and subfields that shape their expectations. "Audience far" is anyone outside of your committee and advisers, people who are not directly involved in the design and development of your research and will read it from a distance. In a dissertation, master's thesis, or research study for a course, you are trying to satisfy both of these audiences, whereas in other forms of writing that are made public through formal publication or other forms of dissemination to broad audiences, you are typically thinking only of audience far.

In Table 9.1, we detail questions to consider as you think about your study's purposes and decide on the audiences for your work.

Articulating Your Study's Purpose

In relationship to writing and the consideration of purpose and audience, Wolcott (2009) stresses being able to cogently articulate the purpose for the study by completing the following sentence, "'The purpose of this study is. . . .'" (p. 34). One of Sharon's qualitative methods professors used to call this "the elevator speech of your research" and stressed the importance of being able to succinctly describe the study's purpose and

Table 9.1 Questions to Consider Related to the Purpose(s) and Audience(s) of Your Study

- For what audience(s) is this study being written? What informs these choices?
- For whom am I writing and why?
- What am I hoping to achieve with this report?
- Do I hope that my research will help solve a problem or address a concern of local interest? How am I viewing "local"—as a monolith or with consideration of multiple, complex ideas of "local"?
- Is my goal to contribute to theory? If so, which ones and in what ways?
- Do I want to publish in an academic journal? Contribute to the improvement of professional practice? Create community awareness? Some combination of these? What informs these goals?
- Do I have overlapping or intersecting goals? Conflicting goals? How will these various goals shape the ways that I write?
- What genre of writing is this: academic report, article, book, policy report or brief, case study for specific audience, program evaluation, ethnography, action research report, and so on? What guided the genre decision?
- Whose voices will be privileged? Why? Why not others?
- Who is heard in my writing and why? Who is not heard and why?
- Are there other audiences who could benefit from my learning and knowledge? Which ones and in what ways?
- How might I structure my writing to fit other audiences' needs?
- What is the relationship between the expectations of writing for my audiences near and far?
- How do my intended audiences shape my sense of how to structure and write my report? Inform which terms I may need to contextualize and define?

primary audiences. (In fact, he used to have Sharon and her classmates engage in the research version of "speed dating," which we refer to as a "Speed Research Exchange" and describe in Recommended Practice 9.2. Sharon and her classmates all found it very useful as students and so we facilitate this with our own students early on in the research development/design stage and then again at the reporting stage.) Being able to cogently describe purpose and audience means that you need to be clear enough about the purpose(s) of the study and the audience(s) you are choosing to speak to in order to make key decisions that set up the structure of your report in important ways. It also means that you need to be able to articulate your goals and purposes in cogent ways that relate them to the ways in which you choose to structure the report.

Like Wolcott, Patton (2015) recognizes the importance of being clear about the purpose of your research. He states, "Purpose is the controlling force in research. Decisions about design, measurement, analysis, and reporting all flow from purpose" (p. 248). In research, purpose is about the goals and objectives of the study going into it as well as those that might have emerged as a result of the process and data collected and analyzed. There are many ways to conceptualize purpose, and thinking about this concept specifically can be helpful when designing, implementing, and analyzing data and writing up studies.

There are five different purposes of research on a theory to action continuum (Patton, 2015, p. 248):

1. *Basic research:* To contribute to fundamental knowledge and theory
2. *Applied research:* To illuminate a societal concern or problem as you explore solutions
3. *Summative evaluation:* To determine if a solution works (i.e., program or policy)
4. *Formative evaluation:* To improve a policy or a program during implementation
5. *Action research:* To understand and attempt to solve a specific problem in as efficient a way as possible

This basic overview is helpful since the goals and audiences, as well as expectations, will vary depending on the purposes of your study (Patton, 2015; Wolcott, 2009). And you, as the researcher, know your data best, know the formal theories to which your data relate, and therefore are the primary determiner of these decisions about the readership of your study. The people who will read your report (in whatever forms) are directly related to your goals for the research and in the report that stems from it. To be even more specific, if the goal of your study is to contribute to the broader academic conversation about a certain topic, then your audience will likely be other scholars of that topic. If the goal of your study is to improve a specific program, then the audience will likely be practitioners who designed or implement that program and/or individuals who fund the program. Depending on the program, you may also be speaking to policy makers as well as other scholars studying similar kinds of programs. This is to say that there may be multiple audiences for any given report.

As you begin to write your report, you should revisit, in a focused way, what the goals and purposes of your study are and then think about the various audiences who will read your report. These audiences may include participants, program administrators and operators, practitioners, other researchers, policy makers, general readers, and mass readers (Miles et al., 2014). A study may reach a different audience (or set of audiences) than intended; however, to achieve the study's purpose, you should be clear about who will read your work. This will help you to make the appropriate decisions as you progress throughout the study and when writing. The study's approach (i.e., ethnographic, narrative, action research) also influences how you write about your research (see Creswell, 2013).

Recommended Practice 9.2: Speed Research Exchange

Preparation and setup: Prepare a small piece of paper with your research questions for as many people as you will meet with during this activity. This should be decided upon ahead of time; we suggest at least three and no more than five rounds. Set up a room with however many small tables you need given if this is in a class or in a smaller inquiry or writing group. Each person should prepare and rehearse a 2-minute overview of their research, timing it to be sure it is in fact 2 minutes and not longer.

Activity: Have a timer (or phone) ready to time intervals. In pairs, share the piece of paper with your research questions on it with each partner. Set the timer for 2 minutes and in this time state your research questions and the context of the research and then share your one to two key purposes for the study and two to three audiences that you envision for the study (this often happens in tandem). The other person takes notes and then has 5 minutes to address goals and audiences, asking questions and sharing feedback that will help you refine your sense of the purposes and audiences of the research. Everyone takes notes.

We suggest that you do this at least twice, once early on in the research development/design stage and then again at the reporting stage. In our classes, we have debriefing sessions for discussion of questions that surfaced through engaging in this process. We encourage that you engage in these debriefing sessions and take notes.

Format and Structure of a Final Research Report

There are many ways to structure a final report, and these depend on the qualitative approach, audiences, goals, research questions, and style of the writer. The rhetorical structure of a final report may differ based on the qualitative approach used (e.g., narrative research, action research, grounded theory) (Creswell, 2013).[3] In addition, the extent to which the presence of the researcher is communicated in the write-up may vary by approach as well, with the researcher's presence being more explicit in ethnography and less in grounded theory (Creswell, 2013). This does not mean that one approach is more or less "objective" than another. As we have discussed throughout this book, the researcher is a primary instrument in qualitative research (Lofland, Snow, Anderson, & Lofland, 2006; Porter, 2010). Whether you are more or less explicit about this does not mitigate this fact. We, like the vast majority of scholars of qualitative research, recommend being as explicit and as transparent as possible about your role, as the researcher, in all research processes, including the analysis and writing choices you make.

One question related to this that we are often asked is if the data analysis and findings sections of final qualitative research reports should be structured directly in response to the research questions (i.e., should the research questions actually be the headings that structure the display of data and analysis). To this, we answer that findings can be organized by and written in direct response to your research questions, but we, like other scholars (e.g., Holliday, 2007), believe that a more holistic picture is painted when findings are organized by and written in the structure of key themes that speak to your research questions but are not directly structured as a response to them. This allows, among other things, for "the emergence of independent realities which may be counter to or hidden by the dominant preoccupations of the researcher"

(Holliday, 2007, p. 94). It also allows for you to articulate the major learnings of your study in ways that are more holistic and inductive rather than determined and shaped by the research questions, which can constrain the creative thinking and analysis by the researcher.

There are no standardized report structures or formats in qualitative research, although a common format includes an introduction, a section that synthesizes relevant literature, a section on research design and methods, a section that presents data and key learnings, a section that discusses the analysis, and a conclusion. The introduction of the written report provides contextual and background information, describes the study's purpose and goals, frames the research questions that guide the study, and provides a conceptual overview for the rest of the text, stating how it will be structured with a clear rationale for why it is structured that way. The conclusion discusses the key learnings from the study and the implications of the research for theory, research, and/or practice. The conclusion should not simply be a summary but rather should be generative: "Summaries reiterate what has been said; conclusions deal with the 'so whats'" (Glesne, 2016, p. 241). While not the final goal, summaries are still useful in qualitative reports, and many final reports end with them. It is important to note that summaries can be incorporated throughout the report as a way to keep readers on target instead of having one long summary at the end (Wolcott, 1990). Wolcott (1990) avoids using the term *conclusion* and suggests avoiding "dramatic but irrelevant endings or conclusions that raise issues never addressed in the research. Beginnings and endings are important. . . . Look for ways to make them better without letting them get bigger" (p. 56). This is great advice. The way you approach the introduction and conclusion, as well as the entire report, involves many factors as described above and is also about your preference and style as a researcher and writer; there is no "right" way to write a report, but there may be audience-specific norms and styles that it is important to be aware of and adhere to in your report. In the subsection below, we detail one way to approach a final qualitative research report.

Sample Qualitative Report Structure

Traditional qualitative reports, at a minimum, should state the purpose and topic of the study, describe the many contextual factors of the study's setting and participants, detail the methods by including how concepts emerged and evolved, present data, contextualize and discuss these data in relationship to relevant literature and/or the research setting, and address the study's significance and implications for its intended audiences and in relation to its expressed goals. To make this process more clear, we provide a sample template for a written qualitative research report in Example 9.1. We adapted this template based on guidelines that Sharon created for our students (who hail from many fields, disciplines, and programs) since this helps them to develop, with their advisers and others within their specific programs and fields, their own customized structures and formats. In Appendix S, we provide an example of a final qualitative report that largely follows this structure.

Example 9.1:
Sample Final Report Template

Introduction

The primary goal of this section is to introduce and contextualize your research topic and research question(s) within relevant field(s) and disciplines as well as in relation to the setting and contexts (e.g., macro, micro, organizational, community) that shape it. You should provide a sense of the overall context of the research study in terms of setting context/location, demographics, and other relevant background information. In this section, which may include subsections, the primary goals are to

- Describe and frame the research questions

- Provide background and context that frames the multiple contexts (including temporal) that shape the study and inform the setting and/or research milieu

- Explain the rationale and significance of the study. This, which may be a subsection depending on how you structure the introduction, introduces the "why" and "so what" of your study. The goal is to help readers, who are typically outsiders to your setting and the research, understand why the study is important and significant to relevant field(s), to you (as the researcher), possibly to the setting itself, and to furthering understanding in your substantive area(s) through the articulation of your focus in the context of the field(s) as a whole.

- You should *integrate literature* throughout the introduction to help frame, contextualize, and justify your study.

Conceptual Framework

The goals of the conceptual framework section are to contextualize the study in terms of key contexts (macro, micro, participants, and setting), framing concepts and influences, study goals, framing theories, and researcher identity and positionality. In addition to exploring and explaining all of the concepts and contexts above and their generative intersection and interaction in your study, the conceptual framework includes the study's *theoretical*

(Continued)

(Continued)

framework, which is a critical integration of empirical and theoretical works that contextualize and create the theoretical framing for your study; positions your research questions within relevant, cutting-edge research within and across fields; and makes an argument for why the proposed study is important, necessary, and what it will contribute to the field. This section should also include how the study approach and goals stem from and relate to the study's overarching methodology to create an understanding of the relationships between study goals, contexts, values, theories, and the methodological choices you detail in the research design.

The theoretical framework, which many refer to as the literature review (but we argue is tighter and more focused than a traditional literature review), comprises a large portion of your conceptual framework because it frames your topic, the research questions, the goals of the study, and the multiple context(s). As such, the review of literature needs to be integrative and meta-analytic. The goal is to position your research questions and focus within relevant, cutting-edge research in field(s) that you identify as salient to the study of your topic. Organize these bodies of literature thematically (through the use of sections and subsections that are conceptually developed, organized, explicated, and narrated) and include a conceptual overview at the beginning of the section that contextualizes each body of literature and discusses how and why it frames your topic and specific research questions. The carefully framed and reviewed literature forms the theoretical framework of your study and as such must contextualize the topic and research questions and their core analytic constructs in your questions (e.g., social mobility, leader professional development, communities of practice).

Methodology and Research Design

In this part of the report, you detail *how* you structured and engaged in the investigation into your research question(s). This section typically begins with an introductory overview to the overall methodological approach, research design, and the section as a whole. Depending on your research approach, you may include the following aspects and subsections:

- A subsection on *Site and Participant Selection* that includes a contextualizing and specific description of the setting and study participants and how/why they were selected given the research questions (with a specific subsection that describes and explains the *selection criteria* used to make these decisions).

- A subsection on *Data Collection* that details each data collection method, including a rationale for the choice of each specific method (e.g., interviews, focus groups, observation and fieldnotes, reflective writing, document review), the nature and scope of the information/knowledge you gained from the method, and the specific structures and procedures of each method with rationales that relate to your research questions and the study goals. As well, describe the strategy/logic behind how methods are timed and sequenced (*Sequencing of Methods*) and be clear about the research design in terms of how data were triangulated in data collection. The data collection section should include not only data collection strategies, such as interviews, focus groups, observation and fieldnotes, member checks, and the like, but also additional forms of data such as research memos, research journal, specific archival/document data, and other kinds of data collected.

- In the *Data Analysis* section, include a specific description of the processes by which you analyzed your data. Topics to discuss include how data were transcribed, how they were analyzed (inductively, deductively, or a combination of both) how themes were developed, what analytical approach and processes were used, how codes were developed, examples of codes, analytical triangulation, and so on. Another way to structure this is by creating subheadings related to each of the three prongs of data analysis that we detail in Chapter Eight, including *data organization and management, immersive engagement,* and *writing and representation*.

- Include a section about *Issues of Validity* in which you discuss threats to validity and how you addressed each one methodologically by structuring validity measures such as triangulation, strategic sequencing of methods, pilot study, piloting data collection instruments, participant validation processes (member checks), and collaborative processes including dialogic engagement (critical inquiry groups). Discuss issues of validity that specifically relate to your identity and positionality as a researcher (this may be included in a subsection titled *Researcher Roles and Positionality*).

- Throughout your Methodology and Research Design section, it is important to cite methods literature to describe and substantiate your choices of methods and processes.

(Continued)

(Continued)

Findings

The goal of this section is to share the major findings from your study. Begin this section with an overview that helps to orient your readers. Organize this section around three to five major themes that emerged from your data analysis and that help you to respond to your research question(s). Each finding should be its own heading/section of the report. In these sections, draw directly upon your data, including and analyzing carefully selected excerpts from the data that both substantiate your interpretations and assertions, bring to life your claims, and make clear the perspectives of the participants in their own words, framing this in well-formed, organized analytic arguments that you make. All data excerpts should be cited with the source (e.g., Participant Name or Pseudonym, Interview, Date). It is important that you use this section to articulate a clear line of argument about the themes and findings that you have culled from your analysis of the data and that you use data and cite theory or theories that you have used to frame the argument.

Discussion and Implications

This part of the report answers the "So what?" of your study as it looks across the three to five major findings discussed in the findings section.

- In the *Discussion* section, discuss the significance of what you discovered broadly as well as in terms of how it sheds light on your research questions. Be sure to link this discussion to how these findings relate to current thinking in the field(s) that contextualize your study (i.e., frame using the literature from your theoretical framework).

- The *Implications* section discusses implications that emerged from your analysis of the data. You may discuss implications for (1) theory development in relevant field(s), (2) future research in this area, and (3) your various audiences such as policy, practice, and so on.

A goal of this chapter is to offer options and suggestions for you to consider as you write and represent your qualitative work. You will incorporate what works for you, your style, your audiences and goals, and the story that you are telling. You can take Example 9.1 and adapt it according to the requirements and needs of your particular study and milieu. Appendix S provides an example pilot study written by Charlotte Jacobs, one of Sharon's doctoral students and a colleague at our applied research center, who wrote the report as her final assignment for our introduction to qualitative methods course. Charlotte adapted a general format in Example 9.1 in relation to her own goals and the research questions while adhering to course requirements.

In addition to the general conventions and structures of writing that we overview here, it is important to note that there are alternative ways of representing qualitative research. Qualitative research can be presented in formats other than a journal article, book chapter, or conference presentation; it can be shared and disseminated in the form of theater, poetic texts, unperformed performance texts, storytelling, alternative presentations, multimodal quilts, films, websites, workshops, and brochures (Keen & Todres, 2007, para. 9).[4] And more traditional and creative forms can be combined in interesting ways as well. It is important to consider that alternative textual forms and representations do not necessarily avert the problem of the authoritative researcher and often place the researcher more prominently in the text (Hammersley & Atkinson, 2007). Furthermore, as in all research, ethical issues need to be considered in writing that relate to issues of representation as well as confidentiality, transparency, and authenticity. We discuss these considerations briefly at the end of this chapter and in depth in Chapter Eleven.

To help you determine how to format and structure your final report, we suggest that you reflect on and discuss the questions presented in Table 9.2.

Table 9.2   Questions to Consider for Determining How to Structure a Final Research Report

- How will my report be structured and organized?
  - As a historical, natural, chronological narrative that re-creates the fieldwork process of exploration and discovery?
  - As a chronology that follows a developmental cycle?
  - As a narrative with analysis following in a separate section?
  - Using a thematic approach in which I present a typology of concepts?
  - Using a biographical approach, in which I chronicle each participant's story?
  - Using a priori theoretical structures as my framework?
  - In some combination or different form than suggested above?
- What will influence the above choices?
- Will I introduce the participants or will quotes be unattributed (i.e., will data be aggregated)?
- Do I report my findings as "objective" (The effects are . . .) or as constructed realities (The participants perceive and report . . .)?
- How can I use a zoom lens, moving recursively between concrete descriptive details and abstract theoretical ideas?
- How do I connect my data to other research in a way that builds trustworthiness, positions my research in the field, and adds to the "so what" of my findings?

Incorporating and Representing Data

Data do not speak for themselves. Readers need you to lead them through the meaning of quotations, tables, and charts. In qualitative research, as in quantitative research, data serve as evidence for the argument you are making (Holliday, 2007). Having said this, the ways data are presented in qualitative texts can vary considerably. For example, in ethnographic accounts, you will tend to include significant amounts of "thick description"—that is, contextual description, based on data collected through observation and fieldnotes, that represents the setting and participants in ways that depict and describe the context in depth and with texture so that the readers can imagine and feel the place and people and will then have interpretative chapters that move from description to interpretation (Wolcott, 2009).

Other kinds of qualitative research, while still describing the scene with care and attention to setting context, will typically spend less time on the thick description since this is not an expressed goal of those approaches and the written reports that stem from them. These examples relate qualitative approach to written text, but there are other influences such as the forum of the written report as well. For example, in academic journals, you will be limited as to the amount and type of data you can include as well as in how much information you can include about the study methodology, which is in contrast to writing a paper for a course or a book or a book chapter. Regardless of the format or structure, as a researcher and author, you should guide your readers in and out of quotations so that you make sure that the meaning is clear and encourage them to read the quote (Richardson, 1990). Quotes are the primary way that you will bring the participants' voices into the written report, and therefore it is vital to make sure that you structure your text in ways that draw the readers into the data as they read your interpretations of the data. Furthermore, it is important to make your interpretation clear and distinct from the data to let the readers understand "how subjectivity has been managed" (Holliday, 2007, p. 106). This is an important point to consider as you structure your report and determine how you will include and integrate the primary data into it.

Using Quotations

There are different ways to approach using quotations in qualitative texts, and the way you incorporate quotations into your written work largely depends on the topic, type of study, the goals, the audience, and your writing style. Three general types of quotations are typically used in qualitative texts: short, embedded, and long (Richardson, 1990). Short quotations allow you to provide examples of a theme or aspect of your data in a focused way. With short quotations, you can present ways that different participants discuss the same topic or illustrate a particular theme. Embedded quotations consist of short phrases or words embedded or interspersed throughout the text. While embedded quotations often do not have the context of other, longer quotations, they can be useful for helping to focus on a certain thematic aspect. Furthermore, it is important to remember that raw data do not constitute reality; they are still the researcher's construction since they have been selected from a larger corpus of data and that means that these choices are shaped by the researcher's interpretations of the data (Holliday, 2007). Long quotes are typically employed when discussing complex phenomena. Because a lot of information may be included in these quotes, readers should be directed to what is important and why, and the quotes need to be embedded in meaningful analysis (Creswell, 2013). The vitality and credibility of texts are enhanced when a variety of different quotes with multiple voices and images are included (Richardson, 1990). Determining which data to include is a subjective process that is often driven by the themes that organize the text and selecting the data that are most illustrative of these themes (Holliday, 2007). It is important to remember that quotes do not make an argument; they support and enliven it.

Generally, quotes should be short, as illustrative as possible, contextualized, interpreted, and integrated with the rest of the text (Kvale & Brinkmann, 2009). While it is important to incorporate quotations in a manner that flows with the manuscript, we do not recommend "cleaning up" quotations. As we discuss in previous chapters and in detail in Chapter Eight, we believe that this can change the meaning of the quotes as they were actually stated by the people whose views you are trying to authentically represent. While any attempt of "speaking" for another person in a narrative is problematic, deliberately changing an individual's words can create a slippery slope of researcher liberties (to change people's words and perspectives) with which we are not comfortable. If you edit quotes in any way, this must be made clear to your audience (each manual—such as APA or the Chicago manual—has specific guidelines for how to do this), and you should provide a clear, grounded, and cited rationale for doing so. One exception to our stance involves editing proper names from quotes for confidentiality. However, even in these cases, it is still necessary to inform your audience if and how quotes were edited. For example, depending on what assurances were made regarding confidentiality or anonymity, identifying information such as individuals' names as well as names of programs and/or places may need to be removed. As with editing all quotes, make sure to inform your reader and edit quotes consistently. As we discussed in Chapter Six, discussing these issues with participants and thought partners can help to determine how and if quotes will be used and edited (Roulston, 2014).

Other Considerations for Representing Data

Representing data can take many forms, including a written report, a presentation, images and displays, a film, and/or many other manifestations. It is essentially how you communicate your study findings to others. As stated in the previous chapter, you must be clear that you are presenting your interpretations and inferences even though these are framed as data-based findings, which they are. Given that data are both real and interpreted, it is important to provide enough information about the context and other relevant features of the study and especially to transparently represent your processes so that your readers can determine the validity of your claims. Including larger quantities of raw data (either as in more or longer data excerpts) does not necessarily make a study better or more participant centered. We often read final reports in which students present a significant amount of raw data in an attempt to allow the data to "speak for themselves." We appreciate the spirit of this idea, but it can easily slide into a situation where there simply is not enough analysis, which is crucial to the representation of the overall learnings.

While data cannot speak for themselves and therefore require that the researcher embed them in an analytic framework, the researcher should include participants' direct statements in ways that speak to the overall arguments and learnings of the research. Furthermore, "raw" data do not represent a participant's reality since these data are being excerpted from a larger corpus and since transcription is a form of analysis. Thus, when presenting your data, the excerpts should be chosen and included in ways that are both purposeful and intentional and that make clear that they have been chosen in support of your analytic points and overall argument. This entails determining the utility and centrality of the data that you include in your study (Marshall & Rossman, 2016). It also requires careful, explicit analysis of the data so that it supports and exemplifies your analytic arguments. Questions that may help guide your decisions about how to include and represent your data are detailed in Table 9.3. These are just a sampling of questions that can serve as entry points into considering these issues of data inclusion and representation in your writing.

Table 9.3   Questions for Considering How to Include and Represent Data

- What choices will I make about how I portray the participants and the site of research?
- How will these choices affect the ways that I include data and the kinds of data I include?
- Will my writing be evaluative or descriptive? Some combination? What would influence these choices? How will that influence how I represent the participants and include data?
- Is there a possibility that readers could identify participants? What are the risks to confidentiality? How can I address these concerns?
- How does the overall approach I am working within shape how I represent participants and include data?
- How can I structure this report so that I achieve the right balance between the inclusion of data and my interpretations?
- What are ways to structure my findings so that it is clear how they relate to the corpus of data?
- How will I structure the report so that data are properly contextualized and triangulated?
- How do I include excerpts in ways that will draw the readers in? Create energy around them?

Finding a Balance Between Description and Analysis

In the report itself, we find that the great majority of our students struggle precisely with the balance between describing and analyzing their data, that is, in figuring out the balance of and the lines between description, interpretation, data inclusion, and analysis in their written reports. Analysis should organize the description of data in a way that makes it clear and accessible, and this organized description allows readers to understand your interpretations (Patton, 2015). It can be useful to distinguish between "particular description," "general description," and "interpretive commentary" (Erickson, 1986, p. 149). This entails that the raw data (data excerpts) are included in the study report as the particular description, the patterns that you discover in your data are written up as the general description, and the next layer of meta-analytic patterns is written in the form of interpretive commentary (Erickson, 1986).

Building on Erickson's (1986) framework, particular description consists of raw data, which include direct quotes from participants as well as cited excerpts from researcher-generated data such as observation fieldnotes, research memos, research journals, and vignettes that the researcher constructs. These excerpts of the data help you to describe the setting, people, and their statements in real-time ways that provide insight into the ethos and feel of the setting and people within it. General description is used to explain and contextualize these data, to articulate their place within the broader corpus of data, that is, to show whether and how the data excerpted are emblematic and representative of the data and setting as a whole. Finally, interpretive commentary helps readers make sense of the connections and broader arguments that you are trying to make (Erickson, 1986). This commentary is "necessary if the reader is not to be lost in a thicket of uninterpretable detail" (Erickson, 1986, p. 152).

Balancing between the particular and general and the description and analysis can be difficult. It is for this reason that our students (and even our colleagues at times) typically, at least in early drafts, present too much general description and in doing so do not write in ways that are sufficiently concrete or abstract. This proves an ongoing, cross-researcher challenge. And there are no set rules to guide these choices and writing processes, which means that we are all left to figure this out as we write each report. When students are facing this dilemma and ask us to help them figure out what to do to address it, we suggest that they find a few well-respected and well-written studies in their field and read these as examples, looking precisely for how the authors achieve these balances. We also suggest that they engage with others, including advisers and peers, in reviewing their early drafts with an eye toward this specific issue. Since the goal is to embed the analysis in the data and to help your readers understand that the data examined are the foundational basis of your argument, it is important to ask others if they find your arguments to be well grounded and plausible.

Voice, Language, and Participant Portrayal

The question is not whether we will write the lives of people—as social scientists that is what we do—but *how* and for *whom*. (Richardson, 1990, p. 9)

While the previous section focuses on how data are included and represented, this section focuses on how choices you make about author voice, language usage, and ways of representing individuals and groups influence your writing in ways that are important to consider before you start writing. Journal editors, advisers, and dissertation committee members may have preferences for style, tone, and pronoun usage. We strongly argue for the use of first person to signify the role of the researcher(s) in the study instead of trying to assume nonexistent objectivity and distance with the third person. Absenting of yourself in the write-up is disanalogous and in conflict with the values of qualitative research. However, the voice that you use depends on the goals and type of research you are conducting. For example, an evaluation report is different from a case study report, and the conventions of pronoun usage and tone differ. Equally as important to how you portray yourself as a researcher in relation to your study is how you represent participants and/or groups in your study. How are people described and characterized, and what role (if any) did participants play in making those decisions? Even if participants did not play a role, which they often do not, you must be very careful to resist the imposition of interpretative authority to the fullest extent possible by paying attention to how you portray individuals and groups as well as to how you consider issues of validity, accuracy, and ethics.

It is important to be intentional and explain why you are making the voice and language choices that you are making. To help you think about these issues, we detail important questions to consider in Table 9.4.

In addition to these questions, we always ask, *If people in my study were reading this, how would they feel? Would this resonate for them?* While of course we cannot actually know the answer to this question unless we ask multiple study participants (and in ways that create the conditions necessary to try to elicit their honest responses and possible pushback), this is the ultimate test for everything we write since we consider fidelity to participants' experiences as the primary consideration in qualitative research. How would we feel if we were portrayed in these ways?

To underscore this point, we provide an example from Sharon's experience as a study participant. She was asked many years ago to be a part of a study about first-generation

Table 9.4 Questions to Consider Related to Participant Voice and Language Choices

- How am I representing people, experiences, and places?
- What kinds of adjectives and specific descriptive language do I use for people, and what influences these choices? What assumptions and biases inform these choices?
- How do I bring in their voices, and what informs these choices?
- How are quotations and other data used in the report narrative? What informs these choices? Do they help me accomplish my goals in terms of how I represent participants? Why or why not?
- Am I making arguments grounded in my data or are they an inferential leap or both?
- What data support my arguments?
- Did I look for disconfirming evidence? In what ways might I try to do more of this? What do these data tell me?
- How do I as the author represent myself in the narrative?
- What forms of authority do I use to make my case? What mediates these choices?
- To what extent will literary devices be used? Which ones and why?
- Would a new writing format change what I know? How and in what ways?
- Do the participants have a say in how they are represented? Should they?
- Will my writing be accessible to the participants? Will they ever know how they are being portrayed? If they did, how might that affect how I represent them?
- Are there terms, concepts, or kinds of language that I am using that have been challenged by thought partners and why? How am I responding to these questions or concerns?
- Are there ways that I might include participants in some of the decisions about voice and language choices? If so, how? What might be some of the challenges and benefits of this?

female college students. Since she and her sister were the first women in their family to attend college, she was excited to be a part of the study. She actively engaged in multiple rounds of interviews and a focus group as well as other forms of data collection. When the study was published, she excitedly looked to see how she was portrayed. She was surprised to read these words ascribed to her: "Sharon's experience of being a Russian immigrant greatly influenced her experience of being a college student." She read these words and the entire passage about her with surprise and disappointment. Sharon is not a Russian immigrant; her family has been in the United States for three generations. The researcher misrepresented her. What Sharon shared in her interviews was that she is Jewish, that her ancestors had emigrated from Russia, and that her college experience was very much shaped by her family's narrative and by their gratitude for the opportunities and freedoms afforded to them in the United States. Not only were the facts inaccurate, but they also blurred a major cultural/historical line that is sensitive to many Jews from Russia since at the time when Sharon's family lived in Russia, Jews were an oppressed minority, and her family, like so many other Russian Jewish families, left Russia to escape severe religious persecution. Sharon's family members (including Sharon) do not even consider themselves Russian despite the fact that their ancestors were born there. Russia was considered an oppressive place to escape, not a homeland or culture to which Sharon or her family feel they belong or would ever claim as a proud part of their identity. Rather, they consider themselves Jewish as both religion and culture and eschew ties to Russian culture. The researcher's carelessness—whether it was caused by sloppiness or manipulation of the facts to create a specific kind of narrative about Sharon—ended up being a problem not only in its inaccuracy but in that it offended Sharon and made her question the researcher's skills and ethics. This is an important example of how researchers need to be very careful about how participants' identities and experiences are portrayed.

There are no easy answers or solutions to these kinds of questions about representation. One way we address this is by incorporating language that accomplishes what we consider *tentative specificity*. What we mean by this is that we do not generalize for a group of people or for all participants. If we make a claim, we state specifically to whom and what this claim is referring. For example, using language such as, "For the participants in this study, I argue that. . . ." or "According to this participant. . . ." We do not consider this being apologetic in our assertions; rather, we consider it ethically and methodologically important to clearly describe whom we are talking about and how we are basing our claims. This language of tentative specificity allows us to make claims with our data but keeps us from generalizing and lumping participants together.

We discuss reflexivity at length throughout this book. As a horizontal in our approach (and in qualitative research more broadly), we consider researcher reflexivity as the connective tissue across all qualitative research processes, including determining analytical interpretations, thinking about representations, and considering ethical issues. It is an ethical responsibility to examine our roles as researchers and challenge our biases, assumptions, and epistemologies. As we have suggested throughout this book, structured reflexive practices and dialogic engagement exercises are two ways that you can attempt to challenge these notions and delve into the many complexities that come with the role of the researcher in qualitative research. The final report, whatever shape it takes, needs to reflect these examinations. This will inevitably look differently depending on the research questions, the study's goals, overall qualitative approach, and the audiences for the product. However, you must engage in due diligence with respect to these issues no matter the venue or format and write with a voice that is simultaneously authoritative and nonimpositional. This can be a delicate balance and one that many novice researcher writers struggle with. In the next section, we detail critical aspects that all writers of qualitative texts should consider regardless of the goals, research approach, or format.

CRITICAL WRITING CONSIDERATIONS

> You write in order to change the world . . . if you alter, even by a millimeter, the way people look at reality, then you can change it. —James Baldwin (cited in Pipher, 2007)

Thinking critically about writing and representing your research involves asking questions such as, "For whom do we speak and to whom do we speak, with what voice, to what end, using what criteria?" (Richardson, 1990, p. 27). As a writer, you must consider these issues no matter what or whom you are writing about because there is power in the production of all texts, and we, as writers, should think about the ways in which our texts reproduce and/or challenge social structures (Richardson, 1990). The power inherent in qualitative research becomes even more salient when you begin to write up your research since this is what takes the personal, the political, and the contextual from your data and transforms it into a message shared with the world. Writing makes your research public in particular ways that are powerful and important to carefully scrutinize. Finding the appropriate authorial voice requires that you make writing decisions that shape the structure, format, and style of your report.

The characteristics of qualitative interviews (relational, contextual, nonevaluative, person centered, temporal, partial, subjective, and nonneutral) that we detail in Chapter Five apply to all aspects of qualitative data collection, analysis, and when writing and representing your study. Understanding that your interpretations are and should be articulated as relational, contextual, nonevaluative, person centered, temporal, partial, subjective, and nonneutral and clearly communicating this in your report is a crucial aspect of critically and ethically approaching qualitative research. We discuss specific aspects for you to consider related to research writing in Table 9.5 on page 293 and continue discussing these issues in Chapter Eleven.

CONCLUDING THOUGHTS: THE ETHICS OF RESEARCH WRITING

> Writing is political act. (Glesne, 2006, p. 191)

> The personal is political. —Hanisch (1970)

Table 9.5   Considerations for Writing and Representing Qualitative Data

Characteristic of Qualitative Data	Considerations for Writing and Representing Qualitative Data
Relational	• How am I presenting the relational nature of my interactions with participants? • How have I discussed with participants how they will be portrayed in the final report? • How have I asked for (and incorporated) participants' feedback? In what ways and was this adequate? Accurate? • How am I dealing with and representing power asymmetries, including the researcher-participant relationship and its possible implications?
Contextual	• How have I represented the study's contextual factors (macro and micro)? • How do I understand the role of context in what emerges from the data? • Does my description of the context differ from participants' descriptions? If so, in what ways? What can this teach me about the context? • Have I provided enough information for readers to understand the contextual factors at play? Including the temporality of mediating contexts near and far?
Nonevaluative	• Am I rendering the realities I depict in ways that are evaluative? If so, in what ways and why? How can I challenge my assumptions underneath these evaluations? • Am I judging various aspects of the participants' choices, realities, lives? If so, in what ways? What shapes this? How did it affect the collection and analysis of data? And how might it affect how I represent the participants and their experiences or perspectives? • How have I represented how my presence affected the data I generated? • How have I described the ways in which my biases may have affected the data I generated?
Person centered	• How am I staying true to each individual even as I try to describe patterns, differences, and/or similarities? • How am I showing respect to participants as I write about them? Since respect is culturally mediated, whose understanding of respect is informing these choices? • How am I viewing and writing about each person as a unique individual? • Am I essentializing participants in any ways? If so, how and why? • How am I involving participants in decisions about representation? • How am I acknowledging my power as I write about participants?
Temporal	• Am I accounting for ways that the temporal quality of the data affects my analysis and interpretations? What does this look like and mean? • Am I clear about the role of historical and current sociopolitics (and micro-politics) and how they affect the data and my interpretations of the data?

(Continued)

Table 9.5 (Continued)

Characteristic of Qualitative Data	Considerations for Writing and Representing Qualitative Data
	• Have I accurately captured the context of the setting and participant experiences as the participants describe them? • How might this context have changed or be changing over time?
Partial	• How am I acknowledging, as I write about individuals, that the data collected provide just a snapshot of people's lived experiences? • How have I communicated the partial nature of the data in my representation of setting and people? • How does the partial nature of the data affect what is knowable? What I can confidently argue in relation to my research questions? • How might I need to qualify or contextualize this in my findings?
Subjective	• How is this product shaped by my subjectivities? Which ones and how? How can I challenge my assumptions and constructions of reality? • How have I communicated my roles in the research processes? • Have I connected this adequately to data collection, analysis, and write-up? • How am I accounting for the ways that my subjective experience and the subjective experience of the participants come together? Shape the research and the data? • How have I challenged my interpretations? How might I do so more vigorously?
Nonneutral	• How am I accounting for the inherent power asymmetry present in the data collection, analysis, and writing processes? • What have I done to address issues of bias, power, and their impact on the participants and readers? • How am I representing the multiple layers of bias (mine and the participants') present in the data? • What is the relationship between the nonneutrality of data collection and data analysis?

In Chapter Eleven, we discuss research ethics as a primary consideration in qualitative research, not only ideologically or relationally but methodologically. We choose to end this chapter by underscoring that writing is not a neutral act; rather, it reflects our biases and assumptions in myriad ways that must be carefully scrutinized throughout the writing processes that comprise an approach to research that seeks complexity and criticality in research reporting. This, we argue, is key to ethical research. As we have discussed throughout the book and in this chapter, representation of participants is at the core of ethical issues in research writing. This includes the following:

- Resisting essentializing individuals and groups (bearing in mind the notions of intra- and intergroup variability discussed in Chapter One)

- Assuming similarity and difference about individuals and groups

- Not inferring beyond what is valid and reasonable as evidenced by carefully analyzed data

- Challenging our biases and how they become embedded in our data sets and interpretations of the data

It also includes issues of confidentiality and anonymity and the writing choices we make related to protecting the identities of the participants in our studies. These issues of ethics in writing all revolve, to some extent, around issues of power and respect, and it is important to remember that writing is a complex endeavor.

There are many complicated issues at play in issues of representation, such as how the researcher is represented and how, "even when 'we' allow the 'Other' to speak, when we talk about, or for them, we are taking over their voice" (Denzin, 2014, p. 573). This issue of the "Other" in research representation is a deeply epistemological and ontological one. There is a long history and legacy of impositional "othering" in anthropological research that is strongly critiqued by postcolonial scholars. These scholars take issue (as we do) with representations of the "Other" since these kinds of representations often exoticize, fetishize, and essentialize study participants and work from underlying deficit orientations that the research ultimately then reinscribes (Valencia, 2010). Those who critique this othering aspect of research processes and products problematize its roots and argue for an interrogation of its outgrowths in both historical and contemporary research accounts (Chakrabarty, 2000; Chilisa, 2012; Said, 1989; Spivak, 1999).

We also want to note in this chapter on writing that there are ethical considerations related to authorship that you should consider. This might mean the ethical decisions of how to include participants in formal written reports—that is, as authors? Coauthors? Mentioned in the acknowledgments? These issues revolve around how to decide what authorship means, and these kinds of considerations must happen as early as possible to set and calibrate expectations. Issues of ethics around authorship might mean how to decide on the order of authors given multiple authors. For example, is this decided by alphabetical order of authors' last names? By who does more of the writing and/or background work? By who conceptualized the research and writing and convened the group or dyad in the first place? By whether and how earlier data collection by some or all of the authors is used in addition to data collection by the researchers or research team? And it might include deciding if one should even publish from their work at all given other considerations or concerns. For example, there have been numerous times when Sharon has decided that it would be a conflict of role or interest to publish on some of her applied research when her role is also as an adviser since once she chooses to publish on the work, it might inform how she advises in ways that she views as a conflict of interest. Issues of authorship should be carefully considered through an ethical lens so that issues of power or privilege do not cloud the ways that fairness and equity in authorship are determined.

QUESTIONS FOR REFLECTION

- How would you characterize the relationship of ongoing research writing and writing of the final research report?

- What are the key aspects of a final qualitative research report?

- How do you determine what structure your final report should take?

- How do you understand and approach the balance between description, interpretation, data inclusion, and analysis in a written research report?

- What kinds of questions, discussions, and processes can help you decide on your study's purpose, goals, and audience(s)?

- What are key considerations related to a research study report's structure and format?

- What primary considerations relating to issues of voice, language, and participant portrayal should you consider?

- What factors affect the ways in which data are included and represented in writing?

- In what ways are your interpretations of data relational, contextual, nonevaluative, person centered, temporal, partial, subjective, and nonneutral?

RESOURCES FOR FURTHER READING

Brodkey, L. (1994). Writing on the bias. *College English, 56*(5), 527–547.

Denzin, N. K. (2014). Writing and/as analysis or performing the world. In U. Flick (Ed.), *The SAGE handbook of qualitative data analysis* (pp. 569–584). London, UK: Sage.

Dzaka, D. (2004). Resisting writing: Reflections on the postcolonial factor in the writing class. In A. Lunsford & L. Ouzgane (Eds.), *Crossing borderlands: Composition & postcolonial studies* (pp. 157–170). Pittsburgh, PA: University of Pittsburgh Press.

Eisner, E. (1997). The promise and peril of alternative forms of data representation. *Educational Researcher, 26*(6), 4–10.

Holliday, A. (2007). *Doing and writing qualitative research* (2nd ed.). Thousand Oaks, CA: Sage.

hooks, b. (1999). Remembered rapture: Dancing with words. In *Remembered rapture* (pp. 35–45). New York, NY: Henry Holt.

Ivanic, R. (1998). Issues of identity in academic writing. In *Writing and identity: The discoursal construction of identity in academic writing* (pp. 75–106). Amsterdam, the Netherlands: John Benjamins.

Keen, S., & Todres, L. (2007). Strategies for disseminating qualitative research findings: Three exemplars [36 paragraphs]. *Forum Qualitative Sozialforschung/Forum: Qualitative Social Research, 8*(3), Art. 17. Retrieved from http://nbn-resolving.de/urn:nbn:de:0114-fqs0703174

Lather, P. (1996). Troubling clarity: The politics of accessible language. *Harvard Educational Review, 66*(3), 525–545.

Marker, M. (2004). Theories and disciplines as sites of struggle: The reproduction of colonial dominance through the controlling of knowledge in the academy. *Canadian Journal of Native Education, 28*(1/2), 102–110.

Patton, M. Q. (2015). *Qualitative research and evaluation methods* (4th ed.). Thousand Oaks, CA: Sage.

Richardson, L. (2009). *Writing strategies: Reaching diverse audiences*. Thousand Oaks, CA: Sage.

Wolcott, H. F. (2009). *Writing up qualitative research* (3rd ed.). Thousand Oaks, CA: Sage.

ONLINE RESOURCES

Sharpen your skills with SAGE edge

Visit edge.sagepub.com/ravitchandcarl for mobile-friendly chapter quizzes, eFlashcards, multi-media resources, SAGE journal articles, and more.

ENDNOTES

1. For detailed advice regarding multiple aspects of qualitative writing, see Saldaña (2011).

2. See Wolcott (1990), Creswell (2014), Glesne (2016), and Saldaña (2011) for specific advice about writing qualitative reports.

3. For more of the different rhetorical structures used in qualitative approaches, see Creswell (2013).

4. For a more detailed discussion and examples of alternative ways of representing qualitative work, see Keen and Todres (2007) and Paulus, Lester, and Dempster (2014).

CHAPTER TEN

Crafting Qualitative Research Proposals

CHAPTER OVERVIEW AND GOALS

Our focus in this chapter is writing effective qualitative research proposals. We describe in detail how to write qualitative research proposals since this is the place where study goals and processes become solidified, and where ideas about audience, purpose, and the relationship of theory, prior research, and methods come together. We also devote a chapter to this process because students tend to have a lot of anxiety related to proposals whether they are for a class assignment or master's or doctoral thesis and because all researchers, new and seasoned, must continuously learn the craft of writing and adapting proposals for a range of audiences. In addition to discussing various considerations in and approaches to proposal writing, we include, annotate, and discuss an example of a research proposal near the end of this chapter. The goal of the chapter and the example that we include is to concretely detail many of the aspects of proposal writing that we discuss throughout this chapter and to do so in a way that helps bring key considerations of research design and research writing together. Furthermore, in Appendixes T to Y, we include a range of different kinds of proposals, including examples of conference proposals and funding proposals for various kinds of academic research.

BY THE END OF THIS CHAPTER YOU WILL BETTER UNDERSTAND

- The ways that the goals and audience shape different kinds of proposals (e.g., master's thesis or dissertation proposal, conference proposal, funding proposal)

- The overall components of a qualitative research study proposal

- The goals and content that should be included in the different sections of a proposal, including the introduction, conceptual framework, and the methodology and research design sections

(Continued)

(Continued)

- How to align data collection methods, site and participant selection criteria, and data analysis processes with your research questions

- How to adapt your proposal to aid in the writing of your final report

- Important considerations during and after writing a proposal

QUALITATIVE RESEARCH PROPOSALS

Writing research proposals is an exciting, if also stressful, process that involves an understanding that proposals frame a study but do not determine it. We have discussed throughout the book how qualitative research is iterative and formative. Thus, while you must be clear and deliberate in what you propose to do in your study, it may likely evolve and change. Given this, it is important to remember that while you are striving for clarity and a clear plan, keep in mind that a proposal is, as Sharon regularly says to our students, a plan, not a contract. In most qualitative research, it is expected that a study will change from its initial proposal. However, writing a sound and detailed proposal is an important part of research design. As vitally important as the actual proposal is, it is also good to remember that there is value in the proposal writing process. Putting your ideas and thoughts into writing helps to clarify them as well as highlight gaps or potential concerns. For these reasons, proposals occupy an important place in the research process.

Despite an understanding that research design will evolve, writing proposals can be a daunting task for novice and seasoned qualitative researchers. Because a proposal explains your study to individuals who may be unfamiliar with the context, writing clearly and concisely becomes very important for this task. As you develop an argument for your proposal, there are a variety of important questions to consider. Creswell (2014, pp. 77–78) presents nine questions that should be answered when developing an argument for a proposal, and we share these here to help you to think through your own possible research study and how you will develop your proposal:

1. What do readers need to better understand your topic?

2. What do readers need to know about your topic?

3. What do you propose to study?

4. What is the setting, and who are the people you will study?

5. What methods do you plan to use to collect data?

6. How will you analyze the data?

7. How will you validate your findings?

8. What ethical issues will your study present?

9. What do preliminary results show about the practicability of the proposed study?

Answering these questions, both by yourself and in dialogue with others, will guide how you structure and write your proposal. These questions highlight key components of research proposals, which we describe in depth in the subsections that follow.

Components of Research Proposals

Proposals can vary greatly based on the audiences, purposes, and goals of the proposal. For example, a proposal for funding your research will be structured according to the guidelines set out by the funder (see Appendix W for an example of this). A conference proposal on your research will vary according to the specific association and norms within it as well as the type of format of the conference and your specific kind of session (see Appendixes T–V for examples). A master's thesis (for an empirical study) or dissertation proposal will reflect the preferences of the academic adviser or dissertation chair and committee as well as the norms of that university and even specific programs within it as well as the field or discipline to which it belongs (see Example 10.1). Despite the range and variation of types of proposals, with their respective formats, proposals tend to include an introduction that includes the context and a rationale for your study as well as the study goals and guiding questions, a discussion of the relevant literature, a detailed description of the research design and methodology, and a timeline for completion.

In Table 10.1 and in the subsections that follow, we describe three main components of a proposal: an introduction, the conceptual framework (which critically integrates context, concepts, goals, and theory), and a comprehensive methodology and research design section. During and after reading this section, we recommend that you refer to Example 10.1 for an example of a dissertation proposal. The structure and methodological content of this proposal can also apply to master's theses and capstone projects as well as many kinds of pilot and research studies done in the context of research methods courses. It is important to keep in mind that there is no specified format for a qualitative research proposal, but the template in Table 10.1 can serve as a general guide.[1]

The Introduction

As outlined in Table 10.1, the primary goal of the *introduction* section is to introduce and contextualize your research topic and research question(s) within the field(s) related to your research and the multiple settings and contexts that shape it. You provide a sense of the *background and context* by describing the overall context of the research study in terms of institutional context/location, relevant demographics, and other background information. Finally, by describing the *rationale and significance* of the study, you are helping readers, who are outsiders to your study/setting, understand why the study is important and significant to the field, to you, and to furthering understanding in your substantive area and

Table 10.1 Qualitative Research Proposal Template

Introduction

- Introduce the proposed study, include a brief **overview of the broad issues and milieu that contextualize the study** as well as the study's intended approach. This section should provide an aerial view of the study and frame it broadly, setting the stage for the rest of the proposal.

- Describe and frame the **research question(s).**

- Provide **background and context** about the multiple contexts that shape and represent the setting/study context.

- Articulate the **rationale for and significance** of the study, including introducing the "why" and "so what" of the proposed study, citing existing research to substantiate the value and importance of the study.

- Describe the **goals and purpose** of the study for you, relevant fields, and for the various audiences and possible beneficiaries.

- Topical literature should be cited throughout this section to ground and substantiate the value and context of the study.

Conceptual Framework

- Provide a **conceptual overview** that orients the reader and ties together the various empirical studies, concepts, contexts, and frameworks that you describe and that positions you as a researcher and in relation to the study.

- Discuss the **relevant contexts,** including the setting and population in focus, aspects of researcher positionality, social location/identity and ideology, framing theories, goals, and bodies of literature and their interrelationships.

- Use headings and subheadings to thematically organize and provide a **critical integration** of relevant literature in your **theoretical framework** (often referred to as the literature review, although we argue that a theoretical framework is tighter and more focused than a broad literature review).

- Tie into methodology (in an overview way) to frame **relationship of concepts, goals, and approach to methodology.**

Research Methodology and Design

- Include an introductory **overview** to the methodological approach, conceptual framework, research design, and the section as a whole.

- Describe the **participant and site selection and criteria** as well as **sampling methods** by including a contextualizing and specific description of the setting (site) and study participants and how/why they have been selected given the research questions (with a specific subsection on the selection criteria). Include a discussion of limitations and choices with transparency.

- Describe the **data collection** methods and processes and include a clearly articulated, cited rationale for the choice of each specific method (e.g., interviews, focus groups, observation and fieldnotes, reflective writing, document review), the nature and scope of the information/knowledge you wish to gain from the method, and the specific structures and procedures of each method should be described. (We encourage students to make each data collection method a subheading.) This section should include not only data collection strategies, such as interviews, focus groups, observation and fieldnotes, and the like, but also additional forms of data such as research memos, research journal, specific archival/document data, and other kinds of data collection planned.

- Describe the **sequencing of methods** (i.e., the strategy/logic behind how methods are timed and sequenced; see Chapters Four and Five).

- For the **data analysis** section, include a specific description of the broad approach and specific process by which you intend to analyze your data. For example, discuss how data will be transcribed and analyzed (inductively, deductively, or a combination of both), how data will be coded, how themes will be developed, and so on. Refer to the questions in Table 10.2 to help you develop the data analysis section of a proposal as well as Chapters Seven and Eight for additional considerations to include depending on the methodological approach and research questions. You might also include a data management and analysis plan; see Chapter Eight and specifically Table 8.1 for more information about this. Another way to structure this is by creating subheadings related to each of the three prongs of data analysis, including *data organization and management, immersive engagement, and writing and representation.*

- Describe in detail how you are considering and dealing with **issues of validity/trustworthiness.** This includes describing threats to validity as well as *how you will attempt to address/mitigate these threats.* Describe measures you will use to address validity (i.e., triangulation in data collection and analysis, sequencing of methods, participant validation processes, and dialogic engagement exercises) and describe the rationale as well as limitations of these measures. (See Chapter Six for a discussion of important considerations related to validity and ways to address these.) As you address validity, you may include a subsection about **researcher roles and positionality** in which you discuss issues of validity related to your social location/identity and positionality and how you will address these issues methodologically.

- Address **ethical considerations,** including, but not limited to, institutional review board (IRB) application and approval, informed consent processes, processes to ensure confidentiality or anonymity, rationale for participant compensation (if relevant), and so on. (See Chapter Eleven for a detailed discussion of ethical considerations.)

- Include a **timeline** for completion of each phase of data collection and analysis as well as writing up the study. For funding proposals, you would include a budget.

Note: It is very important to have a conceptual overview to the entire proposal that lets readers know how the proposal is structured and provides a rationale for that structuring. As well, it is important to be attentive to transitions between major and minor sections so that the proposal coheres and has a clear logic readers can follow. **Finally, it is important to have a conclusion that is not a summary but, rather, gives the reader a sense of closure and a reminder on the way out of the importance and value of the study for its various audiences.**

relevant disciplines and fields. A discussion of relevant literature should be integrated throughout the proposal, including the introduction section, and not just in the theoretical framework section of your conceptual framework.

The Conceptual Framework

The goals of the *conceptual framework* section are to contextualize the study in terms of key contexts (macro, micro, participants, and setting), framing concepts and guiding research questions, study goals, framing theories, and researcher identity and positionality. In this section, you will include the theoretical framework (often referred to as the literature review), which is a critical integration of empirical and theoretical works that contextualize and create the theoretical framing for your study; positions your research questions within relevant, cutting-edge research fields; and makes an argument for why the proposed study is important, necessary, and what it will contribute to relevant fields and disciplines. You will also include how the study approach and goals stem from and relate to the study's overarching methodology to create an understanding of the relationships between study goals, contexts, values, theories, and the methodological choices you detail in the research design.

Methodology and Research Design

There is a difference between methodology and methods. Qualitative methodology includes specific research methods you use to collect data(e.g., interviews, focus groups, questionnaires), as well as the related goals, processes, understandings, theories, values, and beliefs that inform them. Thus, the *methodology and research design* section has two primary goals. First, it describes your overall approach to the research and how your conceptual framework interacts with your research design and topic as well as how the conceptual framework guides the research. Second, it details the specific data collection and analysis methods that you plan to use and how these are situated and sequenced in relation to each other to set up the conditions for rigor and validity. You should provide a level of detail that would allow an outsider to understand the context, methods, processes, and procedures necessary to carry out the study. It is important to substantiate the design and methods decisions you are making and cite relevant research methods literature as well as other empirical studies throughout this section of the proposal as well. Paying close attention to the methodology section in studies written by scholars and/or peers who you respect can help you familiarize yourself with research language and with how other studies are conceptualized and structured.

As stated in Table 10.1, you should begin this section by situating your study within a methodological paradigm (i.e., qualitative, quantitative, or mixed methods) and methodological approach. By methodological approach, we mean what research approach guides your proposed study (e.g., ethnography, participatory action research, case study). If your study is not grounded in a specific approach, you can discuss how you are using a general qualitative approach. You should also discuss your epistemological beliefs, ontological values, and interpretive frameworks. For instance, the proposal in Example 10.1 is situated in critical case study methodology. Susan Bickerstaff, in Example 10.1, further connects her research design to her epistemological understandings when she describes that her critical methodological lens comes from her belief in a constructivist epistemology, which means that knowledge is constructed through social interactions.

The methodology and research design section should have two major subsections, one for *data collection* and one for *data analysis*. In the *data collection* section, you describe, in detail, the data you will collect and how you will structure these processes. As Table 10.1 states, you should explain the rationale behind each and all of your methods choices and substantiate these within appropriate research methods literature. Each data collection method should comprise its own subsection. Depending on the format and audiences of your proposal, you may be asked to include data collection instruments (i.e., the specific interview questions or observation protocols) in the appendixes.

The subsection on *data analysis* is where you detail your overall analytical approach (e.g., narrative analysis, thematic analysis) and how you will analyze your data. For example, if you plan to thematically code your data, you should overview what processes you will use, which might include if you intend to use inductive and/or a priori codes and how. If you plan to use data analysis software, you would discuss how this fits in with your overall analysis plan. You should also include detail on how you will engage in formative data analysis (i.e., at which strategic points and toward which design goals) and your plan for integrating what you learn from early analysis into subsequent study decisions and methods choices. Each data analysis stage or specific method of analysis should comprise its own subsection. For example, we recommend a subsection related to the three prongs of data analysis: *data organization and management, immersive engagement,* and *writing and representation*. As with data collection, you should explain the rationale behind each and all of your analytic methods choices and substantiate these within appropriate research methods literature. Questions to consider related to how you might analyze your data are described in Table 10.2.

Table 10.2   Data Analysis Proposal Considerations

- Will codes be developed inductively, deductively, or in some combination? Why? And if a combination, how will this be structured?

- How (and when) will I incorporate formative data analysis processes? What will these processes entail? How will they affect other aspects of research design and development?

- How (and when) will I refine my data collection instruments based on early data analysis?

- Will any aspects of data analysis be participatory? To what extent? Why or why not? What will this entail?

- How will I engage with others in the process of analyzing my data? Are these processes built into my research timeline?

- When and how will I consult extant literature as I analyze my data? What role will it play and why?

- How will preexisting theories (both formal and informal) inform my data analysis? How can I track and explore this?

- How am I, as the researcher, influencing my data? What will I do to address and reflect on this throughout the data analysis and writing process?

- What analytical memos will I compose? How will these be used? Will I share these with others? If so, in what ways and with what specific goals in mind?

Table 10.3   Matrix for Detailing How Methods Align With Research Questions

	Research Questions	Study Goals	Procedures and Processes	Validity Issues	Validity Strategies	Other Considerations
Data Collection Methods						
Specific data collection method (include an entry for each data collection method)	How will this method help me to answer the research questions?	How will using this method help to achieve the study goals?	What specific processes and procedures will you follow?	Describe threats to validity	What strategies will you incorporate to help address threats to validity?	Address any additional considerations or concerns and how you will try to mitigate them
Participant Selection Criteria						
Specific sampling technique (include an entry for each sampling technique)	How will you use this process to answer your research questions?	How will this participant group help achieve the study's goals?	What specific processes and procedures will you follow?	Describe threats to validity	What strategies will you incorporate to help address threats to validity?	Address any additional considerations or concerns and how you will try to mitigate them
Selection criteria (include an entry for each specific selection criteria)	How will you use this process to answer your research questions?	How will this participant group help achieve the study's goals?	What specific processes and procedures will you follow?	Describe threats to validity	What strategies will you incorporate to help address threats to validity?	Address any additional considerations or concerns and how you will try to mitigate them
Data Analysis Processes						
Data analysis processes (include an entry for each specific process)	How will this process help to answer your research questions?	How will this process help achieve the study's goals?	What specific processes and procedures will you follow?	Describe threats to validity	What strategies will you incorporate to help address threats to validity?	Address any additional considerations or concerns and how you will try to mitigate them

Aligning Methods With Research Questions

Research proposals are influenced by many factors, and central to these factors are your research questions and study goals. Including a matrix similar to the one in Table 10.3 that details how your methods align with your research questions and study goals will help you address how your methods for data collection, participant selection, and data analysis are related to crucial aspects of your study. In addition to helping your audience understand the rationale behind your decisions and to see clearly how you are deliberately aligning your research methods with your research questions, creating such a matrix can be a helpful exercise as you plan your study.

Ongoing Considerations

You should continue to revise and reconsider aspects of your research design after a proposal has been submitted. Being responsive to these iterative processes and maintaining fidelity first and foremost to participants' experiences is an important aspect of conducting critical, ethical, and rigorous qualitative research. One way we have addressed this in our courses is that in the session just after students hand in their proposals, we have them engage in the discussion described in Recommended Practice 10.1 as a way to debrief the proposal writing process and think out loud with a partner or set of thought partners about what they learned through engaging in the proposal writing process as well as to move forward in their thinking about next steps. We offer this here as one strategy for reflecting on the proposal after it is completed, since a study's research design requires attentiveness and refinement after proposals are complete.

Recommended Practice 10.1: Proposal Move-Forward Discussion

Discuss the process of writing the proposal and the next steps you will take to address areas of that need development. In pairs, please consider at least three or four of the following questions:

- What was the process by which I developed and iterated my research questions (e.g., reading existing studies, speaking with experts and/or insiders, vetting the research questions by others)?
    - How did this process influence my thinking?
    - Were there benefits and challenges, and what were they?
    - Are there any additional steps I can take to further refine this once I get feedback on the proposal?

(Continued)

(Continued)

- Who were the primary audiences in my mind as I wrote the proposal and why?

  - How did that shape the arguments I made for the rationale and significance of the study?

  - Are there other audiences I might consider? How would I find that out?

- How am I thinking about the relationship of the theoretical and conceptual frameworks?

  - Where are points of confusion (if any), and how can I address those moving forward?

  - What could help me better articulate/refine/iterate my conceptual framework?

- How do the research questions and methods relate with existing research in the field(s) into which I am researching?

  - If I feel unaware of aspects of this, what would be next steps to inquire into this?

- How did I locate myself in the proposal/research, and what informed those choices?

  - Upon reflection, is there anything I might add or change?

- How did I ascertain if I was using the appropriate methods to address my research questions?

  - What/whom else might I look to/speak with to refine my approach?

- What else might I consider that I did not have time to include in the proposal?

- What are the next steps I need to take to move into the development of instruments for data collection?

This discussion process helps our students, and us, to move forward in our thinking after being immersed in writing our proposals. We strongly suggest revisiting these questions again with a few thought partners and possibly advisers once you get feedback on your proposals from your professors, advisers, and peers.

WRITING QUALITY PROPOSALS

In Example 10.1, we provide an example of a dissertation proposal that can be read as a sample structure and approach to multiple kinds of academic proposals, specifically master's theses. This proposal embodies multiple layers of and approaches to criticality and is well conceptualized and written. While we do not want to be prescriptive, which is always a problem with including an exemplar since each proposal is so unique in various ways, we do believe that strong proposals—be they for funding, for master's theses, for conferences, or for doctoral dissertations—often share certain qualities that Susan Bickerstaff's dissertation proposal exemplifies. These characteristics are described in Table 10.4.

Table 10.4   Characteristics of Strong Research Proposals

A strong research proposal:

- Is written clearly and has a logical, explicated organization and structure
- Clearly articulates the study's goals and relevance to research questions
- Provides roadmaps for readers of the organization and flow of the proposal itself at the beginning and throughout the proposal
- Has solid transitions and relationships between sections that indicate a clear logic in the study (and proposal) construction
- Clearly articulates a conceptual framework and argumentation for the study and its methods
- Critically integrates and demonstrates an understanding of the relevant literatures and their relationship to the study (and has a discerning, carefully explicated theoretical framework and not just a broad literature review)
- Clearly describes the overall methodological approach, research design, and specific methods choices and situates each choice within relevant theoretical and methodological literatures
- Provides a sufficient level of detail on the proposed research methods and provides a cited rationale for each of these methods
- Describes validity issues and how these will be addressed and attended to in detail
- Reflexively positions the researcher within the study and addresses the methodological implications of researcher identity and positionality
- Dynamically discusses the interaction between theory and methods
- Clearly describes the relationship between research questions, study goals, and the overall study design
- Demonstrates a facility with research language and relevant research methods literatures
- Describes possible limitations of the study and their potential implications
- Incorporates sources in a purposeful, balanced, and integrated manner
- Demonstrates engagement in the research and respect for the proposed participants and the context of the study through language and choice of methods
- While being as detailed as possible, allows room for emergent changes
- Includes a brief and integrated conclusion

The proposal in Example 10.1, written by Susan Bickerstaff, is only one example, and it fits within a specific academic context that shaped it in ways that reflect norms in academia as well as the specific demands of her field, program, and dissertation committee. It is important to understand the context, goals, needs, and tasks involved in the specific proposal that you are tasked with writing. Knowing these various norms and expectations within the context of a proposal and understanding the guidelines is an important first step. We have annotated Example 10.1 to draw your attention to aspects of this proposal that we think are especially important. We wish to note here that one of the goals of our annotations is to focus readers both to what is exemplary in the proposal as well as to some alternative possibilities for your own proposals since again, this proposal was written to satisfy the requirements and expectations of a specific program and dissertation committee.

Example 10.1: Dissertation Proposal

Challenging "Dropout": Why and How High School Dropouts Return to School

Susan Bickerstaff

March 5, 2008

Dissertation Proposal

University of Pennsylvania Graduate School of Education

Abstract

High school attrition is a critical issue in the ongoing effort to reduce disparities in access, equality, and equity across race and class. Despite decades of attention to the "dropout crisis," research tells us little about the lived experiences of students who leave high school. The number of youth and young adults who have returned to a wide range of alternative education settings (GED classes, adult literacy programs, charter schools, accelerated high schools) is growing, but likewise, we know little about who these students are and why they persist. How do young men and women make the decision to return to school? How do they understand their lives, their stories, and their futures? This study is grounded in a conceptual framework developed around three strands of work: narrative, identity and possible selves, and youth literacies. Using a critical case-study methodology, the proposed project will follow, for 1 year, a group of 8 to 10 adolescents who have elected to return to an education program after dropping out of high school. A central assumption underlying this work is that dropouts[2] have a variety of experiences, both in and out of high school, and these experiences may not be merely reflective of disengagement or resistance but instead constitute learning, growth, and the development of identity. The proposed project aims to offer a more nuanced understanding of the lived experiences of young people, both and in and out of school, in an attempt to contribute to efforts to make high schools and alternative education programs more accessible and amenable to students with a broad range of needs.

Introduction and Problem Statement

In October 2004, an estimated 3.8 million 16- to 24-year-olds in the United States were out of school without a high school diploma or equivalent degree (Laird, DeBell, & Chapman, 2006). This estimate contributes to a longstanding

Annotation: Writing abstracts is often something that students struggle with. In this abstract, Susan does an excellent job of cogently describing her goals and proposed study while also gaining the reader's interest. As you begin reading the proposal, you will note the headings that Susan uses to structure the proposal. As with many proposals, the specific headings incorporated are often a reflection of the wishes and expectations of the proposal audience.

concern about high school attrition, the role of schools in helping students to persist, the nature of student engagement, and the capacity of students who leave school to thrive in adulthood. The concern over the impact of dropout on the well-being of young adults and their families is an important area of inquiry in education but also in a range of related disciplines and in the on-the-ground efforts of programs. Adults who do not complete high school or earn an equivalent credential face barriers to employment and financial stability (U.S. Census Bureau, 2005; Sherman & Sherman, 1991), fare worse on measures of health and well-being (Laird et al., 2006), and are among the largest groups of people in prison (Haigler, Harlow, O'Connor, & Campbell, 1994). Despite the relatively large body of scholarly work on high school attrition and ongoing and longstanding efforts toward dropout prevention in research, policy, and practice, surprisingly little is known about how and why students make the decision to leave school and in many cases why they decide to return to education settings.

High school attrition is a critical issue in the ongoing effort to reduce disparities in educational and social access, equality, and equity across race and class. In her 2006 AERA Presidential Address, Gloria Ladson-Billings proposed that educational researchers rethink the notion of the "achievement gap" in terms of an "education debt." She identified the historical, sociopolitical, moral, and economic components of persistent underlying disparities along the lines of race and class in the American educational system. High school attrition is clearly a symptom of the growing debt owed to poor students, English language learners, and students of color who are grossly overrepresented in the number of high school dropouts (Balfanz & Legters, 2004; Laird et al., 2006; Orfield, 2004). Data from a 2004 study by the National Center for Education Statistics (Laird et al., 2006) indicate that the dropout rate for students from low-income families was four times higher than for other students and that while White students drop out at a rate of 3.7%, Black students and Hispanic students drop out at rates of 5.7% and 8.9%, respectively.[3]

Looking at graduation rates and dropout rates for over 10,000 schools nationally, Balfanz and Legters (2004) found that while high school noncompletion is widespread geographically, most high school dropouts are clustered in approximately 2,000 schools that graduate less than 60% of incoming freshmen on time. More than one quarter of these schools are located in the metropolitan regions of Ohio, Michigan, Illinois, Pennsylvania,

(Continued)

Annotation: In this opening section, Susan contextualizes her proposed study and articulates a strong rationale for the practical and scholarly significance of her study. Furthermore, she situates her proposed study within relevant literature.

(Continued)

and New York, and the differences across race are striking. According to their data, more than 50% of the African American students in these five states attend high schools in which less than half of students graduate on time, yet they report that the number of White students attending schools with low graduation rates is below the national average. In 2002, in Philadelphia, 20 high schools (61%) graduated less than half of the freshman class within 4 years.

Although graduation rates have been on the rise over the past three decades, these improvements have leveled off since the 1990s, most notably for Black and low-income students. Balfanz and Legters (2004) found that the number of schools that graduate less than half of the incoming freshman class on time grew over the 1990s. Between 2003 and 2004, the number of low-income students leaving school actually increased (Laird et al., 2006). Neild and Balfanz's (2006) study of Philadelphia graduation rates revealed that on-time graduation rates between 2000 and 2005 range from 45% to 52%. They estimate that 30,000 Philadelphia students left high school between 2000 and 2005. Some scholars argue that the recent spread of high-stakes testing, new accountability measures linked to No Child Left Behind (NCLB), and high school exit examinations, like those implemented in New York, Florida, and California, put pressures on schools that may result in the intentional or unintentional "pushing out" of low-performing students. While NCLB has a strict set of guidelines that states must use in regards to what constitutes Adequate Yearly Progress (AYP) for attendance and test scores, little effort has been made to hold states accountable for graduation rates. Losen (2004) reviewed graduation rate accountability by state and discovered that only 10 states have what he terms "a true floor for adequacy in graduation rates." All other states, of which Pennsylvania is one, do not use graduation rate as a measure of AYP and therefore have little incentive under the federal regulation to increase graduation rates. A case study of the Chicago public schools indicated that graduate rates fall after the implementation of high-stakes testing (Allensworth, 2004).

Research Perspectives on Leaving School

In addition to this demographic work, a substantial corpus of research attempts to understand why students leave school in an effort to prevent attrition (e.g., Sherman & Sherman, 1991). Rumberger's (2004) model of school engagement is an often-cited summary of the range of factors related to persistence in high school. Drawing on data from the National Education Longitudinal Study, Rumberger (2004) identifies a number of explanations for leaving school that include school-, family-, and employment-related reasons. One of his primary findings is that school-related reasons, which include academic and social engagement in school, are offered by students most frequently, rather than familial or employment reasons. Unfortunately, survey research has significant limitations for understanding the complexities of an individual's decision to leave school. Qualitative research specifically focused on high school dropouts (e.g., Fine, 1991; Rymes, 2001) suggests that students' decisions to leave high school are nested in a number of social and contextual factors that cannot be captured by a list of school-leaving reasons. Instead, former students share intricate narratives that offer a nuanced picture of their schools, their self-perceptions, their peers, their families, and their communities. Rymes (2001) argues that students frame and reframe their reasons for dropping out of school depending on the context in which they are speaking and the reinforcement they receive from listeners. Her findings provoke questions about whether surveys, administered once, can effectively capture the complexities of dropping out.

In addition, Stevenson and Ellsworth (1993) point out that statistical analyses of large-scale data sets on high school noncompleters typically result in discussions framed around student characteristics, rather than school, community, or societal factors. Thus, the tendency, even if unintentional, is to view dropouts through a deficit lens that focuses on risk factors, motivation, family characteristics, and disengagement. Even the concepts of self-image or self-esteem, which underpin some theories of dropout (e.g., Finn, 1989), conceal institutional and social realities and focus on the traits of individuals.

The notion of student "resistance" to dominant norms or culture is frequently used to explain the so-called achievement gap (e.g., Fordham & Ogbu, 2000; MacLeod 2000), yet theories of resistance can also focus attention away from the schools these students leave, placing the "blame" for disengagement on the students. In general, "resistance" is undertheorized and therefore of uncertain utility as a conceptual framework (Trimbur, 2001). In their ethnography of African Canadians' disengagement from school, Dei, Mazzuca, McIsaac, and Zine (1997) point out that resistance to school can take on many forms, including an increased engagement in student organizations, self-education, or political efforts. Their study is useful because it signals the importance of considering race, class, gender,

(Continued)

(Continued)

and culture as more than just variables but as aspects of identity that are central to the lived experiences of adolescents who leave school.

For several decades, many researchers have worked to counter the tendency to view dropouts as one-dimensional or as failures. Several notable studies have offered an alternative lens for understanding the experiences of students who do not graduate. For example, Farrell (1990) gave voice to high school students and dropouts by employing "at-risk" students as "collaborators" who interviewed their peers about their experiences. Fine's (1991) ethnography of an urban high school spotlights the institutional forces that silence students. Her data suggest that many students who leave school exhibit a "critical consciousness" and question the ideology of meritocracy and social mobility. Similarly, in Willis's (1977) ethnography of working-class British youth, the "lads" rejected middle-class (and by extension school) values of individuality, obedience, social mobility, and theoretical knowledge. Willis's analysis famously posits that these youth "penetrate" dominant discourses and critically read the social and economic realities of their situation. Blake's (2004) study of children of migrant workers and incarcerated youth revealed that some youth felt school was "useless, boring, complicated, dangerous, and even 'evil'" (p. 39). Like Willis and Fine, Blake found that the young men she observed and interviewed saw little connection between school and the realities of their lives. Unfortunately, even students who penetrate dominant discourses or exhibit a critical consciousness remain disadvantaged by societal power structures. For example, Fine (1991) argues in her study that one of the most striking differences between dropouts and graduates was graduates' failure to challenge or question meritocracy and economic opportunity. Yet, ironically individuals without a high school diploma are routinely silenced and largely excluded from conversations around economic and educational reform.

The strength of critical qualitative work is not only its ability to create rich portraits of the lives of students who leave school but also to point to the structural and institutional systems that in Fine's (1991) words "enable, obscure, and legitimate" high school dropout. LeCompte and Dworkin (1991) argue in their analysis of student dropout and teacher burnout that the structure of contemporary schooling is in conflict with current economic and social systems. They provide a compelling theory of alienation centered on a gap between expectations and experiences that leads to a sense of "powerlessness, meaninglessness, and normlessness" (p. 155). Although they do not include the voices of students or teachers, their work and the

Annotation: Susan does an excellent job of connecting her methodological approach of critical qualitative research to extant theoretical and empirical research on her study topic. She does this in a way that both grounds the value of her study in terms of its theoretical framing and its methodological approach in the landscape of studies on this topic. Furthermore, she aptly situates and justifies the use of her approach throughout the proposal and not just in the design section.

work of sociologists like Bowles and Gintis (1976) delineate the relationships between economic trends, social class structure, and educational systems. It is also worth noting that a significant portion of high school dropouts leave school "involuntarily" because they are expelled or incarcerated. For these young men and women, returning to school can prove exceptionally difficult, and data suggest that a majority of adjudicated youth do not reenroll in high school after release (Brock & Keegan, 2007). Young mothers without childcare provisions, students from low-income families who feel pressure to maintain a full-time job, students with special needs that remain unmet, and students who feel unsafe or unwelcome in their schools (e.g., transgendered, gay, and lesbian students; students with disabilities; or immigrant students) may drop out for reasons that have little to do with academic achievement, aptitude, motivation, or resistance.

Research on Students Who Return

For many students, leaving high school is not a permanent decision; instead, it is a well-accepted fact that many high school dropouts cycle in and out of adult education programs, often intermittently throughout much of their lives. Kolstad and Kaufman (1989) reported that 44% of school leavers eventually go on to attain high school diplomas or their GED certificate. Of particular interest are adolescents and young adults who decide to return to education within a few years of dropping out. Several recent studies indicate that the number of 16- to 20-year-olds in alternative schools, GED classes, and other adult education settings is growing (Flugman, Perin, & Spiegel, 2003; Welch & DiTommaso, 2004). The number of school leavers returning to adult education programs is indicative of both greater interest on the part of the young adults who return and enhanced efforts on the part of educational institutions (e.g., community colleges and charter schools) to provide opportunities for individuals who seek a second chance to realize their goals. Some scholars (e.g., Entwisle, Alexander, & Olson, 2004; Rachal & Bingham, 2004) suggest that the prevalence of GED programs and the flexibility they offer in terms of structure and scheduling may actually encourage some students to leave traditional high schools. However, there is little compelling evidence to support this claim. Adult education programs frequently struggle to meet the need of the growing numbers of young students, and the literature suggests that teenage and young adult learners present particular challenges that many adult education programs feel unequipped to handle (e.g., Hayes, 1999).

(Continued)

(Continued)

Although studies indicate that the number of young adults returning to education settings is significant, little attention in the research has been given to who these students are and why they persist. Conventional wisdom suggests that these students return to find a better job, to gain a sense of pride, or to provide more opportunities for their young children. While few studies have looked at the experiences of returning students, some large-scale quantitative studies have attempted to map the characteristics of students who return to education programs (e.g., Chuang, 1997; Entwisle, Alexander, & Olson, 2004; Goldman & Bradley, 1996; Sherman & Sherman, 1991; Wayman, 2001). Salient factors drawn from this literature indicate that dropouts who leave school later, have more work experience, score higher on standardized tests, and are of higher socioeconomic status are more likely to return to school. Unfortunately, these studies tell us little about the experiences of returning dropouts, their decisions to return so soon after leaving, and their reasons for persisting.

This project attempts to uncover something about youth who leave high school and subsequently choose to return. Understanding more about young adults who return to education programs is of great importance for two primary reasons. First, this population can help researchers increase their knowledge of the issues that contributed to students' dropping out of high school. Such information would inform us about the ways that learning environments can be structured and revised to support potential school leavers, sustain their school participation, and perhaps increase graduation rates. Second, understanding who these students are and why they return may help institutions improve programs for returning students. Inherent in this inquiry is an acknowledgment that school systems are inadequately serving large segments of the student population and that adult education programs are increasingly overwhelmed by the number of young students. A more nuanced understanding of the lives of these young people and their lived experiences, both in and out of school, can contribute to efforts to make both traditional high schools and programs for returning students more accessible and amenable to students with a broad range of needs.

To address the enduring "education debt" in our country, we must understand why so many students, particularly poor students and students of color, leave high school. The proposed study attempts to speak critically both to the question of why students drop out of school and why they return. A fundamental tenet of this study is that high school students who leave school have a variety of experiences that may not be effectively captured

Annotation: After discussing the topic in a broad sense, Susan brings the readers into the particular focus of her study. She appropriately contextualizes the background and reinforces the study's importance. Furthermore, she clarifies key concepts and terms and reinforces the rationale and significance of her proposed study.

in the term *dropout*. Students return to school with a range of experiences both in and out of high school, and these experiences may not be merely reflective of "disengagement" or "resistance" but instead constitute learning, growth, and the development of identity.

Research Questions

Recognizing that the inequity in dropout rates across race and class is a serious issue for education research, practice, and policy, this study attempts to speak critically both to the question of why students drop out of school and why they return. This project focuses on adolescents and young adults between the ages of 16 and 21, a period that has come to be considered "youth"; it endeavors to add the voices of former high school students to the discussion in a way that has unfortunately been done infrequently in research on high school leavers. The proposed study focuses on three broad research questions and several subquestions. During the course of this year-long project, these questions may change as data emerge; however, I begin with these questions with the hope that the perspectives of students returning to an education program will add to our understandings of how and why students leave school and under what circumstances they return.

Annotation: Susan clearly frames and articulates her proposed research questions while stating that they may be refined and/or additional ones may emerge.

1) What expectations of learning and life do high school dropouts bring with them when they return to school?

   a) How do they describe their hopes, fears, and expectations for the future?

   b) What identities do they take up in their narratives of leaving and returning to school?

   c) Against and within what master narratives do they frame their stories of leaving and returning to school?

2) Why do students leave school and why and how do they make the decision to return?

   a) How do they talk about their experiences as high school students?

   b) How do they talk about their out-of-school experiences and their decision to return?

   c) What do these narratives of dropout and return reveal about the ways in which students' understandings of themselves and of education, success, and achievement have changed over time?

(Continued)

(Continued)

3) What are the experiences of returning students in a college-based program for adolescent dropouts?

   a) How are students positioned and how do they position themselves in this program?

   b) To what extent and in what ways do they take up and enact identities as "college student" and "dropout"?

   c) What literacy practices and texts are central to the ways in which students represent their experiences and enact various identities and cultures?

Conceptual Framework

Embedded in these questions are multiple assumptions about the experiences of high school dropouts returning to school, including the notion that narratives can yield important insights into student experiences and that students who return to school imagine a life for themselves that is of interest to practitioners and researchers. These assumptions are grounded in literature that crosses multiple domains of scholarship. The proposed study builds on literature written about high school dropouts and uses a conceptual framework developed around three strands of work: (1) narrative, (2) identity and possible selves, and (3) youth literacies.

Narrative

In Metzer's (1997) qualitative study of dropouts returning to school, he reports that most of the dropouts in his study returned to school after "some kind of pivotal event or realization" (p. 6). Metzer writes that while many respondents reported feeling more mature, wanting a better life, or feeling more motivated, he maintains that critical incidents in the respondents' lives (i.e., the loss of a job, the birth of new baby, or the graduation of friends) were essential in precipitating the respondents' return to school. The notion of the "turning point narrative" is pervasive both in scholarship on narrative and in popular media. American culture mythologizes the former gang member who now works to prevent violence, the cutthroat businessman turned philanthropist, the drug abuser turned motivational speaker. These archetypes embody a "redemption-sequence narrative" in which the protagonist makes a life transition from a negative situation or experience to a positive one (McAdams & Bowman, 2001). Metzer's data

Annotation: Susan provides a concise and clear conceptual overview to orient her readers to her conceptual framework that she then presents in detail. In the sections that immediately follow, she critically examines what she has determined to be the key themes from the extant literature and creates the theoretical and broader conceptual framework of her study, including her methodological approach. Susan integrates these literatures and connects them to her proposed study.

analysis draws on a classic redemption narrative in which former high school dropouts experience an epiphany and then begin a new section of their life. Unfortunately, Metzer presents the turning point stories narrated by the participants in his study uncritically. In his conclusion, he writes of students who made the decision to return: "What is required is the critical event, the idea, perhaps from an outside, insight that suddenly emerges and brings about the decision to return to education" (p. 28). In so doing, he presents turning point as "fact" and propagates an individualistic mythology that states students must experience a "critical incident" in order to make the decision to return to school. Metzer offers little analysis of what these stories tell us about the ways in which students are making sense of their world and their position in it.

By contrast, the present study proposes a poststructuralist narrative lens to understand the stories of returning students in relationship to dominant narratives. Work by Bruner (1990), Labov and Waletzsky (1967), and others has laid the theoretical foundation for the use of narrative as a lens of interpretation in education and social sciences. According to Bell (2002), a narrative approach to research "rests on the epistemological assumption that we as human beings make sense of random experience by the imposition of story structures" (p. 207). *Memories, both individual and collective, as well as experiences, norms, values, thoughts, and feelings, manifest in the ways in which people choose to narrate their lives.* The study of narratives gives insight into the way individuals make sense of experiences over time and therefore is a particularly appropriate frame for an inquiry into the experiences of students leaving and subsequently returning to school. Summerfield (1994) argues that "[stories] represent ways of knowing, ways of constructing our lives and values" (p. 180). Thus, an analysis of a participant's story may or may not reveal the "facts" about her experiences but is likely to reveal the way she is positioning herself in the world. In Warriner's (2004) study of the stories of recently arrived refugees, she chose to examine the interviews through a narrative lens because, in her words, "the content of the story and the way it is told contribute to the representation, construction, and enactment of identity" (p. 282). My first research question rests on the assumption that by studying the narratives of returning students, I can learn about dominant narrative structures available to high school dropouts and the ways in which they take up and reject these narratives in their enactment of multiple identities.

(Continued)

(Continued)

Using a poststructuralist lens, Barbara Kamler (2001) calls attention to the limitations of narratives in uncovering a single truth. She writes, "Stories do not tell single truths, but rather represent a truth, a perspective, a particular way of seeing experience and naming it . . . stories are partial, they are located rather than universal" (pp. 45–46). Kamler's discussion of narrative in relationship to critical writing pedagogy offers a useful frame to qualitative researchers. She argues, "The narrator is a position, an angle of vision and not simply the student confessing his or her authentic feelings or truths" (p. 144). Pavlenko (2002) writes that "narratives are not purely individual constructions—they are powerfully shaped by social, cultural and historical conventions as well as by the relationship between the storyteller and the interlocutor" (p. 214). Personal narratives offer a particular representation of memories that is affected by dominant discourses, local contextual factors, and cultural conventions. This is not to say that the narratives of our participants do not offer us any "truths" about their experiences. Wortham and Gadsden (2006), for example, use narrative analysis to understand how young urban fathers position themselves in relationship to the domains of "street" and "home." I frame this study with the assumption that the analysis and interpretation of narratives can reveal important information about participants' enacted identities, implicit belief structures, and lived experiences, vis-à-vis larger structural and cultural forces.

It is worth noting that I understand identity not as static or unitary but as fluid, multiple, and dependent on local context. Julie Bettie's (2003) notion of class identity as both a performance and performative is useful here. Bettie contends that class identity can be performed in a number of symbolic ways, including through dress, makeup, and sexuality. She "[explores class] as a learned position, arguing that class identity comes to be known equally by markers that exist outside of discovering one's position in paid labor, as an identity lived out in private life and personal relationships—in short, class culture" (p. 42). These performances are the result of constant positioning and repositioning in relationship to local and dominant narratives.

I use the term *dominant narratives* or *master narratives* to refer to the stories that circulate in popular, political, and commercial cultures. Pam Gilbert (1992) argues that "the power of storying to regulate and order our cultural life should be a key question for educational research" (p. 211). This study speaks to this charge by looking at how returning students talk about and respond to dominant narratives about students who do not finish school (i.e., dropouts as deviant, lazy, unmotivated, or defiant). The notion of dominant narratives or master narratives is central to Kamler's (2001) discussion of a narrator's production of a story. In her description of a writing workshop with older women, Kamler reflects on the dominant narratives available to aging women that frame them as passive, silent, and invisible. In response, she encourages writers to take up alternate narratives and "do counternarrative work" to reimagine the identities available to them. Warriner's (2004) interviews with refugee women demonstrated that they too worked both within and against master narratives of meritocracy and opportunity. Orellana's (1999) study of elementary school Latino children investigated the way young boys and girls assumed identities in their writing. She discovered that master narratives of love, romance, and "goodness" or "badness" are evident in student writing, even as a few students subvert and invent alternative identities in their writing. Likewise, Walton, Weatherall, and Jackson's (2002) analysis of preteen narrative writing suggests that master narratives of romance in popular culture limited the subject positions available to the writers.

These studies raise the question about what dominant or master narratives high school dropouts are working within or against. Stevenson and Ellsworth's (1993) interviews with White working-class dropouts revealed that these former students generally accepted and internalized mass-mediated images of dropouts as "incompetent" or "deviant." Interestingly, unlike the participants in Fine's (1991) study, these dropouts "self-silenced" their critiques of the school and voiced master narratives of independence, hard work, and opportunity. In Rymes's (2001) work with returning students, she analyzes the common feature of the "turning point" in which the narrator distances himself or herself from a former self who was responsible for leaving school. The turning point in the storyline allows narrators to create two characters within their life story. The first was lured into gangs or other negative behaviors; the second self-rejected those influences and is on the path to success. Unlike Metzer (1997), Rymes analyzes these data critically and writes, "These stories become testimonials to the abandonment of a rejected self, calling upon a particular set of linguistic and indexical recourse to depict the turn to a new life" (p. 74). The notion that dual identities as "dropout" and "successful student" are, in Rymes words, "mutually incompatible" raises questions about the pervasiveness of the turning point storyline. One wonders if dominant narratives of redemption or rebirth may influence students who struggle to reconcile their past and former selves within a cohesive identity.

(Continued)

(Continued)

This study proposes a closer examination of the narratives of returning students with an eye to what these reveal about available identities and subject positions for students who leave school. What sources of popular culture are important to students, and how do they figure into the way these students narrate their lives? How do master narratives about success and failure affect students' identity constructions? In what ways are these students doing counternarrative work and seeking alternatives stories to make sense of the structures that surround them?

Possible Selves

Closely related to work on narrative constructions of self is the theoretical notion of "possible selves" developed by Markus and Nurius (1986): "Possible selves represent individuals' ideas of what they might become, what they would like to become and what they are afraid of becoming" (p. 954). Born out of literature in psychology, the domain of possible selves was developed to expand understandings of self-knowledge and self-concept. Possible selves are temporally constructed ideations of future identity and are frequently categorized in research into "hoped-for selves," "feared selves," and "expected selves." Hoped-for selves might include financial stability, marriage, happiness, and health. Feared selves may commonly include poverty and loneliness. Markus and Nurius characterize possible selves as both "individualized" and "distinctly social" (p. 954). That is, their framework does not define possible selves as abstract archetypes but as facets of identity that are drawn from "past selves," personal history, and past and current social and cultural context. The work on possible selves has been influential in current understandings of the self-concept as multiple and changing, as opposed to earlier frameworks that viewed self-concept as monolithic (Oyserman & Markus, 1990).

Much of the research on possible selves has focused on links between feared or expected selves and motivation or behavior. This has resulted in a substantial corpus of empirical work measuring correlations between the range of expected selves and academic performance, juvenile delinquency, and other self-change behaviors. For example, Oyserman, Bybee, Terry, and Hart-Johnson (2004) explore the ways in which "self-regulatory possible selves" can motivate current behaviors and actions when they are accompanied by self-change strategies. Using an adult development perspective, possible selves have been understood as "blueprints" (Cross & Markus, 1991) or "roadmaps" (Oyserman et al., 2004) for change and growth throughout the life course. In adolescence, positive outcomes related to academic achievement and delinquency have been associated with what Oyserman and Markus (1990) call "balanced selves" or a combination of goals and fears.

Research indicates that hoped-for and expected selves are influenced by various elements of social identities, including race, class, and gender (e.g., Fryberg & Markus, 2003; Oyserman, Bybee & Terry, 2003; Yowell, 2000). Granberg (2006) characterizes theories of possible selves as similar to self-discrepancy theories. That is, individuals must work to reconcile the differences between their hoped-for selves and their current social identities and contexts. In this way, theoretical work on possible selves has been compared to Steele and Aronson's (1995) work on stereotype threat. Self-regulatory behavior, according to Oyserman, Bybee, and Terry (2006), is only likely when likely selves are "made to feel like 'true' selves and connected with social identity" (p. 188). Gadsden, Wortham, and Turner (2003) found that young African American fathers' perceptions of "familial, peer, and legal systems as barriers and resources" were important to the ways in which they created their possible selves (p. 395). These findings raise questions about the ways in which high school leavers are or are not constrained as they construct hoped-for selves, expected selves, and "plausible selves" (Oyserman et al., 2004). Despite its rich potential and the growing body of research on possible selves and adolescents' academic achievement and career goals (e.g., Shepard, 2000), possible selves has rarely been used to investigate how youth reflect on their identities as "dropout" and "student."

One of the affordances of the possible selves framework is its implications for methodology and its connection to the narrative. Although most of the early work in this area was grounded in an objectivist paradigm and marked by quantitative analyses focused on patterns in large-scale data sets, recently theories of possible selves have been applied to a broad range of work representing a diversity of stances and methods. Packard and Conway (2006) categorize the literature on possible selves into two broad categories. The first views the self as "a collection of schemas" and has deep roots in the psychological literature. The second conceptualizes the self "as a story," grows out of the narrative tradition (i.e., Bruner, 1990), and is aligned with the poststructuralist theoretical stance outlined above. These theoretical differences have implications for method. Markus and Nurius's (1986) Possible Selves Questionnaire, a structured survey instrument, continues to be used

(Continued)

(Continued)

in studies on possible selves; however, more recent work has employed a narrative methodology that positions the researcher not as an extractor of information but as a co-constructor of a participant's ever-unfinished narrative. In her study on methods and possible selves, Whitty (2002) argues that an open-ended narrative prompt "allows one to gather a richer portrait of the individual's life goals than would be afforded by questionnaire measures of these constructs" (p. 124).

The proposed project is influenced by the conceptual framing of a number of studies that combine narrative methods and theories of possible selves to uncover the ways adolescents and young adults view their present and future identities. Halverson's (2005) research explored the possible selves of LGBTQ youth through narrative and performance. One of her central questions—how do youth grapple with "taking on the stigmatized identity of homosexuality" (p. 71)—might be applied to youth who have left school and are confronted with the possibility of taking on the stigmatized identity of "dropout." Shepard's (2004) work with rural young women used the possible selves framework to investigate life-career development. Her study revealed that the participants had detailed and elaborated possible selves within the "domains of personal attributes and relationships" (p. 85); however, they described their career possible selves in vague and generalized terms. In her analysis, she connects this finding to the social, economic, and material realities of rural community in which she conducted her study. Similarly, Gadsden et al. (2003) found that the young African American fathers in their study constructed possible selves of fatherhood that were largely "undefined" because of various life experiences (including father absence in their own lives), real and perceived obstacles, and the ways in which they situated their identities within various contexts. In a study conducted by Hunter and her colleagues (2006), young African American men revealed through narrative that they were aware that they may not be able to actualize their desired selves. This research indicates that the constructing of possible selves occurs within and against social contexts that work to restrict the imagined futures of marginalized and stigmatized groups.

Recently, this theoretical construct has also been applied to understanding multiple, contextual, and "continuously revised" (Giddens, 1991) identities. As indicated above, qualitative research on students returning to school that has demonstrated the ways in which students "reinvent themselves" (Hull, Jury, & Zacher, 2007), talk about the "turning point" that provoked their return (Metzer, 1997), and through their discourse reveal their "abandonment of a rejected self" (Rymes, 2001). Viewing these stories through the lens of the possible selves offers an additional opportunity for analysis. Halverson (2005) notes that by definition, possible selves "implies a more fluid notion of identity, one that demands the acknowledgement of both past and future self-conceptions, and an understanding of the interplay between past, present, and future, as the way in which identity is formed" (p. 71). King and Raspin's (2004) research demonstrates this fluidity in a useful way; they asked participants to write narratives about their "current best possible self" and about a possible self that they once imagined but is now impossible or unattainable (a "lost possible self"). Although he does not explicitly use the possible selves framework, Farrell's (1990) analysis of the stories of high school students centers on seven current- and future-oriented "selves." He posits that the struggle to reconcile and integrate these multiple selves is a critical challenge for adolescents. The proposed project seeks to understand more about how young people returning to high school create and enact narratives of changing identities. How do they describe their hoped-for, feared, lost, and expected selves?

Youth Literacies

Implicit in the central research questions driving this study is a concern with identity. In understanding more about students' narratives of dropout and return, this project hopes to uncover something about the way youth who have left school understand themselves as individuals and as members of their classroom, peer group, community, and generation. The third strand of the proposed project's conceptual frame is drawn out of the expansive body of research and theory on literacy as a sociocultural practice as it relates to narrative production and identity construction. In the framing of this project, I self-consciously use the term *youth literacies* to mark a particular theorization of the literacies of adolescence and young adulthood that is grounded in a youth cultures framework. Although the literacy of high school leavers is most frequently considered from a school-based deficit perspective, sociocultural studies of the literacies of youth have revealed the ways in which even "marginalized readers" (Franzak, 2006) engage in intricate literacy practices. Broadened definitions of literacy and text under this framework represent a sharp break from traditional notions of literacy as an autonomous set of skills used to decode and make meaning from decontextualized written text. Drawing on the early work of Scribner and Cole (1981), Street

(Continued)

(Continued)

(1984), and Heath (1983) and more recent pieces like Kress and Van Leeuwen (2001) and Gee, Hull, and Lankshear (1996), this project understands literacy practices as embedded in local context, highly complex and multimodal, instrumental in the maintenance of power and privilege, and inextricably linked to the enactment of identity.

The scholarship on adult literacy and adult education provides a useful lens through which to view the literate lives of young men and women leaving high school. Firstly, adult educators have long pointed to the rich and productive lives of learners who are commonly labeled "illiterate" or "low-literate." For example, Lytle's (1990) four dimensions of literacy development in adulthood—beliefs, practices, processes, and plans—offer a theoretically nuanced way of understanding the literacies of adult learners. This framework calls attention to adult learners' prior knowledge, current activities and strategies, and forward-looking goals and thus prompts researchers and educators to depart from one-dimensional characterizations of students (i.e., functionally illiterate). Similarly, Belzer's (2002) work with adult literacy learners indicates the importance of understanding the past experiences, particularly past school experiences, of learners in adult literacy classrooms. She demonstrates the ways in which schools' messages about literacy have a "destructive" impact on marginalized students who come to see themselves as nonreaders. Freire (1987) challenges adult educators to recognize the ways in which learners enter classrooms with complex ways of reading the word and the world. In Freire's model of adult literacy education, students' questions and concerns become the central text in the classroom; through the critical examination of these texts, students question and examine the status quo and investigate the ways in which language upholds dominant ideologies and the ways particular discourses are favored over others (Shor, 2005). While these principles have been applied to elementary and middle school classrooms, their roots in political education for undereducated adults have particular relevance to young adult learners returning to school. The critical literacy framework is useful because it recognizes the potential for marginalized students, in this case school leavers, to become critical agents and to develop a body of political knowledge (Thompson & Gitlin, 1995).

These insights from the field adult literacy, however, have limitations for work with young adults. They fail to consider, for example, the particular circumstances of young men and young women who are transitioning to adulthood. Research under the umbrella of "adolescent literacy" represents a wide range of work, some of which considers the particular ways in which youth engage with texts to forge identities. For example, in Moje's (2000) study of the literacy practices of gang-connected youth, she reimagines literacy practices as tools. Thus, literacy is used for a purpose: "to claim a space," "to construct an identity," or "to take a social position in [the world]" (p. 651). In her case study of two high school students, Nielson (1998) notes that for these adolescents, "texts hold potential as symbolic resources," which present "opportunities for trying on and taking up often multiple and conflicting roles or identities" (p. 4). For one of her focal students, the reading and rereading of *Catcher in the Rye* is an opportunity to imagine a romantic relationship with a protective figure. For another, the movie *Pulp Fiction* provides an opportunity to imitate powerful masculine archetypes. In Nielson's analysis, texts offer places for youth to harbor dreams and to imagine future selves. Finders's (1997) ethnographic study of the literacies of junior high school girls demonstrates the ways in which reading and writing practices (like magazine reading, graffiti writing, and note passing) are used to solidify group membership and jostle for status within the group. Similarly, studies of teenagers' use of fanfiction, online chat rooms, and weblogs (e.g., Alvermann & Hagood, 2000; Chandler-Olcott & Mahar, 2003) demonstrate the ways in which authors can become part of "imagined communities" of fellow users. Blackburn's (2002) analysis of "literacy performances" among gay and lesbian adolescents offers an example of the ways in which youth use literacy to disrupt imposed identities and cultural expectations.

The appropriation and hybridization of textual forms is another way through which young adults subvert master narratives, resist imposed identities, and create new narratives. In one of the most striking examples of youth appropriation of schooled literacy practices, Schultz (1996) describes a young Filipina graffiti artist who practices her tag repeatedly in a notebook as part of her initiation by a group of more experienced writers. In Wilson's (2003) ethnographic work with incarcerated youth, she finds a cartoon submitted to the Valentine's Day edition of the prison magazine captioned with this: "Nike trainers my one true love, without you I am nothing." Here a young man has appropriated trite literary conventions to express the locally specific importance of brand-name footwear. Similarly, despite their openly hostile opposition to dominant culture, the graffiti artists in Cintron's (1991) study refashion sports' team logos, brand names, and even popular folklore (i.e., the lion as "the king of the jungle") with new meanings

(Continued)

(Continued)

and create what Cintron calls "a distinctive gang register." Lam (2004) profiles a Chinese American adolescent comic book fan whose fascination with Japanese comic books offers him an opportunity to situate himself within a "transnational culture" and to create a borderland discourse available to both American and Chinese peers. Nakkula and Toshalis (2006) assert that adolescent development is a process of creation, authorship, and interpretation. These studies demonstrate the various ways that adolescents use multimodal texts to create new discourses; take up, subvert, and respond to mass-mediated texts; and enact individual and group identities.

Alvermann (2001), Elkins and Luke (1999), Moje (2002), and others point to a continued neglect of the literacies of adolescence except for a persistent interest in "struggling" or "remedial" readers. Moje challenges adolescent literacy researchers to apply a youth cultures framework to the study of adolescent literacy; she defines youth culture as "the experiences and behaviors of youth as part of normed sets of practices, behaviors, and beliefs common to people of a certain age or generation" (p. 213). This conceptualization of youth represents a departure from the notion of adolescents as "developing" or "not fully formed." Instead, as Alvermann (2006) suggests, this framework considers youth as "knowing something that has to do with their particular situation and surroundings" (p. 40). The youth literacies framework also acknowledges youth as valuable cultural producers. A growing body of research on in-school and out-of-school adolescent literacy practices documents the various and creative ways in which youth produce texts that represent a range of styles, genres, and media (e.g., Fisher, 2007; Hull & Schultz, 2002; Mahiri, 2004). Mary Bucholtz (2000) writes that "one of the richest influences on American speech in the new millennium will certainly be youth culture" (p. 280). Youth-produced zines, fan websites, fanfiction, music, and emulative texts challenge the conventional wisdom that mass mediation has a monologic influence on youth culture. Through language and literacy use, youth cultures are sites of stylistic and linguistic innovation; these new texts, styles, and forms of speech are often, in turn, taken up and appropriated in commercial, popular, and mainstream cultures (e.g., Frank, 1997; Hebdige, 1979; Kelley, 1997). Youth literacies offers a conceptual lens for understanding the ways in which young adults manipulate and interpret a wide range of semiotic resources, including locally produced and mass-mediated multimodal texts to enact identities, narrate their lives, and take up social positions in the world.

Research Design

I approach this inquiry into the experiences of students returning to education settings with a critical stance. According to Kincheloe and McLaren (1994), a criticalist researcher is one who accepts the assumptions that (1) "all thought is fundamentally mediated by power relationships," (2) "facts can never be isolated from the domain of values," (3) "language is central to the formation of subjectivity," (4) "certain groups in any society are privileged over others," and (5) "mainstream research practices are generally, although most often unwittingly, implicated in the reproduction of systems of class, race, gender and oppression" (pp. 139–140). My critical lens is rooted in a constructivist epistemology, which maintains that knowledge is largely dependent upon the interaction between the subject and the object. In other words, the thoughts, feelings, and experiences of the researcher are central to the construction of meaning. Crotty (1998) defines the constructivist notion of truth this way: "Truth, or meaning, comes into existence in and out of our engagement with the realities in our world. There is no meaning without a mind" (pp. 8–9). The poststructualist narrative lens, with its emphasis on representation rather than undisputed truth, is compatible with this epistemological stance.

Researcher reflexivity takes on increasing importance within a constructivist framework, because it is understood that the researcher plays an important role in constructing meanings. The interplay between constructivism and criticalism demands that as a researcher, I reflect on my own position as of privilege as a White middle-class graduate student. I must be reflexive about my subjectivities and my assumptions so that I can present these biases transparently and minimize their effect on my representations of the participants and their stories. I enter this setting with the assumptions that larger cultural narratives are an important influence on the lives of youth, that the narratives of participants in this study will reveal complex stories of school leaving and returning that are of interest and importance to practitioners, and that hierarchical social structures, power, identity, race, class, and gender are important to the lived experiences of high school dropouts. These assumptions are based primarily on literature and theory and only in small part on meaningful conversations or collaboration with young adults who have left and returned to school. My study will be conducted in a program located in Philadelphia, comprising former Philadelphia public school students. As a 27-year-old woman who was raised in middle-class

(Continued)

Annotation: Susan connects her methodological approach with her epistemological understandings, including what it means to be a critical researcher as well as how her constructivist epistemology has influenced her proposed study. Furthermore, she relates her constructivist approach to the methodological considerations such as reflexivity and also discusses her positionality.

(Continued)

suburbs in a predominantly White area of New England, I am an outsider in this setting in many ways. Therefore, I take seriously the need to learn *from* the participants in this study. Central to my methodology is careful listening and trust building over time. I do not expect participants in this study will immediately tell me their stories. Nor do I expect that I will immediately understand their experiences. However, over the course of a year, I seek to collaborate with participants, privileging their voices and their experiences as I work *with* them to understand and interpret their narratives and connect their stories to critical theory.

Context

The Stepping Stone[4] Program at the City Community College (CCC) will serve as the site of this research project. The Stepping Stone program serves youth, ages 16 to 21, who have dropped out of school. It is a dual-enrollment program allowing students to simultaneously work toward a high school diploma and earn credits toward an associate's degree. Stepping Stone at CCC is one of 13 Stepping Stone sites across the nation; each is situated within a community college, and each partners with a local school district. The Bill and Melinda Gates Foundation provides funding for the national replication project, but at each local site, program funds come from the community college and partnering school district. With the exception of a 20-dollar application fee and a 50-dollar fee per semester, the program, including books, is free.

The Stepping Stone Program at CCC opened in the spring of 2006. Each semester, the program admits two cohorts of 20 new students. Applicants must read at an eighth-grade level, must have enough high school credits to complete the program by age 22, must live in Philadelphia, and must have attended and dropped out of a Philadelphia public high school. In addition to a battery of assessments in reading, writing, and math, applicants also write three admissions essays and participate in two individual interviews. The first semester of the program is known as the foundation term. During the foundation term, incoming Stepping Stone students take English, math, and a freshman seminar that focuses on study skills. Although these courses are taught by CCC faculty and are nearly identical to the developmental (remedial) courses in English and math required of low-scoring CCC students, each cohort takes all of their classes together and is separated from the general college population. However, upon successful completion of the foundation term, Stepping Stone students enroll independently in a variety of CCC courses depending on their missing high school credits and their educational goals. Stepping Stone at CCC consists of five staff members: a project director, an administrative associate, and three academic coordinators. The staff spends the majority of their time recruiting and screening applicants and providing academic support and advising to current Stepping Stone students.

The Stepping Stone's rigorous admissions standards and its partnership with local community colleges make it unique among programs for returning high school leavers. Therefore, its students, who read at an eighth-grade level and have successfully completed a series of essays and interviews in an effort to enroll in a college program, are not representative of all students who leave high school. However, this site raises interesting questions about students who leave school and within a few years begin working toward both their high school and college degree. Unlike many GED programs that have high turnover and a mix of older and younger students, the Stepping Stone program serves a narrow age range and boasts a national retention rate of 72% between the first and second semesters (Comer, 2007). Therefore, this site may afford an opportunity to observe and better understand the formation of a collective sense of identity, as the students spend their foundation term together. I wonder specifically how students enact and perform identities as college students as they transition into their second term and how they reconcile this new identity with their past- and future-oriented selves.

Over the past 2 years, I have observed and interviewed in a number of developmental English classes at CCC. I have worked with three faculty members, each with a different perspective on "underprepared" community college students and the role of the instructor. I work as a coordinator at an adult education program and I have conducted four ethnographic-style interviews with students, three of whom are high school graduates enrolled at CCC, one of whom is a high school dropout enrolled in an adult education program. These experiences and the data they yielded have informed my assumptions about what I will see and hear from Stepping Stone students. Beginning in February 2008, I began pilot observations in a Stepping Stone foundation term English class. On my first visit, I was struck by how acutely I felt my outsider status, a feeling that was not nearly as palpable during visits to other CCC classes. I suspect this is related to the sense of community I saw demonstrated by the students, who have class together daily from 12:30 p.m. until 3:30 p.m.

(Continued)

(Continued)

Ruth Behar (1996) writes, "Nothing is stranger than this business of humans observing other humans in order to write about them" (p. 5). As I move forward, I wonder how and to what extent I will find a comfortable niche along the participant-observer continuum. Roman (1992) has pointed out that naturalistic ethnographic research, in which researchers take up roles as "intellectual tourists" or "voyeurs," can be as positivist in orientation and potentially exploitative as experimental or behaviorist methodologies. The poststructuralist approach demands an attention to reciprocity in the research setting. Therefore, I carefully consider Roman's questions: "For whom is this research or inquiry conducted? Who decides what the problems of the research agenda are?" (p. 560). I plan to enter the site with a transparency about my research agenda and with a willingness to dialogue openly with students about my research questions and my assumptions. As I continue to negotiate my access to the program, I am working to build a reciprocal relationship so that I can offer something back to the program and the participants in my study. At this point, I hope to offer my services as a volunteer writing coach or tutor, but ultimately the decision about what is appropriate lies with the program staff and the instructors.

Data Collection

I will follow an incoming cohort of Stepping Stone students for 1 year, sitting in their classes with them during their foundation term (summer 2008) and following them for two subsequent semesters (fall 2008 and spring 2009). Using narrative analysis within a case study methodology, I will use a variety of ethnographic-focused methods to learn about the experiences of these students. Audio-recorded individual interviews with students will constitute the bulk of my data. These data will be augmented with ongoing in-class observations, audio-recorded focus groups, informal conversations with program staff and administrators, and analysis of a variety of student-produced documents inside and outside of class. Please see Table 1 for a detailed timeline for data collection and analysis.

I will begin this project with classroom observation. Carspecken (1996) writes of "building a primary record," which is a "thick" recording of in-depth and nonevaluative fieldnotes that serve as the "data anchor" throughout the study (p. 44). According to Carspecken, the primary record is complemented by a field journal that includes journalistic-type observations and researcher thoughts, feelings, and comments. The focus of this project

Annotation: After contextualizing the proposed study context and setting, Susan details her overarching approach to data collection and then her specific data collection plan. She describes how the proposed methods of data collection align with and map onto her research questions, including the sequence in which she will implement each method. As well, she discusses how she has integrated structured reflexivity into her overall validity and engagement strategy throughout stages of the research.

is on the students and their experiences, not on the effectiveness of the program or the nature of instruction or pedagogy. However, my research questions reflect that the participants in this project are situated within a particular space. Therefore, formal and informal program and classroom observations, and informal conversations with program staff and instructors, will help me understand the context in which these students have returned to school. Observational foci will change over time, but initially I will construct a primary record around the questions: What texts do students reference in this classroom space, what narratives are created about dropout and return, and what identities are enacted individually and collectively? In the public classroom space, I wonder what master narratives are circulated by students and by the teacher about school success and failure. To what extent are students positioned by the teacher as "former dropouts" and/or as "college students"? How are they positioned by instructors and staff members? How do they position themselves? Regular classroom observation will serve several other purposes at this early phase, which include building a rapport with students and staff, discovering the daily routines of the program, understanding the literacy practices employed by students and teachers within the classroom space, and identifying potential students to participate in interviews and focus groups.

Early in Phase One of data collection, I will invite a number of students to participate in the project with the goal of interviewing 12 in the first semester. Students will be sampled purposively to achieve a diverse group in terms of age, race, gender, and a variety of observable social tendencies, including attendance, gregariousness, and displayed attitude toward the program. These audio-recorded interviews will be ethnographic in nature and will focus on the participants' stories of dropping out and returning to school. Please see Appendix A for sample interview questions. Ongoing and regular focus groups with a subset of six to eight students will offer a third source of data to accompany interview and observation data. Rymes (2001) used this method effectively with students who returned to an alternative school setting. She writes that narrators strive to "look good," and what constitutes looking good depends on the intended audience (p. 25). Telling a story in an individual interview for a researcher is very different than co-constructing a narrative for a researcher with and for one's peers. I employ this method with the assumption that when students talk about their experiences of dropping out and returning to school with their peers present,

(Continued)

Annotation: Including a timeline is a necessary part of a research proposal. Susan includes a timeline that is concise and yet has some granular detail on the process, which is necessary since these steps take time that needs to be calculated (or at least estimated). It is important to note that you need to understand the relationship and balance between constructing a clear timeline for data collection and analysis and weaving in language to connote (and time to engage in) the iterative nature of this process. Susan does not do as much of this as some other exemplary proposals we see. While Susan indicates that analysis and writing will take place throughout all phases of her study, her timeline does not necessarily capture the way that her analytic process was iterative and emergent (rather than being accomplished through discrete tasks as represented in the timeline) and how she structured it as an ongoing, multifaceted process. However, it is often difficult to capture such nuances in a timeline, and this is why a narrative description of the timeline that stresses the emergent quality of the process is also important.

(Continued)

their stories will be different than when they talk with me alone. These discrepancies may shed light on the ways in which identities are enacted and co-constructed by students with similar life histories. During Phase Two and Phase Three, I will conduct monthly focus group meetings for a total of nine meetings. I remain open to the reality that the composition of the focus groups may change with each meeting.

Finally, the documents I collect and analyze will offer a diverse array of data sources. Merriam (1998) points out that although interviews and observations are collected to answer the research question, many documents exist apart from the research design and therefore do not "intrude upon or alter the setting in the ways that the presence of an investigator often does" (p. 112). Publicity and marketing documents produced by the program will help me understand more about the context within which these students have returned to school. How does this program promote itself? What are its policies and procedures? What is its mission statement? How does this program represent itself, and how does it represent its students? As student literacy practices across contexts are central to my inquiry, student-produced work will provide an important source of data. I will collect classroom assignments and analyze these pieces of writing, looking for themes related to my research questions. For the core group of students, I will request access to samples of out-of-school literacy artifacts. Some examples might include poetry, photography, lyrics, short stories, journals, work-related writing, email, social networking pages (e.g., Myspace, Facebook), and other webpages. I also see potential for researcher-requested student-produced documents like journal entries, literacy logs, activity logs, written versions of their narratives, and self-identifying photographs (Hungerford-Kresser, 2007). To better understand the circulating dominant narratives, I will analyze mass-mediated texts like television shows, movies, music, and webpages that are deemed important by the participants. In addition to analyzing these texts for their representations of youth, dropouts, or students, I can also ask participants to analyze these texts individually or in focus groups. Participants' emic analyses may offer insight into the master narratives available to youth and the extent to which these participants accept or reject these narratives.

My timeline of data collection and analysis includes a total of four interviews with each of focal students over the 12-month period. Student attrition and refusal is built into the plan, ensuring that a minimum of six to eight students will complete all four interviews. As data collection

Table 1 Timeline for Data Collection and Analysis

	Observations	Interviews	Focus Groups	Analysis and Writing
Pilot Phase: January 2008–April 2008	• Classroom observations 1 day per week	• Pilot interviews with three to five students • Informal conversations with instructors and staff	• One pilot focus group with four to six students	• Transcribe selections of interviews/ focus groups as necessary • IRB • Dissertation proposal
Phase One: May 2008–August 2008	• Classroom observation 3 days per week	• Initial interviews with 12 students		• Document analysis of in- and out-of-class writing • Low-level coding • Biweekly researcher memos
Phase Two: September 2008–December 2008	• Classroom observation 2 days per week	• Follow-up interviews with 10 students	• Monthly focus group interviews with six to eight students	• Document analysis of in- and out-of-class writing • Low-level coding • Begin reconstructive horizon analysis on selected sections of data • Biweekly researcher memos
Phase Three: January 2009–May 2009	• Classroom observation 2 days per week	• Third- and fourth-round interviews with six to eight students	• Monthly focus group interviews with six to eight students	• Document analysis of in- and out-of-class writing • Begin high-level coding • Peer debriefing of initial analysis • Biweekly researcher memos
Phase Four: June 2009–May 2010		• Follow-up interviews with students, as necessary • Member checks		• High-level coding • Peer debriefing analysis • Write dissertation

(Continued)

Annotation : Analysis sections of proposals can be difficult for students to write, often because of the iterative and emergent nature of qualitative research. Susan states in the beginning of this section that data analysis will be generative and recursive. However, the analytical processes she describes do not appear to leave room for more flexible ways of analyzing the data and/or for different approaches. As we describe in the beginning of the chapter, a proposal is a plan not a contract. We recommend expressing this flexibility in your proposal, if possible.

Susan's section on data analysis has a strong emphasis on validity throughout. Susan highlights the key influences on her approach to data analysis and focuses on the work of one scholar (Toma) from whom she adapts her approach to validity in analysis within his four components of rigor (credibility, transferability, dependability, and confirmability). Susan articulates clearly how she conceptualizes and intends to use this framework as a guide and set of criteria for valid data analysis in her study. Adopting a validity framework in the way that Susan does can be helpful; however, we want to reemphasize that validity should not be considered in solely procedural terms. It is not something that you achieve; it is an overarching goal and process. We appreciate Susan's critical stance that is reflected throughout the proposal and her desire to transparently represent her research processes.

(Continued)

progresses, the focus of the observations and interviews may shift based on emerging themes from the preliminary analysis.

Data Analysis and Validation

Data analysis will be generative and recursive, beginning during Phase One of the study. In addition to the field journal, in which I will record ongoing comments, thoughts, and hypotheses about what I see around me, twice each month I will write an analytic memo to synthesize my thoughts and comment on themes as they emerge from the data. Using Erickson's (1992) stages of qualitative analysis as an overarching guide, I will begin by reviewing fieldnotes and transcripts holistically and then identify segments and features of organization and participation structures within those segments. Once selections of data have been coded and analyzed in detail, I will, as Erickson suggests, look across the larger data set for themes, paying special attention to discrepant cases. For the close analysis of selected data, I will draw on two analytic traditions.

Carspecken's (1996) critical methodology offers a useful and systematic method of interpretive analysis in which selections of the primary record and transcripts are analyzed for underlying objective, normative, subjective, and identity claims. Reconstructive horizon analysis, as he terms it, is grounded in Habermas's notion of the "horizon of experience" and is used to help researchers outline the breadth of possible meanings within any given speech act. First the researcher creates a "meaning field" in which all possible meanings for the utterance are stated. Then, reconstructive horizon analysis serves as a bridge from low-level to high-level codes and can help researchers validate their inferences. Please see Appendix B for a sample analysis using this method. With its focus on uncovering tacit claims of normativity (what participants think of as normal or proper) and identity, his four-pronged approach is particularly applicable to the notion of "master" or dominant narratives. What implicit claims are speakers making about what is valued, and how do speakers position themselves in relationship to these values? For example: What do Stepping Stone students think a college student "should" be like? What normative claims do they make about high school, education, success, and achievement?

While Carspecken's mode of analysis will be very useful for analyzing segments of data, his framework does not attend to the interactional nature of speech and the emergent nature of meaning in discourse. To complement Carspecken's interpretive heuristic, this project will also utilize discourse analytic strategies from a linguistic anthropological tradition. Drawing on the Bakhtinian notions of double voicing and ventriloquation, Wortham (2001) outlines a number of narrative analysis strategies that can be used to reveal the speaker's positioning with reference to the other people and ideas in the narrative (pp. 70–74). These cues include metapragmatic descriptors (e.g., she "lied" or she "whined"), evaluative indexicals (e.g., "presumably" intelligent), and epistemic modalization (e.g., "I think" or "the evidence proves"). By looking for patterns in pronoun usage, verb choice, and quotations, I will examine how participants position themselves in relationship to others and to images available to them in their communities and in the mass media.

I will assign low-level (or en vivo) codes to the primary record and all transcripts soon after conducting each observation or interview. Although I will hold off on higher level inferences until Phase Three, immediately assigning and categorizing low-inference codes will help focus my observations and interviews more productively. Beginning during Phase Two, I will conduct reconstructive horizon analysis and discourse analysis on selected sections of interview transcripts and the primary record. I will create and keep a master code list that includes all high- and low-level codes. Over time, the categorization and recategorization of the code list will allow themes to emerge from the data inductively.

Qualitative methodologists (e.g., Creswell, 1998; Denzin & Lincoln, 2003) long attempted, in many cases unsuccessfully, to divorce qualitative design from quantitative standards of rigor (i.e., generalizability, validity, and objectivity). In his review of the literature on standards in qualitative research, Toma (2006) identifies four alternative components of a rigorous qualitative research design: credibility, transferability, dependability, and confirmability. I have included several data validation techniques into the proposed research design in an effort to address these four components. "Credibility," writes Toma, "is established if participants agree with the constructions and interpretations of the researcher" (p. 413). Participants in research projects have traditionally been thought of as "subjects," people who can be studied but whose voices are rarely heard. During the final phase of my project, member checks and follow-up interviews will allow me to check my assumptions and enlist participants into a meaningful collaborative relationship in which my inferences are offered for feedback and critique. Johnson (2007) found that the analytic memos she gave to her participants as part of the member checks were too text-heavy and complex, so she distilled the memos

(Continued)

(Continued)

into poems written in the voices of her participants. Her participants then responded to these poems, clarifying and correcting as necessary. During fourth-round interviews with participants, I may also employ a method known as interpersonal process recall (IPR; Carspecken, 1996, p. 163). IPR, originally developed as a therapeutic approach, involves playing a recording of an interview or focus group back to the participant(s) and allowing them to pause the tape to comment on and interpret the data. In addition to serving as a form of member check, IPR offers a rich additional source of data that demonstrate how students' narratives change over time. If discrepancies between my interpretation and my participants' interpretation of the data cannot be resolved, both interpretations will be presented with commentary in the final analysis.

The second component of rigorous qualitative design is transferability. Toma (2006) likens this characteristic to generalizability in quantitative research. He suggests that researchers who connect their methods of collection and analysis to theory offer opportunities for readers to connect the findings to a larger body of work and therefore see the import of the study beyond the local context. I hope that by positioning this study within the literature on poststructural narratives, possible selves, and youth literacies, my work can speak to the issues for a wide range of students returning to education after leaving high school.

Dependability, Toma argues, can be achieved through a number of design features. He suggests that to be dependable, "findings must go beyond a snapshot" (p. 416). Inherent to my theoretical framework is the assumption that narratives are constructed differently depending on the time and place of their construction. By following the focal students over 12 months, my hope is that I can identify which narrative features remain stable over time and which change as the students progress through the program. Triangulation of multiple data sources offers another assurance that a researcher's analysis is dependable and valid. Student interviews, focus groups, observations, and documents offer me a wide variety of data sources, which can be checked against one another. Careful attention to and analysis of discrepant data, in the final report, will lend both credibility and dependability to my project.

Confirmability "is the concept that the data can be confirmed by someone other than the researcher" (Toma, 2006, p. 417). Beginning in Phase Three, peer debriefing will become a regular component of my data analysis. I will work with two fellow doctoral students, sharing sections of my low-level codes, reconstructive horizon analysis, and high-level codes, to solicit feedback on the extent to which my inferences can be validated by the data. Carspecken's (1996) method of reconstructive horizon analysis in itself offers a source of confirmability in that it forces the researcher to record each step she makes between literal-level coding and high inference coding.

Finally, my ongoing reflexivity, during both data collection and analysis, will be essential to the "trustworthiness" (Lincoln & Guba, 2003) of my research design. I enter this site an outsider on many levels. I bring with me assumptions about the experiences of these returning students. My position as a "visitor" in the program, as a researcher and potentially as a tutor, will affect the ways in which the students narrate their stories to me and the ways in which I interpret those stories. As I remain in the site over time and develop a relationship with each participant, those effects will diminish but never disappear. While the experiences and voices of the students are of primary importance, my own experiences in the site cannot be obscured. To do so would imply a positivist orientation in which the findings I report are truth rather than a representation of student experiences. My criticalist stance demands I work to call attention to inequalities, injustices, and power imbalances. Part of that work must be a recognition of and admission of my own position of power. As I embark upon this project, I endeavor to represent the students and their stories as transparently and as responsibly as possible.

Conclusion

Issues of high school persistence and graduation are implicitly issues of equity and social justice, as data suggest that graduation rates continue to fall, particularly in select number of schools in low-income urban communities. Despite decades of attention to the "dropout crisis," the research tells us little about the lived experiences of students who leave high school. The number of youth and young adults in a wide range of alternative education settings (GED classes, adult literacy programs, charter schools, accelerated high schools) is growing, but likewise, we know little about who these students are and why they persist. How do young men and women make the decision to return to school? How do they understand their lives, their stories, and their futures? How do they understand the messages circulating in the media and in their communities about "dropout" and "student"? The proposed project uses a narrative lens to examine the stories and literacies of 16- to 21-year-olds enrolled in the Stepping Stone program. This project has implications for

Annotation: Susan cogently concludes her proposal. Instead of summarizing the entire proposal, she highlights its significance and reminds the reader of her overall topic.

(Continued)

(Continued)

traditional high schools that are working to prevent dropout and for alternative programs like Stepping Stone that are working to serve the needs of young returning students. The public school system is failing the needs of many high school students who desire and value education. A range of theories have been employed to understand why students drop out, but much of the literature continues to frame high school leavers as resistant or somehow lacking the skills or the determination to finish school. The proposed project works to challenge these oversimplified notions of "dropout" and offer a series of nuanced portraits of youth leaving and returning to school.

As stated earlier, there are multiple strengths in this proposal, and Table 10.4 is based on Susan's dissertation proposal since it fulfills all of these. Example 10.1 is particularly powerful in how Susan bridges the conceptual, theoretical, and methodological. We consider this crucial to conducting ethical qualitative research, as these components cannot be separated. Another key strength of the proposal in Example 10.1 is that the author includes how she has set up her study design to include structured reflexivity processes. We have stressed the importance of structured reflexivity and dialogic engagement exercises throughout the book, and when writing a proposal, it is very important to not only vet the proposal by multiple individuals but also to subject the proposal to critical examination. Depending on the type of proposal (i.e., for a dissertation, master's thesis, funding, or a grant), you should ask for feedback from individuals with different perspectives. For example, an individual close to the research setting or context may provide valuable feedback on your research questions, methods, and participant selection. A methodologist or other knowledgeable adviser may push you to refine and/or rethink aspects of your overall research design. Colleagues and peers with whom you have established relationships may also help you to think critically and consider your biases as well as alternative explanations. As you revise and incorporate feedback into your proposal, it is important to keep in mind that a proposal is not your research design: It is what you *propose* to do. Your study can and should evolve, like most aspects of qualitative research, as your understanding of your topic becomes more complex.

ADAPTING RESEARCH PROPOSALS

In the previous chapter, we discussed final research reports. We wish to note that many students use the proposal as the foundation of their reports. We advise our students to do this with some important caveats, which include the following:

- You will need to adjust the text for tense, changing it from future to past tense.

- The proposal was just that, a proposal of a study; it is not a contract. As you revise the proposal to transform it into aspects of the final report, pay careful attention to what changed, got refined, or was added or dropped (and why) and adjust the report accordingly.

- Be sure to refine and add to earlier statements about context, goals, and so on since your understanding of these will have deepened throughout your research.

- Consider if there are audience shifts and how these will affect the report since the proposal is often written for a small group ("audience near"), and the report is typically for a broader audience ("audience far").

- Remember that the conceptual framework will iterate significantly, so pay careful attention to that as you update and revise the proposal.

- You will inevitably add to the review of literature/theoretical framework, so be sure that you take time to do so in the final report (we suggest adding to the literature review ongoingly as you read).

- Remember that proposals tend to lack specific detail on data analysis plans since in a number of broad qualitative approaches much of the analytic process emerges, so be sure that you have carefully documented your analytic processes in the final report.

- Save the proposal in its original form so that you can always refer to it (rather than writing into the proposal to create the report) since the proposal itself is an important artifact of the research.

Building on elements described in your proposal ongoingly can not only save time but also help you to accurately represent and detail your processes.

QUESTIONS FOR REFLECTION

- What are the ways that the goals and audience shape different kinds of proposals (e.g., master's thesis, dissertation proposal, conference proposal, funding proposal)?

- What are the overall components of a qualitative research study proposal?

- How can you make sure that data collection methods, participant selection criteria, and data analysis processes align with your research questions?

- What are the characteristics of strong research proposals?

- How can you adapt your proposal as you begin writing your final report?

- What important aspects should you consider during and after writing a proposal?

RESOURCES FOR FURTHER READING

Creswell, J. W. (2013). *Research design: Qualitative, quantitative, and mixed methods approaches* (4th ed.). Thousand Oaks, CA: Sage.

Becker, H. S. (1986). *Writing for social scientists: How to start and finish your thesis, book, or article.* (2nd ed.). Chicago, IL: University of Chicago Press.

Kumar, R. (2014). *Research methodology: A step-by-step guide for beginners* (4th ed.). London, UK: Sage.

Marshall, C., & Rossman, G. B. (2016). *Designing qualitative research* (6th ed.). Thousand Oaks, CA: Sage.

Maxwell, J.A. (2013). *Qualitative research design: An interactive approach* (3rd ed.). Thousand Oaks, CA: Sage.

Ogden, T. E., & Goldberg, I. A. (2002). *Research proposals: A guide to success* (3rd ed.). Orlando, FL: Elsevier Science.

Punch, K. F. (2006). *Developing effective research proposals* (2nd ed.). Thousand Oaks, CA: Sage.

Robson, C. (2011). *Real world research* (3rd ed.). Malden, MA: Blackwell.

Schram, T. H. (2005). *Conceptualizing and proposing qualitative research* (2nd ed.). Englewood Cliffs, NJ: Pearson Merrill Prentice Hall.

ONLINE RESOURCES

Sharpen your skills with SAGE edge

Visit edge.sagepub.com/ravitchandcarl for mobile-friendly chapter quizzes, eFlashcards, multimedia resources, SAGE journal articles, and more.

ENDNOTES

1. See Creswell (2014) for additional examples of proposal formats.

2. I use the term *dropout* with ambivalence. On one hand, it is a commonly used and widely understood term to refer to the population of high school noncompleters. On the other, it is a one-dimensional term that implicitly points to individual failure, a lack of motivation, and the notion of "giving up." For ease of language in this proposal, I frequently use the term, although I realize that it does not capture the various trajectories and stories of individuals, many of whom have been pushed out or shut out of their high schools.

3. Data drawn from October Current Population Surveys (CPS), Common Core of Data (CCD), and the GED Testing Service statistical reports.

4. Program and institution names are pseudonyms.

CHAPTER ELEVEN

Research Ethics and the Relational Quality of Research

Research in itself is a powerful intervention . . . which has traditionally benefited the researcher and the knowledge base of the dominant group in society. (L. T. Smith, 1999, p. 176)

CHAPTER OVERVIEW AND GOALS

Throughout the book, we have discussed many concepts central to engaging in valid, person-centered qualitative research. In this chapter, we build on these key concepts to incorporate a conversation about research ethics with a focus on the relational and reflexive aspects of ethics in research. We describe methodological choices in qualitative research, framing them within a discussion about relational ethics. We emphasize the integrity of relationships and the choices researchers make about enacting an inquiry stance that is relational and reflexive.

The chapter begins with an overview of the importance of relational ethics in qualitative research and explains why this is so crucial. This overview is followed by a discussion of some of the most commonly discussed—and vitally important—aspects of research ethics, including the institutional review board (IRB) and ethics committees, informed consent and assent, research relationships and boundaries, reciprocity, transparency, and confidentiality. In the last part of the chapter, we connect the concept of relational ethics to various issues that a diligent and ethical researcher needs to address throughout the research process.

BY THE END OF THIS CHAPTER YOU WILL BETTER UNDERSTAND

- What constitutes research ethics
- The connection between relational aspects of qualitative research and research ethics
- The role of institutional review boards and ethics committees in your research

(Continued)

(Continued)

- The ethical implications for relationship building with study participants

- Issues of reciprocity and research boundaries with study participants

- The role and pertinent aspects of informed consent and assent in the research process

- The ideological, methodological, and procedural dimensions of research ethics

- How issues and processes related to research ethics are more than just transactional and procedural

- The concepts of anonymity and confidentiality and how they differ

- Issues of data management and security broadly and in relation to technology

- The role and importance of researcher reflexivity in ethical research

- What it means to push against the expert-learner binary in research

- The role that collaboration plays in conducting ethical research

- Possibilities and limitations of collaboration

- The relationship between responsive and emergent research design and ethical research

RELATIONAL ETHICS: TAKING A RELATIONAL APPROACH TO RESEARCH

Throughout the book, we present a relational approach to research. To push against forces that seek to dehumanize, generalize, and altogether minimize and "other" people, researchers must be fully committed to an inquiry stance that critically examines every facet of research. This requires, among other things, taking a relational approach to and stance on research. The concept of a relational approach to research emerged from feminist research methodology. This approach critically examines and inquires into the relational dynamics between researchers and participants as well as between participants' experiences in relation to the phenomenon at the heart of a study (Fine, 1992; Newbury & Hoskins, 2010; Noddings, 2004; Tolman & Brydon-Miller, 2001; Way, 2001; Zigo, 2001). The relational aspects of the research process (and product), with its methodological attunement to and address of issues of power, identity, and the need to contextualize interaction and data, is at the heart of a relational stance on and approach to research.

A relational approach to research—because it requires that focused and sustained critical attention be paid to the relational aspects of inquiry—is based on an understanding of the need to allow yourself to become vulnerable (in the sense of being open to critical self-reflection and change) and to engage in research with a receptive sensibility, meaning that you are open to changing your opinions, approach to your research, and even critical aspects of the research as you learn with and from the research. This kind of approach also entails that we, as researchers, must not only see and acknowledge differences between ourselves, our critical friends, and our participants but that we view differences as valuable and reflective of the impact of local, national, and international contexts, realities, and tensions as they shape individuals' different "funds of knowledge" (Gonzalez, Moll, & Amanti, 2005) and ways of being in the world (Fine, 1992; Nakkula & Ravitch, 1998). To be clear, we do not mean to suggest a focus on difference in an essentialized way, but rather, we mean an acknowledgment and critical understanding of the natural differences that exist between any two people, which includes an understanding of the shared value, worth, and humanity of all people (Erickson, 2004).

A relational approach to research speaks to the need to allow yourself to become reflexively engaged in interactions with others. For example, this might mean admitting that you do not know the answer to a question and need to check or admitting that something you said or did reflects a bias that is pointed out to you and showing a willingness to hear more, engage in dialogue about it, reflect on yourself, and possibly make changes in your approach and the research itself. We argue here that researchers must be willing to become inwardly reflexive and consciously collaborative in ways that necessitate raising your threshold for discomfort as we engage in and transform our praxis[1] (Freire, 1970/2000; Ravitch, 1998). Perhaps the best characterization of relational approaches to research that we have found states,

> Relational approaches to research are discovery-oriented and emphasize how data emerges out of co-created, embodied, dialogical encounters between researchers and co-researchers (participants). The researcher's attention slides between the phenomenon being researched and the research relationship; between focusing on the co-researcher's talk/thoughts/feelings and exploring the relationship between researcher and co-researcher as it unfolds in a particular context. (Finlay, n.d.)

Paying attention to the relational aspects of research requires a deeply reflective process and ongoing, complex identification of the influences on your sense-making and the influence that has on every aspect of the research and analytic processes that emerge from it (L. M. Brown et al., 1988). It also requires the systematic examination and questioning of what it means to engage in research, to learn from and with people who have different experiences and everyday realities from you (which we argue is everyone), and to understand, as critically as possible, how the data your research generates are embedded in these sets of relationships and power structures. Thus, a relational approach to qualitative research is built on an explicit standpoint of reciprocal transformation (Nakkula & Ravitch, 1998), that is, an understanding that real collaboration means that everyone involved must be willing to be changed in meaningful ways. This stance requires building a receptive sensibility about relational integrity, viewing research as a fundamentally collaborative and dialogic process (as well as a subjective one), and the nature of the research collaboration itself (Chilisa, 2012; Ravitch & Tillman, 2010).

This person-centered, societally contextualized approach to research places a primacy on the authenticity of the relationships between researchers and participants. It examines the roles, power structures, and language used to frame these relationships (such as "subjects," "informants," "interviewees," and "participants"). Within this approach, the active, critical consideration of and thoughtful engagement in relationships is at the very heart of the research process and is seen as part of the methods as well as the findings. Again, the ways methods are inseparable from findings (Emerson et al., 1995) highlights how who we are, and how we engage with the participants in our research (and with fellow researchers when the research is collaborative) has everything to do with the process and outcomes of our research.

Because qualitative research is centered on relationships, it is absolutely vital to frame relational considerations as ethical issues. Throughout this chapter, we articulate ethical dimensions related to a relational approach to research ethics.

DEEPENING THE CONCEPT OF RESEARCH ETHICS: BEYOND IRB AND INFORMED CONSENT

In our teaching and work with students and researchers across fields and disciplines, conversations and questions about research ethics tend to focus almost entirely on procedural issues such as needing to propose research to an IRB at a university and/or engage with site-based ethics committees to gain approval, engaging with various codes of ethics within specific fields[2] (many of which have their own specific rules and applications), and issues of consent forms and confidentiality. We believe that these issues and processes are absolutely central to research ethics. We take this a step further to argue that these concerns and activities are, when thought about critically, not only transactional but also safeguards for our participants and generative to our own understanding of accountability throughout the research process. They are also, to be clear, only the beginning of the necessary considerations and debates about ethics in qualitative research. While ethical codes inform your actions, your communication with research participants is what makes your study ethical (Glesne, 2016).

Ethics in qualitative research are multifaceted, complex, contextual, emergent, and relational; considering ethics critically requires an attention to the procedural and transactional as well as the relational and sociopolitical (Ravitch, Tarditi, Montenegro, Estrada, & Baltodano, 2015). It necessitates that we, as researchers, understand, consider, and approach our roles with humility and an understanding that we must carefully consider these issues collaboratively and relationally in order for our research to be ethical.

Institutional Review Boards, Ethics Committees, and Codes of Ethics

Formalized guidelines for ethical research conduct stem from a legacy of problematic and unregulated research projects in a historically unregulated milieu that caused considerable harm to individuals and groups. Specifically, discussions of the need for ethical regulation, such as through IRB committees at universities and ethics committees at various kinds of organizations, is the outgrowth of research in the medical realm as well as other kinds of experimental research that were impositional, intrusive, and even abusive and that preyed upon historically marginalized and vulnerable populations.[3] The reach of these committees, as well as the codes of ethics that IRB committees and various national and local organizations disseminate, has been extended beyond medical research and is relevant for "all research with human subjects" to develop and enact the ethical regulation of social science research (Hammersley & Traianou, 2012). To be clear, we completely reject the language of research "subjects" because we believe it reinscribes asymmetrical power relationships and view it as dehumanizing. We use the term *research participants* and acknowledge that there is a continuum of participation.

Universities have formed institutional review boards, which are centralized committees of faculty and staff, and various organizations, including school districts and human service agencies, have developed ethics review boards. These appointed committees are responsible for reviewing research proposals and overseeing ongoing research projects to ensure what is referred to as *beneficence*. Beneficence means that researchers should always have the welfare of participants in mind and should not cause harm to research participants in any way. This notion of not doing harm is a central ethical issue in research, and while IRB offices and institutional ethics review boards often get a bad name among students (and even some faculty) as micromanaging their research, we see them as important safeguards. They can be useful to elevating the level of researcher attentiveness and accountability for participants' welfare and can be a support to researchers. At their most effective, IRBs can help to point out possible issues in proposed and ongoing research that help further focused thinking about how to safeguard against anything that could be harmful to participants.

One aspect of the tensions that can exist between researchers and IRB committees is caused by the fact that IRB and ethics committees often do not understand the specific methodologies of qualitative research because of its emergent nature. The biomedical and positivist roots of the IRB shape the ways these committees respond to qualitative proposals. Although certain questions from IRB committees may at times reflect narrow views of qualitative research, feedback and questions from IRB committees are often useful and generative. The feedback received can help the researcher or research team to develop and articulate clearer rationales for research designs and data collection methods and instruments. Furthermore, it can help you conceptualize what "harm"—a key concern of IRB committees—might mean and look like in the proposed study. In this sense, harm can take many forms, including (but not limited to) the following:

- Language and framing of the research that either explicitly or implicitly creates a sense of pressure or coercion to engage in the research

- Engaging in covert or deceptive forms of research and/or deliberately misleading participants about the purposes and goals of the study or how the data will be used

- Asking questions or reacting to participant responses in ways that could make participants feel marginalized, judged, or that cause distress

- Asking "direct reports" (those who work for you, report to you, or over whom you have institutional power) to engage in studies that you conduct in ways that make them feel pressured or coerced

- Asking participants to agree to a certain number of hours of engagement (through various data collection methods) and then adding additional time requirements after they commit (also known as "time creep")

- Sharing data or exposing data that compromise confidentiality or anonymity assurances

- Using evaluative or judgmental language (or body language) about sensitive topics or aspects of identity or choices participants make

- Thinking of and approaching informed consent too casually or as a one-time endeavor

- Placing participants into focus groups or other situations that could expose individuals to discomfort or harm given the topic or relational/professional/ community dynamics without considering the implications for them during or after the group or interaction

- Writing about or otherwise representing participants in ways that are inaccurate, judgmental, essentializing, or deficit oriented

- Issues stemming from how public, Internet-based data interface with promises of confidentiality

Preserving confidentiality can be complicated due to the ubiquity of technology and social media. An ethical dilemma that one of our students faced brings to life the real ethical dimensions of various methods choices. A student in one of our research courses shared a conundrum that emerged in the pilot study she conducted as part of our class. She had full consent from participants in her study to use public data from the Internet as well as inter-view and observation data. She had promised confidentiality to all participants and then realized that if she included excerpts from the blogs that several of the people in her study kept, even though those are publicly available, people reading her study could cut and paste those excerpts into a search engine and identify the participants. As a result, she decided, after much thought and debate, that she could not include those data in her study analysis or report because she viewed this as an ethical issue. Although she had seen the data and they informed her sense of these participants, she felt she could not be transparent about that in order to preserve the confidentiality that she had promised the participants. This ethical dilemma points to the ways in which we, as researchers, must consider and address the various ethical dimensions of our research. It also speaks to this researcher's stance that fidelity to her participants needed to trump a fidelity to the data or findings. This was a diffi-cult choice for her, but she understood the need to keep her promises as a researcher and to protect her participants even though it meant sacrificing doing and writing up her research in ways that she thought were more analytically robust. In the end, she was able to write a very meaningful and important pilot study report that led to a fuller study in which she pro-actively addressed this issue of social media and confidentiality in ways that helped her to avoid this situation. She did this by being careful about the wording in her informed consent forms and the ways she described the research and data collection to her participants as well as in the ways that she approached and structured her IRB proposal.

The IRB process is often site specific. Although the same federal guidelines are used across university settings, the actual local implementation varies considerably. Given that there are standards within and across IRB committees and various other ethics governing bodies, we refer you to these groups for further specifics on how these processes work and how they may affect the research timeline (e.g., gaining approval from an IRB and/ or another ethics board usually takes multiple rounds of proposal development and then response to committee questions and feedback). As a necessary early step in most contemporary research, you should work to understand what is required of you if you are based at a college or university, if your research is within another institution that may have additional guidelines and approval processes, or if you are within any group or community that may have norms about research (this includes online communities). But please do not think that by taking this initial step that you have then entirely addressed the ethical issues of your study, as this is only the beginning. Solely meeting these guidelines, while absolutely important and necessary, is often *insufficient* in terms of thoroughly addressing ethics, especially from a relational perspective (Austin, 2008).

As we end this section on institutional review boards and issues of research ethics, we want to bring up the topic of research with "vulnerable populations." IRBs make distinctions about populations that are considered "vulnerable" (e.g., on Penn's IRB website, these groups include pregnant women, prisoners, and children) so that special safeguards can be put into place to protect these groups (remember that in the case of pregnant women, for example, this was determined in relation to medical research that could harm a fetus). It is important to note that there are special ethical considerations with respect to particular vulnerable populations and that this term can be both defined and interpreted differently (i.e., given the research topic and context) and might be broadened, from an ethical perspective, to include historically marginalized or otherwise underrepresented or underserved groups, groups that are linguistically different from the norm, and/or other ways that groups are minoritized or mistreated. It is important to think carefully about these attributions since there is a fine line between understanding the special interests, situations, and needs of groups and projecting need or deficit onto certain groups in ways that reinscribe deficit orientations (Valencia, 2010). In addition, given group histories and experiences, sensitive topics might create particular vulnerabilities even in groups that are not officially (by IRB standards) considered vulnerable or marginalized. One example is a project Sharon worked on with U.S. military veterans. While not officially considered a vulnerable population, the research focused on issues caused by veterans' time in combat and their reentry into civilian life, topics that can at times bring up emotional and psychological distress in ways that need to be proactively and ongoingly considered. These kinds of issues are caused by the interface between group and topic, and they point to the need for these determinations to be thought about on a case-by-case basis and in deeply contextualized ways. We seek out advice on these kinds of issues from those with more experience since there is much gray in this area, and you need to be thoughtful and careful to resist essentializing, stereotyping, or deficitizing individuals and groups (Valencia, 2010).

GOING BEYOND "NEGOTIATING ENTRÉE" AND "BUILDING RAPPORT"

The oft-discussed qualitative notions of "negotiating entrée" and "building rapport" seem important in a chapter on research ethics since that parlance is common and can be problematic if not fully thought through in ways we discuss here. These concepts, which are common in introductory texts in qualitative research, are very important, and discussions of them can glaze over deeper aspects of introducing yourself and your research to various participants and actors in a research setting. It is important to consider thoughtful, ongoing, authentic, and respectful engagement and relationship cultivation with research participants before, during, and after the research takes place. As we begin to explore some of the language used to describe these processes and offer our critiques of this language, we want to state our goal for doing so at the outset. Our goal here is beyond the specific terminology we critique below; it is that we wish to bring to the forefront of this discussion that the typical research language and ways of thinking about research ethics and relationships are not, in reality, as relational as many who write about research might intend.

The very language used in the phrase *negotiating entrée* has always felt rather transactional and even legalistic to us, and therefore we tend not to use it. While thinking of gaining access and then entering into a site or community as a back-and-forth exchange certainly resonates with us, the language of "negotiating entrée" is not one that we tend to take up given that it sets a tone that we believe undermines the spirit of ethical qualitative research that is grounded in an equity-oriented, relational stance. Our primary critique of this phrase is that once researchers gain "entrée," the negotiation tends to end, and there is the potential for such a negotiation to generate a one-sided and even impositional sense that the researcher will negotiate for what she or he wants.[4] This process should be approached as a dialogic exchange in which all parties feel comfortable to advocate for their ideal research scenario and through which the researcher is thoughtfully responsive to participants' interests, concerns, and limitations. We suggest that you consider thinking about the concept of "negotiating entrée" as an ongoing process of relationship building that does not end once you gain access to a site. However, more important than the language used is that you understand and can aptly address these concerns about ongoing relational integrity. Establishing respectful and mutually beneficial research relationships from the beginning and maintaining them throughout are what is important.

Related to this idea of the early and ongoing need to cultivate relationships in meaningful and respectful ways that resonate for those in the research, the concept/terminology of *building rapport* can make what we view as one of the most important aspects throughout the research process sound transactional and one-sided, as well as a one-time or early-on event, which of course is the opposite of what ethical, relationally oriented research should be. The development of relationships is an incremental, complex, multifaceted, and vital process that is at the heart of qualitative research. One thing that we often say to our students is that you are building multiple kinds of relationships with a variety of people in research and that, however brief, each interaction can and should be conceptualized as a relationship and attended to with care, intentionality, and transparency. That may include leaders and/or members of an organization or a community or affinity group (formal and/or informal); gatekeepers to these groups, who are often "the deciders" of whether you get access; and various stakeholder groups within or related to this group, possibly with subgroups and then finally with each individual.

Building relationships with **gatekeepers** and research participants will look different depending on contextual factors as well as the research questions and study goals. Despite these differences, an approach to relationship building with research participants should

- centralize respect (and understand that concepts such as respect are culturally and contextually mediated);

- address a range of stakeholder concerns, interests, resources, skills, and needs;

- show a careful and thoughtful cultivation of authenticity and honesty (including but not limited to a transparent and authentic presentation of self); and

- critically consider and address issues of reciprocity (discussed in depth in the next section).

Qualitative researchers must focus on going beyond "rapport" to authentic engagement; this requires both a relational mind-set and a number of specific methodological commitments and attendant methods processes described in this chapter and throughout the entire book.

Terms and Concepts Often Used in Qualitative Research

Gatekeepers: *Gatekeeper* is a term/concept used in qualitative research to describe individuals who are in the position to grant or deny a researcher access to the research site in the early stages of the research. At times, this person serves as the key point person for the researcher throughout the research process, which can include dissemination of information from the researcher to key decision makers and participants. Gatekeepers may be formal, such as an organizational leader or governing body, or informal, such as an informal community leader or someone designated to engage with the researcher with no formal historical role in this arena. It is important to note that even once you may have established access to a setting through the gatekeeper(s), you must get consent from all participants in your study. It is also important to get this gatekeeper approval in writing. This can come in the form of an email, formal letter, or other forms of writing. We recommend that students save these documents as a PDF.

With both concepts—negotiating entrée and building rapport—it is important to understand the importance of viewing the development of equitable research relationships as happening incrementally, intentionally, and with thoughtful reflection over time. Of course, how you begin these relationships is important and requires considerable time, thought, and planning, but the cultivation of relationships and trust must happen ongoingly as well. Moreover, how you set up and manage expectations with participants and other stakeholders is something to pay attention to not only during the research but after as well. Students most often spend the majority of their time and planning on the entrance and beginning of their research, which makes sense and is important. All researchers need to attend to "exit strategies" as well, which means that you need to thoughtfully consider and communicate what happens when the research ends in terms of follow-up and further engagement, if any. In Table 11.1, we present questions to ask yourself at the outset of your study and then throughout the process of conducting and concluding your study.

Table 11.1 Considerations for Establishing and Maintaining Healthy Research Relationships

- How do I introduce and frame my research to various participants so that I am transparent and clear about the research and what it means and entails?
- Are there differences in how I frame the research to various participants? And if so, what can that teach me about some of my assumptions within and across these participants?[5]
- How do I make the ideas accessible to a variety of participants without changing the meaning?
- How am I setting and managing expectations within and across stakeholder groups? In what ways might I do this better?
- Is it OK not to share my research questions? Why would I choose to or not to do so?
- Who are the gatekeepers and how do I approach them? What might the implications of their role(s) and approach(es) be for how my research is presented and understood? Are there ways I can address any concerns related to that?
- What materials, if any, do I share with gatekeepers and participants (e.g., research questions, relevant literature, conceptual framework, instruments for data collection)? What informs these decisions? What are possible implications of these decisions? What are the various approaches for doing so?
- How do I set a positive tone as I begin the research? What does that mean and entail? How do I maintain that?
- What does respect mean for individuals within this context? What are differences within and across the participant group in this regard?
- What are other foundational concepts that may be culturally and/or contextually mediated that I should consider?
- What are some ways I can think about engaging from a position of care and respect? And how do I understand how participants conceptualize these things?
- How do I get clear and honest feedback on how the research is being understood by various individuals and/or groups?
- How do I check in over time on how the research is developing? How it is perceived and experienced? By whom? How do I get at the intragroup variability on these perspectives?
- How am I presenting myself? Is this consistent and authentic?
- What mediates some of my choices about how I present myself (e.g., dress, demeanor, body language, etc.)?
- How do I know if I am being given access to all perspectives and roles in an organization or group? What can I do if I believe this is not happening?
- How do I communicate the goals of the research over time?
- How can I engage thought partners and participants in some of these questions and processes?
- What assumptions might be creating blind spots and biases? What are these assumptions and what are their implications for the research? How can I best address and communicate this?
- Have I met the expectations I/we set at the outset of the research? How did the research comply and/or depart from these expectations and why? Should this be discussed with participants?
- Am I setting the right tone? Being open to engagement while setting appropriate boundaries?

This list is certainly not exhaustive but provides examples of the types of questions that an ongoing engagement in thinking about and addressing issues related to access and relationships requires. It is important to note here that even the idea of the development of "healthy research relationships" is subjective and contextual. The structure and engagement in relationship building will vary with context and length of time—some "relationships" may be fairly pragmatic (a one-time interview, for example, where the researcher does not see the participant again) or long term (Sharon's research in Nicaragua involves the same people for the past 6 + years and will continue for at least 4 more). In this sense, the idea of reciprocity in a brief research relationship versus a longer term one will vary. In the next section, we develop the concept of boundaries addressed in the last question in Table 11.1.

RESEARCH BOUNDARIES

The notion of research boundaries is an important one to consider throughout your research and in new ways with each person and each research study. Some qualitative texts talk about the need to avoid being "too friendly" with participants, claiming that it can bias or skew data, and others argue against the researcher being too stoic and removed. Taking a singular stance across the board on these issues misses the key points of qualitative research: that most questions in qualitative research are contextual, and as we often say to our students, context is everything in qualitative research. Given that we so strongly believe that these questions are contextual, where you fall on the friend-to-removed-observer continuum depends on a number of elements, including roles in a setting (e.g., boss and employee, coworkers), social identities (e.g., elder and younger community member, spiritual leader and congregation member) and roles in the research (e.g., outside facilitator, co-researchers, research adviser), as well as sets of wider considerations such as relational style and power dynamics. Furthermore, these elements should be explored and discussed with a variety of thought partners who can help you to think through the relational stance that you take in your research in relation to the contextual specificities. At the same time, we caution you to consider the notion of "boundaries" in your work so that you can examine if the stance you take and choices you make within and across participants is appropriate personally and contextually. Examples 11.1 and 11.2 help us consider these issues.

> ### Example 11.1: On the Need for Proactive Thought on Boundaries With Participants
>
> A graduate student in our course shared a story of her fieldwork with great concern and palpable anxiety. She was engaged in research with a group of 15 women who were incarcerated at the time. When asked within a group
>
> *(Continued)*

(Continued)

interview session for her mobile phone number, she in the moment felt that if she said no, the trust in the group would erode and she would not be able to get the data she needed for her study to be successful. So she shared her mobile number even though she was not sure this was a good move and it made her feel anxious. At least half of the women in this group proceeded to text and call her quite frequently to ask her for favors and to communicate with their loved ones for them since they had restricted access themselves. These calls and texts made her extremely uncomfortable, and she felt pressure and anxiety about how to respond to these requests. She worried a great deal about how all of this would affect the group dynamic, her relationships with the women, and therefore the data and the research process. She also began to worry for her safety both during and after the research.

As we consider this example, which is actually not that uncommon, we would argue had she thought through her relational stance and set clear boundaries in her mind and with the women in advance, she may not have been as taken off guard when asked for her mobile number and may have in the moment had a prepared answer to why she would not give out her private number. But we would argue, as well, that with a different group of women who were incarcerated in the same setting, this might not have posed a problem. So it is on a case-by-case basis that we suggest you discuss and make decisions about these kinds of things.

Example 11.2: On the Need to Set and Manage Expectations With Site and Participants

One of our students worked diligently to gain access to three classrooms in one grade level of students at an independent school. Her goal was to study the relationship of pedagogical style to student learning outcomes in language arts in an independent school context specifically. She had trouble gaining access to several other schools and was feeling a sense of desperation to find a site so that she could move through her doctoral program on the schedule she had set out for herself. To gain access to the three fifth-grade classrooms at a specific school, she promised two things: (1) to share all of her data and findings with the school leaders and (2) to work intensively over a long period of time with the teachers in the school on developing whatever skills her research identified to be areas of necessary growth and professional development. Both promises ended up being problematic for her personally and professionally as well as for the research.

In the case of the first promise to share her data and analysis, this deeply undermined the teachers' and parents' willingness to be more open about the institutional and leadership challenges they saw in the setting (since they knew the school leaders might be able to identify them in the raw data given the small size of the faculty) and also, for this same reason, created issues of her own sense of openness with the questions she asked during interviews and focus groups as well as the ways she analyzed the data.

In the case of the second promise to use her findings to guide her own implementation of intensive, ongoing professional development for the teachers, once her dissertation was finished, she accepted a job in another country and could not keep her promise to work intensively with the teachers, which was viewed by the school as a massive breach of trust and ethics and ended up creating tension and problems moving forward. Even had she not moved, we believe that the second promise blurs the boundaries of the research. Reciprocity, as we discuss, is extremely important in all research, but as this example highlights, you must think carefully about what you promise and the ethics of making too many promises to gain access. In addition, we advise students not to promise to share all of their raw data and full analysis unless circumstances require it and it does not compromise the research. Sharing executive summaries with site-specific recommendations may be a better option.

Both of these examples help us to see that the notion of setting appropriate boundaries is an important one to consider at the outset of research and ongoingly throughout your study. In Example 11.2, the student should have had a conversation with stakeholders and participants to offer alternative options regarding her second promise when she found out she was leaving the country. However, a key lesson is to not promise too much at the beginning (or any time, but most people seem to overpromise primarily as they enter into the research). Ethical issues revolve in large part around beneficence, or doing no harm. Thinking through issues of boundaries and relationships will help to prevent any unintended harm and consequences. In addition, we specifically suggest having multiple conversations with more experienced researchers that focus specifically on the setting of boundaries and calibration and articulation of expectations with participants and gatekeepers.

There are many lively debates in the field of qualitative research about relational ethics and how things such as "rapport" and "friendship" fit into and shape an ethical approach to research. We describe some of these important considerations in Table 11.2.

The way you think about boundaries and relationships in your research is directly related to your understandings of reciprocity, which we discuss in the next section.

Table 11.2   Considering Relationships Between Researchers and Participants

- How might friendships or prior relationships with study participants make individuals feel pressured or coerced to agree to participate in the research?
- How might friendships or prior relationships with study participants bias data?
- How might relationships be impositional or unwanted by the participants?
- How will you address participants' desire to engage in relationship building that may be unwanted by the researcher?
- How do friendships with participants influence the ability to ask tough questions?
- How do various kinds of relationships with participants influence data collection processes?
- How might relational boundaries with participants influence data analysis?
- How might relationships with participants shape the way roles are established in research processes?
- How do I document relationships with participants and the effects these have on data?
- What is the role of relationships in the disclosing (or lack of disclosing) of certain information?
- What considerations have been made for when researchers leave the research setting(s)?
- What does cultivating relationships with participants mean in terms of balance and reciprocity?

Source: These questions were developed from McGinn's (2008, pp. 770–771) discussion of researcher-participant relationships.

Reciprocity: Not as Simple as It May Seem

Reciprocity is a term we hear a lot in conversations about qualitative research broadly and research ethics specifically. Students often ask questions about what reciprocity means and entails with an assumption that, because it sounds positive, it always is. Schwandt (2015) offers a broad framing of issues of reciprocity in qualitative research:

> [Reciprocity] is a kind of social behavior—a mutual give-and-take, an exchange of gifts or services—in which we routinely engage in social life but that is especially important in field studies where the researcher is accorded the privilege of access to the lives of those he or she studies. Paying respondents, informants, and participants and doing small favors like giving them a ride to work, picking up their dry cleaning, babysitting their children, buying them a meal, and so forth are reciprocal social acts that a researcher might perform while doing fieldwork. Reciprocity is part of the larger ethical-political process of building trust, cultivating relationships, and demonstrating genuine interest in those among whom one studies. Which of these kinds of activities to engage in or how far to take the act of reciprocity is a matter of ethical judgment regarding what is appropriate in the circumstances in question and hence requires practical wisdom. (p. 267)

Schwandt's definition of reciprocity captures the way many researchers think about the concept: that reciprocity, or giving back to participants in "exchange" for their time and the sharing of their perspectives and experiences, is always a positive step toward mutuality and equity.

We, however, think that a more critical approach to reciprocity (rather than this more normative one) is vital to conduct ethical research. For instance, the examples that Schwandt mentions above such as paying participants or "doing small favors" are not necessarily a good idea or ethically appropriate. We believe that in many cases, these specific acts are actually inappropriate and that they have the potential to blur the boundaries and roles of researchers. While perhaps in an ethnographic study, in which a researcher is truly embedded in a community for long periods of time, some of these acts may be more appropriate, we would argue that even in those circumstances, these kinds of acts need to be scrutinized and problematized in a variety of ways. For most of our students (who might use ethnographic methods but are not engaging in full ethnographies or in-depth participant observation and are therefore not fully immersed in groups or communities), such personal acts could seem quite strange and even potentially off-putting. We caution our students to be careful about assuming it is positive to engage in such acts and ask them to think about reciprocity more along these lines:

> Reciprocity concerns balanced patterns of giving and taking between people. Research relationships are not necessarily reciprocal, but good research ethics practice requires that researchers consider what they take from research participants as well as what they give to them. There are several dimensions to this issue, including the different conceptualizations of what is given and taken, the rights and responsibilities of each party in research relationships, and the practicalities of building rapport. (Crow, 2008, p. 739)

Conceptualizing the give-and-take as well as rights and responsibilities, as Crow (2008) discusses, demonstrates how these aspects may be conceptualized differently by various people for a host of reasons ranging from individual to cultural to organizational contexts and how they shape people's views of reciprocity and relational exchange.

Noteworthy in the above quote from Crow (2008) is that researchers are always asking participants for their time, seeking to learn from them, and that in turn researchers can engage in reciprocity through a number of processes and acts, including (but not limited to) the following:

- Giving assurance that data will be treated ethically in terms of confidentiality and anonymity as well as respect for how participants are portrayed

- Providing opportunities for participants to reflect on and openly share aspects of their lives

- Affirming or validating people's experiences in contextually appropriate ways

- Creating the conditions for participants to have a voice in the public sphere (if they express that desire, goal, or need)

- Giving financial compensation for costs and/or time (which can be problematic in terms of deciding how much and then understanding if financial compensation motivated participation)

- Sharing their own lives in response to participant sharing (in ways that are appropriate given the circumstances and context)

This is only a partial list, and even with these, there could be possible complications and issues since these assume a great deal about participants and their needs and therefore could lead to problematic actions, interactions, and engagements. Having said that, we do believe that there can be times when compensation of a few sorts can be appropriate. Some examples (with their respective challenges and considerations) are described below.

- Providing a meal or snack when interviews or focus groups are scheduled at mealtimes or after work (we would argue that meals should be kept modest, and this raises issues of sensitivity to religious and cultural food restrictions such as observing the laws of Halal, being vegetarian or Kosher, food allergies, etc., as well as cultural norms about when eating is appropriate)

- Compensating participants for travel costs (this needs to be thought through carefully since at times, giving cash can be problematic, and reimbursement without receipts is often not possible)

- Providing childcare at town hall meetings, interviews, and/or focus groups so that those who cannot access or afford it can still participate (this may come with insurance concerns)

- Providing small tokens of appreciation such as an inexpensive pen or tablet from one's organization

- Some of our students and colleagues will give gift cards of small denominations ($5–$20, but there can be issues with this as discussed immediately below, and we direct our students away from this much of the time)

You must consider, if giving a gift of any size, how those whom are not invited to participate will feel if they see or discover this. That is one unintended way that reciprocity, even with the best intentions, can go awry. There are surely others. Even with the best of intentions, we have seen these acts of reciprocity lead to unintended negative consequences that can harm the participants and/or the relationships as well as affect the data in negative ways. To be clear, these are ethical issues since they affect the participants and can undermine equity and relational boundaries in a host of ways.

As a part of his discussion of reciprocity, Crow (2008) discusses Patti Lather's (1991) views on reciprocity, which resonate very much with our own:

> More fundamentally, Patti Lather described reciprocity as mutual negotiations of meaning and power. This requires a collaborative approach to research where participants are invited to negotiate interpretations of the data and to contribute directly to data analyses. Researchers who are committed to reciprocity try to avoid imposing their own meanings on the research and strive to decenter their roles as 'experts' in the process. Full reciprocity involved a conscious commitment by researchers to using the research to help participants understand and change their situations. (p. 770)

This definition speaks to a deeper understanding of reciprocity that we believe must be the foundation upon which any of what we think of as these more micro-level reciprocity choices are made. This conceptualization of reciprocity helps to foreground issues of equity in a more relational, rather than simply transactional, sense, which we consider absolutely necessary and that we build on in the sections to come. Reciprocity, like all aspects of research, must be approached critically by thinking about issues of power asymmetry as well as potential consequences of such actions. Because of the contextual and relational nature of qualitative research, there are no rules for attempting to make research reciprocal. If something does not "feel" right, then it probably is not. However, it can also be difficult to know how your actions will land on or be experienced by others. Listening to your "gut" and seeking the advice of others can help you navigate these often-murky waters. You also have to be careful since even in Lather's definition, there might be an underlying assumption that the situations of participants need changing and could possibly, depending on the situation and context, position the researcher as "savior."

We provide one example from a recent conversation that Sharon had with one of her doctoral students as a way to bring some of these issues to life. This student had ended her formal data collection, which had her living in a rural community in southern Morocco for 1 year working with local community organizers to develop language centers for culturally based language learning. As her research was concluding, she came to Sharon quite concerned about how the people in the community would maintain language programs she had helped them to develop. While it seemed that she should continue to "help them" because it became evident that as she was pulling back from community involvement, programs were not being attended to, in reality she was leaving the country and there would be no way to be in contact with them given how remote the community is. Sharon tried to help her to see that while she could transition out of her role in a variety of ways that would support the transfer of knowledge and management skills, community members would need to maintain and further develop the programs, so she needed to let go so that the people of the community would take care (or not as the case might be) of things on their own since it is their community and programs. While sometimes difficult, we, as researchers, have to consider that we are rarely insiders in communities, and we need to set up and maintain respectful boundaries so that we are not making ourselves too central or creating dependency on skills or resources that may be present as we conduct our research but that leave when we do. At the same time, we need to make good on our promises. This is a delicate negotiation of variables, and each scenario is specific and needs to be thought through with others who can help you to understand the range of concerns and implications that exist. In the section that follows, we discuss another commonly thought about but often underexamined aspect of research ethics, informed consent, and we suggest that you keep Lather's notions of reciprocity with you as your lens for this section and the rest of the chapter.

INFORMED CONSENT, WITH AN EMPHASIS ON *INFORMED*

Participant consent in qualitative research can refer to situations in which researcher(s) seek to (1) access settings and groups to which they are outsiders, (2) obtain data or documents not publicly accessible, or (3) elicit information or data from research participants through interviews, focus groups, questionnaires, observation, writing, and other means (Hammersley & Traianou, 2012). Our students often ask what consent actually means, which we believe is an important place to start. Consent entails that "other than in exceptional circumstances, participants agree to research before it commences. That consent should be both informed and voluntary" (Israel & Hay, 2008, p. 431). This notion of agreement to the research, or consent to engage in it, necessitates being clear about what you are asking of participants and that you articulate this clearly to them, allow time for questions about your explanations, and allow for push back (or rejection of participation) if some of the conditions are unacceptable or problematic for any reason. This constitutes the *informed* part of informed consent.

In their book, *Ethics in Qualitative Research,* Hammersley and Traianou (2012) do an excellent job of defining consent as a basis for discussing the ethics of informed consent. They state,

> There are a number of questions that must be addressed regarding consent. These include: *whether* it is necessary or desirable to seek consent; *from whom* consent should be obtained; *how* consent ought to be secured; *for what* consent is being sought; and what counts as *free* consent? The answers to these questions are likely to differ according to circumstances. (p. 82)

These are important dimensions of consent to consider, and they parlay into research-specific questions about the goals, roles, and processes of consent in research. Informed consent, or what Hammersley and Traianou (2012) refer to as "free consent," is an important concept and process in qualitative research since transparency and honesty are central to ethical and valid research. The *informed* part of informed consent is where the heart of the matter lies. Informed consent should not only be considered transactional but also be thought of as a particular kind of attention to meaningful dialogue with participants about the research and their involvement in it.

Consent is a truly vital concept and ethic. Informing participants means that you, as the researcher, will give potential participants information about what you are asking of participants, including demands on their time; what participation will entail; potential risks that could occur; how the data will be handled; who will have access to the data; how the final write-up will be disseminated; the purposes, goals, and methods of the research; who supports or funds the research; and any potential benefits. The process of informing participants can vary by study and context. In some cases, this can come in the form of providing and explaining an informed consent form, which typically includes a brief overview of the research goals, a statement about the voluntary nature of the research, the benefits and risks of the research, and a list of the requirements for participation on the part of agreeing participants, as well as contact information should they have additional questions or concerns. In Table 11.3, we provide an overview of the typical components of a consent form. In addition, please see Appendixes Y.1 to Y.5 for sample consent forms. We have included a range of consent forms so that you can understand that there is variation in how these forms can be structured. One important message is that researchers need not always compose consent forms in formal, contractual ways; they can be warm and made personal in a variety of ways even as they cover specific information that is mandatory. You

Table 11.3   Consent Form Overview and Contents

A consent form should

- Establish in clear language that participation is voluntary and that participants have the right to withdraw from the research at any time
- Present the expectations for study participants, including accurate time commitment
- Give an honest analysis of any potential risks to participants' well-being
- Make clear if data and reports will be confidential or anonymous

Generally, consent forms should include the following:

- Description of the study, which discusses the purposes, methods, and timeline
- Description of organization funding the study (if any)
- A statement that participation in the study is voluntary and that participants can withdraw at any time for any reason or not answer specific questions during interviews or focus groups
- List of potential risks to participants
- List of potential benefits for participants
- Description of specific issues and processes regarding confidentiality/anonymity
- Overview of what participation entails (i.e., how will participants be involved in the study)
- Duration of the study and the time commitment required of participants
- Statement of whether there is any sort of compensation for participation and clear information about compensation
- How the results will be used/disseminated
- Contact information and institutional affiliation of the principal researcher
- Lines for printed name, signature, and date of signing

Note: This information can be presented in paragraphs, bullets, or an outline on the consent form or in an attached letter that is given (and discussed) with participants along with the consent form.

can also adjust forms so that they can be sent electronically, without a line for signatures. Since the forms must fit the context in which they are used, IRB and research advisers are the best source for information on how to structure your specific forms.

Researchers typically send a consent form to each participant via mail or email or give it to participants in person and should use it to (at least briefly) discuss these aspects of the research process.[6] At other times, informed consent is more of an engaged and labor-intensive process that requires careful explanation, lengthy discussion, and even perhaps revision of the terms included in the consent forms. Many qualitative researchers argue that the process of informed consent should be a dynamic and ongoing one that is continually engaged in and even recalibrated over time. We believe that the nature of the study and the context inform where you fall between these two ends of the informed consent continuum. In the latter case, this requires ongoing reflection and engagement that is built on a careful and ongoing attention to being transparent about roles, expectations, and processes. It also connotes that there is ongoing access to participants, which of course is not always the case.

What is of primary importance is that participants be informed in ways that are respectful, accessible, and transparent. This means that the form is written and/or communicated in a way that is accessible to participants and that participants know exactly what they are signing on for in terms of time commitment and logistics so as to avoid "time creep." Furthermore, it should be clearly communicated to participants that the research is voluntary and that they have the right to refuse to answer any question and/or withdraw at any time without fear of upsetting the researcher or harming themselves in any way. Participants should also be informed about the goals of the research, who will have access to the data, and where the results will or will not be published or shared. In addition, consent forms can allow participants to decide if and how they will be represented in a final report as well as allow them to state if they do or do not consent to be audio or video recorded.

We guide our students to be completely upfront and to err to the side of caution, not over- or underpromising anything in any way. An informed consent form should be used to engender discussion rather than as a pro forma thing to get the person to sign as if his or her perceptions and questions are tertiary to the research. Too many times we have seen researchers slide the form across a table to participants without ample (or even any) explanation of its contents. It is clear in those moments that an opportunity is lost to really explain the research and connect the dots between goals, methods, and processes. Furthermore, this connotes a lack of respect for participants, which is an ethical issue.

It is important to note that there can be contextual challenges to informed consent and ways that context can mediate choices around the form that it takes. For example, in some populations, formally writing down one's consent is considered unsafe or unwise. This happened in one of Sharon's research projects where despite spending time on consent forms, once in the field, it was clear that participants were worried about their names being attached to the research since it was in a country that is a dictatorship where participation with Westerners can be viewed as disloyalty and punished. There are also times when it would seem culturally strange to ask for that kind of formal written agreement, such as in Sharon's research in the rural parts of Nicaragua where that would undermine the way that relationships are built over time. Since Sharon was not sure how to handle this, she contacted the IRB staff member whom she had worked with on gaining study approval and he told her that she could get verbal approval and record that at the beginning of the interview, capturing it on audio recording. Some of these challenges are contextual and some are cultural. In any case, you need to understand that informed consent is a concept and process born out of Western institutions, and therefore you should think about their cross-cultural, cross-national use.

Assent

Assent is an important consideration in research with minors. For those younger than 18 years, consent must be granted by either a parent or a guardian with legal voice for the minor. *Assent* refers to the process of giving minors the opportunity to agree or disagree to participate in research even if they cannot legally consent. This means informing them in detail about what the research is about, what it entails, and where their information will go once the study ends.

While assent is not legally mandated or binding, we, along with many other researchers, believe that the ethic of informed consent should apply, in a relational sense, to minors as well. Methodologically, this involves giving minors assent forms that can be used to describe and explain the research in ways that are appropriate for the age of the youth. This approach shows respect for minors and an understanding that while they might not need to be informed for legal reasons, for relational and ethical reasons, it is imperative that minors are actively engaged and informed about the research in which they are being asked to engage as well as are given an opportunity to decline even if their parents/guardians have given permission for them to participate. See Appendix Z for sample assent forms.

TRANSPARENCY IN GOALS, EXPECTATIONS, PROCESSES, AND ROLES

In Chapter Six, we discussed transparency as it relates to validity, stating that transparency in all aspects of the qualitative research process not only is necessary to conducting ethical studies but also is an important aspect to achieving validity. We explored two terms within the discussion of the relationship of transparency to validity—what we refer to as internal-facing and external-facing transparency. *Internal-facing transparency* means being clear and transparent about all aspects of a study with our participants, including the goals of the research, our expectations of them, what the process and timeline will look like and entail, what their roles and our roles and related responsibilities include, who else can access the data, whether anonymity or confidentiality is being ensured, and so on. *External-facing transparency* means that threats to validity and the presence of bias, which are present in all studies both quantitative and qualitative, should be clearly articulated, in detail, in a final research project, whether that is a paper, film, or other form of representation. This means including a discussion of study limitations and changes in research design with rationale and related constraints that might stem from them (or benefits) and so on.

Both of these kinds of transparency—the kind that is about our participants and the kind that is about our audiences and readers—are vital to the validity and trustworthiness of our studies, and they require, among other things, a commitment to being clear and honest about the goals, expectations, and processes of your research as well as the roles and responsibilities of all involved: you and the research participants as well as any others involved, including thought partners. While a true informed consent process will set this up at the outset with your participants, we argue that this must happen intentionally throughout (and after) the research as well. Some questions to ask that can guide your sense of and approach to transparency are presented in Table 11.4.

While these questions are not exhaustive, they can help direct your thinking to the kinds of questions you should ask yourself (and possibly your co-researchers) to engage in a truly ethical informed consent/assent and broader research process.

CONFIDENTIALITY AND ANONYMITY

While related, there are important differences between confidentiality and anonymity. *Confidentiality* is related to an individual's privacy and entails decisions about how and

Table 11.4   Transparency-Related Questions to Consider

Questions for thinking about internal-facing transparency:

- Have I described the research accurately in terms of its goals and focus?
- Have I been clear about expectations on participants' time and of their engagement?
- Have I been clear about what the process and timeline will look like and entail?
- Have I clearly articulated what the participants' roles and my roles and related responsibilities include?
- Did I share who else can access the data in their various forms (e.g., images, transcripts, written reflections), whether anonymity or confidentiality is being ensured, and so on?
- Have I shared what will happen with the data and analysis in terms of what forms they will take and for what audiences?
- Are there ways that I was not open about any of these issues? If so, why? How do I address this?

Questions for thinking about external-facing transparency:

- Have I clearly stated the threats to validity of this study and what I did to address them?
- Did I clearly articulate the presence of my specific biases and how they affected various aspects of the research?
- Did I clearly list the study limitations, changes in research design with rationale, and related constraints that might stem from them?
- Did I share clearly a rationale for my design and choices to engage in certain methods or to not engage in certain methods?
- Am I consistent and respectful in how I represent the participants and their experiences, thoughts, and contexts?

what data related to participants will be disseminated (Sieber, 1992). Discussing confidentiality with participants might mean that pseudonyms will be used and/or other identifying facts will be changed or not disclosed. For example, you may not include identifying information such as participants' names, unique attributes, or job titles in a final report. *Anonymity* means that there would be no way for anyone to identify an individual within a sample of participants because data and resulting reports are aggregated and not individually contextualized or displayed. An example of anonymity may look something like this statement: "Of the 100 people interviewed, 32 stated that they believe that the professional development practices supported instruction." In some studies, no names or identifying information are ever associated with data. Depending on the type of research, anonymity is normally only promised in studies with larger samples.

It is difficult to guarantee both confidentiality and anonymity. For example, through a process called deductive disclosure, some participants may be able to be identified as a result of specific traits, circumstances, and/or experiences (Kaiser, 2009). Unfortunately, there are not many guidelines to deal with deductive disclosure (Kaiser, 2009; Sieber & Tolich, 2013), and researchers often change identifying characteristics or leave specific participants out of final reports (Kaiser, 2009). Furthermore, not all participants wish to have their information kept confidential and/or anonymous. What is key here, as we highlight throughout this chapter, is that these decisions need to be *presented to and/ or discussed with participants*. Researchers need to be transparent about the promises they are making and make sure that they can keep these promises. For example, if using focus groups, there is no way to promise confidentiality to participants. While you, as the researcher, may not disclose what was discussed in a focus group, you cannot control what information the other focus group participants will or will not share.

A multitude of issues should be considered related to confidentiality and anonymity assurances that are made to participants. If these assurances are made, you must deliberately plan for how you will go about this. Using pseudonyms throughout the research process and not just at the end of a study is one way to help safeguard participants' identities. To have anonymous data, identifying information should be removed from all study materials, including transcripts and/or coding sheets so that responses cannot be connected to individuals (Ogden, 2008). These are only some examples of ways that researchers can attend to securing confidentiality or anonymity, and there are additional ways to engage in processes that protect either or both. Where, how, and if identifying information is stored is another important consideration. In Table 11.5, we discuss key considerations for managing your data to safeguard participants' identities. In the subsection that follows, we present additional considerations related to confidentiality and anonymity and highlight these considerations with specific examples.

Data Management and Security in the Information Age

It is important to note that there are additional challenges to anonymity and confidentiality brought on by the advent and pervasiveness of social media and new technologies, including various forms of digital photography and video, audio and video recordings online, and virtual materials from the Internet, including blogs, chat room discussions, and other publicly accessible data on individuals that could undermine other de-identification processes (Hammersley & Traianou, 2012).

There is a difference between using "online strategies for eliciting information" and "using naturally occurring online data for research purposes" (Hammersley & Traianou, 2012, p. 117). The ethical issues related to the first are similar to existing concerns of data collection but the second, naturally occurring online data create specific issues of privacy that require additional scrutiny (Hammersley & Traianou, 2012). There are significant debates in the field about how to consider and approach these technology-mediated data collection methods with respect to the issues they engender related to privacy and the conundrum between researcher-generated data and the kind of data that a researcher can access through a more distant (and even voyeuristic) scanning of the web (Hammersley & Traianou, 2012). Issues of consent, privacy, and transparency are central to these debates, and they call up contrasting notions of ethics within and beyond these areas. There are no clear answers to any of these issues, but the goal is to think about them proactively and to set up specific systems that make sense for each study so that the ways to deal with these issues are clear at the outset.

Using the Internet to collect data, while making some aspects of data collection easier, highlights issues related to privacy rights and can make promises of confidentiality difficult to achieve. For example, we remind you of the story we shared earlier about the student who had to adapt her study to the realization that despite using pseudonyms, readers could search for quotations from blogs or other social media in which her participants posted or shared ideas in writing were she to include them in her final report, and thus those reading her work could identify participants' identities. As there are not yet established guidelines for how to treat these data, we recommend being diligent to the participants' wishes and having multiple conversations with them and other researchers who routinely collect and analyze these types of data to make proactive as well as emergent decisions in this realm. Furthermore, we find the staff at various IRBs, across multiple institutions, to be helpful and supportive when discussing these issues with students, as their primary concern is to protect study participants. They also have more experience with these matters and can offer creative solutions. Given the prevalence of online data collection and its possible ethical implications, we suggest a thorough examination of literatures related to the role of social media and the Internet more broadly with respect to your specific study design, context, concerns, and needs. Several resources shared at the end of this chapter address these issues.

While data storage and management have always been concerns to researchers, new cloud technology and transcription services, as well as the incredible mobility of data through email as well as on laptops, smartphones, and electronic storage devices, create a new set of serious ethical issues for researchers to consider related to data security. Data management and security in an age of technology is a central and ongoing concern in the protection of anonymity and/or confidentiality.[7] Careful consideration of all possible ways that data security can be breached must be considered and strategized at the outset of a study (and carefully addressed in your proposal, field-generated data, and final reports). It should then be attended to throughout the course of the research and even once studies have been concluded. In Example 11.3, we provide an example from a doctoral student that underscores this issue.

Example 11.3: Lost Phone Creates Breach in Confidentiality

A doctoral student called Sharon in a panic, with an urgent question relating to her dissertation data collection. The student had audio-recorded an interview with a key research participant on her mobile phone, and her phone was stolen later that day before she had time to download and delete the file from her phone. She was extremely upset and concerned about the implication of this for the confidentiality that she had promised her participant(s). She was also, of course, upset that she lost the data from this interview.

Sharon spoke with her about this and determined that she should check in with our university's IRB staff about this issue because it provoked an unintended breach in confidentiality since she stated the actual name of the participant at the beginning of the interview recording and there was specific context-related information disclosed throughout the interview, and now these data were out of her control.

With the IRB staff member who attended to this issue, they determined that the student should notify the person she interviewed about the issue to let him know what happened. They advised her to call and discuss the issue with the participant, referring him to the informed consent form where she ensured confidentiality and secure data management, and to discuss the unfortunate occurrence and ask if he wanted to withdraw from the study. She was quite worried that this would have negative implications for this person's trust in her and that others in the setting would find out, and the data would be affected in future interviews. However, she knew that it was important to inform him immediately and so she did.

She called the person she had interviewed that same day. He was quite understanding and said that he did not feel he shared anything that he would be concerned about others hearing, and he said that he was not upset or angry about it. And, over time, it did not seem that others had heard about the incident since the rest of the group's involvement and engagement were quite positive throughout the remainder of the research. She did, however, lose the data from that interview, since even though he was quite understanding about the situation with the theft of her mobile phone, the participant said he did not have time to be interviewed again. So the student reconstructed as much of the interview as she could from the notes she took during the interview, treating them as jottings that she turned into fieldnotes, and while not as valid or useful as an interview transcript would be, she had these notes as context. Sharon advised her to write a memo that evening in which she captured as much as she could about the interview while it was fresh in her mind. While not considered interview data in the same way as a transcript, she was able to glean context and learnings. Most important, it appeared as though she did not seriously harm the participant in her study. Had the participant considered the information more "sensitive" or had the participant been angered, there could have been significant consequences and harm.

The example described above is not as unique as it may seem. In an age in which privacy is undermined by social media and technology, you have to consider all sources of data and their storage and security quite carefully. Some say that "the devil is in the details." While the devil may not be, issues of privacy sure are when the details are about how you collect and secure data in the digital age. Planning and safeguards must be put into place to proactively and ongoingly plan for and deal with these issues. In Table 11.5, we present questions that can help you consider issues of data management and security in specific ways.

Table 11.5   Considerations Related to Data Management and Security

- In what ways could my data be compromised? What steps can I take to protect against this happening during and after my study?
- When and how do I de-identify names and other identifying information from my data (i.e., before and/ or during data collection)?
- Do I assign pseudonyms or numbers to participants in the data? What are some challenges that can arise from this?
- Are there politics around pseudonyms in terms of cultural meanings and identity politics? How am I addressing these political issues?
- What are participants' wishes for how they will be represented in the final report? How am I ensuring that I am respecting participants' wishes?
- What have I promised participants (e.g., confidentiality or anonymity)? How am I ensuring that I am keeping my promises?
- Where am I storing my data? Who else has (or could have) access to this?
- What precautions do I need to take to more securely store my data?
- What is my plan in case my data are compromised?
- What are ways that social media might/do play into my research and constitute part of my data set? What are ways that I must consider this in terms of ethical issues?
- How does technology usage play into my data collection plan? And what do I need to consider with respect to data security?
- Who will I seek counsel from if there are these kinds of emergent issues?

THE ETHICAL DIMENSIONS OF THE "RESEARCHER AS INSTRUMENT"

In Chapter One, we discuss that in qualitative research, the researcher is considered the primary instrument[8] of the research throughout the research process, meaning that the subjectivity, identity, positionality, and meaning making of the researcher shape the research in terms of its process and methods and therefore shape the data and findings. Thus, the identity and positionality of the researcher is viewed as a central and vital part of the inquiry itself. We revisit this concept here, focusing on its ethical layers and dimensions.

With the concept of the researcher as instrument as a broad frame for qualitative research, it becomes an ethical imperative to consider your role as the researcher throughout all phases and parts of the research process. One key area we see as central to this is with respect to the relational dimensions of research and the role that you play in the research relationships developed. Because we cannot be separated from our biases, "it becomes unethical for us . . . not to explore our own biases and prejudices and the contexts that shape (and misshape) them" (Nakkula & Ravitch, 1998, p. xiii). Through processes of dialogue and critical self-reflection, researchers can uncover and confront these prejudices, although it must be recognized that there are still infinite possibilities of both understanding and misunderstanding (Nakkula & Ravitch, 1998).

Thus, critically confronting and engaging with/challenging our interpretations and the biases that shape them constitutes an especially important set of considerations given the powerful way that researchers can affect others. Addressing this ethical responsibility requires a reflexive approach to research that includes developing and maintaining a commitment to a specific and holistic openness to critical feedback and change.

As we discuss in depth in Chapters One and Three, researcher reflexivity involves paying careful attention to research processes by asking questions of yourself, participants, and other individuals and noting (and acting upon) their responses (Glesne, 2016). When we, as researchers, acknowledge our biases and prejudices, we must not only acknowledge but also actively monitor our subjectivities and how these influence our research (Glesne, 2016; Peshkin, 1988). Ethically, this entails that you should be attentive and actively invested in formal processes of reflexivity that support the critical exploration and interrogation of your biases and their impact and influence on all aspects of the research process. We, along with many of our students and colleagues, find these processes of self-reflection—discovering, locating and reckoning with our subjectivities, and finding ways to engage address them in research—quite generative and exciting. This is not to say that these processes are easy, but they are crucial to ethical research.

PUSHING AGAINST THE "EXPERT-LEARNER BINARY"

As discussed in depth in Chapter Two and throughout the book, sociopolitical contexts (both macro and micro) influence research dynamics and relationships. Researchers must work to be consciously aware of these contexts and their role in all aspects of research and that doing so is an ethical responsibility at the heart of qualitative research. As a part of an ethical stance as a researcher, you should consider how these larger forces manifest themselves in your roles and relationships as well as your research goals, processes, and interactions. As Patti Lather (1991), a scholar well known for critiquing and challenging research conventions, suggests, when you work with research participants to shift power dynamics and meaning, the development of the research and the relationships at its core requires flexibility since dynamics and relationships are continuously formed and re-formed, negotiated, and reenvisioned. We believe in an approach that situates everyone involved as "experts of their own experiences," meaning that everyone involved brings wisdom and generates knowledge (Jacoby & Gonzales, 1991; van Manen, 1990). While this may sound obvious, it is a departure from how research often happens, and we view it as an ethical necessity in qualitative research.

What we hear from students and colleagues more often than we wish is that people often reinscribe broader power asymmetries in their research because they uncritically accept hierarchies and hegemonic behaviors and practices that are socially constructed. An example is in a project Sharon worked on years ago in which her co-researcher, who was a new colleague, approached a group of rural women in India who were involved with them in participatory research, which sought justice in the realms of education and public health, as if they were not as smart as she was, simply because they were not literate or formally educated. To be clear, this was an unconscious bias playing out in seemingly subtle but perceptible ways, including that she continuously asked if they understood what she was saying (which would be fine except it was done in a condescending tone that reflected her assumption that they did not understand); spoke to them as if they had no relevant skills or knowledge, ignoring their deep and vast knowledge about their culture, community, homes, families, the local economy, and so on; and seemed to assume that they did not know their religious texts simply because they could not read them (although they certainly could hear them and did daily in their place of worship). In addition to a set of patronizing behaviors reflecting the biases above, this colleague also took the liberty of taking photographs of these women without first asking them for their permission, which is an act of power and, to many, is seen as disrespectful. This clearly bothered one of the women, who later asked Sharon where the photographic images would end up. As they engaged in the research, Sharon found that these biases shaped her co-researcher's overall behavior as well as how she approached the research interviews in ways that were deeply ethically troubling and that negatively affected the research since the women were less inclined to trust them or share their genuine opinions. This resulted in a series of difficult but important conversations to raise her colleague's awareness about these biases and their implications on the women, on the research relationships, and on the data themselves.

This kind of bias, with its attendant assumptions about and impositions on individuals and cultural groups, reflects a "deficit orientation" (Valencia, 2010). A **deficit orientation** means viewing people from various groups (e.g., cultural, social, communal) as "less than" or as deficient (as in lacking certain knowledge, skills, or value as framed by the dominant group) and therefore creates a power dynamic and set of equity issues in the research relationship. Taking the stance that everyone is an expert of their own experience and that all of our knowledges are valuable pushes into traditional views of who is the expert and who holds the wisdom. Furthermore, taking a stance in relation to these issues can be a powerful way to interrupt deficit-oriented norms and the status quo. For us, this rests on multiple frames that help to decenter the privileging of Western, academic knowledge over other forms of knowledge. One of these frames, which has deeply informed Sharon's work, is a "funds of knowledge" frame that looks at various forms of knowledge through a resource framework (Gonzalez et al., 2005).

Terms and Concepts Often Used in Qualitative Research

Deficit orientation: A deficit orientation refers to when individuals consider people from various groups (e.g., cultural, social, or communal) as lacking in certain knowledge, skills, or value. This can often be applied to members of dominant groups defining norms within hegemonic confines that valorize White, Western, upper-class ways of knowing and being while deeming those who do not fit these backgrounds and characteristics as less than or deficient. Deficit orientations can, at times, lead to the devaluation, degradation, or pathologization of nondominant groups and group members. Countering deficit orientations means shifting to a resource orientation in which the strengths, skills, knowledges, and value of each individual and various groups are the lens through which one views self and other. In a relational and critical approach to qualitative research, researchers reject and problematize deficit orientations and take the stance that everyone is an expert of their own experience and that all knowledge and ways of being are valuable and worthy of respect.

We strive in our own research, as well as urge our students and colleagues in their research, to become sufficiently comfortable with the uncertainties that arise from engaging in actions and processes that decenter codified knowledge (i.e., knowledge valued in the Western academy) and expertise by allowing time and space for mutual evaluations and conversations that strive for balance and equity in reconciling what are sometimes divergent expectations and interpretations (Chilisa, 2012; Ravitch & Tillman, 2010). Research is often built on what we think of as an "expert-learner binary"—that is, an assumption that researchers are more knowledgeable and expert, and participants are viewed, and sometimes even view themselves, as passive recipients of this information and the research process overall rather than as knowledge generators and experts in their own right. This, of course, is problematic and must be countered. It is crucial to acknowledge that everyone brings to and should benefit from research experiences, and this acknowledgment involves pushing against the expert-learner binary (Ravitch & Tillman, 2010). For example, during Sharon's work in Ecuador, she watched various forms of deficit orientations play out in concerning and even damaging ways between participants and colleagues, some of which were shaped by relationships between indigenous populations and those not from these communities and some caused by her colleagues' lack of understanding that the research was serving to reproduce these divisions by valuing more Westernized versions of knowledge and experience over indigenous knowledges and expertises. The desire to interrupt this problematic status quo leads us to consider the role of collaboration in research. We do so with a specific attention to the ethical layers of collaboration in its variety of sorts. We also examine some of the potentially problematic aspects of collaboration since it is not wholly positive and must be carefully examined in order for it to resist the very asymmetries and problems it seeks to address. Indeed, collaboration is more than a nice idea; it is a complex issue at the heart of qualitative research.

ETHICAL COLLABORATION: "RECIPROCAL TRANSFORMATION AND DIALECTICS OF MUTUAL INFLUENCE"

> An idea does not live in one person's isolated consciousness—if it remains there it degenerates and dies. An idea begins to live, i.e. to take shape, to develop, to find and renew its verbal expression, and to give birth to new ideas only when it enters into genuine dialogical relationships with other, foreign, ideas. (Bakhtin, 1973, pp. 71–72)

Our view on research ethics places emphasis on collaboration of a variety of sorts. Research collaborations are exchanges that ideally foster reciprocal transformation by cultivating an applied reflexivity and collaboratively examining the ways in which individuals and groups are engaged in "dialectics of mutual influence" (Nakkula & Ravitch, 1998). Dialectics of mutual influence refers to the ways that mutuality of influence happens relationally, in the sense that there is an interactive, catalyzing energy that drives both parties to learn, shift, and change in relation to each other. Centralizing collaboration as a methodological and ethical stance on research is central to engaging in ethical qualitative research. There are multiple kinds and layers of collaboration that we view as ideal in qualitative research. These fall into two broad categories: (1) dialogic engagement or collaboration with various colleagues, peers, thought partners, advisers, and teachers who can help you to engage in reflexive and valid research and (2) ways of viewing and approaching relationships with research participants as collaborative. We want to note here that collaboration is not necessarily always useful or positive and therefore that it must be thoughtfully conceptualized and constructed.

Building collaboration that is respectful even as it seeks to challenge hierarchical norms and how we (as people and researchers) enact them in ways that are constructively critical relies on several major beliefs and actions. Collaboration is relational, reciprocal, and dialogic; it should be thought of "as a 2-way, bilateral exchange" that confronts power dynamics, critically understands contexts, and takes "an inquiry stance on practice" (Ravitch & Tillman, 2010, pp. 4–5). Authentic collaboration (rather than relationships that call themselves collaborative but do not uphold the values of collaboration that we discuss throughout this section) is built on a foundational belief that healthy and enduring partnership requires a deep and ongoing reflective process that also focuses on engagement in and exchange of constructively critical feedback and ideas. Furthermore, it requires that you, as a researcher, engage in systematic questioning of what it means and entails to engage in research that comes with everyday realities and challenges that frame and at times push up against the goals and ideals of our studies.

Authentic collaboration is easier said than done since there is no clear roadmap and the collaborative process is co-constructed rather than having clearly delineated steps. Further complicating this is the fact that concepts such as respect, mutuality, and healthy communication are mediated by context, culture, social location, and individual preferences and differences. Thus, ethical collaboration requires building a shared mind-set about relational integrity, meaning that these kinds of issues must be explicitly discussed together, and this includes conversations about the nature and even discomforts of collaboration.

The belief that we, as researchers, need to allow ourselves to become vulnerable in order to engage reflexively in relationships is vital to collaboration and the development of inquiry communities and critical friend partnerships. We believe that this relies on having a receptive sensibility—that is, we should identify and acknowledge our different stances and biases and, moreover, see them as valuable, generative, and reflective of the influences of local, national, and international contexts and tensions as they shape our different funds of knowledge (Gonzalez et al., 2005; Ravitch & Tillman, 2010). It also requires us to allow ourselves to become openly vulnerable, critically self-reflective, and truly collaborative in ways that require us to raise our thresholds for discomfort (Ravitch, 1998). As Nakkula and Ravitch (1998) share,

> The potential for symbiotic development depends, in part, on the acceptance of and interaction with difference, and on the welcoming of misunderstanding and discomfort as part of the practitioner's raison d'etre. Difference is at the heart of dialectical growth. Differences of opinion, worldview, cultural background, and life experience all serve as fuel for the dialogical process. But to recognize and engage with difference requires the willingness to acknowledge misunderstanding and to be misunderstood. . . . Genuine empathy, however, cannot be achieved without an authentic willingness to misunderstand, to strive to connect only to miss the point, and as such to feel disengaged. All too frequently, false connections are maintained in order to salve the discomfort of disconnection. A productive

synthesis of differences requires a grappling with discomfort, a clear recognition of disjunctions. Willingness to reach toward understanding in the midst of such discomforting misconnections is . . . a healthy prognosis for mutual growth. (p. 87)

Reflexive attention to the existential, relational, contextual, and ultimately dialogical nature of research is central to the kind of honest collaboration that we work hard to develop and support throughout our own research. It can be effortful and even quite difficult at times to develop and maintain this level of collaboration. Each person must work to understand how everyone contributes differently to the interaction, how what we bring to this collaboration—culturally, institutionally, professionally—influences the macro collaboration as well as each micro interaction. Taking this kind of relational stance on collaborative work requires actively appreciating and accounting for the myriad—individual, social, institutional—complexities of this sort of work and their impact on our actions and interactions. Reflecting on their work collaborating in participatory action research projects with teachers and students in schools, Sharon and her colleagues, Peter Kuriloff and Shannon Andrus, write,

To be effective, we believe that an ethical stance on research practice must interrogate every issue our teams address in their work in terms of questions of fairness and justice. We must ask ourselves: How do the questions that we raise affect various constituencies at the intersection of race, class, gender, sexual identities, and the like? We must also recognize that those questions inevitably involve interpretations based on individuals' positionalities within the collaborating organizations. In truly collaborative work, we have to be prepared to understand and appreciate those representations—both ours and our collaborators. This means we also have to be prepared to contest positions in a spirit of openness as we recognize that everything each of us is doing is value-laden and contestable: we must be willing to learn and to give ground when we have understood the various positions and conclude we are in error. Such an ethics of practice requires researchers to have a disciplined understanding of our own subjectivities. Peshkin (1982, 1988) has written about this eloquently, describing the ways that researchers must not only recognize and accept the existence of their subjectivities, but sincerely and actively engage in practices to uncover and explore what those subjectivities are. (Kuriloff, Andrus, & Ravitch, 2011, p. 58)

As the passage above addresses, collaboration is not as straightforward as it may seem and can often lead to "messy ethics." Messy ethics entails that there are not necessarily clear answers to all questions, and collaboration can generate a variety of epistemological and methodological issues. Power, context, and people's opinions and concerns mediate collaboration. What may seem innocuous to a researcher may be quite troubling to a participant. Recognizing that ethics can be messy necessitates thinking about, address-ing, and discussing issues of power, context, and people's opinions and concerns. Doing so can prove challenging for a variety of reasons that are contextual and specific to the milieu, people, and processes at the center of each research study. Furthermore, power dynamics and issues of equity mediate collaboration in ways that are both visible and invisible. Therefore, conceptualizing and enacting collaboration evokes unending ethical complications to reckon with. It is vitally important to consider the issues discussed here both in terms of (1) collaboration with various colleagues, peers, thought partners, advisers, and teachers who can help you to engage in reflexive and valid research and (2) ways of viewing relationships with research participants as collaborative.

As we think about ethics in our own work and in the work of our students, we ask some crucial questions: *What constitutes productive collaboration? How does a researcher know if a collaboration is productive or healthy or genuine?* These are difficult questions to answer since they are contextual and rely on the individuals and the consideration and balancing of various viewpoints and beliefs about healthy and productive communication and interaction. For example, two researchers, one with a high threshold for confrontation and one with a lower threshold, might view a heated debate about some aspect of their research quite differently based on their preferred modes of communication, their beliefs about the role of confrontation in communication, and their ability to get beyond the heated affect to the content of the interactions. One of these colleagues might view an interaction or even the whole collaboration as unhealthy because he assumes that confrontation connotes anger and disrespect, while the other one might see the debate as an artifact of trust and comfort in disagreement and therefore as a sign of a healthy, open partnership. We use this example to probe into terms such as *healthy, productive,* and *reciprocal* since these terms are individually, contextually, and culturally mediated and therefore must be determined on a case-by-case, person-by-person basis.

To be clear, collaboration is not always positive, and it is not necessarily unproblematic. For example, there are pros and cons of dialogic engagement, which we espouse to be a cornerstone of qualitative research. Not all feedback is useful or constructive to the ethics and validity of a study. For example, what if your thought partners are enacting their own blind spots and assumptions as they give you feedback or advice and as they analyze your data with you? What if the process of engaging with study participants to seek out their perspectives creates new issues if their opinions and yours are at odds? Who gets to decide? And who gets to decide who gets to decide? How do we, as researchers, deal with being challenged on our opinions—by thought partners and participants—when we truly believe we are accurate or right? How do we know what affect and emotions might be mediating how we hear and engage with feedback and critique from others? For example, in Sharon's dissertation research, a key participant dropped out of her study because she disliked how she was represented on issues of racism. Sharon felt strongly that this study participant was being defensive but also had to question if perhaps her own biases shaped how Sharon portrayed this participant. In the end, the participant withdrew from the study because she did not agree with her portrayal and, Sharon believes, there was a breakdown in communication with both parties being defensive and not finding a way to agree to disagree or to create an alternative that met both of their needs.

In addition to the notion that collaboration is not necessarily always positive and that it must be complicated in a variety of ways throughout the research process, we want to make clear that collaboration is not "proof" of a study's validity; it is one tool (within an overarching process) that can help a researcher or team of researchers to think more carefully from a variety of vantage points about the research at the development, design, implementation, and writing stages. However, it is not a panacea, and it must be reflected on critically with an attention toward the issues we raise throughout this chapter and the book. In Tables 11.6 and 11.7, we present questions that can help guide thinking about ethical considerations to collaboration.

This list of questions for peers and thought partners and for research participants is just the beginning of the kinds of reflexive questions that you should ask about the nature and processes of collaboration in your research. These questions should not simply be asked at the outset of your study. Rather, you should revisit them throughout each and every phase of your research, addressing specific issues within and across the various stages and relationships that constitute and shape your research. It is important to note that in participatory research, including participatory action research, collaboration takes on an added layer and set of processes since the research is co-constructed from start to finish. This creates new and different kinds of issues to consider. Across approaches, we suggest that you create specific memos within your research in which you address issues of ethics in collaboration. In Recommended Practice 11.1, we suggest an ethical collaboration memo as one kind of memo to be used at various points throughout your study.

Table 11.6 Considerations for Collaborating With Colleagues, Peers, Thought Partners, Advisers, and Teachers

1. How do I create the conditions necessary to be challenged on my biases, blind spots, and assumptions?
2. What kinds of processes and experiences would support challenging myself on my biases and assumptions ongoingly?
3. At which points is being challenged and having a space to reflect with others most valuable (i.e., topic and question development, research design, choice of theoretical framework, etc.)?
4. What are some specific techniques and processes I can use for these purposes?
5. What kinds of thought partners do I want (or not want) and why?
6. Am I avoiding getting feedback from specific peers or advisers and, if so, why?
7. Do I have people challenging me in ways that will truly move my understanding? How so specifically? If not, why not?
8. Do I make myself vulnerable in my process of engaging in reflexivity? If so, in what ways? If not, why not?
9. What makes me most defensive and/or uncomfortable and why?
10. Am I asking enough/the right kinds of questions about these issues and how they play out in my research?
11. Are these reflexive questions built into my research design? If so, how? If not, how can they be?
12. Am I being collaborative with these various thought partners? What does collaborative mean to me? To each of them? What assumptions are underneath that?
13. What kinds of issues arise between collaborative data analysis (which requires that others see transcripts) and confidentiality issues? What do I need to consider in relation to this at the design/pre–data collection stages and throughout data collection, with respect to data management and analysis?
14. How can I work with colleagues, peers, and advisers to set expectations for our collaborations that enable us to move forward in our work with clear and transparent goals and processes for collaboration?
15. What is my strategy or approach for dealing with inevitable disagreements or problems that arise throughout these collaborative relationships?

Table 11.7 Considerations for Collaborating With Research Participants

1. How do I present myself in the research context and why?

2. How do I present the research in terms of its goals, influences, and questions and why?

3. How do I define, present, and engage in research collaboration with participants in this study? What assumptions or biases shape these processes?

4. How have I approached and structured the research, and does this allow room for ethical collaboration with participants? What would that look like? How can I/we improve upon this?

5. Should any of these processes be more participatory, and what would that look like and entail?

6. How do I view collaboration with study participants in relation to collaboration with thought partners? What influences these views, and how do they differ?

7. Am I supporting and/or interrupting existing power asymmetries with research participants and, if so, how and why? How do I monitor this and change it if need be? Who gets to determine this and how?

8. Am I willing to actually shape and/or change aspects of the research in relation to participant needs, interests, and feedback? Why or why not? And if yes, what will that look like and who weighs in on it?

9. Are there ways that I can collaborate with research participants that will improve the research? If so, what and how?

10. What does power look and feel like in this study? How is it influencing the research?

11. How can I try to engage in critical feedback exchange in/on the process of the research with participants?

12. Am I receptive to critical feedback? What makes me feel defensive and/or uncomfortable and why? How can I get additional perspectives on this if need be?

13. Do the participants seem to feel comfortable challenging me? If so, in what ways are they or are they not, and how does this look across the participant group?

14. Am I reinscribing an expert-learning binary or a deficit orientation and, if so, how? How can I stop doing that?

15. How do I work with participants to set expectations for our collaboration that enable us to move forward in our work with clear and transparent goals and processes for collaboration?

16. What is my strategy or approach for dealing with inevitable disagreements or problems that arise throughout these collaborative research exchanges and relationships?

Recommended Practice 11.1: Ethical Collaboration Memo

To help you think about and engage with the various ethical considerations broadly and with respect to collaboration specifically, we recommend selecting questions from Table 11.6 and Table 11.7 and mapping out a set of memos that focus on various aspects and moments of collaboration in your research. The goal is to compose thoughtful memos that genuinely challenge you to think critically and write about these issues in focused ways that push your thinking forward. This is best done at both intentional and emergent or strategic moments throughout the research process.

You may write this memo or set of memos and reread them at various points throughout the research process. It may be more beneficial to compose these memos and share them with peers and/or advisers who can help you to continue thinking about and addressing these issues. Ideally, you should compose multiple memos related to systematically interrogating the ethical issues in your study. And the focus on collaboration will help you to investigate if and how your research collaborations are useful, equitable, and authentic.

BALANCE BETWEEN DESIGN FLEXIBILITY AND RIGOR: RESPONSIVE RESEARCH AS ETHICAL STANCE

As stated in Chapter One, a primary value of qualitative research is that it shows a fidelity to participants and their experiences rather than a strict adherence to methods and research design and, in that sense, can take an emergent approach to research design and implementation (Emerson et al., 2011; Hammersley, 2008; Hammersley & Atkinson, 2007; Maxwell, 2013). We also argue in depth in Chapter Three (and throughout the book) that qualitative research must have a solid research design. As Frederick Erickson (one of Sharon's mentors and her academic adviser during her doctoral work) has stated, research is about searching and re-searching, a continuous process of searching for new insights that teach us and help us to develop our understandings of people, research, ourselves, and the world. In a true search, you are not sure what you may find. Thus, there are vast methodological implications of taking an inquiry stance on research, and there are significant ethical dimensions and implications of taking a relational stance in design and fieldwork as well as in data analysis. Ethical issues and implications cannot be separated from interactions with research participants (Glesne, 2016). In this way, all aspects of the research process must be responsive to the daily and overarching realities, contexts, issues, and needs of a given research setting, context, and set of participants.

Responsive research, because it pays careful and ongoing attention to participants and their realities and contexts as well as the realities and multiple, intersecting contexts of the research setting, is an ethical approach to research because of the primacy it places on the identification and incorporation of participants' experiences and feedback into the research itself (not just the findings). Because of its emergent and flexible designs, there are unique ethical issues to consider in qualitative research (Hammersley & Traianou, 2012). The setting and people within a study are not static entities; rather, people and setting are dynamic, ever-changing, and informing of process and technique. You cannot know how you will act in future situations, and how you will act is ethically determined through critical and engaged dialogue with others (Austin, 2008).

Researcher and study responsiveness demands what many now refer to as an *emergent design approach*. Such an approach encompasses, in a variety of ways and on a continuum, all qualitative research. Emergent design, or what we broadly consider responsive, contextually driven research, involves taking a process-oriented, real-time approach to research design, data collection, and data analysis that, with intentionality, develops during a research study in response to what becomes learned and understood. In an emergent design approach, the research questions and other elements of the research design such as data collection methods are reconsidered within the context of emergent understandings of various stakeholder views. Since participants' experiences and mediating contexts are difficult to anticipate, identify, and articulate fully in advance of the implementation of research, you need to respond to these in real time once the research is under way. Within an emergent design approach, researchers and participants "work from local knowledge and interest; bridge to other knowledge domains; and liberate their local knowledge from its specific situated embodiment" (Cavallo, 2000, p. 780). While all qualitative research is emergent in the sense that it can shift and change as you learn in the field and find a need to adjust your design in relation to these learnings, an emergent design approach places a primacy on the incorporation of participants' perspectives on the research process itself; it has an added layer of seeking out data in ways that intend to be formative to the research design and process.

Research design and methods should be flexible so that the implementation of research—including formative analysis that informs the ongoing collection and analysis of data—can respond to the emergent quality of organizational or community realities and the study of these within their organic contexts and naturally occurring processes (Christie, Montrosse, & Klein, 2005). For example, in Sharon's collaborative research in Nicaragua, which works from an action research approach to community and school development that is resource oriented and generated within communities in partnership with multiple collaborators both within Nicaragua and in the United States, the research team has engaged deeply in emergent design research. In the theory of action for that research project, called *Semillas Digitales,*[9] Ravitch and Tarditi's (2011) discussion of emergent design highlights the important ethical issues involved in emergent design, including valuing and *incorporating* local knowledges of and within a research setting. They state,

> An emergent design approach enables the ongoing recognition and incorporation of local expertise, skills, knowledge, resources, and concerns into the structure, strategy, and development of a sustainable educational initiative. Understanding the context and intricacies of the educational and broader community environment at all levels as they iterate over time is essential to the development of a capacity-building approach, one that works from an appreciation for and incorporation of local needs and strengths in relation to educational development and innovation. The design of the development and evaluation of aspects of educational practices, programs and reforms are emergent in that they are co-constructed with stakeholders at multiple levels using a systematic and yet flexible approach. Further, emergent design projects seek out and systematically work to develop the knowledge and skills in the community, teachers, educational leadership, and staff since each of these groups serves as a bridge to the development of new, enriched content and approaches that resonate with local experience, culture, interest, and expertise. (Ravitch & Tarditi, 2011, p. 3)

While Ravitch and Tarditi (2015) emphasize participatory research methods and relationship building, they heed warnings that espoused participatory methods (like all methods) can easily fall into the trap of imposing outside, academic, and Western ideals and assumptions about the needs of local communities and the methods for creating change (Cooke & Kothari, 2001). As university-based researchers, they seek to resist and reject preformulated approaches to participation or even oversimplified understandings of what *local* means since that should not be treated as a monolith. Therefore, their research team members spend considerable time creating and fostering strong foundational relationships with their co-participants even as they endeavor to engage in applied development, action-oriented work. While this is specific to one participatory research study, we argue that these concepts and processes are at the heart of emergent design and that *all* qualitative research is emergent to some degree and should consider these issues.

Because qualitative research takes place in natural settings and does not occur in controlled environments, anticipating the data that will be generated as well as ethical issues that may arise can be challenging and is not always possible (Hammersley & Traianou, 2012). Emergent design may make sense intuitively, but it can mean more work for researchers. This work is viewed as valuable and vital to preserve the integrity and freedom of research. As qualitative researchers, we must push against the confines that seek to constrain our abilities to engage in research that is ethical from relational as well as procedural standpoints. We believe that we must commit to research that can actually respond to what we learn as we learn it and that honestly engages with the complexity of life, people, and setting. In your research, this means that you need to consider how ethics are related to design choices, both in the actual research design of your study and then what Sharon refers to as "micro design choices," or the smaller (but still important) choices you make as you conduct your studies. For example, Sharon often talks about considerations such as scheduling (as in scheduling interviews, focus groups, or observations for a study) as more than merely transactional choices. If you approach your study with a genuine respect for participants and their time, then you will make every effort to schedule around their needs (such as childcare, work schedules, etc.) rather than placing a primacy on your own needs. At least, if it is not always possible to work around participants' schedules, you will work to notice and engage participants' needs rather than simply following a set timeline or schedule. Then there are other kinds of micro design choices you make, such as establishing the conditions necessary to protect the relational quality of research, which might include not only the timing of interviews or focus groups but also possible choices such as having a note-taker in focus groups so that you can be more present and engaged rather than distracted and hurried. Properly planning for these kinds of situations can make research less transactional and can create the conditions that allow for the research to be a more positive, engaged, and enriching experience for all involved.

In this chapter, we have explored what constitutes research ethics in terms of concepts, processes, and methods to consider and engage with throughout the various phases of qualitative research. This discussion included attention to the specific ideological, methodological, and procedural dimensions of research ethics. Furthermore, we examined how issues and processes related to research ethics are not just transactional and procedural but also meaningful and generative. Our hope is that this chapter has provided you with a variety of conceptual and methodological tools to engage in research that is ethical from conceptualization to design through implementation and into analysis, writing, and dissemination.

QUESTIONS FOR REFLECTION

- What does a relational approach to ethics entail?

- What does it mean that it is an ethical responsibility to confront and challenge our biases and assumptions about the world broadly and our research studies specifically?

- What is the role of institutional review boards and ethics committees in your research?

- What are some important considerations in terms of issues of reciprocity and research boundaries with study participants?

- What are some pertinent aspects of informed consent and assent in the research process?

- How are issues and processes related to research ethics more than just transactional and procedural?

- How do anonymity and confidentiality differ?

- What are some key issues of data management and security broadly and in relation to social media and technology?

- How is researcher reflexivity central to ethical research?

- What does it mean to push against the expert-learner binary in research?

- What are the possibilities, limitations, and ethical dimensions of collaboration?

- How would you characterize the relationship between responsive and emergent research design and ethical research?

RESOURCES FOR FURTHER READING

Ethics in Qualitative Research

American Educational Research Association (AERA). (2011). Code of ethics: American Educational Research Association. *Educational Researcher, 40*(3), 145–156.

Alvesson, M., & Sköldberg, K. (2009). *Reflexive methodology: New vistas for qualitative research*. Thousand Oaks, CA: Sage.

Hammersley, M., & Traianou, A. (2012). *Ethics in qualitative research: Controversies and contexts*. Thousand Oaks, CA: Sage.

Josselson, R. (2013). *Interviewing for qualitative inquiry: A relational approach*. New York, NY: Guilford.

Miller, T., Birch, M., Mauthner, M., & Jessop, J. (Eds.). (2012). *Ethics in qualitative research* (2nd ed.). Thousand Oaks, CA: Sage.

Punch, M. (1986). *The politics and ethics of fieldwork* (Qualitative Research Methods Series 3). Newbury Park, CA: Sage.

Smith, L. (1999). *Decolonizing methodologies: Research and indigenous peoples*. London, UK: Zed Books.

Welland, T., & Pugsley, L. (Eds.). (2002). *Ethical dilemmas in qualitative research* (Cardiff Papers in Qualitative Research). Aldershot, England: Ashgate.

Wiles, R. (2013). *What are qualitative research ethics?* London, UK: Bloomsbury Academic.

Zeni, J. (Ed.). (2001). *Ethical issues in practitioner research*. New York, NY: Teachers College Press.

Collaboration

Arnett, R. C., Harden Fritz, J. M., & Bell, L. M. (2009). Dialogic ethics: Meeting differing grounds of the "good." In *Communication ethics literacy: Dialogue and difference* (pp. 79–99). Thousand Oaks, CA: Sage.

Gallagher, K. (2008). *The methodological dilemma: Creative, critical and collaborative approaches to qualitative research*. London, UK: Routledge.

Gershon, W. S. (Ed.). (2009). *The collaborative turn: Working together in qualitative research*. Rotterdam, The Netherlands: Sense Publishers.

Guest, G., & MacQueen, K. M. (Eds.). (2007). *Handbook for team-based qualitative research*. Lanham, MD: Rowman & Littlefield.

Heron, J., & Reason, P. (2001). The practice of co-operative inquiry: Research 'with' rather than 'on' people. In P. Reason & H. Bradbury (Eds.), *Handbook of action research*. Thousand Oaks, CA: Sage.

Ravitch, S. M., & Tillman, C. (2010). Collaboration as a site of personal and institutional Transformation: Thoughts from inside a cross-national alliance. *Perspectives on Urban Education, 8*(1), 3–9.

Rule, P. (2011). Bakhtin and Freire: Dialogue, dialectic and boundary learning. *Educational Philosophy and Theory, 43*(9), 924–942.

Tetreault, M. (2012). Positionality and knowledge construction. In J. Banks (Ed.), *Encyclopedia of diversity in education* (pp. 1676–1677). Thousand Oaks, CA: Sage.

Online Data Collection and Internet Security

Birnbaum, M. H. (2010). An overview of major techniques of web-based research. In S. D. Gosling & J. A. Johnson (Eds.), *Advanced methods for conducting online behavioral research* (pp. 9–25). Washington, DC: American Psychological Association.

Dawson, P. (2014). Our anonymous online research participants are not always anonymous: Is this a problem? *British Journal of Educational Technology, 45,* 428–437.

Denissen, J. J. A., Neumann, L., & van Zalk, M. (2010). How the Internet is changing the implementation of traditional research methods, people's daily lives, and the way in which developmental scientists conduct research. *International Journal of Behavioral Development, 34,* 564–575.

Fielding, N., Lee, R. M., & Blank, G. (2008). *The SAGE handbook of online research methods*. Thousand Oaks, CA: Sage.

Gill, F., & Elder, C. (2012). Data and archives: The Internet as site and subject. *International Journal of Social Research Methodology, 15,* 271–279.

Hesse-Biber, S., & Griffin, A. J. (2013). Internet-mediated technologies and mixed methods research: Problems and prospects. *Journal of Mixed Methods Research, 7,* 43–61.

Hewson, C., & Buchanan, D. (Eds.). (2013). *Ethical guidelines for conducting Internet-mediated research*. Leicester, England: The British Psychological Society.

Hine, C. (2012). *The Internet: Understanding qualitative research*. Oxford, UK: Oxford University Press.

Paulus, T. M., Lester, J. N., & Dempster, P. G. (2014). *Digital tools for qualitative research*. Thousand Oaks, CA: Sage.

Salmons, J. (2014). *Qualitative online interviews: Strategies, design, and skills*. Thousand Oaks, CA: Sage.

Salmons, J. (Ed.). (2011). *Cases in online interview research*. Thousand Oaks, CA: Sage.

Wankel, C., & Malleck, S. (Eds.). (2010). *Emerging ethical issues of life in virtual worlds*. Charlotte, NC: Information Age Publishing.

ONLINE RESOURCES

Sharpen your skills with SAGE edge

Visit **edge.sagepub.com/ravitchandcarl** for mobile-friendly chapter quizzes, eFlashcards, multimedia resources, SAGE journal articles, and more.

ENDNOTES

1. Freire (1970/2000) conceptualizes praxis as the generative intersection of reflection and action toward social transformation. For more on the concept of praxis, see Tierney and Sallee (2008).

2. For example, in the field of psychology, the American Psychological Association has its own ethical guidelines (see http://www.apa.org/ethics/code/index.aspx).

3. For a discussion of this etiology with specific examples, see Hammersley and Traianou (2012).

4. As Glesne (2006) states, "Researchers . . . [in contrast with counselors] . . . traditionally establish rapport to attain ends shaped primarily by their own needs" (p. 110).

5. For example, what language and concepts might I use with academic colleagues, and how do these differ from how I would describe the research to participants?

6. It should be noted that in our own work, with populations that are not literate, these forms are read verbatim.

7. For further discussion of this, see Hammersley and Traianou (2012).

8. For more on the notion of the researcher as instrument, see Lofland, Snow, Anderson, and Lofland (2006) and Porter (2010).

9. For more on the project, see http://www2.gse.upenn.edu/nicaragua/.

Epilogue

Revisiting Criticality, Reflexivity, Collaboration, and Rigor

In this Epilogue, we once again emphasize our central argument—namely, that in qualitative research, the methodological cannot be separated from the theoretical and conceptual. We have discussed throughout the book what we consider to be four of the key pillars of qualitative research—criticality, reflexivity, collaboration, and rigor—so we do not discuss them in depth here. However, we provide examples of alternative ways to think of and enact these pillars from current graduate students to help you think about ways that these key aspects of qualitative research can be applied. We conclude this Epilogue with a brief discussion about what we view as the incredible power and potential of qualitative research.

REVISITING THE HORIZONTALS IN QUALITATIVE RESEARCH

Bias exists in and shapes all research, whether you engage in quantitative, qualitative, or a mixed-methods approach. In qualitative research, understanding and confronting the values and beliefs that underlie your decisions and approaches is vital and should be situated at the heart of the inquiry. This examination of biases, beliefs, and assumptions is an ethical responsibility. There are important implications of your actions (and reactions) as a researcher as well as of how you represent other people's lives, both in terms of honest representation and because others' actions (practice, policy, future research) might be based on the content of your findings and reports. In this regard, the methodological cannot be separated from the theoretical or the conceptual. Thus, qualitative researchers must make deliberate methodological choices to acknowledge, account for, and approach researcher bias and the assumptions that drive, frame, and shape your research. *Criticality, reflexivity, collaboration, and rigor* are necessary as framing concepts for conducting ethical and valid qualitative research.

To be clear, we call these four concepts horizontals not because there is anything linear about them but because we believe that they span throughout and ideally shape all other aspects of the research process. These central concepts and the processes that these horizontals generate are important in all qualitative research regardless of specific approach, study goals, and other specifics of the research in which you will engage. While we would argue that all qualitative research ideally would follow these values in an integrated approach to inquiry, our belief is that it takes a particular kind of focus and intentionality to do this in specific rather than general ways, that is, to engage this methodologically, not just conceptually or theoretically. Thus, our focus on relating these concepts within and across all other aspects, processes, and methods of research is vital since we see their integration as a centralizing ethical dimension of engaging in qualitative research.

Criticality

Our conception of *criticality* in qualitative research entails the wedding of the conceptual and theoretical with the methodological; it holistically frames how theory, methods, process context(s), researcher identity, and positionality intersect with and shape qualitative work. Throughout the book, we discuss ways to critically approach aspects of qualitative research, including developing a conceptual framework and research design, collecting and analyzing data, and representing research. Criticality in qualitative research is a stance, an ideology, and a methodological framework; it is a way to conceptualize and operationalize a research design and set of research practices that engage complexity and help you to enact research with an active, reinvigorated approach to rigor. As Demetri Morgan discusses in Feature Box 12.1, it is not formulaic or prescriptive. When we first presented our conceptualization of criticality in qualitative research to our students a few years ago, Demetri asked us, "How do you know if you are conducting critical research and who gets to decide?" This question stuck with us, and our answer to him then and now is that there is not a checklist or test to determine if you are critically approaching qualitative research. There are processes, practices, and procedures that you can (and should) systematically engage in, including in dialogue with others, to help foster a critical stance. Ultimately, critically approaching qualitative research means that you, as a researcher, are paying careful attention to issues of power and equity so that you are centralizing and faithfully representing participants' experiences.

Feature Box 12.1: Criticality in Qualitative Research

Demetri L. Morgan

My research grapples with the overarching question of how postsecondary institutions can best change in order to promote the success and development of an increasingly diverse group of college students. As has been well documented, higher education institutions in the United States have a long history and continuing existence of exclusionary admissions practices, racism, sexism, and a host of other oppressive forces that make up the contemporary culture of the academy and contribute to inequities in society. Given that I am concerned with how institutions need to address these issues to foster student success and remediate these inequities—the need to take a critical inquiry stance in my research work is beyond paramount. I initially learned about taking a critical stance toward inquiry as

I worked to improve my undergraduate institution's commitment to Black male achievement. Through this effort, I was exposed to a wealth of research on the Black male success issue. As I continued to engage with this body of research over time, what stood out to me was that there were specific scholars who intentionally took an anti-deficit approach to studying Black men in college and sought to look at our assets and not deficiencies, despite a national dialogue that was doing the opposite. This validated me and also helped me think through how I could help my friends and my institution.

I vowed to myself from that point on that my research would seek to attack inequities, be true to my participants, and contain actionable insights for practitioners and other scholars. I would later come to realize that this commitment meant that I needed to work to have a sense of *criticality* run throughout my work. To carry this criticality out in a practical way, I started by reading widely. Although my main field is higher education, I pull from K–12 educational research, political science, sociology, anthropology, communication studies, and psychology to inform my work. This opens me up to different theoretical frameworks and research designs that afford me the necessary tools to remain critical throughout a research project. Furthermore, operationalizing a critical inquiry stance means surrounding myself with people who will hold me accountable to a critical orientation. At times, this means being vulnerable with others so they can call me out on an intellectual blind spot that may affect my criticality. Additionally, this means engaging with nonacademics, practitioners, and students, to shatter the echo chamber that the academy sometimes produces. Last, taking on a critical inquiry stance has meant intentionally practicing being critical across my research projects. Being critical is not formulaic, so thinking about the uniqueness of each project and what that means for my critical disposition serves as a repetition that helps me remain critical in all my research projects. This creates an ongoing process of self-reflection to implementation and back to self-reflection that strengthens my sense of criticality and helps to infuse it in all that I do.

Considerations I have when thinking about criticality in my work:

- What drives me to continuously take on a critical inquiry stance?

- Why is it important to me to be critical in my approach to research?

- Am I reading from a broad collection of sources?

- Do I have three to five scholars whose work engages the type of criticality that I could adapt to help me shape my own sense of critical research?

- Do I regularly interact with people who will be critical of my criticality?

- Do these people have a range of perspectives and ties to my work?

- Am I being critical in all the different projects of which I am involved?

- How do I know this?

- How do the different critical approaches in my work inform each other?

Reflexivity

Reflexivity is an active and ongoing awareness and monitoring of your personal role and significant, ongoing influence on the research. This means recognizing and reckoning with your role in the construction of and relational contribution to the process and content of the research throughout its development, enactment, and write-up. Reflexivity necessitates that you assess and continuously reassess your positionality, subjectivities, and guiding assumptions as they directly relate with and shape your research.

As the researcher, you are the primary instrument of your research, and therefore the importance of systematically considering and methodologically addressing your social location and positionality cannot be overstated. As Jaime Nolan discusses in Feature Box 12.2, to faithfully represent her participants' experiences, she must monitor her own identity and positionality. She discusses how this is often even more difficult when conducting research, which often connotes an aura of objectivity. Embracing the subjectivity inherent in all research, especially qualitative research, is another way of critically approaching it. As we discuss throughout the book, reflexivity is a central means to help you embrace and address your subjectivities.

Feature Box 12.2: Reflections on Reflexivity

Jaime Nolan

The world is a story we tell ourselves about the world.

—Vikram Chandra

I don't believe in charity. I believe in solidarity. Charity is so vertical. It goes from the top to the bottom. Solidarity is horizontal. It respects the other person. I have a lot to learn from other people.

—Eduardo Galeano

My research has facilitated in many ways my ability to sit quietly in Galeano's spirit of coalition building, humility, and learning from others in the liminal spaces between strangeness and familiarity. All of these things for me are central to reflexivity, which I view as mindfulness and presence in my evolving sense of relationship with my participants, myself, the phenomena I study, and the larger social world. And being mindful in this way has become core to my inquiry work. My engagement with a group of American Indian students has been life changing.

I have had to negotiate the limitations in my own frames of reference, my own cultural constructs about story and, as Gadamer (2004) writes, see the "relative significance of everything within this horizon whether it is near or far, great or small" (p. 302). As a mixed-race White/Latina, viewed as both exotic enough to encounter others' quizzical gazes and White enough to move through many institutional spaces with a sense of privilege, I have to confront my own personal complex, history, and its intersection in my work.

The process of building a sense of trust, so important to this research, raised questions of my own identity, positionality, and how I could hold, honor, and tell both the big story, contextualized in cultural history, and the little stories of my study participants' lives. I have had to remain mindful of the complexities of my own identity and positionality in my efforts to remain faithful to my participants as well as representing their stories without misappropriating their ideas, imagery, and language. This is not easy, however, particularly in institutional spaces that often prize distance, objectivity, and the hubris of certitude. Working with these students made visible to me again how much I have to learn from others, who are, as Gloria Anzaldúa taught me, the theorists of their own lives. I am constantly reminded that I have power to do both tremendous good and tremendous harm. I have come to see more clearly and value deeply the process of being reflexive and the way in which reflexivity deepens ethical praxis in a spirit of responsiveness and faithfulness to the stories I hear to the needs of the world and those around me.

Questions I considered in focused ways throughout the process of my study:

- Who am I in relationship—To the research? To participants? With myself? With the phenomena I am studying?

- How am I situated in all aspects of the work?

- What are my assumptions and how do I negotiate them throughout the process?

I also drew heavily from the memos I wrote and made a point to share my research with participants at several points throughout the research process. I found that debriefing with someone at various points in the process to be invaluable to reflexivity.

Collaboration

Collaboration entails engaging with participants, colleagues, advisers, and mentors in thoughtful and deliberate ways. If you are on a research team or are a lone researcher, collaboration is necessary to conducting complex, ethical, and valid research. One way that collaboration can be facilitated is through *dialogic engagement* practices that support a critical stance to qualitative research by helping you confront your biases and challenging you to think about issues of power and equity. Collaboration requires discretion in that you must consider who you are choosing to be thought partners, what their views are, if the collaborations are fruitful, and if they actually help to facilitate criticality broadly and critical self-reflection specifically. Thus, collaboration is central to a reflexive and critical research practice. As Matthew Tarditi discusses in Feature Box 12.3, collaboration is an integral part of how we, as researchers, challenge our subjectivities.

Feature Box 12.3: Thoughts on Collaboration

Matthew J. Tarditi

At the heart of collaboration is the expansion of possibilities, especially given the limitations of individualism. There is no singular conception or type of collaboration, but there are certain characteristics of collaborations that make for more meaningful relationships that are steeped in respect, trust, dialogue, and dynamics that treat all engaged as valued knowers. Over time, I have learned that consensus is not the primary goal of collaboration, and the common pressures to collapse difference and reach a single conclusion are counterbalanced by the enriching qualities and characteristics of diversity, difference, and heterogeneity. Collaboration is not exclusively positive, nor is it inherently beneficial, but it does have the powerful potential to (1) connect individuals through dialogic relationships in which we learn together and (2) expand the horizons for what is possible when we work together and accept difference as an opportunity to collectively grow.

We ascend to the ideals of learning when it becomes a dialogic, social process, one that occurs in and through relationships. In part, this is how we reflect on and challenge our subjectivities, epistemologies, ideologies, and ontologies (our understanding of reality). In dialogue with others, we appreciate and accept individuality (and difference) as we deepen our understanding of humanity as a whole. Life is ideally a collaborative endeavor, and the foundations of any collaboration are the direct relationships among individuals within a dyad or among a network of partners. Although many professions, disciplines, and cultures emphasize individual achievements, collective efforts and collaborative partnerships are returning to prominence in the network age of ubiquitous technology and global interconnectedness. Dialogic and/or polyphonic (multivoice) relationships challenge previously understood boundaries and borders and provide the propitious relational dynamics to enable new opportunities for growth, spaces in which our ideas, concepts, and perceptions intersect and collide.

The unifying foci of my intellectual, professional, and personal pursuits are the parallel and coevolving nature of trust and dialogue as they relate to collaborations. As part of an international educational program operating in Nicaragua since 2009, I have had the great fortune to cross borders (literally, culturally, socioeconomically, etc.) as part of an unrelenting desire to build partnerships and form collaborations among the multitude of stakeholders. It is my contention that (relational) trust and authentic dialogue are two fundamental, indispensable qualities of any collaborative endeavor, and they are especially important in those comprising a diverse range of participants residing across a wide spectrum of power and privilege. From my experience in various partnerships, I believe there are three central characteristics to the development and enactment of inclusive collaborations among diverse stakeholders:

(1) **Respect**—treat people fairly, justly, and as equals, understanding that these concepts are mediated by power dynamics and social structures.

(2) **Active listening**—engage with what people are saying to understand the roots and reasons behind their points of view instead of waiting your turn and preparing your counterargument or unrelated point.

(3) **Authenticity**—Be true to yourself and who you are. It may sound simple, but we too often compromise our own perspectives, values, and beliefs for the sake of consensus or deference. Every individual's opinion is a valuable contribution, and critical and/or opposing views enrich the discussion of ideas and the making of decisions.

As a researcher, educator, professional, and person, I approach life in dialogue with the world around me, and the collaborations that I initiate or participate in depend greatly on the positive personal relations and emotional connections between and among partners. That being said, the complexities are many, and there are no universals; however, I humbly offer some guiding questions to consider when approaching collaborations.

(1) *What is the purpose of the collaboration? What is the relationship between individual and group goals? How do we measure success?*

(2) *What are our individual roles within the collaboration, and how are we organized (e.g., hierarchically, horizontally, in a network)? What are the existing organizational/relational structures of collaborators, and how do these mesh within the newly emerging collaboration?*

(3) *How are decisions made and who gets to decide? Why? Is there room here for other possibilities?*

(4) *What are the communicative pathways and spaces for dialogue and discussion? Are all voices being heard?*

Rigor

Rigor in qualitative research encompasses a variety of concepts, considerations, and actions, including that the researcher (1) develops and engages in a research design that seeks complexity through the structure and strategic sequencing of methods and the mapping of research methods onto the guiding research questions; (2) maintains a fidelity to participants' experiences through engaging in inductive (or what we think of as emergent design) research that is responsive to emerging meanings while at the same time ensures a systematic approach to data collection and analysis; (3) seeks to understand and represent as complex and contextualized a picture of people, contexts, events, and experiences as possible; and (4) transparently addresses the challenges and limitations of your study.

Rigor allows for the validity of your study from its first to its last moments. Adrianne Flack discusses this important aspect of rigor in Feature Box 12.4. She highlights how rigor cannot be achieved through prescribed formulas because it necessitates creativity and responsiveness.

Feature Box 12.4: Reflection on the Role of Rigor in Qualitative Research

Adrianne Flack

As an early career qualitative researcher, achieving rigor in my studies has at times felt daunting. "Growing up" as a researcher in the shadow of debates about perceived differences in rigor between qualitative and quantitative research designs, with paradigmatic privilege being bestowed on "experimental" methods, I have learned over time how to combat such critiques by taking initiative in growing my repertoire of research knowledge and skills. One of the unexpected discoveries I made about conducting qualitative research is that it really is a creative enterprise in which there are many powerful ways to construct a strong study with accurate findings. It just takes effort to learn what you need to learn, integrity, and a good dose of creative thinking (and yes this can be a lot of fun).

My identity as an emerging scholar deeply aligns with the philosophical underpinnings of qualitative research, which privileges the lived experience of participants in developing solutions to problems of practice in education—my own context being higher education. My research agenda, which centers on issues of access to and completion of postsecondary education for educationally and economically disadvantaged students, relies heavily on carefully hearing, observing, and interpreting the experiences of learners who come to postsecondary education with differing needs, abilities, and expectations than more culturally and academically prepared college goers. Hence, privileging the voice of such participants is paramount to developing practices and policies that can facilitate their educational and career success. In doing so, however, it is critical that my findings are indeed an accurate reflection of what is going on in the world and how learners make sense of their experiences in context. Some of the ways in which I try to build and demonstrate rigor in my studies include the following practices:

- Rather than solely focus on the general principles of establishing trustworthiness, I take time to learn about what validity looks like and what constitutes rigor in employing a specific methodological approach. Weaving this specific information into the appropriate parts of my research design is a critical step to using an approach with fidelity.

- I take time to read primary sources pertaining to any theoretical constructs or methodological approaches I desire to use in my research. In the event that it is more advantageous to rely on reading secondary sources, I read across multiple accounts to gain a comprehensive picture of a scholar's work.

- I am reflective about the ways in which differences in language and culture between participants and me might shape the research experience.

- I try to think outside the box regarding the types of data that I can collect to help answer my research questions. It is a big world; think broadly about your options!

- I think about who the potential readers are for my work and anticipate the kinds of questions and critiques they might offer as I design and conduct my research and write my final reports.

- Last, I engage a wide range of interested others in conversation about my work as I am doing it. This has been an extremely helpful way to crystallize my thinking about my procedures and findings and, in a similar vein, illuminate potential blind spots.

CLOSING THOUGHTS: THE POWER AND POTENTIAL OF QUALITATIVE RESEARCH

It is our hope that reading this book and engaging in its many recommended practices, questions for consideration, thought experiments, and related resources has provided you with a solid working understanding of what qualitative research is and how to conduct complex, ethical, and rigorous qualitative studies. We have endeavored to share a framework for how to design and conduct qualitative research that integrates the theoretical, methodological, and conceptual complexities and processes that comprise qualitative research studies. This book was written with the primary and intersecting goals of providing you with a grounded understanding of what qualitative research is and entails, offering a conceptual framework on criticality in qualitative research that you can adapt to cultivate your theory of action for your own work, and helping you to develop the methodological tools, conceptual understandings, and procedural techniques to engage in valid, ethical, engaging, and respectful qualitative research. Throughout the book, we have sought to support you as you cultivate and integrate theoretical, methodological, and conceptual knowledge and skills; to help you see their important and generative interaction; and to help you appreciate and understand the central concepts, topics, and methods you need to engage in your own high-quality, well-conceptualized research.

All research is powerful. With this power also comes great responsibility and potential. It is our hope that you never forget the power of research and remain as respectful and humble as possible by centralizing first and foremost participants and their experiences. We designed this book to help you do this by centralizing the iterative, reflexive, systematic, and recursive ways that you develop and design your research (and your broader methodological approach) to achieve a focused and intentional approach to research rigor and validity. We hope that our passion for research and belief in the power of equitable research to contribute to the world will inspire you to retain an appreciation for the real considerations and steps necessary to engage in rigorous qualitative research. For us, the possibility born out of equitable research that seeks local and contextualized knowledge generation and relational ethics through an attention to rigor and validity is at the heart of qualitative research.

QUESTIONS FOR REFLECTION

- How are criticality, reflexivity, collaboration, and rigor central aspects of qualitative research?

- As discussed in this chapter and throughout the book, what are ways that you can engage in criticality, reflexivity, collaboration, and rigor?

- Which of the four horizontals strikes you personally as the most difficult to get right as you engage in your own qualitative research? Why?

- What should you keep in mind about the power and potential of qualitative research? What applications does this have for your work?

ONLINE RESOURCES

Sharpen your skills with SAGE edge

Visit **edge.sagepub.com/ravitchandcarl** for mobile-friendly chapter quizzes, eFlashcards, multimedia resources, SAGE journal articles, and more.

Appendixes[1]

CONTENTS

Susan Feibelman

- Appendix P: Example Data Display 2
  Mustafa Abdul-Jabbar

- Appendix Q: Example Data Display 3
  Susan Feibelman

- Appendix R: Example Coding Scheme and Coded Excerpts
  Mustafa Abdul-Jabbar

- Appendix S: Example of a Pilot Study Report
  Charlotte Jacobs

- Appendix T: Example of a Conference Proposal
  Demetri L. Morgan, Cecilia Orphan, and Shaun R. Harper

- Appendix U: Example Research Paper Proposal,
  Hillary B. Zimmerman, Demetri L. Morgan, and Tanner N. Terrell

- Appendix V: Example Project Statement for Grant Proposal
  Sarah Klevan

- Appendix W: Example Fellowship Proposal
  Arjun Shankar

- Appendix X: Example Preproposal Letter of Interest
  Arjun Shankar

- Appendix Y: Consent Form Template and Examples

  - Appendix Y.1: Consent Form Template
  - Appendix Y.2: Example Consent Form
  - Appendix Y.3: Example Consent Form
  - Appendix Y.4: Example Consent Form
  - Appendix Y.5: Example Consent Form

- Appendix Z: Assent Form Examples

  - Appendix Z.1: Example Assent Form
  - Appendix Z.2: Example Assent Form

Appendix A

Example Conceptual Framework Memo and Accompanying Concept Maps

This is a good example of an early conceptual framework memo that focuses on the development of the conceptual framework at an early stage. In Brandi Jones's memo, you will see how she has located her research questions in a range of relevant bodies of literature as well as in the contexts of practice that generated the study. In addition, Brandi locates herself in the study, including her personal and professional background and its relationship to her study. Brandi does an excellent job of exploring the formal theories that support the development of her research questions and begins to engage with these literatures as she starts to solidify her research questions and the scope and frames of her study.

Conceptual Framework Memo

Brandi P. Jones
November 16, 2013

INTRODUCTION

The study of psychosocial and identity development of college students fascinates me. Many developmental theories have been used to explore issues in adolescents or undergraduate students. I am interested in exploring the relevance of developmental theories in graduate student populations. As society demographics and academic environments become more diverse, there is a need for doctoral education to evolve (Kim, 2009). The concept of developing the whole student is absent from doctoral training, particularly in fields related to science, technology, engineering, and mathematics (STEM). I would like to explore the relationship between racial identity, self-esteem, and professional identity of African American doctoral students in engineering fields.

Necesitamos teorias [we need theories] that will rewrite history using race, class, gender, and ethnicity as categories of analysis, theories that cross borders, that blur boundaries—new kinds of theories with new theorizing methods. . . . We are articulating new positions in the "inbetween," Borderland worlds of ethnic communities and academies . . . social issues such as race, class, and sexual difference are intertwined with the narrative and poetic elements of a text, elements in which theory is embedded. In our mestizaje theories we create new categories for those of us left out or pushed out of existing ones (Anzaldúa, 1990, pp. xxv–xxvi).

RESEARCH QUESTIONS

What is the relationship between racial identity, self-esteem, and professional identity?

For minority students, self-esteem and ethnic identity are related (Phinney & Alipuria, 1990). There is a general lack of research on self-esteem's relationship to professional identity development. Current research on African American doctoral students in STEM fields largely focuses on recruitment, mentoring, student engagement, retention, and academic preparation. There is a need for more research on how African Americans in STEM doctoral programs develop professional identities. Although existing research indicates that relationships and networks are important to the doctoral student experience (Sweitzer, 2009), it does not discuss racial identity in the construction of professional identity development. Research on professional identity development lacks any in-depth analysis of academic disciplines and racial identity.

Figure 1   My Thinking Around Factors That Contribute to Professional Identity

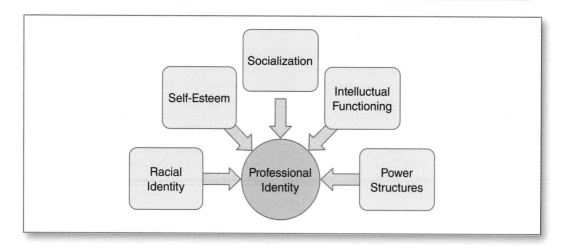

According to Felder (2010), African American PhD students' progress toward degree completion is "a journey wrought with obstacles" (p. 455). The doctoral degree process forces students to juggle multiple and competing identities. At the same time, doctoral students are developing their professional identities; they are constructing an image of who they want to be and how they want others to view them. Their self-images may conflict with their perceptions of who their programs want them to become (Hall & Burns, 2009):

> Becoming a professional researcher requires students to negotiate new identities and reconceptualize themselves both as people and professionals in addition to learning specific skills . . . formal curriculum designs can be used more intentionally to help students and faculty understand the roles identity plays in professional development and to make doctoral education more equitable. (Hall & Burns, 2009, p. 49)

What are the factors that influence professional identity development for African American doctoral students in engineering fields?

Given the nature of research in STEM fields, few opportunities arise for PhD students to have frank discussions about race and ethnicity in academia. I am interested in learning how students who lack these opportunities develop a sense of racial identity alongside professional identity. According to Felder (2010), faculty members exert a strong influence over doctoral student degree attainment. How do faculty, who often lack a background that prepares them to address matters of identity, shape doctoral degree processes in a way that impacts, sometimes adversely, minority students' professional identity? Professors' perceptions of student performance influence their relationships with those students (Felder, 2010, p. 460). How does students' self-esteem impact faculty perceptions of them? Is there a correlation between students' self-esteem and their eventual career paths? How do African American students develop scientific identities? How does racial identity interact with professional identity? Answering these questions can inform doctoral programs' retention efforts in STEM fields as well as how they prepare students for research careers.

How are self-esteem and racial identity related to the way African American engineering doctoral students perceive themselves and are perceived by others? How does racial identity and self-esteem play a role in how African American doctoral students experience graduate studies and navigate social structures?

Carlone and Johnson (2007) described someone with a strong scientific identity as having the ability to demonstrate a meaningful understanding of scientific content, the motivation to understand the world scientifically, and others' recognition as a "science person" (p. 1190). Furthermore, students' likelihood of being recognized as a scientist relates to how they look, talk, and act (Carlone & Johnson, 2007). I am interested in learning how self-esteem and racial identity relate to the ways students perceive themselves and are perceived by others. Lewis (2003) argues that aspiring African American scientists rely on the

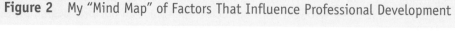

Figure 2 My "Mind Map" of Factors That Influence Professional Development

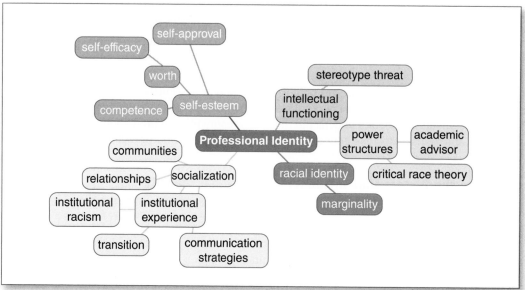

"judgment and invitation of practicing scientists throughout every phase of the educational and career process" (p. 271). Exploring the relationships among racial identity, self-esteem, and professional identity can inform advising, mentoring, retention, and the professional development of doctoral students. "Science career attainment is a social process, and the desire of an aspirant is only one factor in this process. An aspiring scientist relies on the judgment and invitation of practicing scientists throughout every phase of the educational and career process" (Lewis, 2003, p. 371). What factors influence the recognition of being a "science person" for a group of people (African Americans) who have not been recognized historically as "science people" (Carlone & Johnson, 2007, p. 1197)?

COMPONENTS OF EMERGING CONCEPTUAL FRAMEWORK

There are a number of formal and tacit theories that relate to doctoral student development. I am particularly interested in those relating to professional identity of African American students in engineering fields. According to the American Society for Engineering Education (2011), the percentages of African Americans in engineering decrease with each level of education. There are 5.5% at the bachelor's level, 5.3% at the master's level, and 3.5% at the doctoral level. Furthermore, of all engineering faculty in the United States, 2.5% are African American. My research inquiry was born out of my countless conversations with doctoral students about their career plans. The doctoral students who were pursuing careers in academia appeared to have confidence that was lacking in students who seemed to "settle" for other careers. This intrigued me to better understand the factors that influence professional identity (Figures 1 and 2). Professional identity is defined as "the relatively stable and enduring constellation of attributes, beliefs, values, motives, and experiences in terms of which people define themselves in a professional role" (Ibarra, 1999, pp. 764–765; Dobrow & Higgins, 2005). My hunch is that there are African American engineering doctoral students who enter graduate studies with the goal of becoming faculty and decide to pursue other careers along the way. I would like to better understand how racial identity, self-esteem, socialization, intellectual functioning, and power structures influence professional identity.

Self-Esteem

Does racial identity and self-esteem play a role in how engineering doctoral students experience graduate studies and navigate social structures? Studies have shown that there is a correlation between self-esteem and career trajectories (Salmela-Aro & Nurmi, 2007). Self-esteem is critical for factors such as steadfastness and pursuit of certain career paths (Higgins, Dobrow, & Chandler, 2007).

Racial Identity

Racial identity is "a sense of group or collective identity based on the perception that one shares a common racial heritage with a particular group" (Helms, 1990). The multidimensional model of racial identity defines racial identity in African Americans as "the significance and qualitative meaning that individuals attribute to their membership within the Black racial group within their self-concepts" (Sellers, Shelton, Rowley, & Chavous, 1998).

Intellectual Functioning

Is the academic performance of African American doctoral students influenced by stereotype threat? According to Claude Steele (1997), stereotypes influence intellectual functioning and identity. Steele (1997) argues that achievement can be hampered by the threat of negative stereotypes.

Power Structures

According to critical race theory, "racism is a central factor in defining and explaining individual experiences" (Solorzano & Yosso, 2002). I'm interested in understanding how African American doctoral students navigate social and political structures. How is the doctoral experience influenced by race and racism?

Socialization

Socialization is "the process by which newcomers learn the encoded system of behavior specific to their area of expertise and the system of meanings and values attached to these behaviors" (Taylor & Anthony, 2000, p. 186). Faculty careers are shaped by socialization in

Figure 3 Theoretical Framework

Professional Identity	Self-Esteem	Socialization	Racial Identity	Power Structures	Intellectual Functioning
• Theory of Doc Student Professional Identity (Sweitzer, 2009) • Science Identity Model (Carlone and Johnson, 2007) • Possible Selves Theory (Oyserman, Grant, Ager, 1995) • Researcher Identity (Beiber and Worley, 2006; Quaye, 2007)	• Self-Efficacy Theory (Bandura, 1977, 1997) • Research of (Hughes and Demo, 1989; Porter and Washington, 1989)	• Attribution Theory (Weiner, 1985) • Campus Ecology Theories (Moos, 1986; Strange and Banning, 2001) • Socialization Theory (Taylor and Anthony, 2000) • Research on social adjustment (Ostroff and Kozowolski, 1993; Romero and Margolis, 1998; Thompson, 2006; Weidman, et al., 2001) • Social Network Theory (Sweitzer, 2009) • Tools of Bourdieu (1977, 1986, 1990)	• Black Racial Identity Model (Helms, 1990) • Psychological Nigrescence (Cross, 1971) • Multidimensional Model of Racial Identity (Sellers, et al., 1998)	• Critical Race Theory (Yosso, 2005; Solorzano and Yosso, 2002; Harper, 2009)	• Stereotype Threat Theory (Steele and Aronson, 1995; Steele, 1997)

the doctoral process (Sweitzer, 2009). What role do factors, such as transition into graduate school, institutional support, communities, and social networks, play in the development of professional identity? "Doctoral programs that are able to effectively embrace students' differences, rather than viewing them as obstacles, and use those differences to the programs' and students' advantage will likely be more successful in preparing the next generation of faculty" (Sweitzer, 2009, p. 17).

Several bodies of literature frame my thinking around issues related to professional identity development (Figure 3).

Professional Identity

Theory of doctoral student professional identity (Sweitzer, 2009)

- Influence of relationships in professional identity development

Science identity model (Carlone & Johnson, 2007)

- Illustrates the three dimensions of science identity—competence, performance, and recognition

Possible selves theory (Oyserman, Grant, & Ager, 1995)

- Predictors of persistence and the kind of self one might become

Researcher Identity (Beiber & Worley, 2006; Quaye, 2007)

- Graduate student conceptualization of graduate student life
- Graduate student perceptions of faculty life

Self-Esteem

Self-efficacy theory (Bandura, 1977, 1997)

- Beliefs in one's capabilities and factors such as persistence, steadfastness, and pursuit of career paths

Demo and Hughes (1989)

- Social structural processes and arrangements related to racial group identification

Porter and Washington (1989)

- Dimensions of self-esteem among ethnic groups

Socialization

Attribution theory (Weiner, 1985)

- Motivation and emotion in success and failure

Campus ecology theories (Moos, 1986; Strange & Banning, 2001)

- Relationship between the student and campus environment

Social network theory (Sweitzer, 2009)

- Professional identity development and the process by which it occurs

Racial Identity

Black racial identity model (Helms, 1990)

- Various stages of Black identity development

Psychological Nigrescence (Cross, 1971)

- Various stages of Black identity development

Multidimensional model of racial identity (Sellers et al., 1998)

- Dimensions of African American racial identity

Power Structures

Critical race theory (Harper, 2009; Solorzano & Yosso, 2002; Yosso, 2005)

- The role of race and racism in education

Intellectual Functioning

Stereotype threat theory (Steele, 1997; Steele & Aronson, 1995)

- Societal stereotypes about groups and how they influence functioning and identity development

RESEARCHER POSITIONALITY AND CONTEXT

This study will not only contribute to the research on African American doctoral students in STEM fields but also inform my work as the Associate Dean of Engineering and Applied Science at a predominantly White research institution. Maxwell (2013) noted the importance of personal experience as a source of motivation (p. 24). My many interactions with doctoral students have sparked my research questions. It is with a spirit of passionate curiosity that I wish to pursue self-esteem, racial identity, and professional identity of African American STEM doctoral students as a research topic.

As I pursue this research topic, I make a number of assumptions about the development of Black doctoral students at predominantly White institutions based on my wisdom of practice. I've spent countless hours listening to stories and experiences of students. I get angry every time I encounter a Black doctoral student who has low self-esteem. I get angry when I hear Black doctoral students question whether or not they are smart enough to continue. I get angry when I hear Black doctoral students talk themselves out of pursuing a career in academia because they don't think they have what it takes to succeed. I get angry for all of the obvious reasons, but the one that is not so obvious is that I too am a Black doctoral student with shaky confidence in my ability to be a good research scholar. I come from a working-class family. I am a first-generation college student. I am state school trained. My mother grew up on a plantation in Louisiana. It is for all of these reasons that I am not always confident in my ability to succeed as a doctoral researcher.

My assumptions are colored by who I am in the world right now . . . Black; student; STEM professional; researcher.

As STEM graduate programs seek to diversify their doctoral student bodies, it is important that they understand which factors lead to minority student academic and career success. Most STEM departments rely heavily on grants to fund doctoral education. Most national funding agencies, such as the National Institutes of Health (NIH) and National Science Foundation (NSF), have diversity initiatives in place. Funding agencies are interested in the enrollment and completion of underrepresented minorities in STEM fields. For example, NSF grant applications state, "Guided by the Strategic Plan, NSF established a performance area focused on broadening participation to expand efforts to increase participation from underrepresented groups" (National Science Foundation, 2013). This emphasis is forcing STEM departments to increase their numbers of doctoral students, postdoctoral researchers, and faculty from underrepresented backgrounds. The study of ethnic identity and self-esteem in professional identity development can inform how faculty members work with African American doctoral students in STEM fields, particularly at predominantly White institutions. Murphy (2002) and Towler and Dipboye (2003) suggested that efforts in academic and work environments are more effective when tailored to fit populations.

Figure 4   Components That Contribute to My Research Inquiries

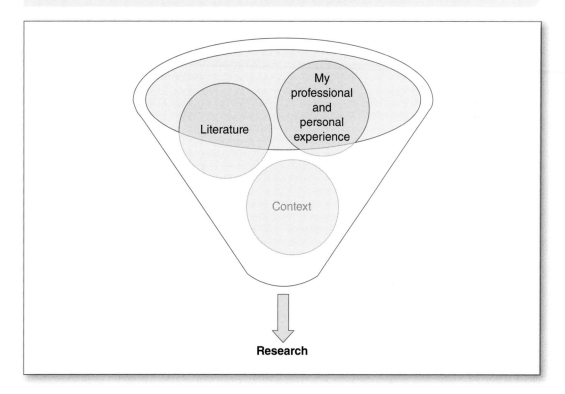

RESEARCH DESIGN

I will collect data via interviews and questionnaires with African American doctoral students in engineering fields at Ivy League institutions. I will likely include Princeton University, University of Pennsylvania, and Columbia University in my research study. Interviews will be conducted according to the three stages into which Tinto (1993) divided the doctoral degree process: transition and adjustment, or the student experience during the first year of study; attainment of candidacy, or the period between the first year and the time a student attains candidacy; and completion of the dissertation, or the time between candidacy and the student's defense. I will explore the use of the following instruments:

- Racial Identity Attitude Scale (Helms, 1990)
- Multidimensional Model of Racial Identity (Sellers et al., 1998)
- Multi-Group Ethnic Identity Measure (Phinney, 1992)
- Model of Nigrescence (Cross, 1971, 1991)
- Rosenberg Self-Esteem Scale (Rosenberg, 1979)
- Anti-Deficit Achievement Framework (Harper, 2010)
- Padilla's Expertise Model (Padilla, 1991)
- Science Identity Model (Carlone & Johnson, 2007)

Appendix B

Conceptual Framework Memo and Accompanying Concept Map

Mustafa Abdul-Jabbar's memo does a great job of laying out the framing formal theories and their relationship to the context, design, and methods of the study. The visual representation of his conceptual framework highlights how all of the constructs in his research come together.

Conceptual Framework: Theoretical Framework and Synopsis of Methodology With Preliminary Findings

Mustafa Abdul-Jabbar
July 24, 2012

Analysis of leadership's role and impact on student achievement concludes that leadership is an important predictor variable of student success. In fact, recent evidence presented by Leithwood, Louis, Anderson, and Wahlstrom (2004) cites that "the total (direct and indirect) effects of leadership on student learning account for about a quarter of total school effects" (p. 5). This finding by the authors precipitated their conclusion that "leadership is second only to classroom instruction among all school-related factors that contribute to what students learn at school" (p. 5).

Other studies have established that trust mediates student learning and achievement (Bryk & Schneider, 2002; Tschannen-Moran & Hoy, 1997, 2000). Bryk and Schneider (2002) add that while "trust does not directly affect student learning · · · trust fosters a set of organizational conditions, some structural and others social-psychological, that make it more conducive for individuals to initiate and sustain the kinds of activities necessary to affect productivity improvements" (p. 116). Thus, in the effort to improve schools and promote student learning and achievement, like leadership, trust matters too.

While both leadership and trust have been found to impact, mediate, and in certain instances predict student achievement, "the relational aspects of trust and leadership have not been fully studied" (Daly & Chrispeels, 2007, p. 31). In particular, Daly and Chrispeels (2007) argue that "studies are needed that investigate how the various facets of trust are related to leadership behaviors" (p. 31). To this end, this dissertation investigates relational trust as it manifests within schools that have recently implemented the distributed leadership framework in their instructional leadership programs.

This dissertation highlights research that connects the distributed leadership and relational trust frameworks in the task of identifying specific behavioral anchors that give rise to specific trust considerations across distributed leadership team members in particular schools. Across specific school sites, these behavioral anchors are further utilized in crafting a typology of leadership practices that inform *respect*, *integrity*, *personal regard*, and *competence* interpretations in the school organization.

The research presented in this dissertation employs mixed methods in its design, featuring a quantitative sampling frame and both quantitative and qualitative data collection methods (i.e., surveys, interview data, observation data, etc.). Perceptual data from interviews involving distributed leadership team members include a growing awareness around an emerging team identity, perceptions of more equality between teachers and administrators on the distributed leadership (DL) team, perceptions that relationships have deepened, and the overall finding that teacher confidence in school administration has grown through collaboration on the distributed leadership team. Also, DL team members (both administrators and teachers) have reported perceptions of increase in trust in one another's capacities, capabilities, and integrity. These perceptions have arisen in the course of sharing work, work space, and team goals. The ambition of this study seeks a strategic framework—a syncretization between distributed leadership and relational trust theory—for documenting a typology of educational leadership practices that influence trust in schools.

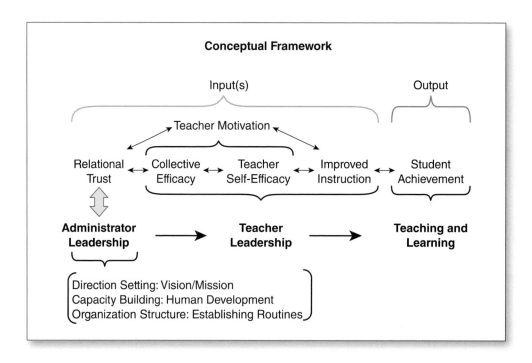

Research Question(s)

1. How do leadership behaviors of a distributed leadership team influence relational trust between members of that team?

2. How do leadership practices of a distributed leadership team at a specific school influence norms of trust, innovation, and collaboration among the greater faculty in the school?

Appendix C

Conceptual Framework Example From a Dissertation

We have chosen to include this lengthy conceptual framework section from Laura Colket's dissertation, which is entitled "Breaking the Single Story: Narratives of Educational Leaders in Post-Earthquake Haiti," because she so thoughtfully and carefully articulates the overall conceptual framework, the theoretical framework within it, and the role of self and context as well as formal theory in the development of the study methodology.

Dissertation Excerpt on Conceptual Framework

Laura Colket
June 12, 2012

The conceptual framework that guides this dissertation research has become like an ecosystem: It is a living, breathing, ever-changing environment in which the various components are constantly interacting with, feeding off of, and growing from one another. Conceptual frameworks are often built from bodies of literature, theoretical stances, philosophical orientations, ethical commitments, and personal experiences (Miles & Huberman, 1994; Maxwell, 2005; Ravitch & Riggan, 2012), and indeed, each of these components, both personal and academic, has influenced this framework in myriad ways. Ultimately, this conceptual framework has also become deeply intertwined with my methodological choices as the researcher. As Ravitch and Riggan (2012) argue, a conceptual framework functions as the guide for research; it "serves to situate the research questions and the methods for exploring them within the broader context of existing knowledge" (p. 136). Accordingly, this chapter combines my conceptual framework and methodology, ultimately creating the basis for this dissertation research. In the section below, I begin with an overview of critical theory and feminist poststructuralism, the two theoretical umbrellas that have most greatly shaped my work.

Theoretical Umbrellas

At a theoretical level, this dissertation has been significantly guided by an orientation toward two broad theoretical umbrellas: critical theory and feminist poststructuralism. For example, I draw on critical race theory, postcolonial studies, and critical historiography throughout this dissertation in order to understand the social and historical context of this research, and each of these three theoretical perspectives has been shaped by the broader umbrellas of critical theory and poststructuralism.

Critical theory. While critical theory as a whole is somewhat amorphous, as it is a self-critical tradition and thus constantly developing and filled with significant disagreement among its scholars, there are some general themes that guide the work of many critical theorists. For example, questions of power—specifically related to systemic inequalities and possibilities for transformative change—significantly shape and influence the work of critical scholars (e.g., Freire, 1993; Giroux, 1986; LeCompte, 1995; McLaren, 2003; Morrell, 2004; Rubin & Silva, 2003; Weis & Fine, 1993). Critical theory has been described by Kincheloe and McLaren (2000) as a means to "disrupt and challenge the status quo" by producing "dangerous knowledge, the kind of information and insight that upsets institutions and threatens to overturn sovereign regimes of truth" (p. 279). This focus on radical transformation supported by deep critiques of the current system is a central component to the work of many critical theorists. And, while critical theory is inherently distrustful and deeply critical of dominant thought paradigms, practices, and systems of power, many critical theorists also maintain a sense of optimism, a belief in the possibility of change.

In addition to seeking to uncover, confront, and ultimately transform oppressive forces, critical theorists also aim to expose and challenge master narratives (i.e., the meta-stories about historical events or experiences that come to be the ostensible "truth"). As such, critical research methodologies work to understand the social, historical, and contextual conditions under which such master narratives take shape and ultimately influence individual and collective interpretations, actions, and ideologies (Kincheloe & McLaren, 2000; LeCompte, 1995; Lather, 1992; McLaren & Giarelli, 1995). Critical theorists, therefore, see "truth" as a relative concept and are thus distrustful of research that claims to have unearthed any sort of final or ultimate truth.

This dissertation is most deeply influenced by the work of Paulo Freire. I draw from Freire's concept of praxis, which he describes as "reflection and action upon the world in order to transform it" (Freire, 1970/2007, p. 51). While this research is not explicitly action research, through this dissertation, I am speaking to an issue that I believe can, and I hope will, change. My experiences with this dissertation have created changes in myself as a researcher and as someone involved in an international partnership, and one of my goals with this work is to inspire similar changes in others who are in the same field. As Freire (1970/2007) argues, "No reality transforms itself" (p. 53), and as such, critical reflection on the field of international development is an imperative first step to creating change. In fact, Freire emphasizes that critical reflection is action.

Freire's work is guided by a deep belief that radical transformation is possible. As Freire (1998) explains, "I have always rejected fatalism. I prefer rebelliousness because it affirms my status as a person who has never given in to the manipulations and strategies designed to reduce the human person to nothing" (p. 103). Through this dissertation, I provide deep critiques of the "manipulations and strategies" that exist in the field of international development, which, intentionally or not, continue to be oppressive forces. As Freire (1970/2007) argues, oppression dehumanizes not only the oppressed but also the oppressors, and the radical transformation that is needed, both at the personal and societal level, cannot happen through acts of kindness (which maintain the oppressive dynamics) but rather must happen through solidarity.

Through this research, I have formed close relationships with my research participants and have come to see them not as participants but rather as partners, colleagues, and mentors. Ultimately, I see this dissertation not just as an intellectual exercise but also as an act of solidarity with the Haitian leaders who have participated in this research and with Haiti more broadly.

Critical race theory (CRT) (Delgado & Stefancic, 2001; Ladson-Billings, 1998; Solorzano & Yosso, 2002), as an outgrowth of critical theory more broadly, also holds a significant place in my conceptual framework—specifically in regards to CRT's focus on counter-storytelling. As one of its main tenets, critical race theory emphasizes the importance of highlighting the voices and experiences of people of color, as their realities and perspectives are particularly likely to be left out of the larger dialogue (Delgado & Stefancic, 2001). Specifically, critical race theorists argue that counter-storytelling—or highlighting the voices, experiences, and realities of people of color—is a powerful tool that can be used to challenge and alter the dominant narrative (Delgado & Stefancic, 2001). Leadership studies, in particular, are based largely on the White male experience of leadership. By sharing the stories and experiences of these seven Black and Mulatto Haitian leaders, I am not only speaking back against the broader discourse on international aid that is often devoid of the perspectives of the Black and Brown people in the global South, but I am also speaking back against the field of leadership studies that has historically ignored the experiences of leaders of color.

Feminist poststructuralism. Similar to critical theory, feminist poststructuralism is also a complex tradition with some internal inconsistency (Scott, 2003). There are, however, certain trends or tenets that shape this theoretical perspective, and, in this section, I will provide a brief overview of these various guiding principles of feminist poststructuralism. I will begin first by looking at poststructuralism more broadly and will then examine the influence of feminism on poststructural thought.

Poststructuralism resides under the larger umbrella of postmodernism, which Kincheloe and McLaren (2000) describe as a state of "hyperreality," an understanding of society as "saturated with ever-increasing forms of representation," which have "a profound effect on our construction of the cultural narratives that shape our identities" (p. 292). Although I create here a distinction between postmodernism and poststructuralism, the differences between the two are not always clearly delineated. In fact, Lather (1992) acknowledges that the terms *postmodernism* and *poststructuralism* are sometimes used interchangeably; even so, Lather describes poststructuralism as "the working out of academic theory within the culture of postmodernism" (p. 90). As a child of a postmodern society, Lemert (2009) argues that poststructuralism is guided by three main principles: decentering, discourse, and difference. Lemert draws heavily from Derrida as he explains,

> In the absence of a Center (whether intellectual or political), one cannot trust reality. In the absence of a trustable reality, one can only rely on language. What a person says is sufficiently true. No one person's saying is more true. Thus, the world is a world of difference. In the absence of a Center, there is difference. (p. 377)

As such, truth, identity, and reality are continually contested and constantly in flux. This creates a deep distrust in any claims to an ultimate "truth" or "reality," which, necessarily, has important methodological implications that will be discussed in greater detail later in this chapter.

Poststructural theory is often traced back to the work of Foucault and is associated with his push for disrupting the "regimes of truth" that shape our society (Foucault, 1980). As Foucault argues, "truth" is created and maintained through systems of power that circulate through academic discourses, the media, and schools. Lather (1992) suggests poststructural theory is in a position to disrupt these regimes of truth by "[producing] an awareness of the complexity, historical contingency, and fragility of the practices we invent to discover the truth about ourselves" (p. 88). We can see here a similarity to critical theorists' parallel commitment to disrupting master narratives and challenging any claims to an ultimate truth.

Foucault's conception of biopower is useful here as well (Foucault & Rabinow, 1984). The story of Haiti is often told through statistics. Illiteracy rates, infant mortality rates, levels of poverty; these numbers are often used to define Haiti, to label Haiti as a "developing country," a country in desperate need of help from the already "developed" world. Foucault argues, however, that the statistical rendering of people can be used as a tool of power and control, and he illustrates the ways in which these statistics ultimately make totalizing discourses about entire populations possible (Foucault & Rabinow, 1984). Indeed, statistics can be used to create narratives from afar and from above; personal narratives, on the other hand, can complicate and fracture those totalizing discourses. Stories have the ability to reveal rich, complex, even possibility-oriented realities that can counter deficit perspectives. While there are undoubtedly deep and significant challenges facing the Haitian educational system, there are also noteworthy inspiring and positive stories about education and leadership in Haiti (Bell, 2001; Racine, 1998; Smith, 2001), and it is the goal of this dissertation to add to this important body of literature that highlights the voices and experiences of the Haitians themselves.

Importantly, a major critique of traditional poststructural thought is the male hegemonic assumptions that are embedded in the work of early theorists, such as Foucault. Feminist poststructuralism stems from this dissatisfaction with the traditional theory and creates an important partnership between feminist theory with poststructural theory. In many ways, this theoretical partnership is quite natural; as Scott (2003) argues, feminism needed a new theoretical lens to help its scholars think in "pluralities and diversities rather than unities and universals" (p. 378). Additionally, as Scott explains, both feminism and poststructuralism are late 20th-century developments, and both "share a certain self-conscious critical relationship" (p. 378).

The merging of feminism and poststructural theory has produced important theoretical developments that are central to this dissertation research. For instance, feminist scholars highlight the ways in which the researcher's positionality, social location, and related biases impact the research process—all the way from the creation of research questions through to the presentation of the research. There is also a particular focus in feminist methodologies on the ways in which power dynamics influence the research relationships. As such, reflexivity becomes a key component of the research process in feminist scholarship. Another central commitment of feminist methodology is the focus on advocacy.

Importantly, however, as Lather (1992) argues, this commitment to advocacy does not make feminist research any more ideological than more traditional forms of research; indeed, positivistic research is no more neutral than a feminist approach to research—both methodological approaches are value laden. Finally, in feminist research methodologies, issues of voice and representation are brought to the forefront (Orner, 1992; Ellsworth, 1989). With this disciplined critique of traditional positivist methods, a keen awareness of the effects of a sexist society, and an openness to internal critique, feminism has supplemented the poststructural theoretical perspective in important and productive ways. As will become clear throughout this dissertation, I have been deeply influenced by feminist poststructuralism, from my initial research design, to the relationships I developed with the participants, and all the way through to the writing of this dissertation.

This section has provided an overview of the two main theoretical umbrellas that have shaped this dissertation research (critical theory and feminist poststructuralism), and throughout this dissertation, the various ways in which these broad perspectives have helped to shape my research will become clear. Specifically, critical race theory, postcolonial studies, and critical historiography (each of which has emerged from the above theoretical umbrellas) have each helped to shape this dissertation research, and as such, each of these theoretical perspectives will be utilized throughout this dissertation. Finally, before moving on to the methodology, it is important to also explicate my understanding of power and discourse, as these two concepts have played a central role in this dissertation research. Kincheloe and McLaren (2000) define power as "a basic constituent of human existence that works to shape the oppressive and productive nature of the human tradition" (p. 283). In this definition, we can see the focus not only on the critique of the dangers of power but also on the more positive possibilities of power. This dissertation is guided by a similar understanding. Power is all around us and has been used to shape much of the general public's understanding of Haiti (pre- and post-earthquake); it has also been used to shape the allocation of money and resources in the post-earthquake reconstruction. Power can also be used, however, to push back and resist the deficit positioning of Haiti and the related modes of "intervention."

Foucault (1980) argues that power is not an entity that some people or institutions have while others do not. In fact, he rejects the powerful-powerless binary and instead suggests that power is all around us: As Foucault explains, power only exists when it is put into action. In other words, power exists in relations and networks between people and groups, through strategies and through acts of resistance. Acts of resistance are particularly useful as a point of analysis, because, as Foucault explains, in any power relationship, the one over whom power is exercised must "be thoroughly recognized and maintained to the very end as a person who acts: and that, faced with a relationship of power, a whole field of responses, reactions, results, and possible inventions may open up" (p. 789). In other words, Foucault argues that power does not exist without resistance.

Discourse is deeply connected to power and can be understood as both "social and ideological," as a way of talking that "represents a particular version of reality and defines the kinds of identities that reality permits" (Wall, 2004, pp. 313–314). Said (1978) illustrates the dangerous and even violent ways in which Western nations essentially discursively created a "reality" about an entire group of people—and, as he argues, this discourse ultimately created the rationale for colonialism. Bhabha (1990) points to an inherent contradiction in this Western discourse; as he argues, the Western discourse is shaped and controlled by a need to prove, a need for logic, a need to access an objective reality; however, the Western discourse is based largely on stereotypes and assumptions about large groups of people—stereotypes that are devoid of logic and cannot be proven. Clandinin and Rosiek (2007) point out, however, that although discourse does shape our understanding of reality, discourses can be productively challenged, and through that, our conceptions of reality can shift. This dissertation, therefore, seeks to productively challenge the discourse about Haiti, the Haitian government, and Haitian leadership. I do so by highlighting the experiences of seven Haitian leaders as they have been working to create systemic change in the post-earthquake context. In the sections below, I elucidate my research methodology broadly and highlight the quite specific methodological decisions that have informed the research design, data collection, and data analysis for this dissertation.

Methodological Choices

It is important to begin by providing context about the broader field of educational research, in order to ground the methodological choices for this research in a particular paradigm. The field of education has experienced a number of important paradigmatic shifts over the years, each with significant methodological implications. Specifically, in the early 20th century, educational research began its longstanding and quite influential relationship with positivism, and the impacts of this relationship have lasted to this day. While positivism has undoubtedly left a strong mark, educational research has been influenced by a number of other factors as well. Over the course of the 20th century, educational research has been shaped by a variety of disciplines (primarily philosophy, psychology, anthropology, and sociology), a number of fields (e.g., women's studies, ethnic and cultural studies), and a number of theoretical perspectives (behaviorism, sociocultural theory, critical theory, and poststructuralism, to name a few). Educational research continues to be shaped by each of these theoretical, disciplinary, and interdisciplinary spaces, and the distinct nature of these spaces has sent the field of education into a variety of directions. As Marshall and Rossman (2006) argue, this has resulted in the lack of internal coherence in the field, and as Denzin and Lincoln (2005) explain, we are left with "an embarrassment of choices [that] now characterizes the field of qualitative research" (p. 20). This plethora of options has created internal debate about research, and the lines between what is and what is not considered valid educational research have remained unclear.

Although positivism has enjoyed a strong hold on the field of education, it has not been met without strong resistance. For example, the discipline of anthropology has had a significant influence on the field of education, specifically by bringing qualitative research into education. The discipline of anthropology brought important methodological alternatives (i.e., ethnography) to the positivistic, experiment-based methods of psychology that had been primarily shaping the field of education. This post–World War II addition to the field of education opened the doors for what has come to be possible. As Gitlin (1994) notes, however, qualitative research dissertations were still rarely accepted even as late as the 1970s; in fact, it wasn't until the early 1980s that there was more of a common, although "begrudging acceptance" of qualitative research (Gitlin, 1994, p. 1). This growing acceptance of qualitative research was supported by the growing popularity of critical, feminist, and poststructural theory in the 1960s and 1970s.

Unfortunately, however, positivism is still very much alive and well in educational research. Kincheloe and Tobin (2009) are very clear about this, as they highlight the dangers inherent in claims that positivism is no longer influential. In fact, Kincheloe and Tobin argue that positivism is so embedded in Western culture that its influences on educational research have easily become invisible. Because these influences have gone unnoticed in many ways, educational research with positivistic underpinnings has come to be seen as objective while other forms of research are often unfortunately considered second-rate. In collecting data for this research, the strong (and invisible) grasp that positivism has on research—even qualitative research—became quite apparent. The interview excerpt that is highlighted at the beginning of this chapter illustrates this well. In this group interview, two of the participants grapple with the notion of objectivity as it relates to subjectivity, and they make noteworthy claims about what (and who) is and can be considered objective. In another conversation with a participant, he asked about the model I would be using to quantify the concept of leadership. When I responded that my research was purely qualitative, he said, "Yes, I know, but everything that is qualitative can be quantified" (Fieldnotes, 6/20/11). Positivism continues to have a strong influence in educational research, and unfortunately, the systems of power behind research methodologies often remain unseen. Foucault's (1980) conception of regimes of truth can be applied here to understand the systems that—almost secretly—guide our sense of what are and are not acceptable forms of research. Over the years, positivism has been so central to research in a number of disciplines, and the idea that objectivity is possible, for example, has essentially become internalized. A commitment to countering and questioning this deep and often invisible influence of positivism has been fundamental in the design, implementation, and writing of this dissertation research.

I now shift into a focus on the specific methodological choices that have guided this dissertation research. Importantly, within the qualitative research paradigm, there are a variety of methodological options, with narrative research being but one of those options (Creswell, 2007; Patton, 2002). For my dissertation, I chose a narrative research methodology for a number of reasons, both political and epistemological, and as this chapter progresses, the foundations behind my various methodological choices will become clearer, particularly in relation to my larger theoretical framework but also in relation to the specific methods that were used to collect and analyze data for this research. I begin, below, with an overview of hermeneutic phenomenology, in order to provide the necessary philosophical grounding for my methodological choices.

Hermeneutic phenomenology. Phenomenology is a philosophical approach to research that uses the perspectives of individuals as the entry point for understanding the larger significance of a specific phenomenon (McConnell-Henry, Chapman, & Francis, 2009). Such an approach has both epistemological and ontological implications, as it presumes that knowledge about phenomena comes from the individuals who have directly experienced it. As Heidegger (1953/1996) explains, "Ontology is only possible as phenomenology" (p. 36). In other words, reality is constructed, and thus must be understood, through the perspectives of individuals.

Husserl (1913) is considered the founder of phenomenology, and he believed that researchers could, and should, bracket—or put aside—their own beliefs and assumptions in the process of phenomenological research. However, Heidegger, a student of Husserl, challenged the notion of bracketing and extended phenomenology by emphasizing the importance of the researcher's perspective and interpretation (McConnell-Henry et al., 2009; Nakkula & Ravitch, 1998). As Heidegger (1953/1996) explains, "The methodological meaning of phenomenological description is interpretation" (p. 38); thus, Heidegger argues that the researcher's interpretation—which is necessarily shaped by his or her past experiences, social location, and the broader historical context—is fundamental to a phenomenological approach.

Importantly, hermeneutic phenomenology (Heidegger, 1953/1996) distinguishes itself from Husserl's phenomenology in an additional way; while Husserl believed that phenomena could be reduced to essential meanings, Heidegger critiqued the notion of "reduction," suggesting instead that phenomenology creates an "expansion of horizons," illustrating there are multiple points of view and that any one perspective can never represent the entire story (Nakkula & Ravitch, 1998). Such an approach to phenomenology is central to this research; rather than attempting to reduce the experiences of the participants to a certain essence, I see this research as creating an "expansion of horizons" in order to see and attempt to understand the various perspectives and experiences of the seven participants.

As Patton (2003) explains, phenomenology "[focuses] on exploring how human beings make sense of experience and transform experience into consciousness, both individually and as shared meaning" (p. 104); as such, it involves a methodological approach that captures how people "perceive [a phenomenon], describe it, feel about it, judge it, remember it, make sense of it, and talk about it with others" (p. 104). In-depth interviews, therefore, play an integral role in phenomenological research in order to access the "lived experience" of the research participants, and as such, this philosophical approach aligns well with a narrative methodology; narrative research has indeed been deeply influenced by the hermeneutic perspective and phenomenology. Below, I will discuss narrative research and illustrate more clearly why I have chosen a narrative research methodology for this dissertation.

Narrative Research

Story, in the current idiom, is a portal through which a person enters the world and by which their experience of the world is interpreted and made personally meaningful.

Narrative inquiry, the study of experience as story, then, is first and foremost a way of thinking about experience. (Connelly & Clandinin, 2006, p. 375)

As the above quotation illustrates, narrative research highlights people's lived experiences and can be used as a starting point for understanding how people make sense of their lives and for understanding how and why people's stories are shaped and reshaped (Clandinin & Rosiek, 2007). Maynes, Pierce, and Laslett (2008) define personal narratives as "retrospective first-person accounts of individual lives" (p. 1). As the authors highlight, however, narrative research moves beyond simply the telling of stories to seeking to understand the broader significance of the stories by analyzing them in their social and historical milieu. In other words, personal narratives are not strictly "personal" but rather necessarily highlight the intricate relationship between the individual and society; as such, narratives allow researchers to examine the ways in which people's narratives are constructed through "historically specific social relationships" (Maynes et al., 2008, p. 2). Pavlenko (2002) agrees, explaining that "narratives are not purely individual constructions—they are powerfully shaped by social, cultural and historical conventions as well as by the relationship between the storyteller and the interlocutor" (p. 214). In other words, both macro and micro forces are constantly shaping people's stories. These intimate connections between the macro and the micro will be illustrated in numerous ways throughout this dissertation.

This dissertation is most deeply influenced by Sara Lawrence-Lightfoot's conception of portraiture, a specific approach to narrative research. For Lawrence-Lightfoot (1997), portraiture creates a bridge between science and art through "a method of qualitative research that blurs the boundaries of aesthetics and empiricism in an effort to capture the complexity, dynamics and subtlety of human experience and organizational life" (p. xv). Similar to the descriptions of narrative research above, Lawrence-Lightfoot also highlights the importance of context in this method of research. As she explains, portraits must be placed in their social and historical context, as context provides the necessary clues for understanding; indeed, context helps to create a lens through which the portraits can be viewed. Lawrence-Lightfoot emphasizes that it is not simply about presenting the portraits, but the portraits must also be analyzed, and having a solid understanding of the context is necessary for such analysis.

Maynes et al. (2008) note that there has been a recent increase in using personal narratives as evidence in social science research for a number of reasons. For one, narratives can be useful in critical research, as they offer an opportunity to include typically marginalized voices, and by doing so, their stories can become counternarratives that dispute universalizing claims. Additionally, there are epistemological reasons, such as the commitment to counter positivistic paradigms and to highlight the existence of multiple truths, subjectivities, and realities. Reissman (1993) also suggests that the act of narrating can be both a political and therapeutic act: The narrator has an opportunity to (re)tell his or her story and, in doing so, reconstruct the self. Sandelowski (1991) argues that stories "do not simply present, but rather (re)construct lives in every act of telling for, at the very least, the outcome of any one telling is necessarily a re-telling" (p. 163). In essence, both the act of narrating and the result of that process hold significant meaning in narrative research.

Finally, narrative research is not simply a reflection of the participants and their context but is also very much a reflection of the researcher as well. As Pavlenko (2002) argues, "The constructed narrative and subsequent analysis illuminates the researcher as much as the participant" (p. 210). And in her presentation of portraiture, Lawrence-Lightfoot (1997) emphasizes the role of the researcher as well. As she explains, portraits are "shaped through dialogue between the portraitist and the subject, each one negotiating the discourse and shaping the evolving image" (p. xv). Indeed, Lawrence-Lightfoot argues that the researcher is more present in portraiture than any other form of research. As she explains, the researcher's voice is omnipresent in portraiture research; Lawrence-Lightfoot describes researcher's voice as overarching and undergirding the text, framing the piece, naming the metaphors, and echoing through the central themes. But her voice is also a premeditated one, restrained, disciplined, and carefully controlled. Her voice never overshadows the actors' voices (though it sometimes is heard in duet, in harmony and counterpoint). (p. 85)

Ultimately, though the narrative is created in dialogue with the participants, it is the researcher who frames the final work. The researcher's perspective and lens necessarily shape how data are analyzed and presented, and as such, the researcher is present in both explicit and implicit ways throughout the research process and the final product. Importantly, however, the researcher must find a balance, so that her voice does not overpower the voices of the participants. With this dissertation, I hope to strike that balance, creating a final product that is simultaneously representative of myself as the researcher and representative of each of the participants in this research.

Ultimately, I have chosen narrative research not in an attempt to illustrate the "facts" about educational reform in Haiti but rather to gain a deeper understanding of the work and experiences of a selected group of seven Haitian educational leaders. And the goal of this research is not to generalize their experiences but rather to add texture and complexity to the current stories that are circulating about the post-earthquake reconstruction in Haiti. Ultimately, the narratives of these individuals not only highlight a variety of experiences related to educational leadership and post-disaster reform that would likely otherwise remain hidden or obscured by the powerful neocolonial discourse, but their narratives also offer insight into the process of leadership and reform in the incredibly precarious, difficult, and uncertain terrain of up post-disaster contexts more broadly. In the following section, the specific methods that were used for this research are discussed.

Appendix D

Example Researcher
Identity/Positionality Memo

Role of the Researcher

Keon McGuire
December 2013

This attempt at providing both a justificatory and explanatory accounting of my arrival to this topic of study, in many ways, offers an illustrative example of the impossibility of an objective researcher's position. That is to say, my entire life and developmental journey casts a haunting shadow onto this empirical investigation, and as such, it is most appropriate to state that this topic, spiritual identity development, found me.

    While my mother did not raise me in church from the time I was born, my spiritual-religious journey did begin very early in my life. Infrequently, as an adolescent, my mom would take my twin brother, Donté, and I to church. Sometimes for an Easter service, decked out in our best K-Mart, two-tone blue suits with clip-on bowties (insert picture?) and other times enrolling us in a week-long Vacation Bible School, we found ourselves learning (and then reciting) biblical parables, the Ten Commandments, and Jesus' Be-Attitudes. This earliest memories were my entrée into formal, institutionalized Christian faith. However, informally, growing up in a small town in southeastern North Carolina—or the Bible belt—Christianity was literally and figuratively everywhere. The physical presence of cathedrals, storefront churches, and ubiquitously sprawling bumper stickers that read "Jesus Saves" or "Jesus is my co-pilot" were cultural mainstays. And although attending church was not an activity we engaged in regularly, I recall gracing meals (*God is great. God is good. And we think Him for our food. Bow our heads as we are fed. Give us Lord our daily bred. Amen*) and saying bedtime prayers (*Now I lay me down to sleep. I pray the Lord my soul to keep. If I should die before I wake. I pray the Lord my soul to take. Amen*). In retrospect—meaning post-postsecondary and graduate training that has taught me to critically, sometimes suspiciously, interrogate everything about how I know what I know—adherence to these acts of reverence and acknowledgment that of (a Christian) God were some of my earliest scenes of instruction in Christianity. More so, the very fact that these activities were routine for a "non-church-going" family speak to the larger community and familial ethos of what a participant in my study referred to as *culturally Christian*. Meaning, even for those of us who were infrequent attenders at weekly religious services, we still participated in practices reflective of and embedded in a theology of after-lives, souls, and a benevolent (re: great and good) supreme being who provided physical sustenance. While I would soon come to view these as vacant acts of religious piety, it taught me a powerful lesson: to be Black and to be southern was to be (at least somewhat) Christian.

For the sake of not allowing the shadow (my story) to speak too loudly and overwhelm, I will discuss three long-moments stretching from adolescent to present. The first long-moment covers ages 8 to 18. It was at the age of 8 that I received salvation, accepted Jesus Christ into my heart, and was baptized. Looking back, this for sure constitutes a major turning point and defining moment in my life. About a year prior to this moment, I spent the summer at Taylor Holmes Recreational Center, in the north-side public housing neighborhood. It was during this summer that I met Luther H. Moore III, who, in addition to directing the recreational center, also was a taekwondo instructor and Christian minister. Little did I know at the time he would soon become my pastor, godfather, and undoubtedly the most important spiritual advisor in my life. In the meantime, after taekwondo class we would head to the second floor of the center where Mr. Moore would lead a Bible study. It was my first experience with such lesson, but it was exciting. Learning Bible verses and singing gospel songs in that game room-turned-sanctuary—with its pool and foosball tables and chalkboards—led me to the realization that "I need to be baptized! I need to be saved!" At least that is what Donté and I exclaimed to our mother, who at the time was cautiously curious and began to inquire into just what we were being taught.

Fast forward about 8 months and myself, Donté, and my mother, Latanya Howard, are standing next to a cold baptismal pool, dressed in all white, as onlookers sang *"Take me to the water. Take me to the water. Take me to the water. To be baptized."* It was in this Pentecostal-Apostolic community of about 50 consistent parishioners that I grew up over the next 10 years. It would be several years before my (step) dad, Derrick Howard, would join the church, Emmanuel Temple. However, once he did he soon became a deacon, and my mother—who did everything from teach Sunday school classes, collect and count offerings, keep financial records, ushered, and planned events—never really had a title worthy of her distinction. Needless to say, in addition to the 30-mile drive to church, we spent countless of hours at Emmanuel Temple. As my brother and I were bridges to (re) introducing our parents to the gospel, we became living examples of the Bible verse "The children shall lead them." Suffice it to say we loved Emmanuel Temple and soon began to serve as armor bearers (re: mini-assistants) to Elder Moore, which only formalized our apprentice-mentor relationship.

More than anything, it was in the private moments in Elder Moore's church and Taylor Holmes's offices, ride-a-longs as he ran errands for work, and working the half-acre garden that were the most formidable in my spiritual development. Further, as my first Black male mentor, he simultaneously modeled what it meant to be a Black man in this world. Eventually, and perhaps expectedly, my brother I would become youth ministers at 16, participating in a range of ministerial activities, including leading prayer, teaching Sunday school and Bible study classes, and praying parishioners during alter call. I refer to this first long-moment as *structured sincerity*. While I will revisit this concept later in Chapter 4/5 (?), essentially what I mean is that while my experiences were in many ways overdetermined by environmental factors and agents, none of it *felt* forced and, as such, allowed for experiences full of depth, meaning, and agency along my quest for a personal relationship with Jesus through, yet beyond, the rituals. For brevity, I left out moments of frustration, moments of inconsistencies, or even moments of questioning the validity of a God—a minor episode that pales in comparison to my third long-moment. Nonetheless, the larger arc holding this long-moment together is aptly reflected in what was shared above.

The beginning of my second long-moment coincided with my first year of undergraduate studies at Wake Forest University and lasted until my senior year. In contrast to the master narrative that students become less religious in college, my tenure exhibited an intensification of previous experiences. In addition to thinking about course selection, potential majors, and meeting new friends, my brother and I were reflecting on ways we would promote the gospel of Jesus Christ. We were to be both students and (evangelical) witnesses. Our first attempt at the latter involved a biweekly informal Bible study in the room of our residence hall. Specifically targeted at Black men, for half an hour a group of four to eight of us would meet to review a Bible passage and then pray. Intimate and personal, this provided a space for those of us who were Christian-identified to maintain a level of commitment to values that we were raised with in a community of support. However, this group lasted through the fall of our first year.

After this group dissolved, I spent the remainder of the semester attending a few churches with several upper-class students. The defining moment of this long-moment arrived at the end of my first year. Before heading home for the summer, I was invited by my friend and big brother, Cassiel Smith, to attend a weekend retreat at a church I visited once before. My only other interaction with the ministry prior to that moment was a campus Bible study I attended where the youth pastor was teaching. I sat in the back for the majority of the meeting and swiftly exited, disagreeing with some portion of the theology being taught. Little did I know, in just a year's time, I would be standing in her position.

Called an Encounter, this weekend retreat launched my long-moment of intensification. To further clarify what I mean by intensification, I will share an illustrative example by way of a scene from that weekend. The 2.5 days were full of mini-lessons that covered topics such as forgiveness, healing, purpose, music, and media—all in relationship to how we should govern ourselves as *true* believers. The Encounter's apex was on Saturday night when all participants gathered for a final lesson aptly titled *The Cross*. The thrust of the message centered on three conclusions: (1) We were all sinners, (2) Jesus died for our sins, and (3) because he died for our sins, we were all somewhat responsible for his crucifixion. Thus, the question placed before us was, how do we respond with gratitude for such a faithful act of sacrifice? Replete with props worthy of a theatrical production (e.g., 6-foot wooden cross, crown of thorns that was passed around to participants), the message was forcefully preached and ended with a clip from *The Passion of the Christ*—itself an *intense* visual (re)dramatization of Jesus' death. In the midst of the clip, deep wells would begin again as we were instructed to not turn away from the gruesome consequences of our savior's love-gift. Laid across the alter were 12-inch-long, 1-inch-wide nails that were painted red three-quarters of the way up the nail, indicative of blood. We were then told grab a nail once we were ready in order to never forget what great price was made on our behalf. For you, the reader, imagine the imprint such an experience branded on one's conscious. This is intensification.

I refer to this second long-moment as my *zealot pursuit of authenticity.* Energized by *my encounter,* I returned to Wake Forest for my second year with both a revitalized commitment to evangelism and dogged dedication to *becoming* a true believer. Alongside my brother, I doubled down on sharing my beliefs in hopes of converting peers on campus—at one point standing on the dining hall table asking my peers to join me in prayer—and street witnessing throughout Winston-Salem. I spent more time reading the Bible and praying then I ever had before with the goal of committing at least 2 hours a day to these activities. My ultimate goal was to be an *authentic* believer—an authenticity based less on my interpretation of the Bible and more on what I was being taught. A goal that was ever elusive, concerned with demonstrating outward expressions to mirror espoused beliefs. At the height of this long-moment, I was doing some form of ministerial work 7 days a week. Beneath the certitude I displayed was a humming anxiousness that just possibly I was out of place. However, in my *zealot pursuit of authenticity,* there was no room for uncertainty, and all doubt was met with more fasting, more praying, more witnessing, and more *doing*.

It was only a matter of time before pure psychological stress of such an approach would expire. Ushered in through a painful experience of disappointment from my youth pastor, my second Black male mentor, to whom I had become extremely close and unwaveringly loyal. The hurt threw everything into disarray and created the distance I needed to engage in serious personal reflection. A year and a half before moving to Philadelphia for graduate school, this moment was a turning point and stands as a developmental land post marking an overlap of my second and third long-moments.

I am presently in my third long-moment, or my *reclaiming the gray* phase, where I have many questions but also many "answers." This long-moment started in large part due to my physical distance away from home and family as well as my first extended stay in a city outside of the South. As much as I did not want to fulfill the young-boy-moves-to-the-north-and-drastically-change stereotype, in fundamental ways, I "lost" the religion of my youth. The process of deeply questioning long-held beliefs was frightening but seemed urgent in the most liberal environment I had ever lived. Some questions I had held at bay for years, afraid where the answers would lead. For instance, as I was raised by my mother and grandmother for the first 8 years of my life, I was always suspicious of a theology that seemed to empower my stepfather and disempower my mother concerning gender relations in the home. This was the prerequisite for facing the question: What value, if any, does a text written only by men hold to me? Other queries I had no language or prior experiences to properly frame or prepare for. For example, how did I explain having sex positive or queer friends who were not Christian-identified? Would my *love the sinner, hate the sin* approach still work?

While there were short stints of church attendance, by and large I have not regularly attended church in the last 5 years. I have carved out spiritual-religious practices and community that consists of brunches and music listening-sessions with close friends. I took part, reading a Bible verse, at my friends' same-sex wedding ceremony—an unthinkable act to my family. In fact, my family does not fully know how to make sense of my present spiritual-religious identity, except for somewhere outside of the bounds of a true believer. Yet, I firmly reclaim the Christian moniker as I still draw powerful life lessons and values from Christian traditions and biblical teachings. Further, fingerprints of my Christian upbringing are forever present, whether acknowledged or not. Now, my Christian ethics are conjoined with Black feminist principles, queer sensibilities, and more refined antiracist and anticolonial groundings. For me, this version of being a Christian is less about chasing a version of authenticity that I did not create. Rather, it is about applying Christian principles alongside other ways of being, in my work for equity and justice.

These three long-moments all brought me to this present study. After reading the works of scholars who discuss issues of faith, spirituality, and religion in the context of student development theory, something felt missing. Absent from the discussions of stages were the messy, nuanced, and complicated experiential realities of students' journeys. To me, the theories read like sterile prognoses and diagnoses, preoccupied with cognitive dispositions and, as such, incapable of capturing a certain depth and gravity I knew to be true of my own story of transitions. A hunch that I believed could be true of others' stories.

In some ways, I am cognizant of how my shadows inform the questions I chose to ask and the interpretations I made. When appropriate, throughout my analysis, I will acknowledge areas where I perceive overlap between narratives of others and my own. Yet, there are many other important divergences from my own experiences. I will attempt to make these points of departure clear as well. Still, there will remain some places in which my own shadow exists in an analytical blind spot. For that reason, I have enlisted the assistance of colleagues as a part of a peer debriefing team to share their sense-making of a select few student narratives. Further, each student received and provided feedback on their coded transcript. Lastly, you the reader may perhaps make certain connections between my story and the final product that elude me regardless of my best efforts at critical reflexivity. That is a hope that I hold out and invite.

Appendix E

Example Researcher Identity/Positionality Memo

Casey Stokes-Rodriguez
February 11, 2013

Since my introduction to research as a sociology undergrad at Wagner College, I've been skeptical and disillusioned. All that I have ever been exposed to (and to be fair, sought out) in terms of research has been the normative approaches that adhere to strictly empirical, quantifying data. Researchers go into communities and get what they need using an extractive imperial model. To be a critical thinker is a skillset, one that I did not yet have, nor was encouraged to have at Wagner. I was resistant to "mainstream" research and never considered thinking critically about ways in which it could be conducted differently. It wasn't until last semester in my "Building Community Capacity" course where I was first introduced to the idea of participatory action research, ultimately prompting me to seek out this course. I find myself wondering if I will ever feel comfortable in the role of a researcher.

As a social work student at Penn, I have developed a concern with how the perpetuation of power imbalances can be manifested at both the clinical and macro levels. I've come to realize and now pay more attention to the idea that we all have biases of which we are not aware. I have also developed a heightened sensitivity to the privilege I carry with me as a middle-class, White, American woman. During my time abroad, the concept of *gringa* constantly reminded me that my appearance alone elicited strong feelings in others. I have spent a lot of time over the past 2 years reflecting on how I am perceived within different communities I am working with. Much of this reflection stems from time that I spent in South America.

After graduating from college in 2009, I spent 5 months traveling through Peru and Ecuador, spending the majority of my time in the small town of Ayacucho, Peru. This trip was my first experience traveling abroad without any connection to an agency. My reasons for this nonaffiliation had to do with economics rather than my realization of the often imperialistic ways in which international volunteer agencies carry out their work. Volunteering internationally is often incredibly expensive, which would not be as problematic if the majority of the funds went to "local" communities instead of circulating back to the sponsoring agency. Nonetheless, I was able to experience what it meant to go into a community as a complete stranger, unattached to an organization. It didn't take long to realize that as an American, I will always be attached by others to the United States. The contradiction of trying as much as possible to disassociate myself from the imperialism associated with the United States while simultaneously clinging to the utility it may bring is a phenomenon that I think about often. I believe these thoughts are part of what Appadurai refers to when he encourages "uncomfortable thought" regarding the "awkwardness" and "creative tension" that comes with being a traveling researcher (Kenway & Fahey, p. 29).

My love of the Spanish language paired with a freeing sense of escape and adventure inspired my initial travels to Peru. I was "attracted like moths to a flame" to the idea of international travel (Kenway & Fahey, p. 29). After months of searching online, I stumbled across a Belgium woman's blog referencing "La Casa Hogar Los Gorriones," a small private orphanage in the town of Ayacucho, Peru. At the time, the casa had no formal webpage or contact information. I emailed the "founder" of the casa, a Belgium man, and after several conversations, he invited me to come and spend time as a volunteer at the Casa Hogar while in Peru. I went with no plan and no real sense of what my role would be but was fascinated at the idea of immersing myself in another culture. My first trip to Peru, particularly time spent at this orphanage, was an incredibly transformative time in my life and continues to affect me in numerous ways in the course of my daily life.

There was a certain sense of unexplained (probably ignorant) guilt on my part after returning from my first trip. The intense connection that I made with the children, the overwhelming humility I left with, the desire to stay connected were all things that weighed heavily on my mind. Explaining to family and friends what I was doing and why I was in Peru was a difficult task. My fanatic Christian uncle repeatedly sang my praises on my Facebook page stating how wonderful it was that I was doing "mission work" in Peru. My grandmother constantly applauded me saying "those kids are so lucky to have you."

I felt an intense need to clarify that I did nothing extraordinary. To this day, I'm very conscious about the way in which I describe my experience to others. When asked, I usually state that I have "spent time at a Casa Hogar in a small town in Peru." I've come to realize I'm not comfortable articulating my experience in South America to others. Through conversation with other friends and colleagues who have traveled abroad, I've learned that I'm not alone in that sentiment.

It's difficult to describe the deeply personal transformation that comes from getting to know people of a different culture in a different country. There is a certain connotation that one associates (at least I associate) when an American student states that he or she volunteered at an "orphanage" in a "Third World" country. Essentially, this is what I did. However, I am keenly aware of the far too frequent exploitation that happens in these situations and the sense of "high morality" that people seek to obtain from "international charity work."

Conversely, I now believe that I unfairly maligned the orphanage in my initial descriptions to others, describing the "awful" conditions and focusing on the deficits versus the strengths. The following are all true statements: Ayacucho is the poorest region in Peru. The children at the casa were almost all found abandoned on the streets left to die. There is a hole in the ground instead of a toilet. The kitchen is in a dirt hut. The women who devote their lives working at the orphanage barely get paid.

These descriptions are no exaggeration; however, it is worth examining why I felt the need to *"justify"* my reason for being there. Leading with these descriptions perpetuated the "power geometries of globalization" (Kenway & Fahey, p. 12). I was coming from the United States to provide assistance to this "poor community in need." Although I genuinely feel that I gained much more than I gave in that experience, by communicating the environment to others in such a deficit-oriented manner, I was reinforcing imperialistic behaviors. I appreciated the caution to "constantly reflect on whose voices you are suppressing when you speak" (Ghose, 2013). In an international context when members of the community are not physically present, you have a heightened ethical obligation to be more sensitive and aware of how you represent the individuals in as authentic a way as possible.

I later had an additional opportunity to travel to the casa to conduct research in 2011. I approached the research coordinator of SP2 to learn more about the process of creating my own research topic. I would have gotten a stipend to cover a large portion of my travel expenses and been able to use my research abroad in lieu of taking the Introduction to Research course at Penn. I specifically remember grappling with the decision to "conduct" research "on" people who I had come to know and love. I didn't have the tools at the time to understand why this felt so wrong, but I ultimately declined the research opportunity and just went to visit on my own. Had I known about PAR at the time, I would have spent more time considering the possibility of beginning a collaborative research process with the people of the La Casa Hogar Los Gorriones.

Upon reflection, my immediate disdain for conducting normative research and lack of exploration to imagine conducting an alternative type of research could be defined as a "compliant imagination." I rejected the opportunity because I was unable to imagine any type of research beyond "the type that claims that reason and rationality alone are what steer research" (Kenway and Fahey, p. 8). In the same vein that I consider myself having a compliant imagination for not even considering that another type of research was possible, I was simultaneously on the brink of a "defiant imagination" for rejecting the proposition of carrying out the normative research (Kenway & Fahey, p. 8).

Since my first trip in 2009, I've returned to the orphanage twice for a substantial amount of time. I feel that after 4 years of knowing, engaging, loving, and learning from the community of La Casa Hogar Los Gorriones, I have developed a sense of trust within the community. These kids ignited a spark within me that has kept me in close contact since my initial visit. There is now Internet access, and I communicate at least once a month with the kids and senoritas via Skype. While I'm away, I've thought about ways to stay connected to Ayacucho. The casa continues to struggle with being able to sustain itself. When asked "What would be helpful?" Papa Gil, the orphanage founder, continuously emphasizes the need for funds. I've held numerous fundraisers for the casa and send a check over at least once per year. I say all of this not to boast but to emphasize that despite my connection to the casa, it is not until recently that I've felt a level of reciprocal trust and communication where I could consider doing research and not feel exploitative.

With my intense love and loyalty to the orphanage comes a level of critical observation. It is important to understand the dynamics of a community before attempting to engage in research. I have control over how I conceptualize my role as an outsider and a researcher and how to ideally, transparently, and openly negotiate that with the community. I have no control over (nor should I) the power dynamics within the organic community. For the purpose of storytelling, I have been romanticizing and loosely referring to the "Casa Hogar" as "the community." I found it helpful to reflect on how I am "essentializing the word community as a homogenous entity where people have egalitarian interests to produce knowledge, work with partners and decide on matters of common good in undisputed manners" (Bowd, Ozerdem, & Kassa, p. 6). The orphanage was originally "founded" by Papa Gil and his wife. There are 27 children living in the casa as well as 10 senoritas and five cooks who are employed there. Over the years, my conversations with the senoritas have revealed that Papa Gil is not always the easiest person to get along with. As an outside volunteer, I perceived the casa was governed in a "horizontal," egalitarian manner. The senoritas alluded to the fact that more times than not, their opinions are not considered when making decisions regarding care of the children or management of the house. Ultimately, Papa Gil has the final say.

It is difficult to view a man I respect and admire through a critical lens without feeling judgmental; however, it is essential to reflect on how complicated the power dynamics of the community are. Using the frame of postcolonial theory, Papa Gil and his wife, although with the best of intentions, came into a community and created the kind of orphanage they wanted. Although I was not around in 2001 when the orphanage was founded, I can imagine that with the initial establishment of this power imbalance via a "savior mentality" came confounding feelings of gratitude and resentment from the native Peruvians living in this particular barrio. How does power shape the personal and collective relationships within the orphanage?

Participation is an integral aspect of PAR. It would be essential to understand who would be able to fully participate in the reflexive process and whom I would be perceived as aligning myself with. If the "method" is the "product" in this research, my assumption would be that all parties affected by the research would want to have a voice in the decision-making process. It's simple to understand that a goal in engaging in this type of research would be to give voice to the community. The crucial and difficult part is having a comprehensive understanding of how the community self-identifies and developing a keen awareness of how the power dynamics play out within that community. Furthermore, how am I perceived by the community? I have made assumptions that a level of trust has been established and my relationships with the members of the orphanage are based on reciprocity. Would my relationships change if a research component was added to my visits? What biases do I enter the community with that I am still unaware of? Am I unknowingly perpetuating imperialistic, colonizing thoughts and behavior?

My goal would be to help facilitate a change at the casa *if* a problem is addressed by the people who live there. Another goal would be to conduct research that identifies the strengths of the work being done with the physically challenged children at the casa in hopes of securing more funding for the work. I feel that strength of mine is that I have a deep gratitude and devotion to this community and would be prepared to make a long-term commitment in hopes of commanding resources. Among numerous other details to carefully examine would be whether or not conducting research for this community is realistic given the close emotional connection and sense of loyalty I have with the people living there.

Much of my reflection and skepticism stems from not knowing where the boundaries lie. How close is too close? The more I learn about techniques and values of PAR and critically engage with more "radical" researchers, the less isolated I feel in my thinking regarding research. In the social work field, and recently in my research class, it's been emphasized that too much of an emotional connection with your work compromises the process and invalidates results.

In discussing this conundrum of complete objectivity and detachment within the social work client-therapist dynamic, an acquaintance shared with me an interaction with her therapist. She related that when dealing with her feelings of abandonment during a divorce, one of the most moving and helpful experiences was "when my therapist cried with me."

I'm sometimes concerned that because I become emotional when thinking about the injustices and poor treatment of individuals I work alongside of, it means that I haven't yet reached the point where I can "do something" about it on a systematic level. As if (in the next "stage") I'll become numb to things that are happening—and when that occurs I'll be able to fully focus on "solutions."

I've been reflecting on this idea in the social work field and normative research field that emotions need to stay disconnected from work. In a recent incident at my internship, I was working with a young man from Guatemala. He fled from Guatemala a year ago, was apprehended at the border, and placed in the foster care system via the Division of Unaccompanied Children's Services (DUCS) program. This teenager is incredibly resilient and optimistic, despite his lifetime of abandonment and struggle. He recently learned that he will again be displaced from his new family. The uncertainty he faces and injustices inflicted on him by the larger system anger and frustrate me. I think it less has to do with irrational thoughts and not knowing how to manage emotions and more to do with the increasingly overwhelming understanding of the role of imbalanced power dynamics in the world. This is something I continue to grapple with as I reflect on the interconnectedness of my current work and work abroad.

Therapy, much along the lines of PAR research, needs to be a reciprocal process. I understand that there are boundaries both within the research process as well as within the therapeutic process. I agree that it is so important to realize boundaries within any relationship, whether it be working with a community conducting research or engaging in a one-on-one therapeutic process. My issue lies in the idea that boundaries are concrete and objective. I believe this becomes especially problematic when working with increasingly marginalized communities, or immigrant communities within this country, or working with communities in other countries that have a specific set of cultural norms. The need to be flexible and creative in one's thinking is critical because many times these specific communities don't fit inside established boundaries.

A strength that I will bring to research is the genuine belief in the feeling of interconnectedness. I strongly believe in the concept of reciprocal transformation and dialectical growth when engaging in this work. With that belief comes an understanding that the "community" must identify you as a collaborator first and foremost for reciprocal transformation to take place (Ghose, 2013).

I recently read an African proverb highlighting the idea of reciprocity. "An anthropologist proposed a game to a group of children in an African tribe. He put a basket full of fruit near a tree and told the children that whoever got there first won the sweet fruits. When he told them to run they all took each other's hands and ran together, then sat together enjoying their treats. When he asked them why they had run like that as one could have had all the fruits for himself they said: 'how can one of us be happy if all the other ones are sad?'"

As I endeavor to become an effective researcher and advocate in my social work practice, I hope to exhibit this same kind of reciprocity.

Appendix F

Example Memo About Refining the Research Question

Susan Feibelman
February 12, 2012

Topic: The development of women educators as school leaders within independent schools

Chair and Committee Members: Dr. Sharon M. Ravitch, Committee Chair

Revised Research Question

How do women prepare themselves to take up key leadership roles in independent schools; how does the gendered nature of school leadership inform this preparation?

Rationale

The pilot study I completed for Qualitative Research Methods I (EDUC 801-107 2010C) examined the ways that women are mentored for leadership roles in independent schools. The findings of this study led me to a broader consideration of the impact gender has on leadership attainment and the positionality of women in independent school settings. The pilot study focused on one form of leadership preparation—the mentor-protégé relationship. Although this professional relationship has been broadly defined in the literature, I discovered the participants in my study were frequently confined to a mentor-protégé relationship that was embedded in the hierarchical, supervision, and evaluation structure of their schools. With the mentor-protégé relationship nested within a patriarchal structure, it has the capacity to reinforce the androcentric culture of independent schools. I will use the research for my dissertation to explore how this patriarchal system shapes access to leadership development for aspiring women leaders.

As already noted, my research question is embedded in a system that promotes social/cultural stereotypes of gender, which are used to define the social power and status of members in a professional community. Thus, women who aspire to leadership positions in independent schools are faced with managing the roles and behaviors that have been culturally assigned to women and that serve as barriers to their professional development/leadership preparation. In addition, independent schools function as quasi-closed systems, which are loosely organized through membership in regional and national professional associations. This organizational framework is both similar to and distinctively different from public school systems that are directly linked to state education agencies.

Research studies that have interrogated the effect of the "glass ceiling" on aspiring female executives has informed a growing body of literature regarding barriers faced by women who occupy leadership roles in public education. This research offers an invaluable roadmap for exploring women's leadership development in independent schools and serves as a point of orientation for my own research.

Literature Review Outline

I. Gendered nature of organizations

a. Statistical narratives and theoretical frameworks from the field of business management and public education describe the enduring impact of the glass ceiling for women.

  i. Describes the pace at which women in the corporate sphere are moving into CEO roles (Barsh & Yee, 2011; Ibarra, Carter, & Silva, 2010; Jackson, 2001; Reinhold, 2005)

  ii. Explores the structures and systems that impact women as they compete against White males for leadership roles (Ibarra, 1993)

b. Administrative organization of schools has been modeled after corporate structure.

  i. Professionalization of roles (principal and teacher) follows gender-based stereotypes (Adkison, 1981; Grogan & Shakeshaft, 2011; Sanchez & Thornton, 2010).

  ii. Patriarchal peer groups serve gate-keeping functions (e.g., Old Boys/New Boys Networks); define appropriate behavior and roles for women (Adkison, 1981; Brunner, 2000; Chase & Bell, 1990).

c. Occupational sex segregation shapes access to school leadership roles (Sanchez & Thornton, 2010; Tallerico, 2004)

  i. Relationship between independent school culture and social structure:

    How does the relationship between the culture of an independent school and its social structure inform the strategy of action used by aspiring women leaders in the school? (Swidler, 1986)

  ii. Call for a feminist poststructural stance to interrogate the gendered nature of school leadership (Chase, 1995; Gilligan, 1982, 2011; Grogan & Shakeshaft, 2011)

Interrogating School Leadership Using a Feminist Epistemology

    a. Moving women from the invisible to visible, as both subject and researcher (Harding, 1987).

    b. Feminist challenges to male-defined pathways to success and androcentric approaches to research methodology and epistemological frameworks supports

        i. The study of women from diverse backgrounds and perspectives (Viswewaran, 2003)

        ii. Documenting women's firsthand experience of becoming leaders (Chase, 2033; Gilligan, 1982, 2011; Harding, 1987; Lather, 1992, 2001; Olesen, 1994)

II. Gendered nature of leadership—women in the role of school leader

    a. Troubling lack of research that examines the experience of White women and women of color as leaders of independent schools; however, an instructive body of research addresses the personal and professional lives of women leaders in public school settings (Gardiner, Enomoto, & Grogan, 2000; Sherman & Wrushen, 2009).

    b. The current body of research paints a picture of aspiring women leaders who

        i. Have limited leadership opportunities (Skrla, 2003; Skrla et al., 2000)

        ii. Encounter professional leadership challenges that map onto institutionalized gender bias (Brunner, 2000; Lord, 2009; Skrla et al., 2000)

        iii. Have few female role models (Lord & Preston, 2009)

    c. Research studies bring to the foreground the experience of women, both White and of color, who aspire to school leadership positions, separating their lived experience from the male-dominated narrative of school leadership (Gardiner et al., 2000; Sherman & Wrushen, 2009)

III. How do women prepare for leadership roles in independent schools?

    a. Subquestions

        i. How do women participate in emergent and prescribed networks of support that further their growth as school leaders? (Coleman, 2010)

        ii. How do aspiring women leaders utilize mentor-protégé relationships? (Gardiner et al., 2000; Jackson, 2001; Mertz, 1987; Noe, 1988; Peters, 2010; Sherman, Muñoz, & Pankake, 2008)

        iii. How does gender bias affect personal and professional choices and strategies? (Brunner, 2000; Searby & Tripses, 2006; Sherman & Wrushen, 2009; Skrla et al., 2000)

        iv. How do aspiring women leaders describe their leadership style? (Searby & Tripses, 2006; Sherman & Wrushen, 2009)

        v. What roles do university programs and professional development through regional and national associations play in the enhancement of leadership skills for women in independent schools? (Killingsworth et al., 2010)

IV. Summary

As discussed by Olesen (1994), qualitative researchers have applied a feminist stance in their interrogation of the ways in which women develop the capacity to form mentoring relationships and construct networks of support that result in greater access to leadership roles in schools (Gardiner et al., 2000; Searby & Tripses, 2006). These qualitative studies have examined variables such as the "lack of networking, few positive role models, and inadequate sponsorship and mentoring among women" (Searby & Tripses, 2006), which are further complicated by the "gender filters" being used along "leadership pathways" that preference males as school leaders (Sanchez & Thornton, 2010).

Extending this interrogation of women's leadership development to independent schools will

i. Open the space for women's voices to become part of the lived experience of independent school leadership (Olesen, 1994)

ii. Build a vocabulary that refocuses the leadership discourse in independent schools

Possible Methods

This will be a qualitative research study that utilizes interviews, observations, and focus groups with the participants in the study, to collect narrative data. These data will be coded and triangulated and subjected to member checks.

Appendix G

Example Critical Incident Memo

Researcher Memo: Critical Incidents

Laura Colket
June 15, 2011

"I am going to speak to you in French from now on." It was during my second trip to Haiti that I started to talk with the leaders about the possibility of participating in my research. The first person I knew I needed to speak with was Dr. Creutzer Mathurin, an extremely accomplished and well-respected leader at the Ministry who has been working at a high level at the Ministry for the past 15 years. His approval and permission was quite important for me; in fact, I do not think I would have proceeded without it. And so, before dinner on a Thursday evening, I had the opportunity to speak with him privately about my dissertation. Our conversation that evening began in English (to that point, the bulk of our communications had taken place in English; this is something I have since come to change), but as I started to tell him about my dissertation that evening, something interesting happened. When I asked him if he would be interested in participating in my research, he paused, looked at me, and switched to speaking in French. He told me that he was going to be speaking to me in French from now on. And then, without my asking, he began to tell me about the days immediately following the earthquake. Tears came to his eyes as he recounted, in French, the confusion, sadness, horror, and chaos of those first few days. He told me about the meetings he had with the minister in the lot where the Ministry of Education once stood. During those meetings, he recounted, they had only a tree for shade and rocks and broken chairs for seats. He told me that they had to wear masks over their faces because the smell of the bodies was starting to take over. Tears started to come to my eyes as well. It was in this context that these leaders had to make decisions about how to try to move forward after the earthquake.

Beyond the details of his story, there was another level of this interaction that moved me. It was in this moment that I began to understand the significant role language would play in my research. His story was so powerful and moving and, though he speaks English fluently, I am not sure he could have conveyed his experience to me with such power if he had told the story in English. His response that evening made it abundantly clear that, although each of the people I was hoping would participate in my research spoke

English fluently, I had to remember that English is not their first language. With the help of my committee, I made the (now seemingly obvious) decision that the participants should choose which language they would like to use for the interviews and that I would have interpreters available as necessary.

Another concern this linguistic moment highlighted for me relates to the everyday reality of these leaders. Although they are working in their home country, working to rebuild the *Haitian* educational system, within a government that conducts its business in French, many of their meetings and interactions with international agencies and organizations take place in English. The intimate relationship between language and power is clear and became even clearer in this moment. I realized I had been taking for granted the fact that all of our partners spoke English and I immediately felt ashamed that the bulk of our interactions had taken place in English. That evening, I made a commitment to myself to no longer fall back on English because it is easier for me but instead to challenge myself to improve my French. And so, that small moment in which he switched from English to French was anything but small to me. His shift in language reverberated and created a shift in my perspective and ultimately in my research design as well.

"I will participate under one condition. . . . " Since I began this research journey, the participants of this research have said certain things that have stuck with me, that have been pushing me forward, constantly reminding me of why I am doing this. One interaction in particular that I will always remember involves another moment in which I was asking a potential participant to participate in my research. Our team had just visited the campus of Quisqueya University, a prestigious university located in Port-au-Prince that was completely destroyed in the earthquake. As we walked past the large tents that were now serving as lecture halls, I was talking to a woman who had worked at the Ministry for years; she was one of the leaders I was hoping would be willing to participate in my research. When I told her about my dissertation and asked if she would be interested, she paused for a moment and then said that she would participate under one condition. My stomach churned. Everyone else had readily said yes; what was she about to say? Maybe I had missed something important; maybe I had done something wrong. But, as she continued to speak, she said, "I will participate under one condition: that you give your dissertation back to Haiti, so that it can become part of the Haitian archives" (Fieldnotes, 2/17/11). I was speechless. In that moment, something important was solidified for me: My dissertation was not simply going to be an academic milestone, it was not something that only I was interested in; my research was incredibly meaningful to the participants, and they believed it would be useful and important to a broader community as well—and, specifically, that it would be important to Haiti. In any moment of stress or frustration I have had since that moment, I have thought back to what she said that day; her words have kept me going through this research.

Interestingly, that moment not only stuck with me, but it apparently stayed with her as well. At the very end of my final interview with her, 5 months after that day, she stopped and said there was something she wanted to add. She went on to explain:

> At some point I said it was very important that you send back the dissertation, that it was my only condition, that we need it in the archives. [But afterwards], I thought, no . . . it made me think of files and drawers and dust. We need the

information to be returned to us, but not just for the archives. For example, when you will be finished, I hope INERE [a Haitian research institute] will be set up, and we can use your research, that you will be able to give conferences here on what you learned. I hope that the knowledge about us will be returned to us. And not just to us as the participants, but to us as a nation. But when I thought about the word *archives*, I just saw drawers and papers, you know, like a morgue, and that is not what I want. (Amelie Samson, interview, 7/3/11)

Whether it was her intention or not, her words have put an additional, very important, pressure on me. This dissertation is not just for myself or for my committee; it is also for each of the participants, and it is for Haiti. And accordingly, this dissertation will be translated into French and Kreyòl so that it can be for Haiti.

Appendix H

Example Researcher Memo

Researcher Memo: The Partnership Between PennGSE and the Haitian Ministry of Education

Laura Colket
May 12, 2011

Shortly after the earthquake, Dr. Sharon Ravitch, a faculty member of the University of Pennsylvania Graduate School of Education (PennGSE), and I, a third-year doctoral student at the time, assembled a team to explore how PennGSE might best assist in post-earthquake education reconstruction in Haiti. After an exploratory trip to Haiti in July 2010, we solidified a relationship with the Haitian Ministry of Education (MENFP) and began working closely with Dr. Creutzer Mathurin, Senior Advisor to the National Education Task Force of MENFP, and Dr. Jacky Lumarque, Chair of the Presidential Commission on Education and Training and President of Quisqueya University in Port-au-Prince, to formalize a partnership and develop the parameters of PennGSE's assistance, both at MENFP and Quisqueya University. Over the past 18 months, we have been working with our partners at the Ministry and at Quisqueya, with our work guided by the goals and requests of our partners in Haiti. And in this section, I provide data to illustrate the participants' perspectives on our partnership with them.

In the following excerpt from an interview transcript, Creutzer Mathurin explains, from his perspective, why he chose to build this partnership with us. Creutzer:

> This partnership with UPenn was fortuitous. After the earthquake, we had many offers from institutions, many universities approached us, saying they were interested in working with us. We had universities from Canada, from the United States, who came, and what struck me was that in the majority of the cases, after we spoke, they asked us where they could find money to work with us. And you know, the situation we were in, we needed institutions that were ready to accompany us in a manner that goes beyond money. Money shouldn't be the first priority. And a second thing that is very important is the chemistry, the personal aspect. You know . . . when we had our meeting, and Sharon [Ravitch] told us about other projects she was already working on. . . . But what struck me the most was that she spent time to understand what did we need, what did we want, what were we already in the process of doing. It was not about telling us what our problems

were. And that was important. She didn't come with something already planned for us. Instead, she told us, this is what we know how to do, this is what we have experience with, this is what we are in the process of doing in other contexts. And so, I listened to her, I spent some time thinking. . . . It was Sharon's approach to really working to understand what we wanted, what we were hoping to do. She had that approach instead of coming to us with something already planned, an idea to sell to us. And also, it was that, well, ok, we knew we would need money to do what we wanted to do, and we knew we would have to find the money, but the others who came focused on needing money at the beginning. So, it was like an opportunity to them. And I understand that, that this can be an opportunity, but it can't be only be that . . . I don't remember exactly when, but I had some exchanges with the former minister and I remember, talking to him about how we need serious partners, partners we can be confident in, partnerships we can be confident in. Because, we weren't in an easy situation. And you know, that was the feeling I had in the exchanges with Sharon. Sharon knows how to gain people's confidence. So, the type of partnership that we are trying to create with Penn, you can't think about it like the others, it is not like the others. (Creutzer Mathurin, Interview, 10/27/11)

It is clear to see from this interview excerpt that this work is deeply personal and highly relational. Even in our first meeting with Creutzer, the Minister of Education, and Jacky Lumarque, Creutzer was struck by Sharon's orientation to her work—and to their work as well; Creutzer was struck by the fact that she started with questions rather than answers, and he was struck by the genuine respect she demonstrated for his and his colleagues' knowledge and experience. Indeed, all of this led Creutzer to instantly feel confident in Sharon, to trust her as a potential partner. After meeting Sharon one time, Creutzer could sense that Sharon was genuine in her approach—that she had come to Haiti as someone to support their work, not as someone seeking "participation" in a project or approach she had already designed.

This first meeting created an important foundation for our work with the Ministry of Education, one that has continued to be based in deep respect and solidarity with our partners in Haiti. Indeed, from the perspectives of the participants in this research, the partnership we have been building with them has made it clear that different kinds of international partnerships can exist. Below, Dr. Sergot Jacob articulates his vision for this new kind of international partnership in Haiti:

You know, we have a dream, and that dream is to devise a new kind of cooperation with Haiti. This is why your university is here, because we want to change international cooperation. NGOs and traditional donors, we have been working with them for years, since the Second World War. . . . And what has been the impact of all that? . . . 76% of the population living on less than $2 a day. I am not trying to say that it is one specific organization or another. What I am trying to say is that Haiti is one of the most assisted countries in the world. . . . And as you can see, development cannot happen without assistance. But this is why the University of Pennsylvania is here, because we want to try to find a new way for collaboration, we want to see what would happen if we changed the model. . . . So, what we need to do is create institutional strength. Capacity building is the master word for us. Once we reconfigure the education system, we will have a new Haiti. (Interview, 7/1/11)

Sergot's words speak to the complicated nature of international involvement in Haiti. As he explains, he is not suggesting that international aid disappear altogether but rather that the focus of that involvement shifts fundamentally to focus on cooperation and capacity building. As Sergot explains,

Mahatma Gandhi once said that everything that you do for me, but without me, is against me. So, the best way is to put our hands together, devise together, you understand what I want, you try to move with me in my way. You cannot develop a country from outside. A country needs to develop itself. (Group interview, 6/24/11)

And so, the vision for moving forward in Haiti (and in similar neocolonial contexts) involves reframing the concept of participation. Participation should not be about the host country participating in projects that are conceptualized, implemented, and assessed by the outside, and international aid should not be based in false generosity that is ignorant of colonial history and blind to current imperialist relationships between the global North and global South.

Appendix I

Contact Summary Form Example

Source: Miles, Huberman, and Saldaña (2014, p. 125)

Contact Summary Form: Illustration (Excerpts)	
Contact type:	Site: Tindale
Visit _____	Contact date: 11/28-29/79
Phone _____	Today's date: 12//28/79
(with whom)	Written by: BLT

1. <u>What were the main issues or themes that struck you in this contact?</u>

   Interplay between highly prescriptive, "teacher-proof" curriculum that is top-down imposed and the actual writing of the curriculum by the teachers themselves.

   Split between the "watchdogs" (administrators) and the "house masters" (dept. chairs & teachers) vis-à-vis job foci.

   District curric. Coord'r as decision maker are school's acceptance of research relationship

2. <u>Summarize the information you got (or failed to get) on each of the target questions you had for this contact</u>

Question	Information
History of dev. Of innov'n	Conceptualized by Curric. Coord'r, English Chairman & Assoc. Chairman, written by teachers in summer, revised by teachers following summer with field testing data
School's org'l structure	Principal & admin'rs responsible for discipline; dept chairs are educ'l leaders
Demographics	Racial conflicts in late 60's, 60% black stud pop; heavy emphasis on discipline & on keeping out non-district students slipping in from Chicago.
Teacher response to innov'n	Rigit structured, etc. at first, now, they say they like it/NEEDS EXPLORATION
Research access	Very good; only restriction: teachers not required to cooperate

3. Anything else that struck you as salient, interesting, illuminating or important in this contact?

Thoroughness of the innv'n's development and training

Its embeddedness in the district's curriculum as planned and executed by the district curriculum coordinator

The initial resistance to its high prescriptive (as reported by users) as contrasted with their current acceptance and approval of it (again, as reported by users)

4. What new (or remaining) target questions do you have in considering the next contact with this site?

How do users really perceive the innov'n? If they do indeed embrace it, what accounts for the change from early resistance?

Nature and amount of networking among users of innov'n.

Information on "stubborn" math teachers whose ideas weren't heard initially—who are they? Situation particulars? Resolution?

Follow up on English teacher Reilly's "fall from the chairmanship."

Follow a team through a day of rotation, planning, etc.

CONCERN: The consequences of eating school cafeteria food two days per week for the next four or five months.

Source: Miles, Huberman, & Saldaña, 2014, p. 125

Appendix J

Contact Summary Form Example

Source: Miles, Huberman, and Saldaña (2014, p. 127)

Contact Summary Form: Illustration With Coded Themes (Excerpt)

CONTACT SUMMARY

Type of contact: Mtg. ____Principles____   ___Ken's office___   __4/2/76__   SITE Westgate
                     With whom, by whom   place   date

Phone _____   _____   _____   Coder MM
       With whom, by whom   place   date

Inf. Int. _____   _____   _____   Date coded 4/18/76
          With whom, by whom   place   date

1. Pick out the most salient points in the contact. Number in order on this sheet and note page number on which point appears. Number point in text of write-up. Attach theme or aspect to each point in CAPITALS. Invent themes where no existing ones apply and asterisk those. Comment may also be included in double parentheses.

Page	Salient Points	Themes/Aspects
1	1. Staff decisions have to be made by April 30	STAFF
1	2. Teachers will have to go out for their present grade level assignment when they transfer	STAFF/RESOURCE MGMT
2	3. Teachers vary in their willingness to integrate special ed kids into their classrooms—some teachers are "a pain in the elbow."	RESISTANCE
2	4. Ken points out that tentative teacher assignment lists got leaked from the previous meeting (implicitly deplores this).	INTERNAL COMMUNIC.

2	5. Ken says, "Teachers act as if they had the right to decide who should be transferred." (would make outcry)	POWER DISTRIB.
2	6. Tacit/explicit decision: "It's our decision to make." (voiced by Ken, agreed by Ed)	POWER DISTRIB/CONFLICT MGMT.
2	7. Principals and Ken, John, and Walker agree that Ms. Epstein is "bitch."	STEREOTYPING
2	8. Ken decides not to tell teachers ahead of time (now) about transfers ("because then we'd have a fait accompli")	PLAN FOR PLANNING/TIME MGMT.

Source: Miles, Huberman, & Saldaña, 2014, p. 127

Appendix K

Example Site and Participant Selection Memo

Susan Feibelman
November 27, 2012

How were participants identified for this study?

I have used the strategy of purposeful selection to identify an initial group of participants from the attendees and faculty who participated in the National Association of Principals of Schools for Girls (NAPSG) Women and Leadership Seminar (October 2011). I had attended the seminar as a participant (October 2005) and knew from this experience that attendees came from a geographically and culturally diverse set of independent schools, self-identify as future school leaders, and are actively pursuing networks of professional support. Through prior arrangement with the executive director of NAPSG, I had the opportunity to solicit participants for my study on the first evening of the conference.

Out of the 40 conference attendees, 12 expressed interest in participating in the study. Approximately 80% of the responses came from women who were employed at girls' schools, although the conference participant roster indicated that only 40% of the attendees were in this category. Because I wanted a diverse sample with respect to age and race, as well as geography and type of independent school, I decided to limit the number of participants from girls' schools. Having selected a preliminary group of five participants from the NAPSG conference, I proceeded to recruit two African American women who had previously participated in my pilot study (Feibelman, 2011). The remaining participants were brought into the study through my professional affiliations and the snowball method of sampling. It is important to note that despite my efforts to recruit a diverse set of participants for this study, 70% of the participants are currently associated with a girls' school (see Table 1 and Table 2).

Irrespective of their current affiliation, the 14 participants have been classroom teachers and/or occupied formal leadership roles in 40 different independent schools, which enables them to draw upon a deep and broad range of experience as independent school educators. Five participants in this study have worked at four or more independent schools, while the average number of schools a participant has been

Table 1   Interview Study Sample[2]

Leadership Role	Head of School	Admissions Director	Division Director	Assistant Division Director	Dean of Faculty	Department Chair	Interim Dean of Students	Diversity Director
	2	4	4	1	1	1	1	1
Age	30–34	35–39	40–44	50–54	60+			
	1	3	5	3	2			
Race	African American	White						
	5	9						
School Population	Single Sex—Girl	Coed						
	8	4						
School Size	PK–12	K–12	6–12	8–12	K–8	5–8		
	4	2	3	1	1	1		
Region	Northeast	Mid-Atlantic	West Coast	South				
	6	1	4	1				

affiliated with is 2.9. Only 3 of the participants are presently employed at the independent school where they began their career, and 2 of the 14 participants have spent their whole career in single-sex girls' schools.

Although more women in the study are currently working at single-sex girls' schools, the majority has multiple years of experience as educators in coed and single-sex environments. It will be worth exploring themes that emerged during interviews in which some participants described a robust mind-set that permeated every aspect of the girls' school culture. They described a school environment that is organized to intentionally develop leadership competencies among women teachers and administrators in order to provide students with accessible authentic role models.

Table 2 Participant Demographic Matrix

Participant Code	Age	Race	Position	School Type	Region	Method of Recruitment for Study
A	50–54	African American	Associate Director of Admissions/Coordinator of Diversity and Multicultural Affairs	Coed K–12	Northeast	NAPSG
K	50–54	White	Lower School Division Head	Girls—Single Sex K–8	West Coast	NAPSG
E	30–34	White	Assistant Middle School Director	Girls—Single Sex 6–12	Northeast	NAPSG
D	35–39	White	Teacher/Department Chair	Girls—Single Sex 8–12	Mid-Atlantic	NAPSG
B	40–44	African American	Middle School Division Head	Coed PK–12	Northeast	From pilot study
C	50–54	African American	Upper School Division Head	Girls—Single Sex PK–12	Northeast	From pilot study
H	60+	White	Head of School	Girls—Single Sex PK–12	South	Professional network
I	60	White	Head of School	Coed School 7–12	West Coast	NAPSG
G	40–44	White	Dean of Faculty	Coed PK–12	Northeast	Snowball
F	35–39	White	Admissions	Girls—Single Sex K–12	Northeast	Professional network
L	40–44	African American	Director of Admissions and Financial Aid	Coed 7–12	West Coast	Snowball
J	40–44	White	Head of Upper School	Girls—Single Sex PK–12	South	Snowball
M	35–39	White	Interim U.S. Dean of Students	Girls—Single Sex 6–12	Northeast	Snowball
N	40–44	African American	Director of Admissions and Financial Aid	Girls—Single Sex 5–8	West Coast	Snowball

Appendix L

Example of Interview Protocols Organized Around Research Questions

May 16, 2012
Susan Feibelman

INTERVIEW PROTOCOL: ASPIRING WOMAN INDEPENDENT SCHOOL LEADER

How do women prepare themselves to take on influential leadership roles in independent schools? In what ways does the gendered nature of these leadership roles inform this preparation?

1. What do you consider to be particularly instructive life-stories that come to mind? Please explain.

2. How did you make the choice to work in an independent school? How many different independent schools have you worked in? Please describe each school (e.g., culture, location, demographics, mission, values, etc.).

3. Prior to your current position, what experiences and/or people have been vital to your development as a leader?

4. In your current position, what experiences and/or people have been or continue to be vital to your development as a leader?

5. What forms of support are available to you in this current position? What are some examples of how you make use of this support?

6. How have you asked for support and how have you developed your own support systems in order to attain your leadership goals?

7. What forms of support have been missing from your leadership development? How has this shaped the choices you are making as an aspiring leader?

What is the relationship between leadership and the school's distinctive culture? How do beliefs about gender and leadership inform this relationship?

8. What do you think is important for an aspiring school leader to understand about independent school culture? What are examples from your current school that come to mind? How would you compare the culture of this school with other independent schools in which you have worked?

9. Is there a formal school leadership team at your school? If so, please describe the composition (e.g., race, gender, age, roles, years at the school, etc.) of the team.

10. How does the team make decisions? What role do you play in this process? What is an example of this?

11. Please describe your day-to-day responsibilities as a school leader at < INSERT NAME OF SCHOOL > . What are some examples? Are these examples similar to or different from those of other leaders with comparable roles in the school? Explain.

12. Have you been encouraged to step into school leadership roles; who encouraged you? What are some examples?

13. Do you have a particular belief system about leadership that guides your decision making in this school? What are some examples of how this is evident in your work?

Do women who are interested in taking on leadership roles in independent schools seek out mentor-protégé relationships, networks of support, and/or sponsorship?

14. Do you have or have you had someone (mentors) who encourages or supports your career aspirations? Or was there anyone who particularly influenced your career development? (Tell me about them and in what ways they have supported you.)

15. How critical was this person (mentor) in helping you attain your current leadership position? What things did they do that were critical in assisting you to secure this position? What was/is the power or networking level of this person?

How do the social identifiers of race, class, and age inform their strategies of action?

16. Please tell me about your family background. How has this shaped who you are today? (E.g., tell me a story about yourself that helps to explain who you are today.)

17. Which of your personal traits has most influenced your leadership style? How do you think these traits contribute to your leadership effectiveness? What are some examples that come to mind?

18. How do your friends describe your leadership style? How do you know this?

19. How do your colleagues describe your leadership style? How do you know this?

20. How would you finish this statement: "If you really understood my leadership ability, you would know_____"?

21. Is there anything else you would like to add to this interview?

INTERVIEW PROTOCOL: WOMAN HEAD OF SCHOOL

How do women prepare themselves to take on influential leadership roles in independent schools? In what ways does the gendered nature of these leadership roles inform this preparation?

1. What do you consider to be particularly instructive life-stories that come to mind? Please explain.

2. How did you make the choice to work in an independent school? How many different independent schools have you worked in? Please describe each school (e.g., culture, location, demographics, mission, values, etc.).

3. Prior to your current position, what experiences and/or people have been vital to your development as a leader?

4. How have you asked for support and how have you developed your own support systems in order to attain your leadership goals?

5. What forms of support have been missing from your leadership development? How has this shaped the choices you make as a head of school?

What is the relationship between leadership and the school's distinctive culture? How do beliefs about gender and leadership inform this relationship?

6. What do you think is important for a head of school to understand about independent school culture? What are examples from your current school that you come to mind? How would you compare the culture of this school with other independent schools in which you have worked?

7. Is there a formal school leadership team at your school? If so, please describe the composition (e.g., race, gender, age, roles, years at the school, etc.) of the team.

8. How does the team make decisions? What role do you play in this process? What is an example of this?

9. Please describe your day-to-day responsibilities as the head of < INSERT NAME OF SCHOOL > . What are some examples? Are these examples similar to or different from those of other leaders with comparable roles in the school? Explain.

10. Do you have a particular belief system about leadership that guides your decision making in this school? What are some examples of how this is evident in your work?

Do women who are interested in taking on leadership roles in independent schools seek out mentor-protégé relationships, networks of support, and/or sponsorship?

11. Do you have or have you had someone (mentors) who encourages or supports your career aspirations? Or was there anyone who particularly influenced your career development? (Tell me about them and in what ways they have supported you.)

12. How critical was this person (mentor) in helping you attain your current leadership position? What things did they do that were critical in assisting you to secure this position? What was/is the power or networking level of this person?

How do the social identifiers of race, class, and age inform their strategies of action?

13. Please tell me about your family background. How has this shaped who you are today? (E.g., tell me a story about yourself that helps to explain who you are today.)

14. Which of your personal traits has most influenced your leadership style? How do you think these traits contribute to your leadership effectiveness? What are some examples that come to mind?

15. Is there anything else you would like to add to this interview?

INTERVIEW PROTOCOL EXPERTS IN THE FIELD/SEARCH CONSULTANTS

How do women prepare themselves to take on influential leadership roles in independent schools? In what ways does the gendered nature of these leadership roles inform this preparation?

1. What are the competencies that school leaders are expected to possess? What are some examples?

2. How do aspiring school leaders acquire this skill set? What are some examples?

3. What if anything might hinder aspiring women leaders from acquiring these competencies? What are some examples?

What is the relationship between leadership and the school's distinctive culture? How do beliefs about gender and leadership inform this relationship?

4. What do you think is important for an aspiring independent school leader or head of school to understand about independent school culture? Who succeeds and are there some examples?

5. Do you have a particular belief system about leadership that guides your decision making as a search consultant? What are some examples of how this is evident in your work?

6. In independent schools, what kinds of work and work styles are valued? Are there types of work styles that are necessary but invisible in the organization? Why? Can you give examples?

7. How does your lived experience in independent schools inform your work as a search consultant?

How do the social identifiers of race, class, and age inform their strategies of action?

8. How do you identify potential candidates for leadership positions? Take me through the process. What are the strengths of this process? Are there any flaws and if so, explain.

9. How is expertise and ability identified during the search process? What qualities are evaluated during the hiring process?

10. What aspects of individual performance are discussed most frequently during this process? Why do you think that is the case?

11. What do you think is the relationship between leadership practices and school culture in independent schools; are there beliefs about gender roles that inform this relationship? What are some examples that come to mind?

Appendix M

Example Dissertation Analysis Plan

Dissertation Analysis Plan

Mustafa Abdul-Jabbar

Total Data Used	**Phase I Study Across Seven DLTs**	**Phase II Probe Into St. Ivy School**
	<ul><li>Presurvey</li><li>Postsurvey</li><li>Weblog recurring survey</li><li>DLT member action plans</li><li>Principal/coach reports</li><li>14 Qualitative interviews (i.e., DLT members)</li><li>Site observations</li></ul>	<ul><li>Presurvey</li><li>Postsurvey</li><li>Weblog recurring survey</li><li>DLT member action plans</li><li>St. Ivy principal/coach reports</li><li>23 Qualitative interviews (i.e., DLT members and diversity of other teachers in school)</li><li>Site observations</li></ul>
Specific Data Used	Evidences for group norms (e.g., collaboration, innovation, relational trust)	
	<ul><li>Presurvey</li><li>Postsurvey</li><li>Weblog recurring survey</li><li>DLT member action plans</li><li>14 Qualitative interviews (i.e., DLT members)</li><li>Site observations</li></ul>	<ul><li>Presurvey</li><li>Postsurvey</li><li>Weblog recurring survey</li><li>DLT member action plans</li><li>23 Qualitative interviews (i.e., DLT members and diversity of other teachers in school)</li><li>Site observations</li></ul>
	Evidences for specific leadership behaviors and practices	
	<ul><li>14 Qualitative interviews (i.e., DLT members)</li><li>Site observations</li></ul>	<ul><li>23 Qualitative interviews (i.e., DLT members and diversity of other teachers in school)</li><li>Site observations</li></ul>
	Evidences for impacts and influences by DL program/program structure	
	<ul><li>Principal/coach reports</li><li>14 Qualitative interviews (i.e., DLT members)</li><li>Site observations</li></ul>	<ul><li>St. Ivy principal/coach reports</li><li>23 Qualitative interviews (i.e., DLT members and diversity of other teachers in school)</li><li>Site observations</li></ul>

Appendix N

Example Data Analysis Memo

Data Analysis and Reporting Memo

Susan Feibelman
August 19, 2012

When constructing my research methods, I chose two interview formats for data collection. The first round was organized as a two-part, individual, semi-structured interview with 12 participants who occupy assorted formal leadership roles in 11 culturally and geographically different independent schools. Following a first interview, which lasted approximately 90 minutes, each participant was provided with an electronic copy of her interview transcript to read prior to scheduling a second, follow-up interview. Utilizing her reading of the interview transcript and subsequent reflections as the starting point for the second interview, each participant was encouraged to suggest topics she would like to pursue during the follow-up conversation. Second interviews lasted approximately 60 minutes and provided an opportunity for the participant and me, the researcher, to pursue themes that emerged during the previous contact.

Within the same time frame, a single interview was completed with two women heads representing two geographically and culturally different independent schools. Although they followed a similar semi-structured framework, they consisted of a single, 90-minute interview. Both participants were provided with an electronic transcript of the interview, but no follow-up conversation is planned.[3]

Each interview was conducted as either a face-to-face meeting, using Skype or meeting in a location chosen by the participant, or by telephone. Approximately one third of the interviews utilized one or more of these three arrangements.

In addition, following my first round of coding and data analysis, I will conduct semi-structured interviews with two participants who work for executive search firms.[4] Employing the principles of emergent design (Creswell, 2009), I will develop a semi-structured interview guide that explores the relevant themes that emerged through interviews with aspiring school leaders and school heads.

I have used the strategy of purposeful selection to identify an initial group of participants from the attendees and faculty who participated in the National Association of Principals of Schools for Girls (NAPSG) Women and Leadership Seminar (October 2011). I

Table 1 Participant Demographic Matrix

Participant Code	Age	Race	Position	School Type	Region	Date of Interview	Method of Recruitment for Study
A	50–54	African American	Associate Director of Admissions/ Coordinator of Diversity and Multicultural Affairs	Coed K–12	Mid-Atlantic	May 1 June 7	NAPSG
K	50–54	Caucasian	Lower School Division Head	Girls—Single Sex K–8	West Coast	June 21	NAPSG
E	30–34	Caucasian	Assistant Middle School Director	Girls—Single Sex 6–12	New England	May 21 July 10	NAPSG
D	35–39	Caucasian	Teacher/ Department Chair	Girls—Single Sex 8–12	Mid-Atlantic	May 17 May 30	NAPSG
B	40–44	African American	Middle School Division Head	Coed PK–12	Northeast	May 6	From pilot study
C	50–54	African American	Upper School Division Head	Girls— Single Sex PK–12	Northeast	May 9	From pilot study
H	60+	Caucasian	Head of School	Girls—Single Sex PK–12	South	May 31	Professional Association
I	60	Caucasian	Head of School	Coed School 7–12	West Coast	June 11	NAPSG
G	40–44	Caucasian	Dean of Faculty	Coed PK–12	Northeast	May 30	Snowball
F	35–39	Caucasian	Admissions	Girls—Single Sex K–12	Northeast	May 24 June 7	Professional relationship
L	??	African American	Director of Admissions and Financial Aid	Coed 7–12	West Coast		Snowball

(Continued)

Table 1 (Continued)

J	40–44	Caucasian	Head of Upper School	Girls—Single Sex PK–12	South	June 27	Snowball
M	35–39	Caucasian	Interim U.S. Dean of Students	Girls—Single Sex 6–12	New England	July 11	Snowball
N	??	African American	Director of Admissions and Financial Aid	Girls—Single Sex 5–8	West Coast	TBD	Snowball
0			Search Consultant				Snowball
P			Search Consultant				Snowball

had attended the seminar as a participant (October 2005) and knew from firsthand experience the attendees came from a geographically and culturally diverse set of independent schools, self-identify as potential school leaders, and are actively pursuing networks of professional support. Through prior arrangement with the executive director of NAPSG, I had the opportunity to solicit participants for my study. Starting with a preliminary group of five participants, I proceeded to build the sample by recruiting two women of color who had previously participated in my pilot study (Feibelman, 2011) that examined how women are mentored for leadership roles in independent schools. The remaining participants were identified through informal professional networks and a snowball technique.

Presently I have completed 10 of the 14 interviews that comprise the first round of data collection. In addition, I have conducted the first interview of the two-part interview format with the four remaining participants. All interviews have been transcribed, and I have completed a close reading of each transcript, making marginal notes, which I then used to construct a contact summary form for each participant (Miles & Huberman, 1994). As I completed each contact summary form, I maintained a running list of possible codes in the margin of the form. I transferred this running list onto each subsequent form, which resulted in a master list of possible codes (see Appendix B).

In the process of reading transcripts and writing contact summary sheets, I began to see the data organizing themselves into distinct categories (e.g., network of support, strategies that led to opportunities, school culture, etc.). In response, beginning with participant "I," I modified the contact summary sheets to include these categories as subheadings for my notes. This adjustment in my approach to reading and writing memos allowed me to notice greater consistency affirming and disaffirming data. I returned to these subheadings when I began segmenting my working list of codes.

Developing a working set of codes was an iterative process that involved writing definitions, culling codes, and refining descriptions (see Appendix C). I first organized my tentative list of codes into a table, which allowed me to view them as a whole, and to identify redundant or ambiguous labels. I then proceeded to write a definition for each code, relying heavily on standard definitions from online dictionaries and synonyms provided by an online thesaurus. Although this first effort at defining terms resulted in some sterile descriptions that were at best loosely connected to my research, it provided me with sufficient raw materials to begin working through a second draft of codes and their definitions.

As part of refining my code definitions, I segmented the codes based upon a set of categories I developed in my contact summary form and cross-referenced them with my research questions:

How do women, who self-identify as aspiring school leaders, understand the strategies they use to prepare for leadership roles in independent school settings?

- Do these women seek out mentor-protégé relationships, networks of support, and/ or sponsorship from "recognized" independent school leaders; if so, then how does this work?

- How does gender influence the strategies women use to prepare for independent school leadership?

This systematic exercise of jig-sawing codes and categories helped me to identify where categories and codes intersected, forcing me to test the definition's fit with each category. For example, the code *Authority,* which I defined as "exercising power or influence over others in a school setting" and placed in the categories of *Independent School Environment* and *Leadership,* I determined to be more applicable to the latter category—*Leadership*— but was not a good fit for the former. Interview data describing the environment of different independent schools touched upon the ways that individuals shape the culture of the place. So I replaced *Authority* with *Influence* for *Independent School Environment* and defined it as "possessing social and/or political capital that enables one to direct the human and material resources of the organization."

Table 2   Theme Matrix[5]

Research Question	Category
How do women, who self-identify as aspiring school leaders, understand the strategies they use to prepare for leadership roles in independent school settings?	Independent School Environment Leadership Strategies Used by Women to Gain Leadership Experience
Do these women seek out mentor-protégé relationships, networks of support, and/or sponsorship from "recognized" independent school leaders; if so, then how does this work?	Mentor-Protégé Relationships Networks of Support
How does gender influence the strategies women use to prepare for independent school leadership?	Participant Self-Description Participant Social Identifiers

In July, my goal was to complete the contact summary sheets for each of my transcribed interviews and to use these notes to build a list of codes, along with writing their definitions. While completing this work, I have gained a deeper understanding of the ways in which emergent design is an essential element of qualitative research. The research methods presented in my dissertation proposal describe a process that moves from in-depth interviews with individual participants to data collection that utilizes a blog as a venue for a focus group discussion with aspiring independent school leaders. This shift in data collection modes was intended to provide multiple opportunities for participants to share perspectives and to ensure the trustworthiness of the data. But pausing to build a set of codes before moving to a discussion group format is helping me to craft discussion questions that more faithfully represent my emerging understanding of the themes criss-crossing the one-on-one interviews.

Appendix O

Example Data Display 1

Susan Feibelman

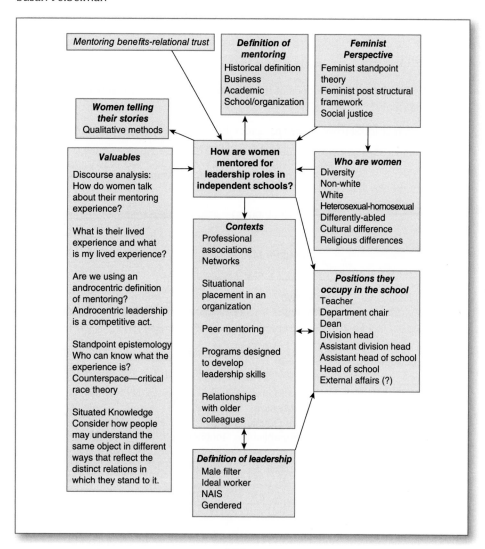

Mentoring benefits-relational trust

Definition of mentoring
Historical definition
Business
Academic
School/organization

Feminist Perspective
Feminist standpoint theory
Feminist post structural framework
Social justice

Women telling their stories
Qualitative methods

Valuables

Discourse analysis: How do women talk about their mentoring experience?

What is their lived experience and what is my lived experience?

Are we using an androcentric definition of mentoring? Androcentric leadership is a competitive act.

Standpoint epistemology Who can know what the experience is? Counterspace—critical race theory

Situated Knowledge Consider how people may understand the same object in different ways that reflect the distinct relations in which they stand to it.

How are women mentored for leadership roles in independent schools?

Contexts
Professional associations
Networks

Situational placement in an organization

Peer mentoring

Programs designed to develop leadership skills

Relationships with older colleagues

Who are women
Diversity
Non-white
White
Heterosexual-homosexual
Differently-abled
Cultural difference
Religious differences

Positions they occupy in the school
Teacher
Department chair
Dean
Division head
Assistant division head
Assistant head of school
Head of school
External affairs (?)

Definition of leadership
Male filter
Ideal worker
NAIS
Gendered

457

Appendix P

Example Data Display 2

Mustafa Abdul-Jabbar

Research Question(s)

1. How do leadership practices of a DL team influence relational trust between members of that team?
2. How is relational trust related to leadership team members' perception of self-efficacy?

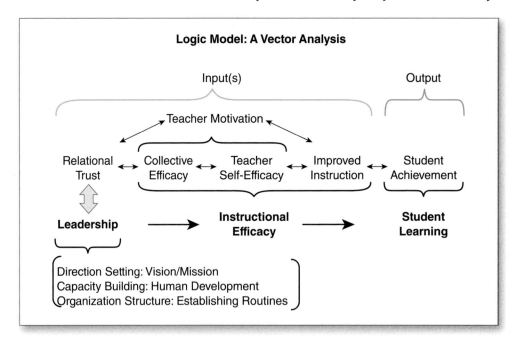

Appendix Q

Example Data Display 3

Emic Categories

Susan Feibelman

Table 1 Connects the Categories That Emerged From the Pilot Study With the Next Set of Research Questions

Categories	Description	Research Questions
Networks of Support	*Some interviewees talked about the benefit derived from meeting with other women who are presently serving in leadership roles. When describing their "ideal" professional network, two interviewees used the term "kinship" to characterize this homophilous group. In addition, some participants expressed their appreciation for having had an opportunity to talk about their leadership journey (Feibelman, 2011, p. 25).*	How do women prepare themselves to take on influential leadership roles in independent schools? In what ways does the gendered nature of these leadership roles inform this preparation? How can networks of support advance the development of women who are interested in taking on influential leadership roles in independent schools? Do women who are interested in taking on leadership roles in independent schools seek out mentor-protégé relationships, networks of support, or sponsorship by "recognized" independent school leaders?
Becoming "visible" as a leader	*Linda Henry (name changed) was the first woman to utter the phrase, but the example she used to define the term was repeated over and over again in anecdotes the women shared. Anne Reynolds (name changed) described her experience as the only woman in the athletic program, which made her a visible choice to serve on schoolwide committees, making it possible for her to gain experience leading;*	In what ways does the gendered nature of these leadership roles inform this preparation? What is the relationship between leadership and an independent school's distinctive culture; how do beliefs about gender and leadership inform this relationship? How do the social identifiers of race and age inform their strategies of action?

(Continued)

Table 1 (Continued)

Categories	Description	Research Questions
	Henry described clerking for the diversity committee at a friend's school, and Lisa Hollis (name changed) described her increasing visibility as an outgrowth of her diversity work in the upper school.	
Pluck or initiative	*One participant described herself as having "pluck," which I interpreted as her way of naming qualities such as ambition, internal drive, and self-confidence. Another interviewee described how she used her initiative as a classroom teacher to gain the recognition and the respect of her peers. This practice earned her the political capital she needed to become acknowledged as a school leader. Yet another participant described how her perseverance resulted in the HOS criticizing her assertiveness.*	In what ways does the gendered nature of these leadership roles inform this preparation? What is the relationship between leadership and an independent school's distinctive culture; how do beliefs about gender and leadership inform this relationship? Do women who are interested in taking on leadership roles in independent schools seek out mentor-protégé relationships, networks of support, or sponsorship by "recognized" independent school leaders?
Pursuit of mentoring	*Without exception, each interviewee's direct supervisor, who is frequently the head of school if the participant is a senior-level administrator, had mentored the preponderance of interviewees. Because independent schools are closed systems, with the division head reporting to the HOS, there is always the possibility of imbalance of power in the relationship. Some women described HOS mentors as being strong, supportive advocates for their growth as school leaders, while other women recognized their school heads to be sources of important information and insights but did not enjoy the benefits of a fully developed mentor-protégé relationship. In the latter situation, the interviewees described the strategies they used to ask for feedback, advice, and direction. A third group of interviewees pursued beneficial mentoring relationships with colleagues.*	How do women prepare themselves to take on influential leadership roles in independent schools? In what ways does the gendered nature of these leadership roles inform this preparation? What is the relationship between leadership and an independent school's distinctive culture; how do beliefs about gender and leadership inform this relationship? How can networks of support advance the development of women who are interested in taking on influential leadership roles in independent schools?

Appendix R

Example Coding Scheme and Coded Excerpts

Mustafa Abdul-Jabbar

DL PROGRAM/THEORY CONNECTIONS
• Connections b/w Distributed Leadership Program/Theory & Greater Faculty or School

LEADERSHIP: TYPES, STRUCTURES, ROLES, ROUTINES, ETC.

A Collaboration/shared leadership (e.g., DL Teams as vehicle for Admin/Teacher collaboration)

B Collective decision making

   ○ By consensus

C Sharing/openness/open floor/part of the process

   ○ "There wasn't a topic that we were afraid to bring up"

D **New DL team structures**

E Team identity & united front

F **Innovation as risk/orientation to risk and change**

G **Innovation as competition/exploitation of hybridity**

H **Role/responsibilities**

I **DL team leadership**

   ○ **Influence**

   ○ **Purpose**

J **DL team vis-à-vis Academic Council**

K **Distinction between DLT and Academic Council**

L **Organizational generational cycle**

M Learning change

N **Action plans**

   ○ **IPADS**

   ○ **PLCs**

RELATIONAL TRUST FACTORS/THEMES

0 Respect

- Genuinely talking and listening to each other
- Inviting, respecting, and valuing others' opinions
- Listening → perceptions of equality
- Honesty (e.g., "I believe he/she is honest")

P Integrity

- Trusting others to keep their word
- Moral-ethical perspective guiding one's work
- Placing the education and welfare of children first

Q Competence

- Fulfilling one's formal role responsibility in a competent manner on behalf of students and staff
- Addressing issue of incompetence

R *Personal regard*

- Willingness to extend one's self beyond formal job definition
- Caring for one another and supporting a climate that fosters personal regard toward each other

---

CODED EXCERPT 1

Interviewer: I'm here with FF. Date is 11/19, 2012. So the first question that I would pose is, what is your overall impression of distributive leadership at the school?

Interviewee: My overall impression is it's one of the programs that has made the biggest impression upon me personally in the ten years of high school education. And I believe it's made the biggest impression upon ##### Catholic High School, and also upon the dynamics of the relationship between faculty and administration that I've ever seen.

Q And I think an obvious example is like the old fashioned faculty meeting after school. No one wants to attend it, including the people running it, the administrators. And generally speaking, are boring. And certainly people will not speak up at meetings.

But through distributive leadership, it's changed the dynamics of one, that relationship between faculty and administration very positively, but it also has impacted the way we approach education, and the whole world of academia. Our teachers, rightfully so, feel much more empowered, much more a part of the educational process.

Interviewer: Thank you. So the next question that I would pose would be, what is your perception of the distributive leadership team members, and I want to say the teachers as leaders here in the school? And based on what you reply, why?

Interviewee:	A	The faculty members at DLT were already leaders in their own right, but just to use that word again, this program empowered them to utilize their talents and gifts as educational leaders. You know, some of them, maybe they felt like they were too young, or maybe some of them felt like they would teach their subjects. But now to be a part of it as a team, #####'s all about success, and now we have a team in place to say let's make our academics even more successful. Let's make our student body, our faculty, more successful. So I think it was just like a natural vehicle that said take off. You know, help us to run the school academically speaking.

Interviewer: Thank you.

A *Interviewee:* And not just help us, but let's run it together.

Interviewer: Thank you. So in terms of the iPad implementation, how do you feel about the idea, and then how do you feel about how that idea is being implemented here?

Interviewee: R I think a big step like that, anytime you take a big leap it's almost like a leap of faith. And it was a little scary and nerve-wracking at the beginning at many levels. Financially. Trying to weigh the positives and negatives. And then would the teachers buy into it?

Interviewer: Have they?

Interviewee: They have. But the teachers bought into it last year through the DLT program, and through running our faculty and development meetings. And trying to get over that nervousness.

Interviewer: Thank you. So I guess another question is, when we look at the ##### Catholic High School, and this is before, before distributive leadership, this is before iPads and that sort of infusion of technology and wiring, and really just sort of the role of the DLT vis-à-vis there's another, as I've been attending the various meetings, faculty meetings, and others, I've run across a group call, they call themselves the Academic Council. And I've been trying to understand what their purpose is. And then trying to understand what the DLT's purpose is, and then understand how they— what's the difference between those two I guess?

Interviewee: I'm not the best person to answer that.

Interviewer: That's good. That's a good answer.

Interviewee: ##### does a good job following the protocol of our positions, so as president, I'm not involved directly in Academic Council, and I'm also not involved directly with DLT. So you know, to the principal and Brian Conway, that might be a better question.

Interviewer: Okay. Perfect. I think that that's all I have for right now. It's really, really short and sweet just to sort of get your overall impression.

Interviewee: No, it's great.

Interviewer: I guess my very last question would be, do you—oh, purpose. That was one last *question on purpose. In your opinion, what is the purpose of the distributive leadership team? That's the last question.*

Interviewee: Well, it's interesting since I've never attended a meeting that I can't answer this. What would I say is the purpose of the distributive leadership team. I would say the purpose is to distribute the leadership of the administration into the faculty. And that's like an age-old conundrum. You know, since some faculty feel maybe some negativity towards the administration. So I really think the purpose is twofold. One is to enhance the relationship generally speaking of the faculty and administration. And then to improve

Q   the distribution of leadership to improve the leadership in the school. And then hence, the natural outcome is to improve teaching, improve test scores, and improve ##### as an academic institution.

Interviewer: Thank you. Thank you very much.

Interviewee: Thank you.

[End of Audio]

CODED EXCERPT 2

Interviewer: This is the part two of ##### #####, October 16th. So you were talking about purpose reason number three. How to make school more competitive.

#####: More competitive. I mean, you know, we're in an open enrollment environment in Philadelphia. Kids can go to any school they want. With us being in the inner city we're obviously attractive to—I shouldn't say attractive, we're available to a lot of different students. What's going to put us aside from them deciding to take two buses and a train to get here rather than at their neighborhood school which they could walk to we have to stay on the cutting edge of you know, what is ##### special? And there's a lot of different reasons. This isn't the only reason. But it doesn't hurt.

Interviewer: Good.

#####: So it definitely gives us a competitive advantage.

Interviewer: Awesome. And my last question, and again, thank you very much, is going to pertain to the connection between the iPad action plan, and distributive leadership. How has the iPad action plan implementation helped your school to carry out distributive leadership in your school?

#####: Definitely. It was such an undertaking that it was something where you were going to need more than just administration as leaders. And we continue to do that. I mean, you know, the way I look at that, we're just on the baby steps of this. I mean, there's stuff here that you know, this

thing has capabilities that I have no idea how to do. And we're going to learn things from the kids on this as well. That's another thing I think about this. Talking about distributive leadership, it goes all the way down to the students.

	Students, today for example, in my AP Bio class, I've been using Drop Box to send out the PowerPoints. One kid says hey, Mr. #####, can you save it as a PDF? Because then if you send it as a PDF, all the other stuff that you say in class, we can write on top of those slides and save it that way. So I said yeah, I can definitely do that. I said, I'm going to have to have someone show me how to do it, but definitely. If it's going to make your lives better, that's a brilliant idea. I like that.
G	So this way I'll save it as a PDF and then send it out, and this is something that a kid came up with. You know, and I think that's something where you know, at a lesser degree I think the faculty is going to come up with things. Hey, this is something I'm doing in my class with this. And I think it just enables a lot of other people to say hey, listen, this is something with the iPads that this is great.
	I just think it definitely, distributive leadership, it's empowered a lot more people to make contributions to making this place a better place. Not just with the team, but with the faculty as a whole.

Interviewer:	Thank you very much. And perhaps we can speak next time and it won't take too long.
##### #####:	Yeah. I mean, I'm always available. That's fine.
Interviewer:	And we can actually follow up how it's hitting the total faculty.
##### #####:	No problem.
[End of Audio]	

Appendix S

Example of a Pilot Study Report

This report is an example of a final pilot study report. Charlotte Jacobs, a first-year doctoral student at the time, conducted this pilot study for an introductory qualitative research class. In addition to the final pilot study report, Charlotte's assent forms are also included in Appendix S.1, her memos in Appendix S.2, a coded transcript in Appendix S.3, and code definitions in Appendix S.4 to help contextualize these processes as well as to document different aspects of the piloting process. References are not included due to constraints on space.

Pilot Study Final Report

Charlotte Jacobs
May 18, 2012

"WHERE MY GIRLS AT?" THE EXPERIENCES OF AFRICAN AMERICAN ADOLESCENT GIRLS AT A PREDOMINANTLY WHITE INDEPENDENT SCHOOL

Introduction: Significance and Rationale

> By insisting on self-definition, Black women question not only what has been said about African-American women but the credibility and the intentions of those possessing the power to define. When Black women define ourselves, we clearly reject the assumption that those in positions granting them the authority to interpret our reality are entitled to do so . . . the act of insisting on Black female self-definition validates Black women's power as human subjects. (Collins, 2000/2009, p. 125)

This powerful and challenging statement is the catalyst for the focus of this pilot study, which seeks to explore and give voice to the academic and social experiences of African American girls who attend elite predominantly White educational institutions. African

American girls occupy a unique place in our society in that their race and gender identities combine to what Collins (2000/2009) defines as the "intersecting oppressions" (p. 26) of being African American and female in a society that historically and currently privileges White males. During the period of adolescence, the identifiers of race/ethnicity and gender become particularly salient. The adolescent developmental stage is characterized by an identity search that seeks to answer the question, "Who am I?" (Nakkula & Toshalis, 2006). For youth, much of that identity search takes place within the school environment where interactions between students, teachers, and parents serve to communicate messages about the meaning and statuses of different identities in our society.

As an African American female who attended a school system where academic tracking began in the fifth grade, I found myself in honors and advanced placement classes that were predominantly White, and I then went on to attend predominantly White higher education institutions. I have often wondered if my academic and social experiences ranging from being "the odd one out," to being a token in the room, to being a role model for younger African American students are specific to my being African American, my being a female, or my combined identity as an African American female. Once I began teaching at an independent school (another predominantly White institution), similar themes of exclusion, academic and social uncertainty, and the positive effects of supportive family and peer groups arose as my African American female middle school students navigated the academic and social terrains of a predominantly White school (Cookson & Persell, 1985; Datnow & Cooper, 1996, 1997).

In response to the growing trend of African American parents seeking forms of education that are alternatives to traditional public schools for their children (Slaughter-Defoe et al., 2012), this study specifically focuses on the experiences of African American females who attend elite predominantly White independent schools. As of the 2009–2010 school year, 9% of the African American student population was enrolled in private primary and secondary school institutions (National Center for Education Statistics, 2009). To this day, independent schools in the United States are still predominantly White institutions. Independent schools are private institutions in which students typically engage in an admissions process and families are required to pay a specific tuition in order for their child(ren) to attend the schools. According to the National Association of Independent Schools (the major organizational and accrediting body for independent schools in the United States) *Facts at a Glance* 2010–2011 statistics, African American students make up only 6% of the total independent school population, while White students comprise 67.6% of the total independent school population.

Past research points out the important influence of schools in the academic and racial socialization of African American students (Arrington & Stevenson, 2012). Since schools serve as "purveyors of sociocultural" knowledge (Brown, 2012, p. 28), it is crucial to explore how African American girls are creating an understanding of what it means to be African American females in environments where, in many cases, they are the extreme minority. The majority of research about African American girls portrays them as one large group, neglecting how the socioeconomic status and the school and home environments can intersect with race/ethnicity and gender to create experiences that are noticeably distinct from one another. That being said, researchers have begun to tease apart the monolithic perspective of African American girls as a single group and have sought to understand the experiences of African American girl subgroups. Janie Ward (1990) and Niobe Way (1998) have contributed much of the scholarship about African American girls in urban settings, but there is minimal research specifically about the daily experiences of African American girls who attend predominantly White primary and secondary independent schools.

The stories of African American middle school girls in elite predominantly White institutions have yet to be uncovered. While the voices of African American boys are gaining presence in the educational research field (Ferguson, 2001; Noguera, 2003), the voices of their female counterparts still remain largely silent. The aim of this small-scale, exploratory pilot study is to serve as a starting point for further research about this particular subgroup of African American girls and their lives as students, peers, and daughters. This paper will begin by introducing the central research questions of the study and then will present an overview of the theoretical frameworks that this study employed to analyze the data. The paper ends with an analysis of the findings and recommendations for future research.

Research Questions

This study takes a phenomenological approach in studying the experiences of African American female adolescents who attend an elite predominantly White independent school. Phenomenology "aims to identify and describe the subjective experiences of respondents. It is a matter of studying everyday experience from the point of view of the subject" (Schwandt, 2001, p. 192). When using a phenomenological lens, the researcher explores "the meaning for several individuals of their lived experiences of a concept or a phenomenon. Phenomenologists focus on describing what study participants have in common as they experience a phenomenon" (Creswell, 2007, p. 57). For the purposes of this study, I am defining the particular experience of being an African American female adolescent in an elite independent school as a phenomenon. For this study, the unique intersectional identity of race/ethnicity and gender combined with the predominantly White school environment creates an experience that is separate from the phenomenon of African American students in independent schools or girls in independent schools. Therefore, a phenomenological approach to this research study is not only appropriate but also necessary.

Following the phenomenological perspective in which participants' lived experiences are the focal point of the research, the guiding research questions for this study are as follows:

1. What are the experiences of African American adolescent girls at predominantly White institutions?

2. In predominantly White institutions, what meaning do adolescent African American girls make about what it means to be to be African American? To be female? To be an African American female?

As part of my methodological approach, I employ the methods of Moustakas's (1994) transcendental phenomenology. These methods included "bracketing" out my own experiences related to the subject of study so that I can view and understand the phenomenon as if "everything is perceived freshly, as if for the first time" (Moustakas, 1994, p. 34). After data collection, I analyzed the data with the goal of finding the "essence" of the experience of what it means to be an African American adolescent girl who attends an elite predominantly White independent school.

Theoretical Framework

Helms's (1986) theory of Black racial identity development, research on adolescent and gender development (Gilligan, 1996; Nakkula & Toshalis, 2006), Collins's (1991/2000) Black feminist thought, and critical race theory (Matsuda et al., 1993; Sue et al., 2007; Tate, 1997) provide the theoretical framework for this study. Together, these theories combine to capture a developmental and critical perspective of the African American female adolescent identity as it is situated within U.S. society.

Black Racial Identity Development

Helms's Black racial identity development theory. Helms's (1986) Black racial identity development theory is an amended version of Cross's (1971, 1978) four/five-stage Black racial identity development model. At the center of the model is a search for what Cross (1971, 1978) terms the "nigriscence identity." "Nigriscence" is defined as "the developmental process by which a person 'becomes Black' where Black is defined in terms of one's manner of thinking about and evaluating oneself and one's reference groups rather than in terms of skin color per se" (Helms, 1990, p. 17). Specifically, the Nigiscence or Black identity models intentionally make a distinction between the aspects of Black identity development that are influenced by racial oppression and the aspects that develop as a part of the typical human self-actualization process (Helms, 1990). This distinction is particularly important in light of this study, for the extreme minority status of the middle school African American female participants at an elite predominantly White school could influence the ways in which their identities are shaped both by experiences of racism and by the typical experiences of adolescence.

Another aspect of Helms's (1986) Black racial identity development model is that she defines each of the four/five stages of the model in terms of the "cognitive templates that people use to organize (especially racial) information about themselves, other people, and institutions" (Helms, 1990, p. 19). Taking this perspective, one's identity is not just about how he views himself in the racial sense but also how he understands and categorizes the races of others. This bimodal aspect of Helms's (1986) theory is evident across the four stages of the model: preencounter, encounter, immersion/emersion, and internalization.

The Black racial identity development model is an important framework to keep in mind concerning the focus of this study, which aims to explore how the African American female adolescents understand their racial identity within the context of their particular school environment.

"Raising resisters." Another aspect of Black racial identity development arises from parental input and socialization around racial identity and issues of race. Ward's (1996) research with urban African American adolescent girls reveals that many African American parents have intentional conversations with their daughters that teach them to resist the prevalent negative stereotypes associated with being African American. African American parents teach their daughters to know when to attribute failure to a lack of individual effort and when to attribute it to societal forces (Ward, 1996). Robinson and Ward (1991) characterize two distinct forms of resistance that African American girls typically display as a result of their focused racial socialization: "resistance for survival" and "resistance for liberation."

"Resistance for survival" offers short-term solutions for how to cope with and understand situations of racism and discrimination. The strategies employed in the "resistance for survival" can take the form of making decisions based purely on emotion rather than a close examination of the underlying factors of the racist or discriminatory experience (Robinson & Ward, 1991). "Resistance for liberation" strategies, on the other hand, provide "solutions that serve to empower African American females through confirmation of positive self-conceptions, as well as strengthening connections to the broader African American community" (Robinson & Ward, 1991, as cited in Ward, 1996, p. 95). The focus of "resistance for liberation" is that it encourages African American girls to critically examine and affirm themselves and their status in society in safe and supportive environments, known as "homespaces" (Ward, 1991). The formation and development of a resistance stance toward racism and oppression in society is a significant concept regarding the focus of this study. As the study participants reflect on their experiences of being African American girls in their school, it is important to note both the messages that they are receiving/interpreting about their identities and who are the primary figures in their lives who are presenting them with these messages.

Adolescent and Gender Identity Development

Adolescent development. Adolescent and gender identity development are also two significant frameworks that connect with the research focus of this study. The fifth stage of "identity vs. role confusion" of Erikson's (1950/1993) identity development model is the stage most closely linked with adolescence. During this stage, " The adolescent expansion of cognitive abilities occasions a heightened awareness of how one's self-concept is linked to such personal characteristics as race, gender, class, and sexuality" (Nakkula & Toshalis, 2006, p. 26). As youth struggle to find answers to the questions "Who am I?" and "Who do I want to be?" they also experience anxiousness in trying to balance the expectations of their friends, family, and society in terms of who they are expected to be (Nakkula & Toshalis, 2006).

Gender identity development. Gender identity is particularly salient during the time of adolescence as youth navigate the messages that they receive from their friends, parents, teachers, and media about societal gender roles (Nakkula & Toshalis, 2006). This finding is supported by the gender intensification hypothesis, which holds that during adolescence, gender-differential socialization escalates in intensity and pushes girls to display stereotypical feminine characteristics (Buckley & Carter, 2005; Hill & Lynch, 1983). In particular, Gilligan (1996) found that girls tend to "go underground" around the time of adolescence by intentionally hiding or downplaying certain parts of themselves and their identity in order to fit into the dominant gender narrative of women as quiet and nonaggressive.

Interestingly enough, this behavior pattern was found to be more common with middle-class White girls than with African American girls. In fact, research shows that the gender roles for African American girls may be more flexible in comparison to White girls (Collins, 1991, 2000/2009; Ward, 1996). Ward (1996) found that African American parents often socialize their daughters to take on both traditional and nontraditional feminine roles and to value traits such as perseverance and strength. Ward (1996) also cites these elements of socialization as an explanation for why African American girls have been found to maintain higher levels of self-esteem than their White counterparts (American Association of University Women, 1992). The self-image concerning gender and overall self-esteem of African American adolescent girls is important regarding this study in light of how the study participants view their experiences in school and how they choose to navigate the academic and social environments of their school.

Black Feminist Thought

Collins's (1991, 2000/2009) Black feminist thought lies at the center of this study because of the way it intentionally addresses the unique intersections of gender and race that are present in the African American female identity. Collins (2000/2009) maintains that the experiences of African American women cannot be viewed through two separate lenses of race/ethnicity and gender, but rather it is the intersectionality of the two identifiers and the way in which society responds to and interacts with African American women that creates a specific way in which the experiences of African American women should be taken into account. Specifically, Collins (1991) holds that "all African-American women share the common experience of being Black women in a society that denigrates women of African descent . . . in spite of differences created by historical era, age, social class, sexual orientation, or ethnicity, the legacy of struggle against racism and sexism is a common thread binding African-American women" (p. 22).

As a critical theory, Black feminist thought goes beyond simply naming the ways in which society oppresses African American women. While Black feminist thought acknowledges the ways in which African American women are oppressed, it also calls for action as a way to dismantle the inequitable systems present in society and seeks to empower African American women to better their status in society through resistance and critical thought (Collins, 2000/2009). In doing so, Black feminist thought values the experiential knowledge of African American women above all else and insists that the sharing and understanding of the lived experiences of African American women is critical to their survival. In line with Black feminist thought, this study places the voices of the study participants at the center of its analysis. The study participants serve as experts of their own experiences, and through the lens of Black feminist thought, the goal is that by sharing their stories, the girls will find ways to empower themselves and others.

Critical Race Theory

Critical race theory (CRT) (Matsuda et al., 1993; Tate, 1997) is another critical lens through which the experiences of the participants in this study will be viewed. As its core principle,

CRT holds that "racism is endemic in U.S. society, deeply ingrained legally, culturally, and even psychologically" (Tate, 1997 p. 234), meaning that racism is a fundamental part of the structural and institutional composition of U.S. society. In light of the fact that schools serve as places of socialization about racial and other societal identities, is it crucial to take a critical view as to what messages students may be receiving in schools. Experiences of race and racism are not uncommon for African American students who attend independent schools (Alexander-Snow, 1999; Arrington & Stevenson, 2012; Horvat & Antonio, 1999). In the Success of African American Students (SAAS) study by Arrington and Stevenson (2012), African American students described that one of the most challenging aspects of attending independent schools was the negative race-related experiences and stress that were even more apparent because of their minority status within their schools. The most prevalent racial stressor for the students in the SAAS study were experiences when it was clear that their peers at school held beliefs that the they would conform to specific stereotypes related to being African American (Arrington & Stevenson, 2012).

Sue et al. (2007) classify these particular types of interactions between White people and people of color as "racial microaggressions." Racial microaggressions are defined as "brief and commonplace daily verbal, behavioral, or environmental indignities, whether intentional or unintentional, that communicate hostile, derogatory, or negative racial slights and insults toward people of color" (p. 271). Following the research on African American students in independent schools, it is expected that the participants have also experienced racial microaggressions at the hands of their teachers and peers. What is important to this study are the ways in which the study participants make meaning of those particular experiences.

Methodology and Research Design

Setting and Participants

This study was conducted over a period of 2 months at Grace School,[6] a coeducational, pre-K through 12th-grade independent school in the greater Philadelphia area. The annual tuition of the middle school at the time of this study (2011–2012) was $28,600 (Grace School website). Currently, the school has 838 students total, and students of color make up 20 % of the overall student body (Grace School website). The middle school has approximately 196 students, but it is unclear the percentage of middle school students who identify as students of color or as African American. At the time of the study, 13 African American girls were enrolled in the middle school: 3 African American girls in the sixth grade, 6 in the seventh grade, and 4 in the eighth grade.

My work as a researcher for a separate project at Grace School presented me with the opportunity to conduct research at this specific school site. I had not met any of the study participants prior to my leading the focus group. In the selection of the participants, I relied on my established relationship with the assistant middle school principal and the Director of Community and Diversity to recruit participants for the study. The participants for the study were selected via letters and personal phone calls from the middle school assistant principal and the Director of Community and Diversity at the school to the families of the participants of interest.

Based on their responses to an introductory questionnaire, three of the study participants were in eighth grade, five were seventh grade, and two were in sixth grade. All of the participants identified primarily as African American, though some identified by additional ethnicities as well. Three of the participants were "lifers," meaning that they have attended Grade School since prekindergarten or kindergarten, three have attended the school between 3 and 4 years, and four participants are new to Grace School and began attending the school at the beginning of the current school year. Of the participants who had attended other schools previous to being enrolled at Grace School, four had attended other independent schools and four had attended public school. Most of the participants belonged to two-parent households where both parents had either graduated from college or had a graduate or professional degree. One participant lived in a single-parent household, and the remaining two participants lived in two-parent households where one parent was a high school graduate and the other was a college graduate.

In addition to the parental/guardian informed consent that was required in order for each participant to become involved in the study, participants were also asked to read and sign a form acknowledging their assent to participate in the study (Miles & Huberman, 1994) (see Appendix S.1 for the Pilot Study Letters, Consent Form, and Assent Form). Due to the fact that the study would require the participants to share their personal experiences, I believed that gaining their assent to participate in the study was essential to the integrity of the study and in creating a relationship with the study participants. Participants were told that they were free to withdraw from the study at any time and that the study presented minimal risk to them (Miles & Huberman, 1994).

Data Sources and Data Collection Processes

Demographic questionnaire. In line with Creswell's (2007) purposeful sampling strategy, all participants in the study completed an introductory questionnaire that asked them to list their age, grade, how many years they had attended Grace School, and other schools that they had attended before Grace School (if applicable). The questionnaire asked participants to first describe their racial or ethnic background in an open-ended format and then asked participants to identify the broader racial/ethnic categories to which they thought they belonged according to the U.S. 2010 census categories. In order to assess the relative socioeconomic statuses of the participant group, the participants were asked to identify the highest level of education of their parents/guardians (Ensminger et al., 2000).

Focus group. The initial source of data collection was an hour-long focus group of all the study participants. The focus group method fits with the overall phenomenological methodological framework in that the study seeks to understand the experience of what it means to be an African American middle school girl in an elite predominantly White independent school. The focus group was also used as a "warm-up" to deeper discussions about topics to which participants could have been sensitive. Researchers have found that focus groups are a useful way to collect data when "the interviewees are similar and cooperative with each other, when time to collect information is limited, and when individuals interviewed one-on-one may be hesitant to provide information" (Creswell, 2007 p. 133; Krueger, 1994; Morgan, 1988; Stewart & Shamdasani, 1990). Since I did not have a previously existing relationship with the participants and because of their relatively young age, I believed that a focus group setting would garner more information at the beginning of the study than structured one-on-one interviews.

When conducting the focus group, I employed Moustakas's (1994) method of asking a broad general question that follows along the lines of "What have you experienced in terms of the phenomenon? What contexts or situations have typically influenced or affected your experiences of the phenomenon?" (Creswell, 2007, p. 61). In relation to the focus of the study, my initial questions were

- What is it like to be an African American girl at Grace School?

- What have your experiences been at this school?

These questions served as gateways for other questions that focused on the interactions between the participants and their teachers and with whom they decide to share the stories of their experiences (teachers, family, friends). I transcribed the focus group discussion from its audio recording in its entirety, verbatim, before beginning data analysis.

Interviews. As a follow-up to the focus group, I conducted one-on-one interviews with three of the study participants that were 15 to 35 minutes in length. According to Creswell (2007), 5 to 25 in-depth interviews usually serve as the primary sources of data in a phenomenological study so that researchers are able to get a full picture of the phenomenon that they are studying from various vantage points (p. 61). Due to the scale and timeline of this study (as an exploratory pilot study in this introductory-level qualitative research methods course), I decided that three in-depth interviews were appropriate. I selected interview participants according to their grade level, how long they had attended Grace School, and their responses during the focus group. In order to attempt to capture the range of experiences present in the study sample, I conducted an interview with a participant from each grade level. Additionally, two of the participants were new to Grace School, and the remaining participant was a "lifer." I also specifically selected two of the study participants because the stories that they shared during the focus group were compelling and I wanted to know more about their experiences. The third participant I selected because she was relatively quiet during the focus group, and I wanted to give her an opportunity to share her stories in a more intimate setting.

The general questions for the interview centered on remaining thoughts that the participants had from the focus group, a discussion of the similarities and differences between their home communities and the Grace School communities, and whether they talked with their families about issues of race. The remaining interview questions were based on the specific information that each participant shared during the focus group. Once I completed the interviews, I transcribed each interview in its entirety, verbatim, from an audio recording before beginning data analysis.

Memos. As a part of my data collection and analysis process, I wrote and reviewed several kinds of researcher memos that focused on my thought process in developing and conducting the study, my ongoing identity and positionality as a researcher, my questions about aspects of the research framing and fieldwork, and the stories and themes that were beginning to emerge from the data. Three of the memos were assignments for the Qualitative Modes of Inquiry course. These memos are as follows: (1) a memo focusing on my researcher identity and positionality as an African American woman working with African American adolescent girls, (2) a fieldwork memo reflecting on the status of the fieldwork of my pilot study halfway through the course, and (3) a vignette memo in which I selected a vignette from my data and then wrote a memo analyzing that vignette according to two of four guiding topics connected to the research process (please see Appendix S.2 for the actual memos). Given that these memos were course assignments, I benefited from feedback on the memos from the teaching team. The feedback helped me to gain insight into my own sense-making processes.

Document review. In order to get a better sense of the context of the research site, I reviewed several artifacts from the school that provided context relevant to my guiding research questions. This included the Grace School website and the school's mission statement and school demographic information found on Grace School's website as well as other related documents from the students themselves such as texts, emails, and other communication artifacts.

Data Analysis

Focus groups and interviews. The analysis of the data from the focus group and participant interviews was based on Moustakas's (1994) phenomenological data analysis and approach. As the first step in data analysis, I read through the transcripts from the focus group and interviews, the memos that I had been writing throughout the study, and the feedback that I received from my colleagues who served as external auditors for the study. As I read through the data, I looked for initial emerging themes and questions that I had in order to begin to gain a deeper understanding of the experiences of the study participants while continuing to reflect on the potential biases that I could be employing as I read the data (Emerson et al., 1995; Maxwell, 2005; Miles & Huberman, 1994). The next step in data analysis was what Moustakas (1994) calls horizonalization. I read through the data and highlighted the "significant statements" or phrases that the study participants made regarding their experiences. Following that, I organized the significant statements into "clusters of meaning" that evolved into the major themes of the data (Creswell, 2007; Moustakas, 1994). The following themes emerged as a result of these first steps of data analysis:

- Thoughts on being African American

- What it means to be African American

- Us vs. them mentality

- Push-back/resistance

- Community

- School environment—isolation

- Money/privilege

- Family

As the last steps of data analysis, I then wrote textural (what the participants experienced) and structural (how the context or setting influenced what the participants experienced) descriptions of the participants' experiences (Creswell, 2007; Moustakas, 1994). I then wrote a final descriptive passage reflecting the combination of the textural and structural descriptions to capture the "essence" of the experience of the participants (Creswell, 2007; Moustakas, 1994) (please see Appendix S.3 for sample pages from a coded transcript and Appendix S.4 for the Pilot Theme Code Descriptions).

For the purposes and given the constraints of this report, I will only focus on two of the seven themes that emerged from the data in terms of describing the overall experience of what it is like to be an African American adolescent girl who attends an elite predominantly White independent school. In a future report, I plan to incorporate the five other themes into the final analysis and findings.

Validity

Researcher bias. Due to the fact that this is phenomenological study, it was imperative that I approach the data with as little a bias as possible so that I could better identify the essence of the participants' experiences (Creswell, 2007). In order to "bracket out" my personal experiences and view and analyze the experiences of my participants with a fresh look, I wrote a researcher identity memo as a way to highlight the biases that I may have brought into the study (Creswell, 2007; Maxwell, 2005). Merriam (1988) recommends that the memo include the researcher's reflections on "past experiences, biases, prejudices, and orientations that have likely shaped the interpretation and approach to the study" (as cited in Creswell, 2007, p. 208). In addition to the researcher identity memo, I also continued to write memos throughout the data collection and analysis process in order to continue to reflect on my own position as an African American woman and researcher in relation to the focus of my study (Miles & Huberman, 1994).

External audit. Another way in which I attempted to engage with and minimize the personal biases that I may have brought to my interpretations of the data was to perform an "external audit" on the data from the focus group. The job of the auditor(s) was to examine a sample of the data and then assess whether the data supported my interpretations (Creswell, 2007; Erlandson et al., 1993; Lincoln & Guba, 1985; Merriam, 1988; Miles & Huberman, 1994). The external auditors for the focus group data were doctoral-level students and professors in the field of education who have experience with analyzing and interpreting data. The external audit was helpful not only in that I was able to share my initial thoughts about the data but also that the auditors offered further questions and points to consider concerning the data.

Triangulation. I also validated the results from the study by triangulating the data from the focus group and the one-on-one interviews so that I could view and analyze the data sources in relation to each other. In creating the questions for the interview, I was sure to include questions that served as follow-ups to some of the information shared in the focus group. It is my hope that by engaging in the triangulation process, the information from each source will reinforce the other so that a clear and honest picture emerges about the phenomenon at the center of this study (Ely et al., 1991; Erlandson et al., 1993; Glesne & Peshkin, 1992; Lincoln & Guba, 1985; Merriam, 1988; Miles & Huberman, 1994; Patton, 1980, 1990).

Analysis and Findings

The themes that I selected that contribute to the findings of this study are *thoughts on being African American/what it means to be African American*, and *push-back/resistance*. The findings reported below represent 13 different voices that blend together to tell the story of what life is like for them as middle school African American girls attending Grace School.

Thoughts on Being African American/What It Means to Be African American

Racial microaggressions. When answering the initial focus group question of "What is it like to be an African American girl at this school?" many of the study participants tended to describe situations in which they experienced, overheard, or responded to stereotypes that others (usually White students) ascribe to them. These stereotypes generally included the assumption that they lived in "the hood" or "ghetto" and that talking or acting a certain way was equated with being African American. Bethany,[7] a seventh grader, explains this general trend:

> Yeah, it's just kinda like they're a little, they seem kinda like some kind of stereotypical. "I'm sure that you know like a lot about it [referring to the hood]. I'm sure that you go there a lot" you know, and just because of the color of my skin, they think that we, you know, either live in the hood or we visit the hood a lot or we know a lot about it.

This experience describes what Sue et al. (1997) define as a racial microaggression. Racial microaggressions are "brief, everyday exchanges that send denigrating messages to people of color because they belong to a racial minority group" (p. 273). Sue et al. (1997) hold that in addition to the surface-level meaning, racial microaggressions also communicate underlying messages to people of color. For example, the belief that all African American people live in the ghetto or a dangerous neighborhood communicates the idea that all African Americans are criminals or are dangerous (Sue et al., 1997).

While experiences such as this are certainly insensitive and sometimes hurtful, researchers have found that repeated experiences as the target of racial microaggressions can create "psychological dilemmas that unless adequately resolved lead to increased levels of racial anger, mistrust, and loss of self-esteem for persons of color; prevent White people from perceiving a different racial reality; and create impediments to harmonious race relations" (Sue et al., 2007, p. 275; Spanierman & Heppner, 2004; Thompson & Neville, 1999). These findings show that not only are the study participants at psychological risk in attending a school with such an environment, but White students suffer as well in that they have no reflection about the negative impact of their words on their African American female classmates.

Otherness. Another overall message of a racial microaggressions is that African American girls and their modes of thinking, doing, and being are outside of the norm. This theme was apparent in that the first topic that the study participants raised in the focus group discussion was about their hair and how the White girls in the school are preoccupied with touching, questioning, and commenting on African American girls' hair. The intense interest and curiosity attached to African American girls' hair creates an "other" status. Collins (2009) comments on the fact that the prevailing standard of beauty of the United States, which holds White women with blonde hair and blue eyes as the norm, creates and maintains an oppressive power through comparing those characteristics to features that are *not* those—those of African American women.

The extreme minority status of the study participants at school (13 African American girls out of a school population of 196) could also potentially contribute to their "other" experiences. Due to the fact that there are so few African American girls in the school, it is easy for them to be "othered" because it is clear that they do not fall into what is the norm for their particular school environment. Additionally, the messages that are prevalent in the media that serve to reinforce negative stereotypes of African Americans could explain why the study participants have experienced so many racial microaggressions and why the African American identity is equated with the hood or being ghetto.

Self-definition. Another finding related to this theme was the confusion associated with defining what it means to be African American. At certain points in the focus group, the participants prided themselves on what they viewed to be positive aspects of being African American: The seventh graders are considered the "fashionistas" of their grade, have rhythm, and know how to dance. Interestingly enough, these points of pride are also characteristics that are often associated with stereotypes of African Americans. Even as the study participants resist the negative stereotypes of being ghetto and from the hood, they seem to embrace the positive stereotypes related to their race. Collins (2009) offers a potential explanation for these contradictory beliefs by stating that

> Black women's lives are a series of negotiations that aim to reconcile the contradictions separating our own internally defined images of self as African-American women with our objectification as the Other. The struggle of living two lives, one for "them and one for ourselves" (Gwaltney, 1980, p. 240) creates a peculiar tension to construct independent self-definitions within a context where Black womanhood remains routinely derogated. (p. 110)

The study participants' rejection of some stereotypes and the acceptance of others reflects their attempt to find positivity in the midst of an environment where negative messages about aspects of their identity are ever-present. Their struggle is also a reflection of the adolescent developmental period in which youth try out different identities in order to find the characteristics that work for them (Nakkula & Toshalis, 2006).

Push-Back/Resistance

Peer-to-peer interactions. Another defining characteristic of adolescence is that youth consistently challenge and question different aspects of society as they try to find where they belong (Nakkula & Toshalis, 2006). The core of the push-back/resistance theme for this study is that participants openly question their experiences at Grace School and challenge the messages that they are receiving as a result of interactions with parents, teachers, and other students. When referring to the assumption that all African Americans live in the ghetto, Angela, an eighth grader, states,

> Honestly, what is it, like the 10 of us right here, so how would all these White kids be if we . . . took them to a school in Philly. We took ten of them to a school in Philly and let them go to school there for five years? I wanna know how they would feel. . . . Maybe they would understand, in terms of racist and the smart remarks made towards them. Maybe, just maybe they would feel how we felt.

Angela's challenge is an example of what Ward (1996) would define as "resistance for liberation." Instead of relying on emotion to address the problematic situation, Angela's statement reflects a critique of not only her personal experience but also the world in which she lives where schools are becoming increasingly segregated by race.

Another example of resistance is the act of confronting and questioning of acts of racism and discrimination in the moment in which they occur:

> Um, I was in um, art class, and me and my friend were sitting on one side of the table and these two boys were sitting on the other side. And the boy sitting next to me, he's like, "Except for Jasmine." So I'm like, "What are you talking about? You know, saying my name, you're talking about me." And he was like, "Oh nothing, you're going to get mad." And I'm like, "Well, how bout first you tell me, and then if I get mad maybe you'll regret it." So then he tells me, and he's like, "Well, so-and-so thinks, says that Black girls are dumb, and I said except for you." And I was like, I asked him, "Did you really say that?" He was like, "Yeah" and I was like, "Why do you think that?" and he's like, "I don't know, they just—the Black girls here don't seem smart." And that really like . . . I was like, wow . . . that's how *you* feel.

Bethany's persistent questioning of her White classmate who unashamedly admits that he thinks African American girls are dumb reflects her strong sense of self and high self-esteem. Instead of engaging in the nonassertive actions associated with "going underground" (Gilligan, 1996), Bethany's response reinforces the findings of previous research that African American girls are socialized to demonstrate traditionally masculine traits such as strength and assertiveness (Buckley & Carter, 2005). Bethany knows that she and other African American girls do not deserve to be labeled "dumb," and she is not afraid to challenge someone who believes that they do.

School environment. The participants in the study also critiqued the larger school environment as contributing to their experiences as African American girls in the school. The participants specifically question Grace School's commitment to diversity by pointing out that the school claims to be diverse even though the African American student population is extremely small and that there are no African American teachers in the middle school. As Angela points out, "They try and propose getting more a diverse community, but I don't think they're, you know, doing so good at that. For one of the reasons, they may have, like, some Black kids, but there are no Black teachers," and Bethany describes the level of diversity at Grace School by stating, "Yeah. You know when you come here you see . . . White people, and then maybe, like, a few specks of color."

Angela and Bethany were two of the most vocal participants in challenging and questioning their experiences at Grace School and also were the two participants who reported that they frequently engaged in conversations with their families at home about issues of race. This finding is explained by previous research that points to the significance of family support in "preparing the students to deal with his or her doubly marginalized status at the independent school" (Cookson & Persell, 1985, as cited in Cooper & Datnow, 2000, p. 189). Part of the comfort level of Angela and Bethany in challenging the status quo of their school is that their families have taught them how to recognize, process, and respond to situations of racism and discrimination.

Another factor that could contribute to the ways in which the participants critiqued their experiences at Grace School, especially regarding the commitment of the school to diversity, is the length of time that the participants have attended the school. Of the two students who critiqued the school the most, one student is practically a "lifer" at the school (has attended Grace School since first grade), and the other participant just started at Grace School this year and specifically paid close attention to the school's levels of diversity when she was assessing schools to which she wanted to apply.

Study Implications

The findings from this exploratory pilot study highlight the importance of the role of the school environment in communicating messages about race, status, and identity to their students. When analyzing the messages that the study participants receive on a daily basis, they are often negative and reinforce their feeling "othered" in the school environment. It is the responsibility of independent schools (and other schools that have students who have an extreme minority status) to be thoughtful and proactive about how to create a welcoming and affirming environment for African American girls. Datnow and Cooper's (2000) study on the experiences of African American students who attend predominantly White independent schools found that the schools that were most successful in creating a positive environment for African American students were those that expressed "an explicit institutional commitment to racial diversity." This took the form of open communication about issues of race in the different places of the school and demonstrating an awareness about race in the admissions process (p. 199). In the case of this study, the small numbers of African American students at the school and the lack of African American teachers is a large contributing factor to the feelings of isolation and frustration that the participants expressed throughout the focus group and their interviews.

Another significant finding from this study is the equally important role that parents play in preparing their children for experiences of racism and discrimination. As the stories of the participants reflect, the messages and advice they receive from their parents almost serve as a form of protection against the psychologically detrimental effects that repeated experiences of racism and discrimination could have on an individual. In the case of this study, this protective role could also be embodied in the presence of African American teachers within the school. Datnow and Cooper (2000) found that another contributing factor to the success of African American students in predominantly White independent schools was the fact that African American teachers within the school could serve as role models in the school and potentially share common experiences with the African American students.

The findings from this exploratory pilot study serve as a starting point from which schools, teachers, and parents can work to support African American adolescent girls who attend elite predominantly White independent schools. In recognizing the complex system of which schools are a part and the multiple relationships that contribute to and form the schooling experience, future research could focus on the parent and teacher influences on students' experiences as well as the education of non–African American students around issues of race and diversity.

Though this study is small in nature, the importance of honoring the voices of the study participants as experts in their experiences cannot be overlooked. While their stories are a reflection of the struggles that arise from being an African American female in society that historically and traditionally privileges the voice of White males, the girls' recognition of their power to define themselves and their willingness to question their place in society shows that they are no longer willing to be overlooked, because, as the participants shouted at the end of the focus group, "Black girls rock!"

APPENDIX S.1: PILOT STUDY LETTERS, CONSENT FORM, AND ASSENT FORM

Consent Form

February 16, 2012

Dear Families,

My name is Charlotte Jacobs and I am a doctoral student at the University of Pennsylvania in the Teaching, Learning, and Teacher Education program. I am also a researcher for the Center for the Study of Boys' and Girls' Lives (www.csbgl.org), a research organization at the University of Pennsylvania whose goal is to address the intersection of gender, race/ethnicity, and class and how these factors play out in the daily school lives of students and faculty members. I am currently working on three different action research projects with student research teams at [SCHOOL NAME].

My research interests lie in exploring the academic and social experiences of African American girls, particularly in the middle school and high school years. My specific focus is to study what factors contribute to the academic and social success of African American girls and women. As part of these larger research goals, I would like to begin by studying the academic and social experiences of African American girls who attend independent schools.

Last month, I approached [MIDDLE SCHOOL HEAD] and [DIVERSITY DIRECTOR] about conducting a pilot study with African American middle school girls at [SCHOOL NAME], and they are supportive of my study and also offered to help me to recruit students at [SCHOOL NAME] who may be interested in participating in the pilot study.

The study will consist of one group meeting where middle school students who identify as African American girls will participate in a focus group that will center on what it is like to be a student at [SCHOOL NAME]. The second part of the study will consist of single or two-person interviews that will follow up on some of the points that the girls mention in the focus group. Both the focus group and interviews will be audio-recorded. I hope to schedule the focus group for Wednesday, March 7th, and then conduct the follow-up interviews in late March or early April.

This study presents minimal risk to your daughters, and participation in this study is completely voluntary. If your daughter wishes to stop participation in the study at any point, she is free to do so. In the final report, all responses will be kept anonymous. I will also make the final report available to your families, [MIDDLE SCHOOL HEAD], and [DIVERSITY DIRECTOR].

This study has important implications for informing the school and larger community about the unique experience of African American girls who attend independent schools. If you are willing to have your daughter participate in the study, please complete the attached consent form and return it to [MIDDLE SCHOOL HEAD] by Friday, March 2nd. If you have any questions about the study, please do not hesitate to contact me either via email (xxx) or phone (xxx).

I look forward to working with your families in the future!

Sincerely,
Charlotte Jacobs

[SCHOOL NAME] MIDDLE SCHOOL
AFRICAN AMERICAN GIRLS PILOT STUDY

Consent Form

After reading the introduction letter that details the study, I,

_____ give permission for my daughter _____

(Parent/Guardian Name)   (Students' Name)

to participate in the [SCHOOL NAME] Middle School African American Girls Pilot Study.

_____   _____

(Signature)   (Date)

Parent/Guardian Contact Information

Name(s): _____

Preferred Method of contact (email or phone): _____

Email address and/or phone number: _____

*** Please return this signed consent form to [MIDDLE SCHOOL HEAD] by Friday, March 2nd.

Thank you for your interest in participating in the [SCHOOL NAME] Middle School African American Girls Pilot Study!

Assent Form

March 7, 2012

Dear Participant,

Thank you for your interest in participating in the [SCHOOL NAME] Middle School African American Girls Pilot Study! My name is Charlotte Jacobs and I am a doctoral student at the University of Pennsylvania in the Teaching, Learning, and Teacher Education program. I am also a researcher for the Center for the Study of Boys' and Girls' Lives (www.csbgl.org), a research organization at the University of Pennsylvania whose goal is to address the intersection of gender, race/ethnicity, and class and how these factors play out in the daily school lives of students and faculty members. I am currently working on three different action research projects with student research teams at [SCHOOL NAME].

My research interests lie in exploring the academic and social experiences of African American girls, particularly in the middle school and high school years. My specific focus is to study what factors contribute to the academic and social success of African American girls and women. As part of these larger research goals, I would like to begin by studying the academic and social experiences of African American girls who attend independent schools.

The study will consist of one group meeting where middle school students who identify as African American girls will participate in a focus group that will center on what it is like to be a student at [SCHOOL NAME] School. The second part of the study will consist of single or two-person interviews that will follow up on some of the points that the girls mention in the focus group. Both the focus group and interviews will be audio-recorded. The focus group is scheduled for Wednesday, March 7th, and the follow-up interviews will be conducted in late March or early April.

This study presents minimal risk to you and participation in this study is completely voluntary. If you wish to stop participation in the study at any point, you are free to do so. In the final report, all responses will be kept anonymous. I will also make the final report available to you and your families.

This study has important implications for informing the school and larger community about the unique experience of African American girls who attend independent schools. If you are willing to participate in the study, please complete the attached assent form. If you have any questions about the study, please do not hesitate to contact me either via email (xxx) or phone (xxx).

I look forward to working with you and your families in the future!

Sincerely,
Charlotte Jacobs

[SCHOOL NAME] MIDDLE SCHOOL
AFRICAN AMERICAN GIRLS PILOT STUDY

Assent Form

After reading the introduction letter that details the study, I,

_____ agree to participate in the [SCHOOL NAME] Middle School African American Girls Pilot Study.

(Student First/Last Name)

_____   _____
(Signature)   (Date)

_____ I am interested in participating in a follow-up one-on-one interview if needed.

Parent/Guardian Contact Information

Name(s): _____

Preferred Method of contact (email or phone): _____

Email address and/or phone number: _____

Thank you for your interest in participating in the [SCHOOL NAME] Middle School African American Girls Pilot Study!

APPENDIX S.2: MEMOS

RESEARCHER IDENTITY MEMO: WHO AM I AND
HOW DOES MY IDENTITY CONNECT TO MY RESEARCH INTERESTS?

The purpose of this memo is for me to write about my identity, my past experiences, my goals, and how all three components potentially intersect to frame my future research about the academic and social experiences of adolescent Black girls in predominantly White institutions (PWIs). From the beginning, my identity and experiences as a Black woman have informed my interest in both applying to graduate school and my research interests. From the time I was about 5 or 6 years old, I knew that I wanted to be a writer. I loved creating stories and creating my own worlds and realities on the page. But inevitably, I would experience writer's block and would complain to my mom that I had nothing to write about. My mom always gave me the same advice each time, "Write what you know." Her advice communicated to me that my life experiences were just as interesting, "deep," and inspiring as anything that I could cook up in my imagination. And so, I take the same stance when approaching my research interests: by choosing to research Black girls and the factors that contribute to their continual success, I am, in essence, studying myself.

But who am I? I am a Black woman who grew up in Shaker Heights, Ohio, a racially diverse suburb of Cleveland. I attended public school for my entire primary and secondary school careers and was tracked into the Honors/AP route beginning in fifth grade. By the time I entered high school, two things had happened: (1) Even though my high school was roughly 50% White and 50% Black, I was typically one of five Black students in my AP/Honors classes, and (2) the other four Black students who were typically in my classes were the same four Black girls that I had been tracked with since the fifth grade. Though on the outside my school was a racially diverse environment, in reality I had been attending a PWI since the age of 10 by the sheer makeup of my AP/Honors classes. Long before graduate school was on the horizon, I began to wonder (a) why there weren't more Black students in AP/Honors classes and (b) if my experience in forming a tight-knit friendship with the other four Black girls in my classes was typical and if the experience somehow contributed to my academic and social success. Those questions continued to follow me throughout my undergraduate experience at Columbia University and now at the University of Pennsylvania.

I consider my ongoing friendship with the four other Black girls in my AP/Honors classes a defining moment in my life. Although none of us ever talked directly about the formation of our group, the implicit message that drove us was that in order to survive, we had to stick together. We supported each other academically by pushing one another to do our best on every assignment that we received. We debriefed with each other when a teacher or another classmate made a comment that was racist or discriminatory. We shared stories about how hard it was to find guys who wanted to date us. Our friendship was more than a friendship—it was an informal support group.

I contrast my experiences with that of my sister, who is 2 years younger than me. We attended the same schools, and she was also tracked beginning in the fifth grade. But the difference for her was that she was the only Black student in her AP/Honors classes. There was no real opportunity for her to form a social group of Black girls because there were none in her classes. Though my sister excelled academically, socially, she struggled. She was very much the lone wolf throughout high school, and even when she attended Princeton for her undergraduate degree, she felt that she could not relate to the other Black students that were on campus. My sister's experience makes me wonder even more about the influence that a Black peer group has on the academic and social success of Black students in PWIs.

I realize that I am going into this work with some major assumptions and biases. For one, I think there's value in members of an ethnic group (especially if they are a minority group) to be able to relate to and support one another. I think back to my sister and that fact that she has never had any Black friends, and subconsciously I feel sad for her. Though she has lived and continues to live a fulfilling and successful life, my personal bias leads me to think that something is missing from her life because she does not have a built-in support group (outside of me). I know that as I interview and learn more about the experiences of Black girls at PWIs that I am going to have to tame my perspective that having no Black friends is somehow an inherent deficit.

I also need to be cautious about mapping my own experiences onto that of my subjects. I think because I identify so strongly with my potential subjects and could see elements of myself reflected in their experiences, I have to be careful to listen to *their* stories and not filter them through the lens of my own experiences. Not all Black girls are alike, and neither are the academic and social situations in which they find themselves. I have to be sure to pay attention to the nuances that could arise in their experiences as a result of their family lives, school environment, and simple personality traits.

For my QMI pilot study, I intend to focus my research specifically on the experiences of Black girls who attend (or have attended) predominantly White institutions (PWIs), but in the future (most likely for my dissertation) my goal is to create a comparison study of the social and academic experiences of Black girls who are in PWIs and Black girls who are in predominantly minority institutions. In addition to learning about the Black girl experience in these spaces, I'm curious in exploring the potential power or influence that Black women peer groups have on academic and social experiences. Do these kind of networks/support systems only form in places where Black girls are the extreme minority, or do they also appear in environments where Black girls are in the majority, and if so, do these support groups take the same form?

Moving forward, I can already predict that another one of my challenges could be that my study has the potential to rock the boat in institutions that are steeped in traditions that may be problematic for Black female students. What would their response be if my findings are such that call into question their practices? I also wonder about institutions that may fall prey to the fallacy of the "white liberal" wherein they either have a color-blind approach to diversity at their school (which in itself is contradictory) or they feel that they offer enough support for students of color and that my findings are simply the experience of my particular subjects and not of other Black girls who will walk through their doors in the future. For example, when I pitched my QMI study idea to a middle school principal of an independent school in the Greater Philadelphia area, she was happy to open her doors to my study and help in any way with the facilitation of the project, but she did ask me if in addition to the Black girls that I planned to interview if I was going to interview White girls at the school as well as a way to compare the experiences of two different racial groups at the school. My initial response (in my head) was "Wow, that's a lot of work" and also "Well, maybe it makes sense that I should compare the two groups." When I related this conversation and overall research idea to the small group in my RAC, they questioned the fact that I was asked to create a comparative study—wasn't my study of the experiences of Black girls enough? Why should I feel the need or fulfill the expectation that their stories would be or should be compared to a normative White standard? From this experience, I have realized that as I complete this study and progress with my dissertation, I will have to be able to defend the fact that I have made the conscious decision to only study Black girls and their experiences without using their White counterparts for comparison.

I hope that my research will draw attention to a subpopulation that for a long time has been overlooked in education research. Black girls deserve to have their needs met so they can flourish academically and socially as they progress through school. For the longest time, they have been either lumped into the "girls" category or the "Black students" category without special attention being paid to how these two identifiers intersect to create a unique experience for Black girls. While my research will most certainly not be definitive, it will hopefully serve as a starting point for getting Black girls on the map.

FIELDWORK MEMO

Research Questions: In predominantly White institutions, what meaning do adolescent African American girls make about what it means to be African American? To be female? To be an African American female?

- What are the experiences of African American adolescent girls at predominantly White institutions?

Right now I feel that I am at a somewhat comfortable point in my pilot study. At this time, I have collected data, performed an initial external audit with the data that I have, and I have begun to collect sources that I will use as part of my literature review for the final report. At the beginning of March, I conducted a focus group with 10 middle school African American girls. The focus group lasted for about an hour, and I opened the session with the initial request of asking the girls to tell me what life was like for them as African American girls at their school. I barely got the tape recorder on before the girls began relating their experiences. The first girl to speak opened the discussion by talking about her hair and the experiences that she has had with White girls at the school not understanding that her hair was different from their own. The conversation then proceeded to cover the topics of the White students at the school assuming that all of the African American students live in and are from the "ghetto," different discriminatory situations that they have experienced from students and teachers, the challenge of feeling like they don't have enough money to do the things that they want compared to their peers, and the acknowledgment that they feel like they don't have many people who they feel like they can talk with these experiences about.

What struck me about this experience was how open the girls were when sharing their stories. It seemed like they simply needed someone to ask this one question, and then they were off. For the most part, their discussion covered all of the areas that I was curious about related to the experiences of African American middle school students in a PWI. I was also taken aback by how aware the girls were about how their status as African American girls possibly influenced the daily experiences that they had at their school. Going into this study, one of my concerns was that by working with middle school students, I would be running the risk of their not having any awareness yet of how different aspects of their identity influenced their experiences. Another thing that I noticed was the immediate sense of trust and camaraderie among the girls in the group. The participants were from the sixth, seventh, and eighth grades and from a variety of income levels and have attended the school for a varying amount of time (all of these data were collected via a preliminary questionnaire). I point this out to say that most of the girls seemed at ease with sharing personal, emotional, and sometimes hurtful stories in front of the entire group. Something that I would be interested in following up on are their particular social relationships at the school. Do they only hang out with each other (African American girls)? Do they spend time talking about race and/or gender and their experiences at school? Is there a pressure to hang out with each other?

The content of the focus group has also got me thinking about the larger elements at play that factor into the girls' experiences. How much does the influence of social media play into the stereotypes that these girls combat on a daily basis ("Pat your weave!")?

What education needs to happen on the part of the White students at the school? What about education for the teachers? As the girls shared their stories, I also wondered what the school's commitment to diversity truly is. Do they pat themselves on the back for recruiting African American students (13 African American girls, 3 African American boys in the entire middle school of 196 people), yet not provide the structure and support that these students need to thrive (as in not having any African American faculty in the middle school)?

I walked away from the focus group feeling happy that I had PLENTY of data but slightly intimidated by how I would go about analyzing the data. There are so many themes and lenses that I feel I could use to analyze the data that I am slightly paralyzed about what to do next. One of the themes that I could hear emerging even as the girls spoke was the idea of racial microaggressions. Sue et al. (2007) define racial microaggressions as "brief and commonplace daily verbal, behavioral, or environmental indignities, whether intentional or unintentional, that communicate hostile, derogatory, or negative racial slights and insults toward people of color." I wonder what kind of story would arise from viewing their experience by categorizing the different microaggressions that they have experienced. I also plan to employ critical race theory as part of my theoretical framework, but I haven't decided what tenets of the theory I will use.

As a way to unpack the focus group data and find my footing a bit, I presented an audio excerpt and transcript of the focus group data to some of my fellow graduate student colleagues and professors who meet in our monthly Inquiry into Practice Collaborative meetings. This group data analysis session also served as part of the external audit that I planned as part of my methodology for the study. The feedback that I received from the group was helpful and overwhelming at the same time. Many of the comments focused around the sense-making that the participants in my study seemed to be doing about their unique identities with their specific school space. Key questions/themes that arose that I am still mulling over are

- How do they (the African American girls) influence other cultures at the school? How are the African American girls themselves being influenced?

- The concept of an asymmetrical space and the impact of that on the way that African American girls view their experiences

- Are there times at school when people bring up race or being Black that is positive or helpful?

- The girls tended to trail off in their speech at certain points during the discussion—perhaps analyze where in the discussion that pattern occurs

- The girls seem to be in the midst of defining for themselves what it means to be African American girls

  - Resisting the broad brush stokes that their White peers use to describe them and focus on the nuances of being African American

  - Their conceptualization of Whiteness—is there as much nuance used when describing the White students at their school?

- The desire to be an individual vs. looking for a sense of community

- The repetition of particular words—"annoying," "offensive," "frustrating"

I'm now at the point where I am preparing for follow-up interviews with three of the participants who were a part of the focus group. While I have general questions that serve as a follow-up to the focus group, I also plan to look through the focus group data once again to create questions that are specific to experiences that they shared in the focus group.

Another interesting development that has come out of the focus group is that a couple of the girls went to the middle school principal to tell her how nice it was to have a space where they could share their stories, and they asked her if the group could meet again or become a regular thing. I am now in conversation with the middle school principal and the Director of Community and Diversity at the school about setting something like this up—a cultural group or affinity group for African American students. It would be interesting to have this focus group be the start of a longer study exploration into the lives of these girls at their school.

VIGNETTE MEMO

Date: 4-23-12

Research Questions

1. In predominantly White institutions, what meaning do adolescent African American girls make about what it means to be to be African American? To be female? To be an African American female?

2. What are the experiences of African American adolescent girls at predominantly White institutions?

* This memo will address Topic 3 (*the roles of being a researcher*) and Topic 4 (*other central issues about which I have questions/concerns*).

There are two key issues that I find myself grappling with as I continue my study about middle school African American girls who attend an independent school. The first issue is my role as researcher in this study and my own particular positionality in relation to the research topic. The second issue/growing concern is how I plan to tell the story of my findings and craft my final report with the knowledge that this report will be read by the girls in the study, their parents/guardians, and interested parties in the school administration.

Last week, I conducted follow-up one-on-one interviews with three of the girls who had participated in the focus group back in March. I selected the interviewees for a few different reasons: (1) I want to make this study as representative of the middle school girls at this school as possible, so I selected one girl from each of the middle school grades (sixth, seventh, and eighth) to interview. (2) Similar to my first point of trying to establish a representative sample, two of the girls I interviewed began attending Grace*[8] School this year, and the other girl I interviewed has been at Grace School since first grade. (3) I chose girls who I wanted to get to know in more detail—two of the girls I selected had shared compelling stories during the focus group that I wanted to know more about, and one of the girls only spoke once or twice throughout the whole focus group, a comparatively small number of responses compared to the other girls in the group.

The vignette for this memo focuses around Student B's interview and her retelling of a specific incident that she experienced while at Grace School. As I mentioned in my previous memo, many of the stories that the girls shared while in the focus group were stories that were (in the view of the girls) painful, frustrating, and shocking. Student B's experience is no exception. While in the focus group, she made the comment that she had overheard one of her (White male) classmates make the statement that he thought that "Black girls are dumb." The conversation in the focus group moved on from there with no real chance for me to probe deeper about the situation. I was intrigued and disgusted at the same time. Someone actually said that?? Out loud?? One of the reasons I chose to interview Student B was because I wanted to know more. Below is an excerpt from her interview where she recounts her experience:

Student B: Um, I was in um, art class, and me and my friend were sitting on one side of the table and these two boys were sitting on the other side. And the boy sitting next to me, he's like, "Except for [Student B]." So I'm like, "What are you talking about? You know, saying my name, you're talking about me." And he was like, "Oh nothing, you're going to get mad." And I'm like, "Well, how bout first you tell me, and then if I get mad maybe you'll regret it." So then he tells me, and he's like, "Well, so-and-so thinks, says that Black girls are dumb, and I said except for you." And I was like, I asked him, "Did you really say that?" He was like, "Yeah" and I was like, "Why do you think that?" and he's like, "I don't know, they just—the Black girls here don't seem smart." And that really like . . . I was like, wow . . . that's how *you* feel.

Interviewer: Yeah . . . [laughs and shakes head]. So what were you feeling in that moment?

Student B: I-I was just shocked. I didn't have any feeling, I was just like, "Wow." I can't even believe he said that. And was able to admit to it and not be like . . . he just, you know, big and brave, "Yeah I said that." It's like then, like, oh, ok.

The next few lines of the interview transcript show me fumbling to make sense of the situation. I ask Student B if the offending student has a lot of social power in the school (yes, but she still couldn't imagine that he would ever say something like that); I ask her if there were other students around who heard the comment (yes) and what their reactions were (they didn't do anything). Student B says that if "the other girls" (referring to the other African American girls who she hangs out with) had been there that they would have said something but that she knew in that situation that none of her other classmates were going to say anything.

This part of the interview was challenging for me. As I saw Student B sitting before me, on the verge of tears, I felt angry. I was angry that this was part of her middle school experience. I was also angry that in that moment when the boy made the comment, she seemed utterly alone—none of her classmates backed her up in what was an uncomfortable and (in my eyes) traumatizing situation.

I relate this story because it brings up for me my struggle with my role as a researcher in this particular study. As a Black woman, I can't help but feel very deeply what these girls seem to be experiencing on a daily basis. In my effort to theorize Student B's and the other girls' stories that are of a similar vein, I analyzed their experiences through the lens of critical race theory, whose key tenet is that race is an endemic part of the fabric of our society (Matsuda et al., 1993). Part of that analysis resulted in my categorizing their experiences as examples of different racial microaggressions (Sue et al., 2007). But when I looked through the transcripts from the focus group and student interviews, I realized that there is another story that could be a part of these experiences. Using Student B's story as an example, I realized that her story is not only about her being a victim of some student's racial microaggression but also about a demonstration of power and self-confidence. Rather than quietly let the comment by, Student B confronts and questions her classmate and pushes him to support his comment with evidence. This is a side of the story that I initially missed because of my emotional response when rereading the transcript for the first time. This revelation more than ever highlights the biases that I am carrying as I conduct this research. It's hard to read through the transcripts and not react emotionally (negatively or positively) to what is on the page. I am slowly learning how to work past the initial emotional response so that I can authentically analyze the data with the goal of drawing out a story that is accurate.

In a similar vein, another concern that I have concerning the data is what my findings and final report will look like. In thinking through the data (which I will do again and again in these next 2 weeks), many of the stories that the girls share are overwhelmingly negative in relation to their school experiences. During Student B's interview, I even asked her if she could think of any positive experiences that she had had at Grace School, and she struggled to come up with an answer, and finally said, "I can't really—yeah, I can't really think of a positive thing rather than like, I guess, like being here. I don't know. I never really thought about it" (interview on 4-17-12). The knowledge that my final report will be made available to the parents/guardians and school administrators makes me really think about how I plan to frame their stories. My intention with my research is not to act as a whistleblower, nor is it to sugarcoat life at Grace School, but rather paint a picture of what is happening in school with these girls. I expect that as I go through the data again, I will find some positive things about their experiences, but I am worried that (1) the negative will outweigh the positive and (2) that my intentions will be misread by my audience.

Any advice that you could give me about writing a "sensitive" report in relation to its audience would be extremely helpful (especially because the relationship that I have with this school is one in which it would be in my best interest to stay in their good graces)!

APPENDIX S.3: TEXTURAL AND STRUCTURAL DESCRIPTIONS OF THE THEMES/CODES, BEGINNING ANALYSIS

Grace School African American Middle School Girls Pilot Study

May 2012

TEXTURAL AND STRUCTURAL DESCRIPTIONS OF THE THEMES/CODES, BEGINNING ANALYSIS (QMI REPORT)

Textural: Description of what they experienced

Structural: Description of the context or setting that influenced how participants experienced the phenomenon

Theme 1: What It Means to Be African American/Racial Microaggressions

Textural Description: Many of the girls describe situations in which they are responding to stereotypes that others are ascribing to them such as living in the ghetto and having others talk or act a certain way that they equate with being Black. The descriptions focus on the way others see them, rather than how they seem themselves. There are also descriptions of "otherness" in the way that people are surprised because they are Black and attend Grace School and how a student singles out the "Black girl" category as being dumb. Many of the experiences that the girls describe would fit under the Sue et al. definition of racial micro-aggressions. There is an instance where they make a delineation between a girl who is African American, but there are some girls who don't consider her to be Black. There are also positive aspects of being Black that the girls point to—having rhythm, knowing how to dance, being fashionistas. The girls have also adopted a mantra of "Black girls rock" that seems to be a positive "state of mind."

Structural Description: The extreme minority status of the girls at the school (13 out of 196) could potentially contribute to some other their experiences—because there are so few of them, it is easy for them to be "othered." The messages that are prevalent in the media that serve to reinforce negative stereotypes of African Americans could explain why the girls are experiencing so many racial microaggressions and why the African American identity is equated with "the hood" or "being ghetto." The adoption of the BGR mantra is also reflective of messages from the media—the phrase was taken from the name of an organization that a few of the girls saw on a BET awards show.

Theme 2: Push-Back/Resistance

Textural Description: The core of the push-back/resistance is that the girls are questioning what they see around them and the messages that they are receiving. They question the belief that all Black people live in the ghetto and challenge those at the school (non-Black students) to define what they mean by "ghetto" and to put themselves in an environment where they are one of the few of their group there. They also question the racial experiences that they have in the moment and confront the person making racist remarks face to face. They also question the school's commitment to diversity by pointing out that the school says it's diverse but that the African American student population is small and that there are no African American teachers in the middle school.

Structural Description: The fact that the girls were in a focus group where they were with (who they assumed) like-minded people potentially made them more comfortable in analyzing and questioning their experiences. Two of the girls who were the most vocal in questioning and challenging acts of racism and discrimination were also the two girls who talked most about having conversations about race at home with their parents. The length of time that the student has attended the school could also contribute to their critique and questioning of the school and its commitment to diversity—of the two students who discussed this topic the most, one student is practically a "lifer" at the school (has attended Grace since first grade), and the other girl just started at Grace this year and initiated the search for a new school and took diversity into account when she was assessing schools to which she wanted to apply.

Theme 3: Community

Textural Description: The girls discuss that the appeal of the focus group was that it gave them time to be together and talk about subjects away from non–African American students. They mention how there aren't that many opportunities to be together while in school; some of the girls are the only African American girl in their entire section—they go through their core classes each day being the only one. They also share the fact that the focus group felt like a "family" to them, which made it easy to share their stories. They enjoy being with people who are like them. There was also a comparison between their home communities or former schools and Grace School concerning race relations. The girls described how people of different races in their neighborhood or school get together and interact with each other. In Grace, the girls do not see that same ease of interaction.

Structural Description: Their extreme minority status could make them want to search for a sense of community in the school and reflect nostalgically about their former schools or their neighborhoods. The fact that they are sometimes isolated throughout the day (being the only one or two in their classes) also makes them seek community. It also does not seem like they have time to come together and discuss issues of race or their experiences in the school.

APPENDIX S.4 CODED TRANSCRIPT

Bethany*[9] Interview Transcript (Seventh Grade), 4-17-12

(Underline = Community; Gray = Push-Back/Resistance; Italics = Thoughts on Being African American/Racial Microaggressions)

Interviewer:	Ok, so this is Bethany. Ok, so one of the first things I want to know is have you . . . was there anything that kind of stuck out to you about the focus group? Things you hadn't really thought about or things that you heard that you thought were interesting? Or anything else that you wanted to share that you didn't get a chance to share at the focus group? Because there were a lot of voices—
Bethany:	Yeah.
Interviewer:	There. [laugh]
Bethany:	One thing, um, they were saying like how, um, they thought that one of the Spanish teachers was kind of racist? Mr. Ames? I really, um, don't agree with that. She's saying, that like he doesn't call on her all the time, but I think like— I was in his class yesterday with the same people and I don't think that is, that he's racist. I think that those people, they raise their hands all the time. He wants to get different people and not that he doesn't, you know, like Black people.
Interviewer:	*Right. Ok. And how many Black people are in that class?*
Bethany:	*Umm . . . all of them?*
Interviewer:	*Really? [laugh] In one section?*
Bethany:	*Yeah. All the— yeah, cuz we're all in middle math, so we're all in that Spanish class.*
Interviewer:	*Ok. So that means. . . .*
Bethany:	*Oh, it's . . . yeah.*
Interviewer:	*So how many would that be? Five or six?*
Bethany:	*Yeah, something like that.*
Interviewer:	*Ok. And then how big is your class?*
Bethany:	*Um, nineteen.*
Interviewer:	Nineteen. Ok. So you, you would say from your perspective that he does call on people pretty fairly? Or he tries to?
Bethany:	Yeah, you just— I think it's more of, um, "I don't wanna call on the same person over and over, I wanna see what other people are saying."
Interviewer:	Mmm, ok.
Bethany:	Not "I don't like Black people."
Interviewer:	Alright.

Bethany:	Yeah.
Interviewer:	Ok. Um, was there anything else that you were thinking about after the focus group or anything else that you wanted to share or say that you didn't get a chance to?
Bethany:	Um, yeah. One thing they were talking about, like their parents and [clears her throat] like, um, "our parents this" and "our parents that" and one thing about my family that's a little different from everybody, my grandma, she is . . . like, she's . . . wealthy—
Interviewer:	Um-hmm.
Bethany:	And she lives in um, New York, and um, you know, I don't— I guess . . . I don't, like, take advantage of that. Like she's always tells me if I ask for something, she's like, "No, I'm not giving it to you" I mean, I know she can get it—
Interviewer:	[laughing]
Bethany:	But, you know, she tries to still sustain that, not wanting me to be spoiled because I know that I have someone to go to that can give me something that I want, you know, yeah. Things like that.
Interviewer:	And what made you think of that in relation to what the other people were saying?
Bethany:	Um, because one of the girls, Tracy, she was talking about how her grandparents give her, like everything, and she gets like new stuff every day, and it kinda just made me think of that.
Interviewer:	Um, so one of the things that I was also struck by in the focus group was the fact that, so all of you were together and you, when I asked you different questions in the focus group, you were all just very open about sharing—
Bethany:	Yeah.
Interviewer:	And so that made me wonder, you know, are all of you friends? Do all of you hang out here at school? and um . . . or . . . so that's the first question. So yeah, do all of the Black kids kind of hang together at school, um . . . and . . . then the other thing is that then are there mixed race sort of friends groups? Um, and then, the other thing, following that is, are there, do you all spend time talking about, like issues of race and gender, like amongst yourselves?
Bethany:	Well, um, [clears throat], we do . . . sometimes like say at lunch, like, like Ashley, and um us, we'll sit together, and um, not always. I sit . . . mmm . . . with some, um, White kids and um, me and the girl Tina, she's not White, she's Hawaiian, um, close [laughs]
Interviewer:	[laughs]
Bethany:	But, uh, yeah, I don't really sit with, like, Yolanda and Jeanelle a lot, only because, I'm really, um really don't know why. I just . . . I dunno, those

are like the other people those are just the people, there in my section and so we talk a lot. Yeah, those are the people I sit with at lunch and talk to and stuff.

Interviewer: Ok. And is there pressure do you think for— for the Black girls to all hang out together? Or for Black, or for the . . . students of color at the school to hang out together? Or it is pretty— does it seem like it's pretty open, you can be friends with whoever you want to be friends with?

Bethany: I mean it's fairly open, you can be friends with whoever . . . you want to be friends with, I mean there's a certain group of people that, like, we don't really hang out with, just cuz . . . they're not really people you wanna hang out with. [laughs] So . . .

Interviewer: And why is that? So just give me a quick, kind of like, overall description. It doesn't have to be very detailed, but why wouldn't you want to hang out with them?

Bethany: They're nasty.

Interviewer: Oh, ok.

Bethany: Yeah . . .

Interviewer: To other people?

Bethany: No, not like nasty like that. Like . . . well, I'm so sorry about saying this, like slutty nasty.

Interviewer: Ok.

Bethany: Yes.

Interviewer: Fair enough. Fair enough. [laughs] So, then do you, so when you were hanging out with the other Black girls, do you all end up talking about things that you experience here at Grace in terms or race and gender or do you talk about other stuff?

Bethany: Um . . . mm . . . a little like after the group we did talk about stuff like that, but then after that, no, not really, we just talk about, you know. Stuff.

Interviewer: [laughs] Other stuff. Other important middle school stuff!

Bethany: Yes [laughs]

Interviewer: Um, let's see. Oh! So one of the things I also thought that was really interesting was that at the end of the focus group. Um, there was cheer—

Bethany: Oh . . .

Interviewer: That you all said, "BGR" Black Girls Rock. And I was like, "Wow! That's really cool!" Um, and so I wondered, you know, one: where did that come from? You know, was it something that was spur of the moment? or had, did, you know, people already know about it?

Bethany: Yeah, um, it's um, it's actually like an organization called Black Girls Rock and they had, like, it's on, like um . . . they had a, um like event thing last year, and it was on like BET, and yeah, that's where that came from, and I guess they just decided to say it.

Interviewer:	Um, and so, following up on that, because it was really interesting to kind of have that cheer at the end. Um, do think, can you think of times here at Grace where race or being Black has been seen as being positive? I mean, helpful in some way? Because, um, in the focus group, a lot of what ended up coming out was—
Bethany:	Was negative
Interviewer:	Yeah.
Bethany:	Yeah.
Interviewer:	*And so I wonder if there is, if they are positive aspects . . . here where there's a positive spin on talking about race or positive spin on being Black?*
Bethany:	*Hmm, I don't really have a forward answer for that question. All I'm going to say is like, sometimes like in class they'll be talking about like Black stuff and it will be really weird. Cuz in my section I'm the only Black person in my section because I'm the only Black person who plays an instrument here.*
Interviewer:	Oh, ok.
Bethany:	*So, I'm like. I'm with like the Asians, so I know that when we talk like about Koreans and stuff, I know they feel it too because I can like see it in their face. Like it's weird. Everyone's like looking at you. It's like you know everything about that topic just cuz your Black or Asian. It's not, you're not—*
Interviewer:	*Mmhmm.*
Bethany:	*It's not like that.*
Interviewer:	*Right.*
Bethany:	*Like, just because we're watching something about Africa doesn't mean I've been to Africa and visited ancestors.*
Interviewer:	*[laughs] Right. And so that. So it's uncomfortable?*
Bethany:	*Yeah, yeah.*
Interviewer:	Ok. Um, so . . . so you don't really, you can't really think about a time when there are the positive things? It's more the negative things that kind of weigh in your mind?

Appendix T

Example of a Conference Proposal

True Lies: Diversity Promises Made to Students in Mission Statements at Large Universities

Demetri L. Morgan, Cecilia Orphan, and Shaun R. Harper

ABSTRACT

Using a transdisciplinary concept from legal studies and business, this paper focuses on diversity promises made to students in mission statements at the 101 not-for-profit U.S. universities enrolling 20,000 or more undergraduates. Zhang and Wildemuth's (2009) seven-step qualitative content analysis method was used to identify themes in the mission statements.

INTRODUCTION

Mission has been called the "life force" of institutions (Scott, 2006), and organizational theorists generally agree that a clear sense of institutional purpose and mission leads to organizational success (Hartley, 2002; Kotter, 1996). This is called "mission agreement" and occurs when organizational operations, articulations of mission, member beliefs about the purpose of the organization, and historical evolution of mission are consistent and reso-nant (Fjortoft & Smart, 1994). Mission agreement is believed to be particularly important for public and ethnically concerned organizations (Berg, Cziksentmihalyi, & Nakamura, 2003). One way to ensure mission agreement is to juxtapose institutional policies, pro-grams, and practices with articulations of mission—also called mission statements (Lang & Lopers-Sweetman, 1991; Meacham & Gaff, 2006).

Mission statements are "shaped by a kind of garden-variety philosophic idealism" and an articulation of what an institution's purpose and mission is (Davies, 1986, p. 85). Articulating mission has two primary benefits. First, articulation allows an institution to communicate its essential character, operations, and values to its own members as well as external audiences (Davies, 1986; Lang & Lopers-Sweetman, 1991). Second, these mission statements can serve to inspire and motivate institutional members to commit to the essential work of the institution (Davies, 1986; Kanter, 1972). Despite their role in conveying the essential purpose of institutions, mission statements are often vague and imprecise. These statements are not typically thought of as static; hence, they may change over time as the institution's mission evolves (Davies, 1986; Morphew & Hartley, 2006; Scott, 2006). While mission statements may evolve, organizational theorists generally agree that they are important because they can serve as a guide for institutional evolution, serving as "smoke screens for opportunism" as institutions determine appropriate programs and policies in light of historic purpose and mission (Lang & Lopers-Sweetman, 1991).

How well institutions embody written missions within institutional policies and practices relates to mission alignment (Eckel & Kezar, 2002; Fjortoft & Smart, 1994). The way in which institutional leaders claim and take up the institution's written mission varies widely, though, leading some to criticize mission statements as "rhetorical pyrotechnics" (Morphew & Hartley, 2006, p. 456). In their study of 158 public and 141 private universities and colleges, Morphew and Hartley found that mission statements varied widely. While there was variation in the specific language used, diversity, service, civic engagement, and liberal education were frequently mentioned across institutions (Wilson, Meyer, & McNeal, 2012).

Scholarship points to the structural and ideological changes and features that predict mission agreement and the creation of inclusive environments for all students (Chang, Milem, & Antonio, 2011; Harper & Hurtado, 2007; Hartley, 2002; Hurtado, Milem, Clayton-Pedersen, & Allen, 1998). Specifically, a blend of institutional policies, procedures, and resource allocations is required to ensure mission agreement and the creation of "diverse learning environments" for students. Additionally, attitudinal and behavioral changes must occur in order for inclusive environments to be fostered. At the heart of these efforts are transformations to the institution's fundamental culture with regard to inclusion. The structural and ideological dimensions of this work raise questions about the feasibility of enacting the promises made with regard to diversity, racial inclusion, and cultural awareness within large, complex, and ossified institutions.

This study builds on the work of Morphew and Hartley (2006), as well as Lang and Lopers-Sweetman (1991), by exploring the use and content of mission statements as they relate to diversity. Specifically, we are concerned with the feasibility of enacting diverse learning environments at the largest colleges and universities in the United States. We purposely situate our analysis in institutions that enroll 20,000 or more undergraduates for two reasons. First, Hurtado (1992) found that reports of racial conflict were higher among students at large institutions. Second, the 101 institutions that we have characterized as large enroll more than two million undergraduates and produce large shares of college graduates. And third, existing research on mission statements has not focused deeply or exclusively on these colleges and universities.

CONCEPTUAL FRAMEWORK

Puffery, a transdisciplinary concept used in legal studies and business (primarily marketing), is the analytic framework used in this study. Essentially, puffery is used to assess the feasibility of claims made in advertisements and elsewhere about the features, performance, and utility of goods and services (Xu & Wyer, 2010). "Legally, the most significant characteristic of 'puffery' is that it is a defense to a charge of misleading purchasers of goods, investments, and services, or to a charge that a promisor has made a legally cognizable promise" (Hoffman, 2006, p. 106). Hoffman further notes that legal applications of puffery share two characteristics. First is the encouragement of purchase or consumption without sufficient evidence to support optimistic claims marketed to prospective buyers. Second is that purchase decisions are made based on so-called facts that sellers communicate. Given this, puffery analyses necessarily entail the scrutiny of false facts (which include half-truths).

The Federal Trade Commission (FTC) aims to protect consumers from deceptive advertising claims. As such, courts use puffery to decipher "falsifiability" and "claims not capable of measurement" (Hoffman, 2006, p. 108). At the center of analysis are questions about whether fact can be ascertained from advertising speech, as well as whether consumers were capable of being deceived by immeasurable claims. This involves a measure of reasonable believability. While the FTC has long held that reasonable consumers are capable of recognizing puffery and exaggerated statements, decades of empirical research in which the concept has been applied suggest otherwise (Kamin & Marks, 1987; Richards, 1990). Puffery analyses are used in this study to review claims that large postsecondary institutions make concerning diversity and to assess the feasibility and reasonableness of such claims.

METHODS

The following research question was explored in this study: What diversity commitments and outcomes are conveyed in mission statements of universities that enroll 20,000 or more undergraduate students? Qualitative content analysis methods were used to systematically determine content categories or themes that represent promises made in mission statements. Qualitative content analysis is a "research method for the subjective interpretation of the content of text data through the systematic classification process of coding and identifying themes or patterns" (Hsieh & Shannon, 2005, p. 1278). This method is particularly useful for taking large amounts of textual data and, through reduction and sensemaking efforts, identifying core consistencies and meanings that help answer a research question (Patton, 2002). Although originally introduced as a quantitative technique, content analysis has evolved as a qualitative method, and examples abound of its effective application in studies in the social sciences (e.g., Mazaheri et al., 2013; Melender, Sandvik, Jonsén, Hilli, & Salmu, 2012; Storlie, Moreno, & Portman, 2014) and specifically in higher education research (e.g., Creamer & Ghoston, 2013; Hartley & Morphew, 2008; Philipps, 2013).

Hsieh and Shannon (2005) introduce three types of qualitative content analysis studies: conventional, directed, and summary. Although some steps in the application of each approach are similar, the conventional approach was most appropriate since the focus of our study was on promises conveyed to current students via mission statements. Prior research on mission statements in higher education has been concerned with what statements communicate to external stakeholders and prospective students. Thus, the strength of the conventional approach to qualitative content analysis is that content categories inductively emerge from the data rather than being fit into preestablished codes (Mayring, 2000). The key to trustworthy and dependable findings based on qualitative content analysis is systematic and transparent procedures. Therefore, the steps we followed, which were adapted from Zhang and Wildemuth (2009), are detailed below.

1. **Creating the Database.** Data for this study were mission statements from not-for-profit colleges and universities that enroll 20,000 or more students in the United States (n = 101). Institutional enrollment figures were verified using IPEDS. Once the list was built, one research team member visited websites for each institution; IPEDS College Navigator has a direct link to the mission statement page of every university website. A document was then created that included the name of each institution and all corresponding text data from the search results. If a values or vision statement was included on the mission statement website, this information was kept as the research team agreed that this information was intended to be taken as a whole since the universities assembled their pages that way.

2. **Defining the Unit of Analysis.** As a result of the large corpus of textual data generated in Step One, defining the unit of analysis becomes one of the most important decisions in a content analysis study since the unit of analysis helps guide researchers as they attempt to make sense of all the data (Weber, 1990). Rather than focusing on frequency of words or the linguistic structure of sentences, we decided that language pertaining directly to students' experiences and outcomes would be our unit of analysis.

3. **Developing Categories and a Coding Scheme.** Initially, an inductive coding approach was used to generate themes that were as close to what institutions convey to students and as free from researcher interpretation as possible. No literature has been published on how to handle issues of trustworthiness when using conventional content analysis. Therefore, following the recommendation of Elo et al. (2014), one member of the research team did all the analysis while the other two team members followed up on the process; this allowed us to avoid issues of interrater reliability. The research team member first conducted multiple readings of the entire body of the data in order to become immersed in the data and gain a better understanding of potential codes. Then the researcher uploaded the text document to Dedoose Qualitative Analysis Software and read the data set again, this time applying thematic codes to segments of data that answered the research question. To facilitate this read through, the researcher employed the constant comparative method in order to make clear distinctions between emerging codes (Glaser & Strauss, 1967). Many of the codes in this phase of coding were in vivo codes, or codes that used the exact language used in the institutional mission statements (Miles, Huberman, & Saldaña, 2013).

4. **Testing the Coding Scheme and Coding All Text.** The researcher responsible for analysis initially coded 50 mission statements. As text segments became redundant across mission statements, the researcher began to assign previously established

codes. Through this process, 90 unique themes emerged from this phase of the analysis. The research team met to discuss the initial codes, and after hearing definitions of the emergent codes, the research team agreed that the codes were germane to the research question and that the coding scheme was appropriate and effective. Following the meeting, the researcher continued to code the rest of the data. Unique themes were added to the coding list throughout the process if a segment of data did not fit into a previously used code. Once all the data were coded, there were 140 distinct codes.

5. **Assessing Coding Consistency and Drawing Conclusions From Coded Data.** Following the completion of the data analysis, the research team met again to discuss the coding scheme and whether the themes that emerged were relevant to the research question. Also, as an added mechanism of trustworthiness, the researcher had a group of five colleagues unfamiliar with the research topic but familiar with qualitative analysis code a subsample of mission statements in order to determine if people outside of the research team also found the emerging themes. All the codes found by colleagues were included in the initial phase of data analysis, and thus confirmability of the codes was determined to be met (Guba, 1981; Lincoln & Guba, 1985). To further the sense-making process, the researcher read all the emergent codes and merged similar codes together while continuously returning to the raw data to make sure meanings were not lost in the process. This phase produced 97 codes. Following that step, the researcher grouped similar emergent codes into more broad themes culminating with 12 categories, 3 of which are related to the research question explored in this paper on diversity.

SUMMARY OF KEY FINDINGS

Diversity-related promises were often conveyed in three categories: (1) the promise of inclusive campus environments in which all students are valued and respected regardless of background, (2) the promise of preparing students for a diverse democracy and workforce, and (3) the promise of equipping students with a global outlook and preparing them for global engagement. In the full version of this paper, we quantify the occurrence of statements within each category, as well as present verbatim examples from universities in our database.

We also use puffery analyses to pose several questions about the feasibility of diversity statements conveyed in mission statements, especially given the size of these institutions. For example, in light of findings presented in Hurtado (1992), Harper and Hurtado (2007), and other published campus climate studies, can Arizona State University actually ensure an inclusive environment to all 59,382 undergraduates—even those who are gay, undocumented, and/or first in their families to attend college? Will all 26,259 undergraduates leave the University of Georgia with a global outlook, prepared to effectively engage with peoples from other nations across the globe—how can UGA be sure of this? Can an institution that has only one diversity requirement in its formal curriculum, few faculty of color, no written plan for making good educational use of diversity in the student body outside the classroom, and no response to visible patterns of racial segregation on campus ensure White undergraduates that they will be effectively prepared for future workplace settings in which the majority of their colleagues will be people of color? These are just three of several questions our data and conceptual framework permit us to raise in the full version of this paper. We ultimately determine the diversity promises that institutions make are actualized for some students, yet are largely incapable of measurement and likely unreasonable exaggerations that do not reflect the realities of most undergraduates attending large universities.

SCHOLARLY SIGNIFICANCE

This paper is important for at least four reasons: (1) It advances the scholarship on mission statements by focusing squarely on diversity promises and on large institutions, (2) it relies on a transdisciplinary framework from law and business for conceptual analysis, (3) it is based on data that are inclusive of every not-for-profit postsecondary institution in the United States that enrolls 20,000 or more undergraduates, and (4) it engenders a critical conversation about the feasibility of promises colleges and universities make in a society that is becoming increasingly diverse. Practically every postsecondary institution in the United States, regardless of size, has a mission statement. Therefore, findings and critical questions emerging from this study have utility across the landscape of U.S. higher education.

Appendix U

Example Research Paper Proposal

ASHE 2014 Research Paper Proposal

Hillary B. Zimmerman, Demetri L. Morgan and Tanner N. Terrell

"ARE WE REALLY NOT GOING TO TALK ABOUT THE BLACK GIRL?" THE INTERGROUP RACIAL ATTITUDES OF SENIOR, WHITE, SORORITY WOMEN

Abstract: Despite the positive effects of cross-racial interactions for students (Chang, Astin, & Kim, 2004), predominantly White sororities remain segregated (Sidanius, Van Laar, Levin, & Sinclair, 2004). Utilizing focus group methods, this study investigates the racial attitudes of White sorority women to understand the influence of sororities on racial attitudes.

Purpose/Objectives: With the changing racial composition of college campuses, institutions have made efforts to understand more about race and the dynamics it causes among college students. However, there are differing views on the influences of a diverse student body. Some believe that diversity leads to cross-racial interactions that have "positive effects on students' intellectual, social, and civic development" (Chang et al., 2004, p. 529). While others point out that even with a diverse student body, "students from different ethnic groups remain relatively segregated and isolated" from White students (Sidanius et al., 2004, p. 96), and the intended positive interracial interactions inside and outside the classroom do not occur or keep pace with the increase in the number of students of color on campuses (Gurin, Dey, Hurtado, & Gurin, 2002; Milem, 1998). Among students who do interact across race, there are many encouraging outcomes, including increased student retention, satisfaction with college, and improved student intellectual and social self-concept (Astin, 1993; Chang, 1996; Gurin et al., 2002; Hurtado, 2001). Research also points out that racially diverse living and work arrangements (Chang et al., 2004; Hurtado, 2005) and peer group influences provide essential conditions for student cross-racial interaction and racial understanding (Antonio, 2001).

Fraternities and sororities are highly influential peer groups or subcultures on college campuses. They are more likely to influence the behavior of their members as attitudes and interests are similar, membership is highly valued, and isolation occurs in regards to interaction with people, not within the group (Kuh, 1990). Media attention over the last year has shed light on the racial tension that exists in the recruitment practices (Crain & Ford, 2013; Grasgreen, 2013), party themes (Jaschik, 2013a, 2013b, 2013c; Lederman, 2014) and behavior (Jaschik, 2014) of fraternities across the nation, drawing attention to the racial homogeneity and lack of racial understanding of these student organizations. A quantitative study of fraternity and sorority in-group and intergroup racial attitudes revealed a host of potentially undesirable outcomes associated with sorority membership for the students, including increased opposition to affirmative action and interracial marriage as well as more tolerance for symbolic racism, which all reinforce intergroup bias (Sidanius et al., 2004). However, the study does not address the nuances and subtleties of the assumptions and attitudes sorority members have about the intersection between race and their sorority experience and the resulting effect of those dynamics on the chapter and campus climate.

Thus, our present qualitative study seeks to answer the following research questions: *What are the intergroup racial attitudes of senior, White, Panhellenic Council (PHC) women, and in what ways does their particular chapter culture affect their intergroup racial attitude?*

Theoretical and Conceptual Framework

The diverse learning environments framework enables policy makers and scholars to categorize existing practices, emerging practices, and policies on college campuses concerning race into three dimensions: structural diversity, psychological dimensions, and behavioral dimensions. The behavioral dimension of campus climate, which is of most interest to this study, deals with "(a) actual reports of general social interaction, (b) interaction between and among individuals from different racial/ethnic backgrounds, and (c) the nature of intergroup relations on campus" (Hurtado, Milem, Clayton-Pedersen, & Allen, 1999, p. 291).

Studying campus climate related to White students can be challenging as the investigation calls for the illumination of racial beliefs and assumptions about one's own race (i.e., in-groups) and other races (i.e., out-groups), which are often unconscious or thought about infrequently. Social identity theory (SIT) has often been employed as a theoretical tool to help scholars better understand the privileging of in-groups and the discrimination against those in out-groups (Tajfel, 1986). The main tenet of SIT is that in-group identification is causally related to intergroup bias (Kelly, 1993).

Sidanius et al. (2004) applied SIT to the college context by performing a quantitative study between minoritized racial or ethnic student organizations and predominantly White fraternities and sororities, testing to see how the affiliation of members in a racially homogeneous group affected their in-group and intergroup attitudes. The study created four conceptual clusters that were utilized to measure the in-group and intergroup racial attitudes of students: racial policy attitudes, social identity attitudes, ethnic prejudice, and perceived group conflict. The racial policy attitudes cluster sought to determine how participants' form attitudes toward policies that have overt or covert ethnic or racial overtones, such as affirmative action or policies on campus that promote ethnic diversity. The social identity attitudes cluster focused on the participants' assumptions and beliefs about their level of ethnic identity within their context and their sense of belonging to their in-group. The ethnic prejudice cluster focused on the positive or negative experiences participants' had with different ethnic groups (including their own). Finally, the perceived group conflict cluster focused on assessing the participants' beliefs about the level of discrimination that their particular ethnic group may or may not face, their views on the cause of segregation on campus, and how their context contributes to it.

Furthermore, Zimmerman et al. (forthcoming) utilized constructivist epistemology and the Sidanius (2004) conceptual framework in their qualitative study of racial attitudes of senior, White, interfraternity councilmen. Although not an initial focus of the study, one of the significant findings was the way in which IFC members drew stark distinctions between their intergroup racial attitudes and their perception of PHC members' intergroup racial attitudes. Therefore, in order to investigate their claims and determine if the PHC environment engenders similar or different racial attitudes, the research team set out to use the same methodology and methods with PHC organizations.

Research Design

Bryant and Charmaz (2007) assert that constructivism assumes that both researchers and participants construct multiple realities that produce useful data. Accordingly, the research team was interested in the co-constructed reality of the intergroup racial attitudes created by the participants and then interpreted by the research team. Focus group methodology was selected for its usefulness in identifying areas of agreement and disagreement within a group (Carey & Smith, 1994; Reed & Payton, 1997; Sim, 1998). Furthermore, the setting allows participants to challenge each other and seek clarification on contentious issues, potentially revealing the sources of complex behaviors and motivations (Morgan & Krueger, 1993).

Participants

Researchers used purposeful (Patton, 2005) and criterion sampling (Creswell, 2012) to recruit organizations to participate in the focus groups. The research site was located at a large, public, research university in the Midwest. The sampling criteria for the study focused on PHC organizations that have large chapter houses at the research site with living accommodations for over 75 people due to research that suggests that members living in chapter facilities possess attitudes that are more racially hostile than those that do not live in chapter facilities (Hughey, 2007; Morris, 1991). The organizations also had to be established at the institution prior to 1950 to ensure that the groups have had a substantial amount of time to develop a pervasive and unique chapter culture (De Los Reyes & Rich, 2003). Additional criteria stipulated that focus group participants had to self-identify as senior (fourth year or above), White, and female. These criteria were used based on research that suggests that senior sorority members are significantly affected by the fraternity/sorority experience on a range of outcomes, including openness to diversity (Pike, 2003; Zhao & Kuh, 2004). In combination, these criteria presented the opportunity to glean information-rich data about their intergroup racial attitudes (Patton, 2005).

In order to recruit organizations and participants, a member of the research team contacted the presidents of all the PHC organizations that met the sampling criteria via email and asked them to participate in the study. After the research team gained consent, they worked with presidents to recruit members to participate in the focus groups, in alignment with purposeful sampling (Patton, 2005). The final sample consisted of four organizations. Each focus group had five to seven participants, totaling 29 participants.

Data Collection

Data were collected during four 1-hour focus groups utilizing a semi-structured interview protocol (Fontana & Frey, 1994). The focus groups were held at the participant's chapter houses in order to foster a sense of comfort so that the participants might feel more open to speaking candidly about their fraternity/sorority experience. To create the interview protocol, the research team used the conceptual clusters from the Sidanius et al. (2004) study as guides to formulate interview questions that would facilitate discussion about how the essence of each cluster functions in their fraternal experience and in the larger chapter culture. Follow-up questions were asked to better understand the overall climate of the chapter concerning aspects of race, racism, and cross-racial interactions. Focus groups were digitally recorded, and members of the research team transcribed each session verbatim.

Data Analysis

Transcripts were independently coded and analyzed for consensus and disagreement among the participants regarding the intersection of their chapter experiences, intergroup racial attitudes, and how these experiences and attitudes were influenced. Coded transcripts were compared, and similar codes were collapsed into parent codes. Parent codes were defined and arranged into overarching themes that were illustrative of the participants' experiences and racial attitudes (Miles, Huberman, & Saldaña, 2013).

Supplementary Analytic Perspective: Color-Blind Racism Framework

Some scholars suggest that many of the epistemologies typically used in educational research, including constructivism, may suffer from epistemological racism (Harper, 2012; Scheurich & Young, 1997). Given the team's interest in race and racism in this study, we also utilized a sociological perspective known as color-blind racism to analyze the data (Bonilla-Silva, 2009). This framework is helpful in revealing the ways in which "dominant racial groups express racial prejudice . . . without appearing racist" (Berry & Bonilla-Silva, 2008, p. 217). Data were analyzed through the color-blind framework to disrupt and challenge the participants' voices and the researchers' interpretation of their voices in order to be mindful of the way that racism is potentially enacted in their chapter and subsequently influencing their racial attitudes.

Trustworthiness

In order to ensure trustworthiness and credibility in research findings, the team members individually coded each transcript and then compared codes throughout the analysis to confirm consistency and triangulate the data. The team shared findings with an outside researcher familiar with the context of the study as a form of peer debriefing, and participants were provided a summary of the themes as a form of member-checking (Creswell, 2012, 2013). The research team engaged in a dialogue during data analyses to challenge each other's biases and assumptions that might have stemmed from the researcher's positionalities (Britten, Stevenson, Barry, Barber, & Bradley, 2000).

Findings

[Below is a summary of the parent codes that are emerging in the analysis. A full theoretical storyline, with supporting quotations, will be included in our ASHE paper, if accepted.]

Minimization of Race

Participants in the focus groups frequently diminished and detracted from the racialized questions and prompts instead focusing on other identities by highlighting the diversity in their academic majors or the hometowns from which they hailed. Participants felt that focusing on race would marginalize the accomplishments of the individual and would run the risk of sending a message that certain members were only accepted because of their race. Participants also felt that raising the group consciousness concerning race could make individuals less comfortable in an organization, with some participants pointing to the fact that they would not feel comfortable joining a predominantly non-White sorority because of its perceived value on a particular racial identity or group of racial identities.

Out-Group Perception

There was a tendency among the participants to perceive predominantly non-White fraternities and sororities in overly generalized and communally negotiated ways. This phenomenon allowed the White sorority women to create one-dimensional views of students of color that could inform their overall racial attitudes. The participants held generalized and simplistic perceptions of students of color and the predominantly non-White fraternities to which they often belong, such as predominantly non-White fraternities and sororities are much more unified, cohesive, and harmonious; predominantly non-White fraternities are more inclined to reach out and collaborate with predominantly White sororities than their male counterparts; and there are significant but elusive differences in the values that unite predominantly White sororities versus their non-White counterparts.

Traditions and History

Participants often pointed to the past when justifying many observations of the contemporary climates of their organizations. Some participants pointed to stories shared by their parents and other family members when certain characteristics could be traced back to the organization's origins. Other participants alluded that legacy within the organization could explain the demographics of the group, as many of the sororities studied had recommendation policies that place higher value on pledges that have connections to current or alumnae members. Participants also contrasted the perceived allegiance to the national organization on behalf of culturally based fraternity and sorority members to their own allegiance to the local chapters. Members also used tradition to explain why collaboration did not occur between their organization and any of the culturally based fraternities and sororities, explaining that because they did not have an established relationship with an organization, they would not be likely to reach out to that group to cooperatively develop an event.

Relative Diversity

Participants generalized about fraternity and sorority life and made contrasting claims regarding their specific organizations. All organizations that completed focus groups alluded to the lack of diversity in sorority life. However, each organization also made efforts to explain ways in which their specific organization was relatively more diverse than other similarly situated organizations. Utilizing some of the characteristics outlined in the Minimization of Race section, participants would highlight diversity in personality types, academic pursuits, and passions to demonstrate ways in which their organization was more diverse than other predominantly White sororities.

Significance

On many college campuses, fraternity and sorority life represents one of the largest unified student groups, wielding significant power as a shaper of campus culture. While there is a body of literature on fraternity and sorority life, a majority of that work investigates topics such as alcohol use, hazing, academic performance, and sexual assault. However, media attention has revealed racial attitudes deeply entrenched with racial meaning. Recently, the campus or national organization's response to these issues has been to suspend or shut down the local organizations in an effort to demonstrate that the actions of these young adults do not reflect the mission of the organization or institution. Unfortunately, this type of response leaves little opportunity to create meaningful learning opportunities from these regrettable situations. Thus, an important occasion to address the racial attitudes of fraternity and sorority life in a meaningful way is missed. The response on behalf of the institution and/or national organization could be explained on the basis that very little qualitative research exists on how the intergroup attitudes allow a group of one hundred or more to host a race-based stereotype-themed event. Therefore, little empirical knowledge exists on how these attitudes are precipitated and how they propagate through the organization. With our study, we hope to provide data on how these attitudes are shaped so that campuses and organizations can formulate more meaningful responses that effectively address the attitudinal origins of these headline-grabbing actions.

Appendix V

Example Project Statement for Grant Proposal

Sarah Klevan, New York University

I. PROJECT DESCRIPTION

Negative schooling outcomes experienced by minority boys have attracted attention from a broad range of researchers and education practitioners. Indeed, concern about the school trajectories of minority boys has even attracted attention from the White House, which introduced My Brother's Keeper in 2014, a program that encourages cities across the country to improve schooling outcomes for minority boys.

One of the most troubling findings regarding minority males' school experiences is their overrepresentation in disciplinary consequences. The past four decades have seen a doubling in the use of suspensions in schools. While the number of suspensions has risen for all segments of the population, the rate of suspension for Black students has increased by a far greater degree (Wald & Losen, 2003). In New York City, the site of the dissertation study, Black students currently represent 33% of the student population but account for 53% of suspensions over the past decade. Numerous studies demonstrate that boys receive disciplinary consequences at a rate higher than girls (Skiba, 2002), thus placing minority males at an especially high risk of receiving disciplinary sanctions. Removal policies, such as suspensions and expulsions, result in a critical loss of instructional time for offending students and are correlated with a host of negative outcomes for students ranging from disengagement in school to involvement in the juvenile and criminal justice systems.

Though a large body of research confirms that school removal policies do not successfully deter misbehavior or maintain safety and order in schools (Fabelo et al., 2011; Kupchik, 2010; Nolan, 2011), little research has investigated alternative disciplinary practices. An emerging alternative approach to removal policies is a practice called restorative justice. Experts in the field of restorative approaches note that the focus of these programs is on rehabilitating relationships in school communities (Evans & Lester, 2013; Karp 2001). This places restorative approaches in contrast with traditional disciplinary practices, which undermine core school relationships and remove students from the school community (Arum, 2003). Restorative approaches take many forms, including peer mediation, restorative circles (in which groups of students resolve community conflicts), restorative conferences (which bring together individuals who have caused harm to their communities with those who have been harmed), and others.

This dissertation will contribute to the conversation around school discipline by producing an ethnographic account of a New York City high school's implementation of restorative justice approaches to discipline and the ways in which these practices shift the disciplinary experiences and outcomes for minority males at the school. Understanding how school actors implement restorative approaches is essential for scaling up these practices to a larger number of schools. Only by gaining a nuanced understanding of restorative approaches, their implementation, and their effectiveness in promoting positive school outcomes can this new approach for managing discipline receive the notice it deserves.

Research on restorative justice approaches is scarce, and many available studies were conducted abroad or in settings other than schools. While findings from existing studies suggest that restorative approaches may have positive outcomes for communities and students (McMoris et al., 2013), available research leaves much to be learned about the implementation of restorative approaches and how they might benefit students most at risk for disciplinary consequences.

The current study is guided by the following research questions: (1) What forms do restorative justice approaches take in an urban high school setting? How are these approaches implemented, and what challenges are encountered in the implementation process? (2) How is discipline managed differently when restorative justice approaches are utilized? (3) How do restorative justice approaches alter the disciplinary landscape for boys in comparison with girls? This study will provide the first in-depth, ethnographic account of restorative justice practices and how these practices alter outcomes for minority males.

III. RESEARCH DESIGN AND METHODS

Dissertation Site

The dissertation site, referred to here as the School for Restorative Justice (SRJ), is a New York City public high school serving approximately 400 students. I recruited SRJ because the school has incorporated restorative justice approaches for the past 3 years. This makes the school an ideal place to study practices that are established and understood by the school community. Additionally, SRJ's male population is approximately double that of its female population (28% female, 72% male). By locating this research at SRJ, I am able to examine restorative justice approaches at a school that serves a majority of male students, the segment of the student population that is most at risk for disciplinary removals. Because of their commitment to improving their disciplinary practices, the administrators at SRJ are enthusiastic about hosting my research. They have introduced me to their staff and have encouraged staff members to participate in this study. This type of access is crucial for completing an ethnographic study, as it allows the researcher to pursue relevant lines of inquiry even if he or she uncovers tensions, challenges, or struggles within the school.

Research Design

This project incorporates extensive observation, document collection and analysis, and informal and formal interviews with key school actors. Much of the fieldwork will take place in the Youth Development Office at SRJ. This is the location in which many restorative practices occur and the space in which school staff often discuss student behavior and disciplinary approaches. However, initial fieldwork has led me to understand that "discipline" is located at numerous sites within the school—in the hallways, in the school entryway, in professional development sessions, and in classrooms, among others. I will collect data in all of these locations over the course of 1 academic year.

I will conduct 40 formal interviews with teachers, the principal and assistant principal, the restorative justice coordinator, the school dean, school safety agents (SSAs), and students. Interviews will be essential for understanding how discipline was managed at SRJ prior to the implementation of restorative approaches. In addition to formal interviews, my research will incorporate numerous informal interviews with members of the school community.

IV. TIMELINE

Thus far, I have secured approval from the New York City Department of Education Institutional Review Board and New York University's Committee on Activities Involving Human Subjects. I have recruited SRJ's participation and have spent 40 hours at the school site, which has resulted in approximately 100 pages of typed fieldnotes. I have also conducted interviews with the principal and the restorative justice coordinator at SRJ. I expect to finish my data collection and analysis by August 2015, and I expect to write and defend my dissertation by May 2016.

V. STATEMENT OF NEED

I currently work 30 hours a week as a research analyst at the Research Alliance for New York City Schools. The time that this work requires will prevent me from completing my dissertation within the timeframe outlined above. This fellowship will enable me to reduce my workload substantially, allowing me to devote the majority of the 2015–2016 academic year to completing my dissertation. I have utilized all other funding opportunities within my department and am in need of funding that will support the writing of my dissertation. This funding will also offset costs such as interview transcription and travel to my dissertation site, which are essential to the completion of this project.

Appendix W

Example Fellowship Proposal

Fellowship Proposal: Fulbright-Hays Doctoral Dissertation Research Abroad Fellowship Program

Arjun Shankar, University of Pennsylvania

PROJECT OVERVIEW

During a visit to a rural school in Kanakapura, India, I interviewed students about their educational aspirations. One student, Radhini, responded that she wanted to *"[be] a doctor . . . [so that] after finding out what each of the villager's problems are they can come back to the village and work again."* Radhini attends one of over 1,000 schools in rural Karnataka that are "developed" by INDIAN NATIONAL-NGO[10] Foundation, an education nongovernmental organization (NGO) headquartered in Bangalore. INDIAN NATIONAL-NGO's founder and former engineer, Ramaswamy, told me in an interview that rural education could be fixed by bringing the notion of "quality management" to bear on educational programming. In narrating his story, he included global thinkers such as Edward Deming and Dale Carnegie and cosmopolitan travels ranging from São Paulo to Munich. Ramaswamy and Radhini are connected in an emerging "zone of engagement" (Tsing, 2005), which places education NGO personnel from cities in relation to teachers, farmers, and their children in rural schools.

In 21st-century scholarship on globalization and development, cities in the global South, such as Bangalore, have been characterized as nodes in an emergent capitalist network, where new infrastructures are developed to "tap into" global economic flows (Sassen, 2001). Such scholarly works emphasize the *transnational* character of the global city (Appadurai, 2001) and the urban inequality that these global networks have produced (Caldeira, 2000; Davis, 2006). However, in illuminating the relationship between the city and the globe, scholars have failed to adequately investigate the effects of global values on rural locales. My research investigates this zone of engagement and asks, How do cosmopolitan values—circulating globally—get reanimated in urban-rural interaction? How do these urban-rural interactions produce new conceptualizations of development and change the character of Indian rural education?

Bangalore's current development agenda serves to contextualize this research. Like many of India's cities after economic liberalization in 1991, Bangalore experienced incredible economic and infrastructural growth. Many have termed it the "Silicon Valley" of India, representative of the emerging trend toward high-tech global "knowledge" or "information" cities (Nair, 2005). Rural areas adjacent to the city—once the agrarian heartland of Karnataka—face encroachment from developers seeking new land for commercial use (Goldman, 2011). For example, under the Karnataka Areas Act of 1966, the state has procured land for the building of the Bangalore-Mysore Nandi Infrastructure Corridor (NICE), a highway project that displaced over 200,000 farmers (Saldanha, 2007). Roads such as the NICE road serve as "self-worlding practices" (i.e. methods by which Bangalore re-creates itself as a world city) (Ong, 2011).

Another emergent self-worlding practice is the proliferation of Indian education NGOs. In India today, there are over 13.3 million NGOs, or 1 NGO for every 400 Indian citizens (Shukla, 2010), the most NGOs in any nation-state the world over. As the government continues to move away from direct educational intervention (Hill & Kumar, 2009), as poverty alleviation takes on new significance in remaking the "world-city" (Roy, 2010), and as particular skills—such as English and computer training—are exceedingly necessary to join a changing workforce (Heitzman, 2004), the education NGO has become one of the most significant new actors within the human development space. Moreover, after the construction of the NICE road, education NGOs headquartered out of Bangalore, like those run by Ramaswamy, now have access to previously isolated rural communities, like Kanakapura. In other words, infrastructural development, like the NICE road project, has necessitated a concomitant attention to particular new forms of human development as implemented by NGOs.

This multisited ethnography focuses on two education NGOs, INDIAN NATIONAL-NGO Foundation and INDIAN NATIONAL-NGO 2, that work in communities along the NICE road. These NGOs have chosen to work with the government, partnering with state and local bureaucracies to improve measurable outcomes. However, in the current historical moment, Indian education NGOs cull their foundational values from discourses that circulate globally (Ferguson & Gupta, 2002). Specifically, NGO personnel utilize the values they have gathered from management, finance, information technology, or infrastructural development in their interventions in Karnataka state schools. Yet, when these values are animated in rural educational interventions, they are necessarily challenged and create emergent systems of meaning (Ong, 2011). What values do NGO personnel believe should be taught in schools? What values do farmers seek to gain from NGO personnel? What do farmers value independent of the formal education system? How do farmers' values change the particular ways that NGO personnel must intervene?

I hypothesize that interactions between NGO personnel and rural community members will reveal, what I term, the *affect of development,* a moral sentiment that pervades the self- and society-making practices of those residing in postcolonial contexts (Fassin, 2010; Gupta, 1998). These self-making practices position members of India's society, both urban and rural, as exemplified by Radhini and Ramaswamy above, making self-development a primary prerogative as India changes. This emerging affective disposition is tied closely to changing values, or "value migrations" (Slywotzky, 1996) currently occurring in rural areas of Karnataka, as traditional forms of subsistence are being rendered untenable. In the case above, Radhini values skills generally associated with the city (i.e., medical training) rather than agricultural skills. Critically, the change in values *does not connote* a physical migration away from the village. Rather, Radhini's aspiration reveals how values circulate and are reengaged in rural contexts. Like Radhini, Ramaswamy's reply shows how values normally associated with cities—in this case globally circulating values regarding quality management—are being reengaged in rural contexts as moral prerogatives toward action.

THEORETICAL FRAMEWORK

An important corpus of current anthropological scholarship concentrates on the *human* and *moral* realms of global development as opposed to the emphasis on market-driven or "neoliberal development" (Harvey, 2005). For example, Nussbaum and Sen's (1993) "capabilities approach" to development includes indicators for both education and health care. The justification for these development approaches is *moral,* resting on social justice agendas to alleviate poverty (Roy, 2010). Fassin (2010) has called this global moral sentiment *humanitarian reason,* "a language that inextricably links values and affect" (p. 1). My research follows Fassin's calls for ethnographic study of these humanitarian morals and their implementation. Such globally circulating moral sentiments have traditionally been thought to drive NGO action (Kilby, 2011). Yet, recent anthropological scholarship has shown how NGOs sit at the nexus of market and moral economies, drawing funds from corporations (Gill, 2000) and deriving their principles from finance and management (Roy, 2010). My research will build upon these studies by showing how NGOs merge humanitarian and neoliberal principles in their educational interventions, a process commonly termed the "neoliberalization of education" (Hill & Kumar, 2009). Further, anthropologists have shown how development becomes an *internal moral prerogative* (i.e., a self-making process) in postcolonial contexts (Gupta, 1998; Pandian, 2008). I extend these arguments by showing how this internal moral prerogative manifests in *both* NGO personnel from Bangalore and rural community members in disparate and overlapping ways.

Within the economic and human frameworks for understanding development, cities, especially newly developing cities in the global South, have become primary sites for anthropological research (Heitzman, 2004). These cities have been characterized as nodes in an emergent capitalist network, where material processes and new infrastructures are developed to "tap into" global economic flows (Sassen, 2001). Such scholarly works emphasize the *transnational* and *cosmopolitan* character of the global city (Batliwala & Brown, 2006). Other scholars have focused on the inequality that these global networks have produced (Davis, 2006; Holston, 2009). However, in order to combat monolithic representations of globalization's effect on the city, some scholars have called for a "regional modernities" approach to cities, focused on the particular manifestations of the global in the local (Sivaramakrishnan & Agarwal, 2003). Ong (2011) proposes the concept of *worlding practices,* "an array of often overlooked urban initiatives that compete for world recognition in the midst of inter-city rivalry and globalized contingency" (p. 3). Anthropologists concerned with cities in South Asia have utilized this framework to show, for example, how the making of world-class Delhi relies on the development of a world-class aesthetic effecting both middle-class and slum dwellers alike (Ghertner 2011), how the making of world-class Kolkata relies on protest and blockades simultaneously (Roy, 2011), and how the making of world-class Bangalore relies on road building projects, like the Mysore-Bangalore Nandi Infrastructure Corridor or NICE Road (Goldman, 2011).

In focusing on global cities, many scholars have reified a separation between the urban and the rural (Lipton, 1977). Scholars have suggested that this conceptual separation has resulted because of the severe differences in modes and standards of living in rural and urban areas outside the city. I will spend the months of June and July with INDIAN NATIONAL-NGO 2 and the months of August and September with INDIAN NATIONAL-NGO Foundation. Both NGOs are working with populations that were directly impacted by the NICE road building project. Both have also chosen to partner with the government, rather than work outside of the government system. I am particularly interested in the two NGOs specified above—INDIAN NATIONAL-NGO 2 and INDIAN NATIONAL-NGO Foundation—because NGO personnel from both organizations have held previous management, information technology, or infrastructural development occupations that have allowed them to travel to cosmopolitan centers, both within and outside of India.

The principal methods of investigation for this part of my research are qualitative, utilizing traditional ethnographic methods for data gathering, including participant observation, document collection, the writing of fieldnotes, and the recording of speech interactions with a digital recorder. I will follow NGO personnel from each organization both in their interactions at their Bangalore headquarters and as they intervene in rural school contexts (Tsing, 2005). By paying close attention to the ideas NGO personnel reference when describing their work and observing their actions in the field, I can begin to see how global values are uniquely engaged in their rural interventions. How do globally circulating values generally associated with the city get engaged in urban-rural interaction? What power asymmetries do these engagements reveal (Habib, 1963; Williams, 1973) and have tended, therefore, to focus on the particularities of rural practices, economic systems, and resistance efforts (Bernstein & Byres, 2001; Guha, 1983)? By creating this conceptual separation, anthropology has yet to thoroughly engage with the ways in which worlding practices *create new linkages between the urban and the rural.* A few anthropological studies do, however, explore urban-rural linkages, showing how rural ideas and practices exert influence on urban areas (Ferguson, 1992), how urban labor migrations impact rural families (Murray, 1981), how students from cities forge new relationships with the countryside (Tsing, 2005), and how villagers no longer migrate to cities but rather have cities migrate to them (Guldin, 2001). In the Indian context, anthropologists have shown how changing migration patterns have begun to blur the line between the urban and the rural itself (Nair, 2005) and have shown the effects of media and consumer culture on rural communities (Rajagopal, 2001). I build upon this corpus of literature by investigating the emergent value systems that arise in the interactions between NGO personnel and rural farmers.

METHODOLOGY AND RESEARCH TIMELINE

I am requesting funding through the Fulbright Hays Doctoral Dissertation Research Abroad Fellowship for research taking place over a 12-month period beginning in June 2013 and continuing through June 2014. The study is separated into two interrelated parts, the first focusing on NGO values and practices in rural school interventions, the second focusing on the values and practices of other members of rural school communities.

The first part of my study, spanning 4 months, will excavate the global values that filter through two education NGOs, INDIAN NATIONAL-NGO 2 and INDIAN NATIONAL-NGO Foundation, headquartered in Bangalore, but working to improve the government school system in rural communities. The second part of my study will focus on students and their families as they negotiate a moment of displacement and NGO intervention. From October 2013 to January 2014, I will focus on two school sites in which each NGO works, one in Kanakapura village and one in Mandya village, both of which have been affected by the NICE road project. I will interview members of the school community, including teachers, headmasters, and students, to get a sense of how they perceive NGO intervention. I will also follow at least eight students as they travel away from the school to their homes. By paying close attention to the ideas students reference and their actions outside the school space, I can begin to see how they engage value systems not associated with the formal school system. I will interview their parents and attend School Development and Management Committee (SDMC) meetings—a monthly school meeting of local community members—to see what values parents wish to impart on their children and how parents view schooling in relation to their steadily changing lives.

Second, in one of my focal school sites, from January 2014 to June 2014, I will work directly with a group of secondary (high school) students on a series of film workshops. Students will learn how to use small handheld video cameras and will create short films using these cameras (Ruby, 2000). This participatory project will provide an opportunity to create *reciprocal* relationships between myself and those who are considered my "research subjects" (Said, 1989). This part of the project will give students the opportunity to represent their communities and the values within these communities. During this time, I will also shoot my own film footage, interviewing farmers and their children directly while capturing their interactions with NGO personnel who come to school sites. I will use both the student footage and my own video footage to create an ethnographic film that will show how rural communities negotiate their changing lifeworlds (Jackson, 2012). Importantly, the filmic medium allows the visual capture of embodied practices (Grimshaw, 2001) and provides an alternative to the written fieldnote, which many anthropologists have claimed can be distanced from phenomenological experience (Ruby, 2000). Using the filmic medium, I will be able to show how particular values are embodied in the everyday life practices of those living and working along the NICE road.

RELEVANT RESEARCH EXPERIENCE AND RESEARCH SKILLS

This research project is the outgrowth of an exploratory research engagement funded by the Center for the Advanced Study of India (CASI) in the winter of 2010 and 3 months of predissertation fieldwork conducted in Bangalore, Mysore, Mandya, and Kanakapura in the summer of 2011. During this time, I was able to cultivate relationships with the key personnel of both the INDIAN NATIONAL-NGO 2 (APF) and INDIAN NATIONAL-NGO Foundation. In collaboration with a research colleague, I conducted an action research project for

INDIAN NATIONAL-NGO 2's field team in Mandya, a village approximately 100 km outside of Bangalore. The project focused on developing research questions that would help NGO personnel identify spaces for intervention in the formal education bureaucracy. The members of the research team were made up of both those who lived in Mandya and those who traveled from the main city center. Besides providing invaluable insights into the formal structures of the Indian education system, the experience also revealed to me the complex and emerging link between the rural and the urban in contemporary India. In my work with INDIAN NATIONAL-NGO, I conducted qualitative research at their school sites in Kanakapura and participated in an organizational development project that articulated methods for INDIAN NATIONAL-NGO to scale up its programming without reifying its practices (Shankar, Dattatreyan, Ravitch, & Ramamurthy, in press). My work with INDIAN NATIONAL-NGO revealed the challenges Indian NGOs face and the discourses they draw from in making organizational decisions. I was also able to observe how NGO personnel interact with those within school communities, including teachers, administrators, students, and parents.

In preparation for my return in June 2013, I have also cultivated relationships with personnel at multiple school sites and have been given the opportunity to work in a school on a film workshop. I have already established housing options in Bangalore, Mysore, Mandya, and Kanakapura as well as affiliations with the University of Mysore and X University.[11]

As language preparation, I have taken 3 years of Hindi through the University of Pennsylvania's South Asia Studies Department. Toward this end, I was given a Foreign Language and Areas Studies Scholarship (FLAS) for the 2011–2012 academic year and also received a Summer FLAS in 2010. Hindi is an invaluable language for my particular project as it allows me to interact with NGO personnel who, many times, are not from Karnataka state but rather are from cosmopolitan locales across India. Therefore, they do not speak the regional language, Kannada, but speak English and Hindi interchangeably. Over the 2011–2012 academic year, I have also taken one-on-one Kannada classes with a visiting scholar from the University of Mysore, Dr. Rallapalli Sundaram, dedicated exclusively to honing my speaking skills. I will continue my Kannada training through the spring of 2013.

To prepare for my fieldwork, I have also taken a number of relevant courses focused on South Asia, including "Society and Public Culture in South Asia," "Neoliberalism and Development in South Asia," and "Urban South India." Moreover I have taken 2 years of coursework focused on film, including "Ethnographic Film" and "Documentary, Ethnography, and Research," in order to develop my understanding of film technique, film as research, and how to incorporate film into my particular project. Finally, previous to joining the University of Pennsylvania as a doctoral student, I spent 3 years as a high school teacher in New York City. My experiences as a teacher greatly influenced my current work and are the basis for my insights into the relationship between education, social structures, and economic development.

Appendix X

Example Preproposal Letter of Interest

05/04/2013

To Whom It May Concern:

I am writing with regards to the SASgov and Dean's Award for *Research and New Media*. I am a fourth-year doctoral student at Penn receiving a joint degree in anthropology and education. My current dissertation research, titled "The Affect of Development," focuses on the relationship between infrastructure projects (road and school building) and human development projects (NGO interventions in schools) in the peripheries of Bangalore, India. While the field component of my research has been funded through a Fulbright-Hays fellowship, the fellowship does not cover equipment costs. Thus, I am seeking the *Research and New Media* award to cover equipment costs for both the participatory film and the web component of my research, described here.

Similar to many Indian cities after economic liberalization in the early 1990s, Bangalore experienced incredible economic and infrastructural expansion. It has been referred to as the "Silicon Valley" of India, representative of the emerging trend toward high-tech global "knowledge" or "information" cities (Castells 2000; Heitzman, 2004; Nair, 2005; Ong & Roy, 2011). This multisited ethnography investigates how roads, NGOs and NGO personnel, schools, students, teachers, and corporations are reorganized as part of India's contemporary project of liberalization.

Specifically, I am conducting my study in schools in two sites outside of Bangalore—Hoskote and Ramanagara—each of which has differing relationships to Bangalore's infrastructural expansion but in which the same education NGO intervenes. Hoskote, approximately 25 km east of Bangalore, a rapidly industrializing factory town, houses a growing number of migrant workers from Uttar Pradesh and Bihar whose children attend its government schools. Ramanagara, approximately 45 km south of Bangalore, on the other hand, is the home of many Kannada-speaking farmers, whose primary industry is silk production but whose children also attend local government schools. Together, the contrast between changing industrial and agricultural occupations in these locales vis-à-vis Bangalore's expansion and an attention to the changing aspirational trajectories of students in these two communities provide a method by which to study the changes wrought as a by-product of "development," at the intersection of the infrastructural and the human.

A central concept in my theoretical framing is that of *affect*. The philosopher Benedict Spinoza writes about affect in his Ethics: "I understand affections [affect] of the body by which the body's power of acting is increased or diminished, aided or restrained" (Spinoza, 1996, p. 70). Following this framing, I ask, Which types of actions diminish an individual's power and which types of actions increase power within and across a social domain? As material processes change—new roads or school buildings are built—a different set of affects is produced, which changes how populations can and do act. Yet in the study of development in the 21st century, the affective dimension has been largely understudied. My task, therefore, is to understand the link between new infrastructural projects and the types of affects—hopes, aspirations, shame, guilt—produced within the education domain that are differentially produced across populations. I argue that a farmer in Ramnagara, or his child in a school there, will have a different set of affects vis-à-vis changing infrastructures than the NGO personnel who work in these areas or those who have the authority to approve such projects, and my project seeks to demonstrate these differences through both textual and visual research products.

In order to study and represent these aspects of current development, I utilize two media strategies: first, a participatory film methodology; second, a multimodal web installation.

As part of the participatory film project, approximately eight students in school sites at both Hoskote and Ramnagara will learn how to use small handheld video cameras and will create short films using these cameras (Ruby, 2000). This part of the project will give students the opportunity to represent their communities and the values within these communities. A central feature will be a "Memory Map," in which students will interview various members of their communities—parents, friends, shop owners, teachers—and ask how community life has changed over the past 20 years. Simultaneously, I will also shoot my own film footage, interviewing members of each community and their children while capturing their interactions with NGO personnel. I will also capture the infrastructural dimensions of the school sites and the surrounding area. I plan to use both the student footage and my own video footage to create a series of short films that will show differing dimensions of how communities negotiate their changing lifeworlds during India's 21st-century development (Jackson, 2012).

Using the filmic medium, I will show how particular values are embodied in the everyday life practices of those living and working in Hoskote and Ramanagara. In this experiential context, *affect* can be utilized as a unit of analysis, allowing the viewer to connect with the emotional investments of those they see on screen (Pink, 2006). How do students depict their communities? What do these depictions reveal about their emotional connection to place? How do NGO personnel sound, look, and act at the moment when they discuss their projects? In what ways do their movements reveal their particular position and investment?

Second, I will develop a web installation to disseminate my research to the larger university community. The web installation will be multimodal, utilizing a mix of audio, video, text, and still photographs. The website as a whole will allow other members of the university community to explore the differential affordances of media as scholarly knowledge production. The onset of the multimedia generation has opened the possibilities for presentation strategies such as the project I have proposed here, which truly embraces nonlinearity (Seaman & Williams, 1992, 1993) and multiperspectival and multisensory processes (Buccelatti, 2002; Nijland, 2001), and it reveals interactive, transparent relations between researchers, research subjects, and audiences (Read, 2001).

The website pages would be designed as "chapters" of my dissertation, each one providing a different reading on the *affect of development*. Thus far, I have outlined three preliminary chapters of the website. First, student films will compromise at least one chapter of the installation, and I envision juxtaposing their representations with my own representations of similar phenomenon. This strategy would give users a multiperspectival understanding of the research context, reveal the complexities of researcher positionality in field research, and provide opportunities for users to interact directly with the student-films, even speak directly with the student-filmmakers.

A second chapter will spatially trace networks of relationships as they impact schools and schoolchildren. For example, one network would connect Coke, NDTV (a national media network in India), the "Support My School Campaign," education NGOs, and school infrastructure projects in Ramanagara. By juxtaposing visual representations on websites with the images and sounds from the actual schools that have been constructed, I will be able to show how visual representations skew perceptions of educational intervention as part of marketing strategies.

The third chapter would be based on a fictionalized account of school life in one of my field sites. Following Wolf (1992), I hope to use fiction as a way of sharing ethnographic insights by having text scroll horizontally across the screen and images of the schools, students, student work, and soundscapes pop up both above and below the text. In this sense, the "affect of development" is experienced by users at an affective level, and at the end of the reading, users can offer their own responses to the piece, interacting with "development" on the site.

In sum, it would be a privilege to continue refining my research method and dissemination strategy under the SASgov and Dean's Award for *Research and New Media*. The grant will go a long way in helping me toward my goal of producing a dissertation both rigorous in its theorizing of context and innovative in its use of media toward these ends.

Thank you for your time,

Arjun Shankar

Appendix Y

Consent Form Template and Examples

In this appendix, we have included several different examples of consent forms. The examples are presented without study information to protect the confidentiality of the participants who were a part of these studies.

APPENDIX Y.1: CONSENT FORM TEMPLATE

TITLE OF STUDY

Part 1: Research Description

Description of the research: [Briefly describe the study in accessible language.]

Risks and confidentiality: [Discuss how and if participants will be promised confidentiality. Describe any potential risks that could result from individuals' participation. Detail who will have access to the data.]

Time involvement: [Describe the time commitment and duration.]

How will results be used: [Describe how the results will be used.]

Contact: [Include a way for participants to contact you if he or she has any questions about the study.]

Part 2: Participants' rights [Include something similar to the below that informs participants of their rights.]

- I have read and discussed the Research Description with the researcher. I have had the opportunity to ask questions about the purposes and procedures regarding this study.

- My participation in research is voluntary. I may refuse to participate or withdraw from participation at any time.

- Any information derived from the research project that personally identifies me will not be voluntarily released or disclosed, except as specifically required by law.

- I DO ( ) DO NOT ( ) consent to be audio recorded.

- Audio recordings will be erased once they have been transcribed.

- My signature means that I agree to participate in this study.

Name: _____

Participant's signature: _____ Date:____/____/____

Researcher's signature: _____ Date:____/____/____

APPENDIX Y.2: EXAMPLE CONSENT FORM

VOLUNTARY INFORMED CONSENT FOR INTERVIEW/OBSERVATION

Interview/Observation Protocol

You are invited to participate in this study because you are a leadership team member of a distributed leadership team (or there is a distributed leadership team active on your campus) in the Archdiocese.

We ask that you read this document and ask any questions you may have before beginning the interview.

This study is being conducted by Mustafa Abdul-Jabbar, a Graduate Research Fellow of the University of Pennsylvania.

PURPOSE: The purpose of the study is to understand more about the Archdiocese implementation of distributed leadership, team decision making, team culture, and other team dynamics.

PROCEDURES: If you agree to participate, the interview will be conducted at a time and place that is convenient for you. Interviews will last approximately 30 to 45 minutes. With your permission, the conversation will be audiotaped.

RISKS AND BENEFITS: Your participation will help to inform feedback reports for the Archdiocese superintendent and help the distributed leadership program to improve services being rendered to the Archdiocese through the current program and ongoing professional development.

COMPENSATION: There will be no financial compensation for participation.

CONFIDENTIALITY: You understand that all information collected in this study will be kept strictly confidential, except as may be required by law. If any publication results from this research, you will only be identified by a pseudonym, and other information that could reveal your identity will be disguised.

WITHDRAWAL: Your decision whether or not to participate will not affect your current or future relations with your school or the researchers. If you decide to participate, you are free to withdraw from the study at any time without affecting those relationships. You are also free not to answer any question during the interview and to end the interview at any time.

Statement of Consent

I have read the above information. I have asked any questions I had, and I have received answers to my satisfaction. I consent to participate in the study. I have received a copy of this consent form.

Signature of Participant _____ Date _____

Name of Participant (Please print) _____

Signature of Researcher _____ Date _____

Contact Person:

Mustafa Abdul-Jabbar

[phone number] [email address]

APPENDIX Y.3: EXAMPLE CONSENT FORM

Title of the Research Study: Breaking the Single Story: Narratives of Educational Leaders in Post-Earthquake Haiti
Principal Investigator: [Laura Colket, address, email]

You are being asked to take part in a research study. Your participation is voluntary, which means you can choose whether or not to participate. If you decide to participate or not to participate, there will be no loss of benefits to which you are otherwise entitled. Before you make a decision, you will need to know the purpose of the study, the possible risks and benefits of being in the study, and what you will have to do if you decide to participate. The researcher is going to talk with you about the study and give you this consent document to read. You do not have to make a decision now; you can take the consent document home and share it with others before making any decisions.

If you do not understand what you are reading, do not sign it. Please ask the researcher to explain anything you do not understand, including any language contained in this form. If you decide to participate, you will be asked to sign this form and a copy will be given to you. Keep this form; in it, you will find contact information and answers to questions about the study. You may ask to have this form read to you.

What is the purpose of the study?

The purpose of the study is to learn more about your experience as an educational leader working with the Ministry of Education in Haiti in the 18 months following the earthquake.

Why was I asked to participate in the study?

You are being asked to join this study because you have been working in or with the Ministry of Education on reforming the educational system in Haiti following the earthquake.

What will I be asked to do?

Over the course of this research, you will be asked to participate in three individual interviews lasting between 1 and 2 hours each and possibly one or two focus groups (lasting 1.5 hours each) as well. In the interviews and focus groups, you will be asked to share your experience as an educational leader in this historical moment. Finally, you will also be asked to let the researcher observe you as you engage in your daily practice. If there are any public documents written about you, the researcher may use those as data as well.

How will confidentiality be maintained and my privacy be protected?

While it is typical practice to make every effort to keep all the information you share during the study strictly confidential, you have the option to decide to let your names be used in this research project. If you decide to use your name in this research, you may, for any reason, at any time, discuss your options for changing your decision with the researcher, and you can always remove yourself from the study if you so decide. If you decide to no longer participate in this research, all data you provided will be destroyed.

Even if you choose to have your name used in this research, the raw data from our interviews will still be protected. All interviews will be digitally recorded, and the audio files will be directly transferred to my personal computer for transcription. The original files will all be deleted from the digital recording device.

While the interview data will be protected, it is important to note that if you choose to participate in any focus groups, you will be sharing information with the other focus group participants. I will request that all information shared during the focus group remain confidential and is not shared beyond the group.

What are the risks?

The researcher believes there will be minimal risk to you, as a participant in this study. However, as your identity will not be confidential, there may be unforeseen complications when your stories are made public. Before making any of your stories public, you will have sufficient time to review the narrative that has been created to represent your story, to ensure you are comfortable with what is being made public. Additionally, your participation in this research is completely voluntary and you are allowed to withdraw at any time.

How will I benefit from the study?

There may be potential benefits for you as a participant in this research. This research will create an opportunity for you to share your experiences and stories from the past 18 months with a broader audience. This research will help to shed light on the process of educational leadership and reform in a postcrisis context, and your participation will be instrumental in that process.

What other choices do I have?

If you would not like to participate in this study, you do not have to. Your involvement is completely voluntary, and your decision will have no impact on your employment.

What happens if I do not choose to join the research study?

Your participation is voluntary, so there is no penalty if you choose not to join the study.

When is the study over? Can I leave the study before it ends?

The research process is expected to last through the summer of 2011. You have the right to drop out of the research study at any time. There is no penalty if you decide to do so.

If you no longer wish to be in the research study, please contact [name], at [email address].

The study may be stopped without your consent for the following reasons:

- The PI feels it is best for your safety and/or health—you will be informed of the reasons why.

- The PI, the sponsor or the Office of Regulatory Affairs at the University of Pennsylvania, can stop the study any time.

Will I have to pay for anything?

No, you will not have to pay anything to participate in this study.

Will I be paid for being in this study?

No, you will not receive any monetary compensation for participating in this study.

Who can I call with questions, complaints, or if I'm concerned about my rights as a research participant?

If you have questions, concerns, or complaints regarding your participation in this research study or if you have any questions about your rights as a research participant, you should speak with the principal investigator listed on Page One of this form. If a member of the research team cannot be reached or you want to talk to someone other than those working on the study, you may contact the Office of Regulatory Affairs with any questions, concerns, or complaints at the University of Pennsylvania by calling (215) xxx.xxxx.

When you sign this document, you are agreeing to take part in this research study. If you have any questions or there is something you do not understand, please ask. You will receive a copy of this consent document.

Signature of Participant

Print Name of Participant

Date

APPENDIX Y.4: EXAMPLE CONSENT FORM

Information and Consent Form

University of Pennsylvania Graduate School of Education Research Participant Information and Consent Form

Protocol Title: Becoming Visible: Strategies Utilized by Female Educators to Gain Access to Influential Leadership Roles in Independent Schools

Principal Investigator: Susan L. Feibelman

Consent and Confidentiality Agreement

You are being invited to participate in a research study. Before agreeing to participate, Susan Feibelman, an investigator named above, will discuss the following points with you:

- Participation in the research study is strictly voluntary.

- There are a maximum of 14 participants in the study.

- The study will take place commencing on or before April 30, 2012, and will conclude no later than March 31, 2013.

- Each participant will be given a copy of the research study questions and a summary of the study proposal.

- The findings from the study conducted will be submitted to [names of dissertation committee members and institutional affiliation].

- The information gathered and the study conducted will not be used for any nonresearch purpose.

- The study will maintain the privacy of all participants; actual names, school names, and other identifying information will remain confidential.

- Susan Feibelman will conduct audio-recorded interviews, which will be destroyed following the completion of the transcriptions.

- Participants may be asked to contribute written comments to a computer-mediated, online discussion board.

- Susan Feibelman will take all reasonable precautions to ensure that participants are in no way harmed or adversely affected as a result of participation in the study.

Your participation in this study is voluntary. If you decide not to participate, you are free to leave the study at any time. If you have any questions about your participation in this research study, make sure to discuss them with the study investigator (Susan Feibelman).

You may also call the Office of Regulatory Affairs at the University of Pennsylvania at (215) xxx.xxxx to talk about your rights as a research participant.

You will be asked to sign this form to show that

- The research study and information above have been discussed with you.

- You agree to participate in the study.

You will receive a copy of this signed form and the summary of the study proposal that will be discussed with you.

Name Signature Date

Name Signature Date

APPENDIX Y.5: EXAMPLE CONSENT FORM

QUALITATIVE RESEARCH
STUDY CONSENT FORM*

Research Questions

- How do women prepare themselves to take on influential leadership roles in independent schools? In what ways does the gendered nature of these leadership roles inform this preparation?

- Do women who are interested in taking on leadership roles in independent schools seek out mentor-protégé relationships, networks of support, and/or sponsorship from "recognized" independent school leaders?

- What is the relationship between leadership and an independent school's distinctive culture? How do beliefs about gender and leadership inform this relationship?

- How do the social identifiers of race, class, and age inform their strategies of action?

Consent and Confidentiality Agreement

I, _____ agree to participate in a research study being conducted by Susan Feibelman, a student at the University of Pennsylvania's Graduate School of Education, under the following conditions:

- Participation in the research study is strictly voluntary.

- Each participant will be given a copy of the research study outline detailing the research questions and research study process.

- The information gathered and the study conducted will be submitted to Dr. Sharon Ravitch and Dr. [committee member name] of the University of Pennsylvania's Graduate School of Education and Wharton, as well as [committee member name] of [university name] and Dr. [committee member name] of [college name].

- The information gathered and the study conducted will not be used for any nonresearch purpose.

- The study will maintain the privacy of all participants, and actual names and schools will remain confidential.

- Susan Feibelman will conduct taped interviews, which will be destroyed following the completion of the transcriptions.

- Susan Feibelman will take all reasonable precautions to ensure that participants are in no way harmed or adversely affected as a result of participation in the study.

Name _____ Signature _____ Date _____

Susan Feibelman _____ Date _____

*Based on the World Association of Opinion and Marketing Research Professionals Rights of Respondents

Appendix Z

Assent Form Examples

In this appendix, we have included two different examples of assent forms. The examples are presented without study information to protect the confidentiality of the participants who were a part of these studies.

APPENDIX Z.1: EXAMPLE ASSENT FORM

Research Project Title: "Understanding Youth Television Watching Habits and Patterns"
Researcher:

I am doing a research study about **youth television watching habits.** This research study is a way to learn more about how young people think about television and what your habits are in terms of how often you watch, what you watch, and how you think about television watching.

If you choose to be part of this research study, you will be asked to **share some things about how often you watch TV, what shows you watch, and other habits related to television. The research will require that you participate in filling out a brief survey that will take about 15 minutes and will be completed online and that you participate in two 1-hour interviews to be conducted at your home during the month of December.**

There are some things about this study you should know before you agree to take part in it. First, this is completely voluntary, and you should not feel any need to participate. And you can withdraw from the study at any time. Your parents know about the study too. Second, the interviews will be recorded and transcribed; only I and two of my research assistants will see the data, and your name will be changed to protect confidentiality. You will be asked to share your TV watching habits and to respond to emails about scheduling both interviews.

There should be no risks to you in this research. Not everyone who takes part in this study will benefit from participation. A benefit means that something good happens to you because you participated in the study. Benefits might be **that you learn about your own daily habits and how TV influences your life. You will also be able to read a summary of the findings to see how your habits compare to 100 other youth in the United States.**

Once I finish with this study, I will write a report about what I learned. This report will not include your name or that you were in the study.

If you or your parents have any questions, feel free to email me: [email address]

If you decide you do want to be in this study, please sign your name below and write today's date.

I, _____, agree to be in this research study.

_____ _____ (Sign your name here) (Date)

APPENDIX Z.2: EXAMPLE ASSENT FORM

My name is [include your name]. I am a graduate student researcher from [include your college or university name]. In this form, you are being asked if you would like to participate in a research study called "Understanding Youth TV Watching Habits and Patterns," which is about the TV-watching habits of people ages 12 to 18 who live in the United States.

If you agree to participate in this study, you will be asked to fill out an online survey that will take about 15 minutes. You will also be asked to participate in **two 1-hour interviews to be conducted at your home during the month of December.**

Please talk about this with your parents before you decide whether or not you will agree to participate. I have asked your parents to give their permission for you to take part in this study. But even if you parents said "yes" to this study, you can still decide to not take part in the study, and that will be fine.

If you do not want to participate in this study, then you do not have to participate. This study is voluntary, which means that you decide whether or not to take part in the study. Being in this study is up to you, and no one will be upset in any way if you do not want to participate or even if you change your mind later and want to stop.

You can ask any questions that you have about this study. If you have a question later that you did not think of now, you can email me [email address] or call me at [phone #], or ask me next time we see each other. Your parents are also welcome to call or email me with questions or concerns they or you have about the research.

Signing your name at the bottom means that you agree to be in this study. You and your parents will be given a copy of this form after you have signed it.

Name of youth (please print)

Signature of Youth Date

ENDNOTES

1. The references have not been included in the appendixes due to constraints on space.

2. Two different pairs of participants are located at the same school, which explains the discrepancy between the number of participants and the number of schools represented in this study.

3. At the conclusion of the interview, I discussed the possibility of returning with follow-up questions that may emerge from my data analysis.

4. Previous research examining women's access to highly influential leadership roles in public schools has described the role of search consultant as that of a gatekeeper, regulating who has entrée to the most prestigious leadership positions in a school system.

5. Appendix [Data analysis part 1]A provides a complete display of this segmentation.

6. For reasons of confidentiality, the site of study will be referred to by the pseudonym "Grace School."

7. For reasons of confidentiality, all participants in the study have been given pseudonyms.

8. This is a pseudonym for the school name. Also, the names of the study participants will be changed in the final report, but for now they are assigned letter names.

9. All names used in the study are pseudonyms. The school name is also a pseudonym.

10. Name of the two focal foundations has been changed to preserve confidentiality.

11. Name changed to preserve confidentiality.

Glossary

Binaries: In qualitative research, binaries can refer to dichotomies such as polarized notions of insider and outsider positionalities, of practitioners and scholars, or of racial, cultural, or gender categories, and so on. Binaries serve to reduce complexity and impose an either/or frame to typically more complex and multifaceted lived experiences.

Codes: In qualitative research, codes are tags or labels that researchers use (through a process called *coding*) to organize data into manageable units or chunks so that you can find, group, and thematically cluster various pieces of data as they relate to your research questions, findings, constructs, and/or themes across the data set. All data can be coded, including transcripts, fieldnotes, archival data, photographs, videos, research memos, and research journals. Once a researcher develops codes through specific processes of reading and organizing the data, codes are then defined succinctly (usually in a phrase or brief sentence). We discuss codes and coding processes in depth in Chapter Eight.

Core constructs: Core constructs (sometimes called core analytic constructs) refer to the central aspects—concepts, phenomena, and topics—that guide and are at the center of a research study. These are typically articulated in your research questions; they are the key concepts, phenomena, and topic areas you seek to study and understand more fully through the research. These need to be carefully considered at the outset of and throughout research question development and research design processes since they generate the areas of theory and conceptualization that drive your research study.

Criticality in qualitative research: Criticality in qualitative research, as we define it, is the meeting place of the theoretical (formal theory), the conceptual and contextual (conceptualizations and enactments of everyday life and social arrangements), and the methodological (where ideology and epistemology meet research methods). It is the researcher who acts as the translator and mediator of the spaces between the realms of the theoretical, conceptual, and methodological. One aspect of criticality in qualitative research is a recognition and address of the power inherent in research and in society more broadly. Criticality in qualitative research also involves transparency, intentionality, and striving to present as complex and contextualized a picture as possible. Criticality is achieved through the methods you use as well as how you engage with all aspects of research. We discuss other aspects of criticality in qualitative research throughout this chapter and the entire book.

Data saturation: In qualitative research, data saturation refers to the point at which you are no longer finding "new" themes in your data. This can be a problematic term, as there can be infinite themes or ways or looking at data. Furthermore, this concept implies that there is a reality that you can find, which is something that qualitative research pushes against. However, based on your research questions, you can reach a point of data saturation in that you are continuing to see recurring patterns and concepts in your data or have enough data to sufficiently answer your research questions.

Data set: A data set refers to the data that have been compiled and organized to answer your research questions. This may include, for example, data from interviews, archival documents, and/or focus groups. The data in a data set have been organized and pulled from your entire corpus of data, which can include all of your data sources such as archival data/documents and artifacts; interview data; focus group data; observation and fieldnote data; survey and questionnaire data; emails and other online data; participant-generated data such as journals, reflective writing, professional documents, photos, videos, and other digital media; and any other forms of existing data.

Deficit orientation: A deficit orientation refers to when individuals consider people from various groups (e.g., cultural, social, or communal) as lacking in certain knowledge, skills, or value. This can often be applied to members of dominant groups defining norms within hegemonic confines that valorize White, Western, upper-class ways of knowing and being while deeming those who do not fit these backgrounds and characteristics as less than or deficient. Deficit orientations can, at times, lead to the devaluation, degradation, or pathologization of nondominant groups and group members. Countering deficit orientations means shifting to a resource orientation in which the strengths, skills, knowledges, and value of each individual and various groups are the lens through which one views self and other. In a relational and critical approach to qualitative research, researchers reject and problematize deficit orientations and take the stance that everyone is an expert of their own experience and that all knowledge and ways of being are valuable and worthy of respect.

Design complexity: Design complexity refers to the ways that you strategically plan, design, and structure your research processes so that you can answer your research question(s) in the most complex, rigorous, and nuanced ways possible. Design complexity necessitates data triangulation and the strategic sequencing of methods as well as adopting a reflexive approach to design and data collection, which we describe throughout the book and especially in this chapter. To think about achieving design complexity, see the questions described in Table 4.1 and the questions in Table 4.2, which help you to think about reflexivity in data collection.

Dialogic engagement: We refer to the collaborative, dialogue-based processes that qualitative researchers engage in throughout a research study as dialogic engagement. These processes, which we discuss and provide examples of throughout the book, focus on pushing yourself to think about various aspects of the research process (and products) through talking about them with strategically selected thought partners. Thought partners are people who can challenge you to see your self and your research from a variety of angles at various stages throughout the research process. These people can include colleagues, advisers, peers, research team members, inquiry group members, and/or research participants.

Emergent: This term is often used in relationship to qualitative research design to signify that qualitative research does not typically follow a fixed design. Based on multiple factors, qualitative research can evolve and change. Researchers can refine and revise their research questions, data collection methods, and other aspects of a qualitative study. This aspect of qualitative research is often described as being emergent. Qualitative researchers also use the term *emergent* to mean aspects or understandings that arise from data. For example, an emergent theory is a working assumption that you are building from analysis of your data.

Emic: Emic refers to culturally and contextually embedded conceptualizations and descriptions of beliefs, behaviors, and ways of being that are meaningful and resonant to individuals and/or groups. In qualitative research, emic refers to being inductive in your approach so that rather than imposing your own concepts and terms, you create the conditions necessary for participants to offer and articulate their own conceptualizations in language that is organic to them as they describe an account, event, or phenomenon.

Epistemology: Epistemology concerns the nature of knowledge, including how it is constructed and how it can be acquired. The epistemological assumption underlying qualitative research is that knowledge is developed "through the subjective experiences of people. It becomes important, then, to conduct studies in the 'field,' where the participants live and work" (Creswell, 2013, p. 20).

Etic: Etic refers to descriptions of beliefs, behaviors, or ways of being or belief that are attributed by an outside observer (the researcher) that are not culturally embedded or necessarily organically conceptualized, articulated, or contextualized. In qualitative research, etic means that the language used to describe an account is not organic to the participants. We discuss the concepts of emic and etic in relationship to validity in Chapter Six and data analysis in Chapter Eight.

Fieldwork: In qualitative research, fieldwork entails the process of collecting data in a natural setting. This means in a setting in which the phenomenon would naturally occur (e.g., a neighborhood, organization, institution, or workplace). The term *fieldwork,* which comes from the ethnographic tradition of participant observation, is often used in qualitative research to refer to the located process of data collection.

Gatekeepers: *Gatekeeper* is a term/concept used in qualitative research to describe individuals who are in the position to grant or deny a researcher access to the research site in the early stages of the research. At times, this person serves as the key point person for the researcher throughout the research process, which can include dissemination of information from the researcher to key decision makers and participants. Gatekeepers may be formal, such as an organizational leader or governing body, or informal, such as an informal community leader or someone designated to engage with the researcher with no formal historical role in this arena. It is important to note that even once you may have established access to a setting through the gatekeeper(s), you must get consent from all participants in your study. It is also important to get this gatekeeper approval in writing. This can come in the form of an email, formal letter, or other forms of writing. We recommend that students save these documents as a PDF.

Hegemony: The concept of hegemony, developed by scholar and activist Antonio Gramsci, refers to the social, cultural, ideological, and economic influence imposed by dominant groups in society. The dissemination of dominant ideologies is enacted and maintained through ideological, social, cultural, and institutional means in such a way that dominant ideas, values, and beliefs appear to be "normal" and neutral (Agnew & Corbridge, 2002) because "the ideas, values, and experiences of dominant groups are validated in public discourse" and represented in and through public process and structures, including education, politics, law, and social institutions (Lears, 1985, p. 574).

Interpretive authority: Interpretive authority in qualitative research refers to the power of the researcher to be the interpreter and translator of people's lived experiences and perspectives. This becomes especially problematic when researchers believe that they have the "true" or "correct" version of someone's story. There is not a single truth or reality in qualitative research, and out of respect to participants and their lived experiences, it is important to acknowledge that there is inherent power in all forms of research. Thus, qualitative researchers should systematically acknowledge and reckon with this power and attempt to resist it as much as possible both through their own individualized reflexive engagement and through creating the conditions and processes of dialogic engagement in which assumptions, biases, and interpretations are rigorously challenged. We discuss this throughout this chapter, present specific recommendations in Table 7.3, and revisit the topic again in Chapter Nine.

Iterative: Qualitative research is often described as iterative, signifying that it (a) involves a back and forth of processes and (b) changes and evolves over time as you engage in these processes. Ideally, these back-and-forth processes lead to a progressive, evolutionary refinement of your research at conceptual, theoretical, and methodological levels.

Methodology: Qualitative methodology refers to where ideology and epistemology meet research approach, design, methods, and implementation and shape the overall approach to the methods in a study, including the related processes, understandings, theories, values, and beliefs that inform them. This also refers to the ways in which your overall stance and approach to research broadly and your study specifically shape your specific research methods to collect and analyze data (e.g., interviews, focus groups, specific analytic processes).

Ontology: Ontology concerns the nature of being or reality. In qualitative research, an ontological assumption is that there is not a single Truth or reality. Researchers, participants, and readers have differing realities, and a goal of qualitative research is to engage with, understand, and report these multiple realities (Creswell, 2013).

Participant validation strategies (member checks): Participant validation strategies, commonly referred to as *member checks,* are processes by which researchers "check in" with participants about different aspects of the research to see how they think and feel about various aspects of the research process and the parts of the data set that pertain to them. These strategies can be technical, including having participants verify the accuracy of statements and/or transcripts. These strategies can also include more relational approaches to engaging with participants to elicit their thoughts and responses to your interpretations and analytical concepts in more in-depth ways at various points throughout the research process. A goal of true participation validation processes is to create the conditions that help you to explore and ascertain if you are or are not understanding participants' responses, how you are understanding them, and to be challenged on your data collection processes and your interpretations of the data.

Positionality: Positionality refers to the researcher's role and social location/identity in relationship to the context and setting of the research. For example, you could be a practitioner in the setting, located as an expert, an insider or outsider to the setting, a supervisor of employees, a member of community involved in the research, share a cultural or ethnic relationship with the participants, and so on.

Positivism: Hughes (2001) describes the key aspects of the positivist paradigm through its view of the world as comprising unchanging, universally applicable laws and the belief that life events and social phenomena are/can be explained by knowledge of these universal laws and immutable truths. Within this paradigm, the belief is that understanding these universal laws requires observation and recording of social events and phenomena in systematic ways that allow the "knower" to define the underlying principle or truth that is the "cause" for the event(s) to occur. Positivist research also assumes that researchers are able to be objective and neutral.

Purposeful sampling: Purposeful sampling, which is sometimes referred to as purposive sampling, is the primary sampling method employed in qualitative research. It entails that individuals are *purposefully* chosen to participate in a research study for specific reasons that stem from the core constructs and contexts of the research questions. These reasons can include that individuals may have had a certain experience, have knowledge about a phenomenon, live or work in a particular place, or some other specified reason related to your research questions.

Qualitative data analysis: Qualitative data analysis encompasses the processes that qualitative researchers employ to "make sense of their data." Broadly speaking, data analysis is understood to include a variety of structured processes for looking across your data set to identify and construct analytic themes and, ultimately, turn these themes into what are commonly referred to as "findings" that help you to answer your research questions. Qualitative data analysis is the intentional, systematic scrutiny of data that occurs ongoingly throughout the research processes. This analysis of data often involves the specific processes of *data organization and management, immersive engagement with data,* and *writing and representation*. These processes are discussed in depth in Chapter Eight. An integrated approach to qualitative data analysis, which we discuss in this chapter, involves understanding that qualitative data analysis is iterative, formative, and summative; should be based on data and theory triangulation; addresses issues of power; and seeks out alternative perspectives. A challenge when analyzing qualitative data is to engage with and make sense of a significant corpus of data in a process that carefully reduces the amount of data, identifies significant patterns in those data, and does so in a way that allows you to construct an analytic framework for communicating the essence of what your data reveal (Patton, 2015).

Reactivity: Reactivity refers to ways that the presence of a researcher affects the research setting and the behavior of the participants. This most commonly refers to when participants change behaviors as a result of being observed but can also refer to when interview participants change responses due to perceived reactions from the researcher. This reaction to researcher presence impacts the data, thereby affecting the validity of the study.

Recursive: Qualitative research is recursive in that it builds and depends on all of its component parts. For example, your research questions are often informed by your personal and/or professional experiences, literature you have read, and the way you understand the world. Furthermore, as you begin to implement your research, the preliminary data you collect will also inform (and possibly lead you to refine) your research questions.

Research instruments: In qualitative research, research instruments refer to the tools that you develop and use to collect the study data. These data collection instruments, which can also be called protocols or guides, include the questions, prompts, and/or procedures that guide data collection. For example, an interview instrument will have a sequenced list of specific questions or prompts that researchers will ask participants during interviews. Other examples of instruments include observation templates or protocols. These are often used on research teams to make sure that observations, conducted by multiple researchers, capture similar information. The extent to which instruments, guides, and protocols are structured depends on the overall approach that guides the methods as well as the guiding research questions.

Social location: In qualitative research, *social location* is often used synonymously with *social identity.* Social location/identity includes the researcher's gender, social class, race, sexual identity/orientation, culture, and ethnicity as well as the intersections of these and other identity markers such as national origin, language communities, and so on (Henslin, 2013; Neyrey, 1991).

Tacit theories: Tacit theories refer to the informal and even unconscious ways that you understand or make sense of the world that are not explicit, spoken, or possibly even known to you without intentional reflection. These might include working hypotheses, assumptions, or conceptualizations you have about why things occur and how they operate. Unlike formal theories, tacit theories are not directly situated in academic literatures. Rather, they are an outgrowth of the attitudes, perspectives, ideologies, and values into which you have been socialized, often without knowing it. Everyone operates from tacit theories about people and the world, and without attention and reflection, they can constrain your research in a variety of ways. One example of this could be if you have a tacit theory that respecting authority is always good and therefore that all challenges to authority reflect badly on people. This theory might lead you to deem anyone who confronts authority as problematic versus seeing that context mediates whether an authority figure should be respected and what that means and looks like. We discuss how tacit theories relate to your conceptual framework throughout this chapter.

Thick description: Thick description, which we discuss in depth later in this chapter, is the way in which qualitative researchers describe a research setting in writing; the goal is for a researcher to accurately and thoroughly describe the important contextual factors. Thick description is an important aspect in increasing the complexity of your research by thoroughly and clearly describing the study's context, participants, and related experiences so as to produce complex interpretations and findings that allow audiences to make more contextualized meaning of your research. Thick description connotes a depth of contextual detail, usually garnered through multiple data sources, including observation and field-notes; it allows readers to have enough information and a depth of context so that they can picture the setting in their minds and form their own opinions about the quality of your research and your interpretations.

Thought communities: Thought communities refer to the different realms or "communities" to which your research may speak or relate. These may be actual communities in a social geography sense of the term (i.e., a school community, a geographically based community) and/or they may be groups of people such as affinity groups or cultural communities. For example, if you are studying a particular aspect of teacher education, different thought communities your research may speak to or influence could include scholars of teacher education, teacher educators, and/or teachers and other educational practitioners.

Unit of analysis: In qualitative research, your unit of analysis refers to the primary focus of your research study and is most often reflected in the core constructs in your research questions. The unit of analysis is the major entity that you are analyzing in your study. It is the "who" and/or the "what" at the center of your study. Units of analysis can be focused on people (e.g., individuals, groups), perspectives (e.g., individuals who share a common experience), structure (e.g., projects, programs), geographical units (e.g., city, neighborhood, state), artifacts (e.g., books, films, photos, newspapers), and/or social interactions (e.g., marriages, births) (Patton, 2015).

References

Adichie, C. N. (2009). The danger of a single story. TEDtalk. Retrieved from http://www.ted.com/talks/chimamanda_adichie_the_danger_of_a_single_story/transcript?language = en

Agnew, J., & Corbridge, S. (2002). *Mastering space: Hegemony, territory and international political economy*. London, UK: Routledge.

Alvesson, M., & Sköldberg, K. (2009). *Reflexive methodology: New vistas for qualitative research*. Los Angeles, CA: Sage.

American Educational Research Association (AERA). (2011). Code of ethics: American Educational Research Association. *Educational Researcher, 40*(3), 145–156.

Anderson, G., Herr, K., & Nihlen, A. (2007). *Studying your own school: An educator's guide to practitioner action research* (2nd ed.). Thousand Oaks, CA: Corwin Press.

Anderson, L. (2008). Reflexivity. In R. Thorpe & R. Holt (Eds.), *The SAGE dictionary of qualitative management research* (pp. 184–186). London, UK: Sage.

Anfara, V. A., & Mertz, N. T. (2015). *Theoretical frameworks in qualitative research* (2nd ed.). Thousand Oaks, CA: Sage.

Angrosino, M. (2007). *Doing ethnographic and observational research: The SAGE Qualitative Research Kit*. Thousand Oaks, CA: Sage.

Argyris, C., & Schön, D. A. (1978). *Organizational learning: A theory of action perspective*. Reading, MA: Addison-Wesley.

Atkinson, P., Coffey, A., & Delamont, S. (2001). A debate about our canon. *Qualitative Research, 1*(1), 5–21.

Austin, W. J. (2008). Relational ethics. In L. M. Given (Ed.). *The SAGE encyclopedia of qualitative research methods* (Vol. 2, pp. 748–749). Thousand Oaks, CA: Sage.

Bakhtin, M. (1973). *Problems of Dostoevsky's poetics* (R. W. Rotsel, Trans.). Ann Arbor, MI: Ardis.

Bakhtin, M. (1981). Discourse in the novel. In M. Holquist (Ed.), *The dialogic imagination: Four essays by M. Bakhtin* (C. Emerson & M. Holquist, Trans.). Austin: University of Texas Press.

Bakhtin, M. M. (1984). From Rabelais and his world (H. Lswolsky, Trans.). In P. Morris (Ed.), *The Bakhtin reader* (pp. 195–244). New York, NY: Oxford University Press. (Original work published 1965)

Barbour, R. (2007). *Doing focus groups: The SAGE Qualitative Research Kit*. Thousand Oaks, CA: Sage.

Barbour, R. S. (2001). Checklists for improving rigour in qualitative research: A case of the tail wagging the dog? *British Medical Journal, 322*, 1115–1117.

Barbour, R. S. (2014). Quality of data analysis. In U. Flick (Ed.), *The SAGE handbook of qualitative data analysis* (pp. 496–509). London, UK: Sage.

Bernard, H. R., & Ryan, G. W. (2010). *Analyzing qualitative data: Systematic approaches*. Thousand Oaks, CA: Sage.

Blaikie, N. W. H. (1991). A critique of the use of triangulation in social research. *Quality & Quantity, 25*, 115–136.

Bogdan, R. C., & Biklen, S. K. (2006). *Qualitative research for education: An introduction to theories and methods* (5th ed.). Boston, MA: Pearson.

Braun, V., & Clarke, V. (2006). Using thematic analysis in psychology. *Qualitative Research in Psychology, 3*(2), 77–101.

Brinkmann, S. (2012). *Qualitative inquiry in everyday life: Working with everyday life materials.* Thousand Oaks, CA: Sage.

Brinkmann, S., & Kvale, S. (2015). *Interviews: Learning the craft of qualitative research interviewing* (3rd ed.). Thousand Oaks, CA: Sage.

Brown, B. (2010). The power of vulnerability. TEDtalks. Retrieved from http://www.ted.com/talks/brene_brown_on_vulnerability/transcript?language = en

Brown, L. M., Argyris, D., Attanucci, J., Bardige, B., Gilligan, C., Johnston, K., . . . Wilcox, D. (with Brown, L. M.). (Eds.). (1988). *A guide to reading narratives of conflict and choice for self and moral voice* (Monograph 1). Cambridge, MA: Center for the Study of Gender, Education and Human Development, Harvard University, Graduate School of Education.

Cannella, G. S., & Lincoln, Y. S. (2012). Deploying qualitative methods for critical social purposes. In S. R. Steinberg & G. S. Cannella (Eds.), *Critical qualitative research reader* (pp. 104–114). New York, NY: Peter Lang.

Cavallo, D. (2000). Emergent design and learning environments: Building on indigenous knowledge. *IBM Systems Journal, 39*(3–4), 768–781.

Centre for Critical Qualitative Health Research. (n.d.). *What is critical qualitative research?* Retrieved from http://www.ccqhr.utoronto.ca/what-is-critical-qualitative-research

Chakrabarty, D. (2000). Domestic cruelty and the birth of the subject. In *Provincializing Europe: Postcolonial thought and historical difference* (pp. 117–148). Princeton, NJ: Princeton University Press.

Chilisa, B. (2012). *Indigenous research methodologies.* Thousand Oaks, CA: Sage.

Cho, J., & Trent, A. (2006). Validity in qualitative research revisited. *Qualitative Research, 6*(3), 319–340.

Christie, C. A., Montrosse, B. E., & Klein, B. M. (2005). Emergent design evaluation: A case study. *Evaluation and Program Planning, 28*(3), 271–277.

Clarke, A., & Dawson, R. (1999). *Evaluation research: An introduction to principles, methods, and practice.* London, UK: Sage.

Cochran-Smith, M., & Lytle, S. L. (1993a). *Inside/outside: Teacher research and knowledge.* New York, NY: Teachers College Press.

Cochran-Smith, M., & Lytle, S. L. (1993b). Teacher research: A way of knowing. In M. Cochran-Smith & S. L. Lytle (Eds.), *Inside/outside: Teacher research and knowledge* (pp. 41–62). New York, NY: Teachers College Press.

Cochran-Smith, M., & Lytle, S. L. (1999). The teacher research movement: A decade later. *Educational Researcher, 28*(7), 15–25.

Cochran-Smith, M., & Lytle, S. L. (2001). Beyond certainty: Taking an inquiry stance on practice. In A. Lieberman & L. Miller (Eds.), *Teachers caught in the action: Professional development that matters* (pp. 45–60). New York, NY: Teachers College Press.

Cochran-Smith, M., & Lytle, S. L. (2009). *Inquiry as stance: Practitioner research for the next generation.* New York, NY: Teachers College Press.

Cooke, B., & Kothari, U. (Eds.). (2001). *Participation: The new tyranny?* London, UK: Zed Books.

Connelly, F. M., & Clandinin, D. J. (1990). Stories of experience and narrative inquiry. *Educational Researcher, 19*(5), 2–14.

Corbett, J., & IFAD Consultative Group. (2009). Good practices in participatory mapping: A review prepared for the International Fund for Agricultural Development (IFAD). Rome, Italy. International Fund for Agricultural Development.

Corbin, J., & Strauss, A. (1990). Grounded theory research: Procedures, canons, and evaluative criteria. *Qualitative Sociology, 13*(2), 3–21.

Corbin, J., & Strauss, A. (2008). *Basics of qualitative research: Techniques and procedures for developing grounded theory* (3rd ed.). Thousand Oaks, CA: Sage.

Corbin, J., & Strauss, A. (2015). *Basics of qualitative research: Techniques and procedures for developing grounded theory* (4th ed.). Thousand Oaks, CA: Sage.

Cornish, F., Gillespie, A., & Zittoun, T. (2014). Collaborative analysis of qualitative data. In U. Flick (Ed.), *SAGE handbook of qualitative data analysis* (pp. 79–93). London, UK: Sage.

Coyne, I. T. (1997). Sampling in qualitative research. Purposeful and theoretical sampling; merging or clear boundaries? *Journal of Advanced Nursing, 26,* 623–630.

Crenshaw, K. (1991). Mapping the margins: Intersectionality, identity politics, and violence against women of color. *Stanford Law Review, 43*(6), 1241–1299.

Creswell, J. W. (2003). *Research design: Qualitative, quantitative, and mixed methods approaches.* Thousand Oaks, CA: Sage.

Creswell, J. W. (2007). *Research design: Qualitative, quantitative, and mixed methods approaches* (2nd ed.). Thousand Oaks, CA: Sage.

Creswell, J. W. (2013). *Qualitative inquiry and research design: Choosing among five approaches* (3rd ed.). Thousand Oaks, CA: Sage.

Creswell, J. W. (2014). *Research design: Qualitative, quantitative, and mixed methods approaches* (4th ed.). Thousand Oaks, CA: Sage.

Creswell, J. W., & Miller, D. L. (2000). Determining validity in qualitative inquiry. *Theory into practice, 39*(3), 124–131.

Creswell, J. W., & Plano Clark, V. L. (2007). *Designing and conducting mixed methods research.* Thousand Oaks, CA: Sage.

Creswell, J. W., & Plano Clark, V. L. (2011). *Designing and conducting mixed methods research* (2nd ed.). Thousand Oaks, CA: Sage.

Crow, G. (2008). Reciprocity. In L. M. Given (Ed.), *The SAGE encyclopedia of qualitative research methods* (Vol. 2, pp. 739–740). Thousand Oaks, CA: Sage.

Denzin, N. K. (1994). The art and politics of interpretation. In N. K. Denzin & Y. S. Lincoln (Eds.), *The handbook of qualitative research* (pp. 500–515). Thousand Oaks, CA: Sage.

Denzin, N. K. (2001). *Interpretive interactionism* (2nd ed.). Thousand Oaks, CA: Sage.

Denzin, N. K. (2009). *The research act: A theoretical orientation to sociological methods.* Piscataway, NJ: Transaction Publishers. (Original work published in 1970)

Denzin, N. K. (2014). Writing and/as analysis or performing the world. In U. Flick (Ed.), *The SAGE handbook of qualitative data analysis* (pp. 569–584). London, UK: Sage.

Denzin, N. K., & Lincoln, Y. S. (2011a). Introduction. In N. K. Denzin & Y. S. Lincoln (Eds.), *The SAGE handbook of qualitative research* (4th ed., pp. 1–19). Thousand Oaks, CA: Sage.

Denzin, N. K., & Lincoln, Y. S. (2011b). Preface. In N. K. Denzin & Y. S. Lincoln (Eds.), *The SAGE handbook of qualitative research* (pp. ix–xvi). Thousand Oaks, CA: Sage.

Denzin, N. K., & Lincoln, Y. S. (Eds.). (2011c). *The SAGE handbook of qualitative research.* Thousand Oaks, CA: Sage.

Dooley, L. M. (2002). Case study research and theory building. *Advances in Developing Human Resources, 4*(3), 335–354.

Dudwick, N., Kuehnast, K., Nyhan Jones, K., & Woolcock, M. (2006). *Analyzing social capital in context: A guide to using qualitative methods and data.* Washington, DC: The International Bank for Reconstruction and Development/The World Bank.

Eisenhardt, K. M. (1989). Building theories from case study research. *The Academy of Management Review, 14*(4), 532–550.

Emerson, R. M., Fretz, R. I., & Shaw, L. L. (1995). *Writing ethnographic fieldnotes.* Chicago, IL: University of Chicago Press.

Emerson, R. M., Fretz, R. I., & Shaw, L. L. (2011). *Writing ethnographic fieldnotes* (2nd ed.). Chicago, IL: University of Chicago Press.

Erickson, F. (1973). What makes school ethnography 'ethnographic'? *Council of Anthropology and Education Quarterly, 4*(2), 10–19.

Erickson, F. (1986). Qualitative methods in research on teaching. In M. C. Wittrock (Ed.), *Handbook of research on teaching* (3rd ed., pp. 119–161). New York, NY: Macmillan.

Erickson, F. (1996). Going for the zone: The social and cognitive ecology of teacher-student interaction in classroom conversations. In D. Hicks (Ed.), *Discourse, learning, and schooling* (pp. 29–62). Cambridge, UK: Cambridge University Press.

Erickson, F. (1998). Qualitative research methods for science education. In B. J. Fraser & K. G. Tobin (Eds.), *International handbook of science education* (pp. 1155–1173). Dordrecht, Netherlands: Kluwer Academic Publishers.

Erickson, F. (2004). Culture in society and in educational practices. In J. A. Banks & C. A. M. Banks (Eds.), *Multicultural education: Issues and perspectives* (pp. 31–60). Hoboken, NJ: Wiley.

Erickson, F. (2011). A history of qualitative inquiry in social and educational research. In N. Denzin & Y. Lincoln (Eds.), *The SAGE handbook of qualitative research* (pp. 43–59). Thousand Oaks, CA: Sage.

Fals-Borda, O., & Rahman, M.A. (1991). *Action and knowledge.* Lanham, MD: Rowman & Littlefield.

Fine, M. (1992). *Disruptive voices: The possibilities of feminist research.* Ann Arbor: University of Michigan Press.

Fine, M. (1994). Working the hyphens: Reinventing self and other in qualitative research. In N. Denzin & Y. Lincoln (Eds.), *The handbook of qualitative research* (pp. 70–82). London, UK: Sage.

Fink, A. (2006). *How to conduct surveys: A step-by-step guide* (3rd ed.). Thousand Oaks, CA: Sage.

Finlay, L. (n.d.). *Relational research.* Retrieved from http://www.lindafinlay.co.uk/relationalresearch.htm

Flick, U. (2007). *Managing quality in qualitative research.* London, UK: Sage.

Flick, U. (2014). Mapping the field. In U. Flick (Ed.), *The SAGE handbook of qualitative data analysis* (pp. 136–149). London, UK: Sage.

Fontana, A., & Prokos, A. H. (2007). *The interview: From formal to postmodern.* Walnut Creek, CA: Left Coast Press.

Freire, P. (2000). *Pedagogy of the oppressed* (M. B. Ramos, Trans.). New York, NY: Continuum. (Original work published 1970)

Gaskell, G., & Bauer, M. W. (2000). Towards public accountability: Beyond sampling, reliability, and validity. In M. W. Bauer & G. Gaskell (Eds.), *Qualitative researching with text, image and sound* (pp. 336–350). London, UK: Sage.

Gearing, R. E. (2004). Bracketing in research: A typology. *Qualitative Health Research, 14*(10), 1429–1452.

Gee, J. P. (2011). *How to do discourse analysis: A toolkit.* New York, NY: Routledge.

Gibbs, G. R. (2014). Using software in qualitative analysis. In U. Flick (Ed.), *The SAGE handbook of qualitative data analysis* (pp. 136–149). London, UK: Sage.

Gibran, K. (1923). *The prophet.* New York, NY: Knopf.

Gibson, W., & Brown, A. (2009). *Working with qualitative data.* London, UK: Sage.

Gilligan, C. (1996). Centrality of relationship in human development: A puzzle, some evidence, and a theory. In G. Noam & K. Fisher (Eds.), *Development and vulnerability in close relationships* (pp. 237–261). Mahwah, NJ: Lawrence Erlbaum.

Gilligan, C., Spencer, R., Weinberg, M. K., & Bertsch, T. (2003). On the *Listening Guide*: A voice-centered relational model. In P. M. Camic, J. E. Rhodes, & L. Yardley (Eds.), *Qualitative research in psychology: Expanding perspectives in methodology and design* (pp. 157–172). Washington, DC: American Psychological Association Press.

Glesne, C. (2006). *Becoming qualitative researchers: An introduction* (3rd ed.). New York, NY: Pearson.

Glesne, C. (2016). *Becoming qualitative researchers: An introduction* (5th ed.). Boston, MA: Pearson.

Golafshani, N. (2003). Understanding reliability and validity in qualitative research. *The Qualitative Report, 8*(4), 597–607.

Golden Institute. (2011). *Community-based oral testimony: A different approach to knowledge.* Retrieved from www.goldeninstitute.org

Gonzalez, N., Moll, L. C., & Amanti, C. (2005). *Funds of knowledge: Theorizing practice in households, communities, and classrooms.* Mahwah, NJ: Lawrence Erlbaum.

Goulding, C. (2005). Grounded theory, ethnography and phenomenology: A comparative analysis of three qualitative strategies for marketing research. *European Journal of Marketing, 39*(3/4), 294–308.

Greene, J. C., Caracelli, V. J., & Graham, W. F. (1989). Toward a conceptual framework for mixed-method evaluation designs. *Educational Evaluation and Policy Analysis, 11*(3), 255–274.

Guba, E. G. (1981). Criteria for assessing the trustworthiness of naturalistic inquiries. *Educational Resources Information Center Annual Review Paper, 29,* 75–91.

Guba, E. G., & Lincoln, Y. S. (1981a). The evaluator as instrument. In *Effective evaluation: Improving the usefulness of evaluation results through responsive and naturalistic approaches* (pp. 128–152). San Francisco, CA: Jossey-Bass.

Guba, E. G., & Lincoln, Y. S. (1981b). *Effective evaluation: Improving the usefulness of evaluation results through responsive and naturalistic approaches.* San Francisco, CA: Jossey-Bass.

Guba, E. G., & Lincoln, Y. S. (1989). *Fourth generation evaluation.* Newbury Park, CA: Sage.

Guba, E. G., & Lincoln, Y. S. (1994). Competing paradigms in qualitative research. In N. K. Denzin & Y. S. Lincoln (Eds.), *Handbook of qualitative research* (pp. 105–117). Thousand Oaks, CA: Sage.

Guba, E. G., & Lincoln, Y. S. (2005). Paradigmatic controversies, contradictions, and emerging influences. In N. K. Denzin & Y. S. Lincoln (Eds.), *The SAGE handbook of qualitative research* (3rd ed., pp. 191–215). Thousand Oaks, CA: Sage.

Guest, G., Namey, E. E., & Mitchell, M. L. (2013). *Collecting qualitative data: A field manual for applied research.* CA: Sage.

Guthrie, S., Wamae, W., Diepeveen, S., Wooding, S., & Grant, J. (2013). *Measuring research: A guide to research evaluation frameworks and tools.* RAND Corporation. Retrieved from http://www.rand.org/content/dam/rand/pubs/monographs/MG1200/MG1217/RAND_MG1217.pdf

Hammersley, M. (2008). *Questioning qualitative inquiry: Critical essays.* London, UK: Sage.

Hammersley, M., & Atkinson, P. (2007). *Ethnography: Principles in practice* (3rd ed.). Hoboken, NJ: Taylor & Francis.

Hammersley, M., & Traianou, A. (2012). *Ethics in qualitative research: Controversies and contexts.* Los Angeles, CA: Sage.

Hanisch, C. (1970). The personal is political. In S. Firestone, & A. Koedt (Eds.), *Notes from the second year: Women's liberation: Major writings of the radical feminists* (pp. 76–78). New York, NY: Radical Feminism.

Hart, C. (2001). *Doing a literature review: Releasing the social science imagination.* London, UK: Sage.

Headland, T. N. (1990). A dialogue between Kenneth Pike and Marvin Harris on emics and etics. In T. N. Headland, K. L. Pike, & M. Harris (Eds.). *Emics and etics: The insider/outsider debate* (Frontiers of Anthropology, Vol. 7, pp. 13–27). Newbury Park, CA: Sage.

Henslin, J. M. (2013). *Essentials of sociology: A down-to-earth approach.* Upper Saddle River, NJ: Pearson.

Hidalgo, N. M. (1993). Multicultural teacher introspection. In T. Perry & J. W. Fraser (Eds.), *Freedom's plow: Teaching in the multicultural classroom* (pp. 99–106). New York, NY: Routledge.

Hill-Collins, P. (2000). Toward a new vision: Race, class, and gender as categories of analysis and connection. In J. Ferrante & P. Brown Jr. (Eds.), *The social construction of race and ethnicity in the United States* (2nd ed., pp. 478–495). New York: Longman.

Holliday, A. (2007). *Doing and writing qualitative research* (2nd ed.). London, UK: Sage.

Holstein, J. A., & Gubrium, J. F. (1995). *The active interview* (Qualitative Research Methods Series 37). Thousand Oaks, CA: Sage.

hooks, b. (1994). *Teaching to transgress: Education as the practice of freedom.* New York, NY: Routledge.

Hughes, P. (2001). Paradigms, methods & knowledge. In G. MacNaughton, S. Rolfe, & I. Siraj-Blatchford (Eds.), *Doing early childhood research: International perspectives on theory & practice* (pp. 31–55). Buckingham, UK: Open University Press.

Israel, M., & Hay, I. (2008). Informed consent. In L. M. Given (Ed.). *The SAGE encyclopedia of qualitative research methods* (Vol. 2, pp. 431–432). Thousand Oaks, CA: Sage.

Jackson, A. Y., & Mazzei, L. A. (2009). *Voice in qualitative inquiry: Challenging conventional, interpretive, & critical conceptions in qualitative research*. London, UK: Routledge.

Jacoby, S., & Gonzales, P. (1991). The constitution of expert-novice in scientific discourse. *Issues in Applied Linguistics, 2*(2). Retrieved from http://escholarship.org/uc/item/3fd7z5k4

Josselson, R. (2013). *Interviewing for qualitative inquiry: A relational approach*. New York, NY: Guilford.

Kaiser, K. (2009). Protecting respondent confidentiality in qualitative research. *Qualitative Health Research, 19*(11), 1632–1641.

Keen, S., & Todres, L. (2007). Strategies for disseminating qualitative research findings: Three exemplars. *Forum Qualitative Sozialforschung/Forum: Qualitative Social Research, 8*(3), Art. 17. Retrieved from http://nbn-resolving.de/urn:nbn:de:0114-fqs0703174

Kelle, U. (2014). Theorization from data. In U. Flick (Ed.), *The SAGE handbook of qualitative data analysis* (pp. 554–568). London, UK: Sage.

Kelly, N., & Zetzsche, J. (2012). *Found in translation: How language shapes our lives and transforms the world*. New York, NY: Perigee.

Kim, Y. (2011). The pilot study in qualitative inquiry: Identifying issues and learning lessons for culturally competent research. *Qualitative Social Work, 10*(2), 190–206.

Kincheloe, J. L., & McLaren, P. (2000). Rethinking critical theory and qualitative research. In N. K. Denzin & Y. S. Lincoln (Eds.), *The SAGE handbook of qualitative research* (3rd ed., pp. 303–342). Thousand Oaks, CA: Sage.

Kindon, S. L., Pain, R., & Kesby, M. (2007). *Participatory action research approaches and methods: Connecting people, participation and place*. London, UK: Routledge.

Kowal, S., & O'Connell, D. C. (2014). Transcription as a crucial step of data analysis. In U. Flick (Ed.), *The SAGE handbook of qualitative data analysis* (pp. 64–78). London, UK: Sage.

Kress, G., Jewitt, C., Bourne, J., Franks, A., Hardcastle, J., Jones, K., & Reid, E. (2005). *English in urban classrooms: A multimodal perspective on teaching and learning*. London, UK: Routledge Falmer.

Kuriloff, P. J., Andrus, S., & Ravitch, S. M. (2011). Messy ethics: Conducting moral participatory action research in the crucible of university–school relations. *Mind, Brain, and Education, 5*(2), 49–62.

Kvale, S. (1995). The social construction of validity. *Qualitative Inquiry, 1*(1), 19–40. © 2011 The Authors. Journal Compilation © 2011 International Mind, Brain, and Education Society and Blackwell Publishing, Inc.

Kvale, S. (1996). *InterViews: An introduction to qualitative research interviewing*. Thousand Oaks, CA: Sage.

Kvale, S. (2007). *Doing interviews: The SAGE Qualitative Research Kit*. Thousand Oaks, CA: Sage.

Kvale, S., & Brinkmann, S. (2009). *Interviews: Learning the craft of qualitative research interviewing* (2nd ed.). Thousand Oaks, CA: Sage.

Lather, P. (1986). Issues of validity in openly ideological research: Between a rock and a soft place. *Interchange, 17*(4), 63–84.

Lather, Patti, 1991, "Deconstructing/Deconstructive Inquiry: The Politics of Knowing and Being Known," Educational Theory, 41(2), 153–173.

Lather, P. (1992). Critical frames in educational research: Feminist and post-structural perspectives. *Theory Into Practice, 2*, 87–98.

Lather, P. (1993). Fertile obsession: Validity after poststructuralism. *Sociological Quarterly, 34*(4), 673–693.

Lears, T. J. (1985). The concept of cultural hegemony: Problems and possibilities. *The American Historical Review, 9*(3), 567–593.

LeCompte, M. D., & Goets, J. P. (1982). Problems of reliability and validity in ethnographic research, *Review of Educational Research, 53*, 31–36. Retrieved from http://tinyurl.com/o4mo3z5

Lee, R. M., Fielding, N., & Blank, G. (2008). The Internet as a research medium: An editorial introduction to *The SAGE handbook of online research methods*. In N. Fielding, R. M. Lee, & G. Blank (Eds.), *The SAGE handbook of online research methods* (pp. 3–20). Los Angeles, CA: Sage.

Lewin, K. (1946). Action research and minority problems. *Journal of Social Issues, 2,* 34–46.

Lester, S. (1999). *An introduction to phenomenological research.* Retrieved from www.devmts.demon.co.uk/resmethy.htm

Lett, J. (1990). Emics and etics: Notes on the epistemology of anthropology. In T. N. Headland, K. L. Pike, & M. Harris (Eds.), *Emics and etics: The insider/outsider debate* (Frontiers of Anthropology, Vol. 7, pp. 127–142). Newbury Park, CA: Sage.

Lichtman, M. (2014). *Qualitative research in education* (4th ed.). Thousand Oaks, CA: Sage.

Lillis, T. (2003). Student writing as academic literacies: Drawing on Bakhtin to move from critique to design. *Language & Education, 17,* 192–207.

Lincoln, Y. S. (1995). Emerging criteria for quality in qualitative research and interpretive research. *Qualitative Inquiry, 1,* 275–289.

Lincoln, Y. S., & Guba, E. G. (1985). *Naturalistic inquiry.* Beverly Hills, CA: Sage.

Lincoln, Y. S., & Guba, E. G. (2003). Paradigmatic controversies, contradictions, and emerging confluences. In N. K. Denzin & Y. S. Lincoln (Eds.). *The landscape of qualitative research: Theories and issues* (pp. 255–286). Thousand Oaks, CA: Sage.

Lofland, J., Snow, D., Anderson, L., & Lofland, L. (2006). *Analyzing social settings: A guide to qualitative observation and analysis* (4th ed.). Boston, MA: Wadsworth.

Mantzoukas, S. (2004). Issues of representation within qualitative inquiry. *Qualitative Health Research, 14*(7), 994–1007.

Marshall, C., & Rossman, G. B. (2016). *Designing qualitative research* (6th ed.). Thousand Oaks, CA: Sage.

Mason, J. (2002). *Qualitative researching* (2nd ed.). Thousand Oaks, CA: Sage.

Mathers, N., Fox, N., & Hunn, A. (2009). Surveys and questionnaires. The NIHR RDS for the East Midlands/Yorkshire & the Humber. Retrieved from http://www.rds-yh.nihr.ac.uk/wp-content/uploads/2013/05/12_Surveys_and_Questionnaires_Revision_2009.pdf

Maxwell, J. A. (1992). Understanding and validity in qualitative research. *Harvard Educational Review, 62*(3), 279–300.

Maxwell, J. A. (2013). *Qualitative research design: An interactive approach* (3rd ed.). Thousand Oaks, CA: Sage.

Maxwell, J. A., & Chmiel, M. (2014). Notes toward a theory of qualitative data analysis. In U. Flick (Ed.), *The SAGE handbook of qualitative data analysis* (pp. 21–34). London, UK: Sage.

Maxwell, J. A., & Miller, B. A. (2008). Categorizing and connecting strategies in qualitative data analysis. In S. N. Hesse-Biber, S. Nagy, & P. Levy (Eds.), *Handbook of emergent methods* (pp. 461–477). New York, NY: Guilford.

Maxwell, J. A., & Mittapalli, K. (2008). Theory. In L. Given (Ed.), *The SAGE encyclopedia of qualitative research methods* (pp. 876–880). Thousand Oaks, CA: Sage.

McCall, L. (2005). The complexity of intersectionality. *Signs, 30*(3), 1771–1800.

McGinn, M. K. (2008). Researcher-participant relationships. In L. M. Given (Ed.). *The SAGE encyclopedia of qualitative research methods* (Vol. 2, pp. 767–771). Thousand Oaks, CA: Sage.

McMillan, J. H., & Schumacher, S. S. (1997). *Research in education: A conceptual introduction.* New York, NY: Longman. Retrieved from http://tinyurl.com/kbw99vk

Merriam, S. B. (2009). *Qualitative research: A guide to design and implementation.* San Francisco, CA: Jossey-Bass.

Miles, M. B., & Huberman, A. M. (1994). *Qualitative data analysis: An expanded source book* (2nd ed.). Thousand Oaks, CA: Sage.

Miles, M. B., Huberman, A. M., & Saldaña, J. (2014). *Qualitative data analysis: A methods sourcebook.* Thousand Oaks, CA: Sage.

Millward, L. (2012). Focus groups. In G. M. Breakwell, J. A. Smith, & D. B. Wright (Eds.), *Research methods in psychology* (4th ed., pp. 411–437). London, UK: Sage.

Morgan, D. L. (2010). Reconsidering the role of interaction in analyzing and reporting focus groups. *Qualitative Health Research, 20*(5), 718–722.

Morris, M. W., Leung, K., Ames, D., & Lickel, B. (1999). Views from inside and outside: Integrating emic and etic insights about culture and justice judgment. *Academy of Management Review, 24*(4), 781–796.

Morse, J. M., Barrett, M., Mayan, M., Olson, K., & Spiers, J. (2002). Verification strategies for establishing reliability and validity in qualitative research. *International Journal of Qualitative Methods, 1*(2). Retrieved from http://www.ualberta.ca/ ~ ijqm/

Moustakas, C. E. (1994). *Phenomenological research methods*. Thousand Oaks, CA: Sage.

Mukherjee, N. (2002). *Participatory learning and action: With 100 field methods* (Studies in Rural Participation 4). New Delhi, India: Concept Publishing Company.

Nakkula, M. J., & Ravitch, S. M. (1998). *Matters of interpretation: Reciprocal transformation in therapeutic and developmental relationships with youth*. San Francisco, CA: Jossey-Bass.

Newbury, J., & Hoskins, M. (2010). Relational inquiry: Generating new knowledge with adolescent girls who use crystal meth. *Qualitative Inquiry, 16*, 642–650.

Neyrey, J. H. (1991). *The social world of Luke-Acts: Models for interpretation*. Peabody, MA: Hendrickson Publishers. Retrieved from http://tinyurl.com/nhsp6lc

Noddings, N. (2003). *Caring: A feminist approach to ethics and moral education* (2nd ed.). Berkeley: University of California Press.

O'Connor, H., Madge, C., Shaw, R., & Wellens, J. (2008). Internet-based interviewing. In N. Fielding, R. M. Lee, & G. Blank (Eds.), *The SAGE handbook of online research methods* (pp. 271–289). Thousand Oaks, CA: Sage.

Ogden, R. (2008). Confidentiality. In L. M. Given (Ed.), *The SAGE encyclopedia of qualitative research methods* (Vol. 1, pp. 111–112). Thousand Oaks, CA: Sage.

Olsen, W. (2004). Triangulation in social research: Qualitative and quantitative methods can really be mixed. *Developments in Sociology, 20*, 103–118.

Omi, M., & Winant, H. (1994). *Racial formation in the United States: From the 1960s to the 1990s* (2nd ed.). New York, NY: Routledge.

Patten, M. L. (2001). *Questionnaire research: A practical guide* (2nd ed.). Thousand Oaks, CA: Pyrczak Publishing.

Patton, M. Q. (2002). *Qualitative research and evaluation methods* (3rd ed.). Thousand Oaks, CA: Sage.

Patton, M. Q. (2015). *Qualitative research and evaluation methods* (4th ed.). Thousand Oaks, CA: Sage.

Paulus, T., Lester, J., & Dempster, P. (2014). *Digital tools for qualitative research*. London, UK: Sage.

Peshkin, A. (1982). *The imperfect union: School consolidation and community conflict*. Chicago, IL: University of Chicago Press.

Peshkin, A. (1988). In search of subjectivity—one's own. *Educational Researcher, 17*(7), 17–21.

Pipher, M. (2007). *Writing to change the world*. New York, NY: Penguin.

Polit, D. F., Beck, C. T., & Hungler, B. P. (2001). *Essentials of nursing research: Methods, appraisal & utilization* (5th ed.). Philadelphia, PA: Lippincott Williams & Wilkins.

Ponterotto, J. G. (2006). Brief note on the origins, evolution, and meaning of the qualitative research concept "thick description." *The Qualitative Report, 11*(3), 538–549.

Porter, M. (2010). Researcher as research tool. In A. J. Mills, G. Durepos, & E. Wiebe (Eds.), *Encyclopedia of case study research* (pp. 809–811). Thousand Oaks, CA: Sage.

Rambaldi, G. (2005). Who owns the map legend? *URISA Journal, 17*, 5–13.

Ravitch, S. M. (1998). Becoming uncomfortable: Transforming my praxis. In M. J. Nakkula & S. M. Ravitch (Eds.), *Matters of interpretation: Reciprocal transformation in therapeutic and developmental relationships with youth* (pp. 105–121). San Francisco, CA: Jossey-Bass.

Ravitch, S. M. (2006a). *Multiculturalism and diversity: School counselors as mediators of culture.* Alexandria, VA: American School Counselor Association.

Ravitch, S. M. (2006b). *School counseling principles: Diversity and multiculturalism.* Alexandria, VA: American School Counselor Association Press.

Ravitch, S. M. (2014). The transformative power of taking an inquiry stance on practice: Practitioner research as narrative and counter-narrative. *Perspectives on Urban Education, 11*(1), 5–10.

Ravitch, S. M., & Lytle, S. (2015). *Becoming practitioner-scholars: The role of practice-based inquiry in the development of educational leaders.* Unpublished manuscript.

Ravitch, S. M., & Riggan, M. (2012). *Reason and rigor: How conceptual frameworks guide research.* Thousand Oaks, CA: Sage.

Ravitch, S. M., & Tarditi, M. J. (2011). Semillas Digitales overview and theory of action. Retrieved from http://www2.gse.upenn.edu/nicaragua/

Ravitch, S. M., Tarditi, M., Montenegro, N., Estrada, E., & Baltodano, D. (2015). Learning together: Reflections on trust, collaboration and reciprocal transformation from a Nicaraguan action research program. In L. L. Rowell, C. D. Bruce, J. M. Shosh, & M. M. Riel (Eds.), *Palgrave international handbook of action research.* New York, NY: Palgrave.

Ravitch, S. M., & Tillman, C. (2010). Collaboration as a site of personal and institutional Transformation: Thoughts from inside a cross-national alliance. *Perspectives on Urban Education, 8*(1), 3–9.

Richardson, L. (1990). *Writing strategies: Reaching diverse audiences.* Thousand Oaks, CA: Sage.

Richardson, L. (1991). Postmodern social theory: Representational practices. *Sociological Theory, 9,* 173–179.

Richardson, L. (1997). *Fields of play: Constructing an academic life.* New Brunswick, NJ: Rutgers University Press.

Rilke, R. M. (2001). *The book of hours: Prayers to a lowly god* (A. S. Kidder, Trans.). Evanston, IL: Northwestern University Press.

Robson, C. (2011). *Real world research* (3rd ed.). Oxford, UK: Blackwell.

Robson, K., & Robson, M. (2002). Your place or mine? Ethics, the researcher and the Internet. In T. Welland & L. Pugsley (Eds.), *Ethical dilemmas in qualitative research* (pp. 94–107). Burlington, VT: Ashgate.

Rossi, P. H., Lipsey, M. W., & Freeman, H. E. (2004). *Evaluation: A systematic approach.* Thousand Oaks, CA: Sage.

Roulston, K. (2014). Analysing interviews. In U. Flick (Ed.), *The SAGE handbook of qualitative data analysis* (pp. 297–312). London, UK: Sage.

Rule, P. (2011). Bakhtin and Freire: Dialogue, dialectic and boundary learning. *Educational Philosophy and Theory, 43*(9), 924–942.

Sadler, B., & McCabe, M. (Eds.). (2002). *Environmental impact assessment training resource manual.* London, UK: United Nations Environment Programme.

Said, E. (1989). Representing the colonized: Anthropology's interlocutors. *Critical Inquiry, 15*(2), 205–225.

Saldaña, J. (2009). *The coding manual for qualitative researchers.* London, UK: Sage.

Saldana, J. (2011). *Fundamentals of Qualitative Research.* New York: Oxford University Press.

Saldaña, J. (2013). *The coding manual for qualitative researchers* (2nd ed.). London, UK: Sage.

Sampson, H. (2004). Navigating the waves: The usefulness of a pilot in qualitative research. *Qualitative Research, 4*(3), 383–402.

Sanday, P. (1976). Cultural and structural pluralism in the U.S. In P. R. Sanday (Ed.), *Anthropology & the public interest: Fieldwork and theory* (pp. 53–73). New York, NY: Academic Press.

Saumure, K., & Given, L. M. (2008). Data saturation. In L. Given (Ed.), *The SAGE encyclopedia of qualitative research methods* (pp. 795–796). Thousand Oaks, CA: Sage.

Schein, E. (2004). *Organizational culture and leadership* (3rd ed.). San Francisco, CA: Jossey-Bass.

Schram, T. H. (2003). *Conceptualizing qualitative inquiry: Mindwork for fieldwork in education and the social sciences.* Upper Saddle River, NJ: Merrill/Prentice Hall.

Schwandt, T. A. (2015). *The SAGE dictionary of qualitative inquiry* (4th ed.). Thousand Oaks, CA: Sage.

Shulman, L. S. (1987a). The wisdom of practice: Managing complexity in medicine and teaching. In D. C. Berliner & B. V. Rosenshire (Eds.), *Talks to teachers: A festschrift for N.L. Gage* (pp. 369–386). New York: Random House.

Shulman, L. S. (1987b, February). Knowledge and teaching: Foundations of the new reform. *Harvard Educational Review,* pp. 1–22.

Shulman, L. S. (2004). *The wisdom of practice: Essays on teaching, learning, and learning to teach.* San Francisco, CA: Jossey-Bass.

Sieber, J. (1992). *Planning ethically responsible research: A guide for students and internal review boards.* Newbury Park, CA: Sage.

Sieber, J., & Tolich, M. (2013). *Planning ethically responsible research* (2nd ed.). Thousand Oaks, CA: Sage.

Smith, J., Flowers, P., & Larkin, M. (2009). *Interpretive phenomenological analysis: Theory, method and research.* London, UK: Sage.

Smith, L. T. (1999). *Decolonizing methodologies: Research and indigenous peoples.* London, UK: Zed Books, Ltd.

Spivak, G. C. (1999). *A critique of postcolonial reason: Toward a history of the vanishing present.* Cambridge, MA: Harvard University Press.

Stake, R. E. (1995). *The art of case study research.* Thousand Oaks, CA: Sage.

Steedman, P. H. (1991). On the relations between seeing, interpreting and knowing. In F. Steier (Ed.), *Research and reflexivity* (pp. 53–62). London, UK: Sage.

Steinberg, S. R. (2012). Preface: What's critical about qualitative research. In S. R. Steinberg & G. S. Cannella (Eds.), *Critical qualitative research reader* (pp. ix–x). New York, NY: Peter Lang.

Steinberg, S. R., & Cannella, G. S. (Eds.). (2012). *Critical qualitative research reader.* New York, NY: Peter Lang.

Stenbacka, C. (2001). Qualitative research requires quality concepts of its own. *Management Decision, 39*(7), 551–555.

Stewart, D. W., Shamdasani, P. N., & Rook, D. W. (2006). *Focus groups: Theory and practice* (2nd ed.). Thousand Oaks, CA: Sage.

Strauss, A., & Corbin, J. (1998). *Basics of qualitative research: Techniques and procedures for developing grounded theory.* Thousand Oaks, CA: Sage.

Stringer, T. (2014). *Action research* (4th ed.). Thousand Oaks, CA: Sage.

Tanggaard, L. (2009). The research interview as a dialogical context for the production of social life and personal narratives. *Qualitative Inquiry, 15*(9), 1498–1515.

Tetreault, M. (2012). Positionality and knowledge construction. In J. Banks (Ed.), *Encyclopedia of diversity in education* (pp. 1676–1677). Thousand Oaks, CA: Sage.

Thomas, R. M. (2003). *Blending qualitative & quantitative research methods in theses and dissertations.* Thousand Oaks, CA: Corwin.

Tierney, W. G., & Sallee, M. W. (2008). Praxis. In L. M. Given (Ed.), *The SAGE encyclopedia of qualitative research methods* (pp. 676–682). Thousand Oaks, CA: Sage.

Tisdell, E. (2008). Feminist epistemology. In L. M. Given (Ed.), *The SAGE encyclopedia of qualitative research methods* (pp. 332–336). Thousand Oaks, CA: Sage.

Tolman, D. L., & Brydon-Miller, M. (Eds.). (2001). *From subjects to subjectivities: A handbook of interpretive and participatory methods.* New York: New York University Press.

Toma, J. D. (2011). Approaching rigor in applied qualitative research. In C. F. Conrad & R. C. Serlin (Eds.), *The SAGE handbook for research in education: Pursuing ideas as the keystone of exemplary inquiry* (2nd ed., pp. 263–280). Thousand Oaks, CA: Sage.

Valencia, R. R. (2010). *Dismantling contemporary deficit thinking: Educational thought & practice.* New York, NY: Routledge.

van den Hoonaard, D. K., & van den Hoonaard, W. C. (2008). Data analysis. In L. Given (Ed.), *The SAGE encyclopedia of qualitative research methods* (pp. 186–188). Thousand Oaks, CA: Sage.

van Hardenberg, F. (n.d.). http://alistair.cockburn.us/Making + the + strange + familiar, + and + the + fa miliar + strange

van Manen, M. (1990). *Researching lived experience.* New York: State University of New York Press.

van Teijlingen, E. R. (2002). The importance of pilot studies. *Nursing Standard, 16*(40), 33–36.

van Teijlingen, E. R., & Hundley, V. (2001). The importance of pilot studies: University of Surrey social research update. Issue 35. Retrieved from http://sru.soc.surrey.ac.uk/SRU35.pdf

Wang, C. C., & Burris, M. A. (1997). Photovoice: Concept, methodology, and use for participatory needs assessment. *Health Education & Behavior, 24*(3), 369–387.

Wang, C. C., & Pies, C. A. (2004). Family, maternal, and child health through photovoice. *Maternal & Child Health Journal, 8*(2), 95–102.

Way, N. (2001). Using feminist research methods to explore boys' relationships. In D. L. Tolman & M. Brydon-Miller (Eds.), *From subjects to subjectivities: A handbook of interpretive and participatory methods* (pp. 111–129). New York: New York University Press.

Weiss, R. S. (1994). *Learning from strangers: The art and method of qualitative interview studies.* New York, NY: Free Press.

Welland, T., & Pugsley, L. (Eds.). (2002). *Ethical dilemmas in qualitative research.* Aldershot, England: Ashgate.

Wenger, E. (2000). Communities of practice and social learning systems. *Organization, 7*(2), 225–246.

Willig, C. (2014). Interpretation and analysis. In U. Flick (Ed.), *The SAGE handbook of qualitative data analysis* (pp. 136–149). London, UK: Sage.

Willis, A. I., Montavon, M., Hunter, C., Hall, H., Burkle, L., & Herrera, A. (2008). *On critically conscious research: Approaches to language and literacy research.* New York, NY: Teachers College Press.

Wolcott, H. F. (1990). On seeking—and rejecting—validity in qualitative research. In E. W. Eisner & A. Peshkin (Eds.), *Qualitative inquiry in education: The continuing debate* (pp. 121–152). New York, NY: Teachers College Press.

Wolcott, H. F. (1994). *Transforming qualitative data: Description, analysis, and interpretation.* Thousand Oaks, CA: Sage.

Wolcott, H. F. (2009). *Writing up qualitative research* (3rd ed.). Thousand Oaks, CA: Sage.

Wong, E. D. (1995). Challenges confronting the researcher/teacher: Conflicts of purpose and conduct. *Educational Researcher, 24*(3), 22–28.

Yin, R. K. (2009). *Case study research: Design and methods* (4th ed.). Thousand Oaks, CA: Sage.

Yin, R. K. (2013). *Case study research: Design and methods* (5th ed.). Thousand Oaks, CA: Sage.

Zeichner, K., & Noffke, S. (2001). Practitioner research. In V. Richardson (Ed.), *Teaching* (4th ed.) (pp. 298–330). New York, NY: Macmillan.

Zeni, J. (Ed.). (2001). *Ethical issues in practitioner research.* New York, NY: Teachers College Press.

Zigo, D. (2001). Rethinking reciprocity: Collaboration in labor as a path toward equalizing power in classroom research. *International Journal of Qualitative Studies in Education, 14*(3), 351–365.

Zuberi, T., & Bonilla-Silva, E. (Eds.). (2008). *White logic, white methods: Racism and methodology.* London, UK: Rowman & Littlefield.

Index